Theory to Countermeasures Against New Radars

Xixiang Zhang · Kaiqi Xiao · Jie Gu

# Theory to Countermeasures Against New Radars

北京理工大学出版社
BEIJING INSTITUTE OF TECHNOLOGY PRESS

Springer

Xixiang Zhang
Southwest China Research Institute
of Electronic Equipment
Chengdu, China

Kaiqi Xiao
Southwest China Research Institute
of Electronic Equipment
Chengdu, China

Jie Gu
Southwest China Research Institute
of Electronic Equipment
Chengdu, China

ISBN 978-981-16-6714-5 ISBN 978-981-16-6715-2 (eBook)
https://doi.org/10.1007/978-981-16-6715-2

Jointly published with Beijing Institute of Technology Press
The print edition is not for sale in China (Mainland). Customers from China (Mainland) please order the print book from: Beijing Institute of Technology Press.



This Springer imprint is published by the registered company Springer Nature Singapore Pte Ltd.
The registered company address is: 152 Beach Road, #21-01/04 Gateway East, Singapore 189721, Singapore

# Preface

The author has been engaged in radar countermeasure research for 60 years, and this monograph is his third book. The first monograph, *Theory to Countermeasures Against New Radars*, was published by National Defense Industry Press in February 1998, and it mainly introduces the jamming technology of conventional pulse radar and improved or partially added anti-jamming radar. The second monograph, *Electronic Warfare Technology and Application in Peacetime-Radar Countermeasure*, was published by Electronic Industry Press in January 2005, and it mainly introduces the manifestation of electronic warfare in peacetime. A host of facts are presented to show that electronic warfare is not only widely used in wartime, but also broadly and secretly used in peacetime, and countries all around the world attach great importance to it. Electronic warfare is "fighting every day" essentially. It should be particularly reminded that some hegemonic countries have put forward a "preemptive strike" strategy. In time of peace, when people are not paying attention, the hegemonic countries use electronic equipment and other techniques to make clear reconnaissance of their target and then make a sudden attack, which will cause huge losses to the opposite side or make the opposite side unable to resist. Therefore, whether in peace or war, we must be vigilant at all times to prevent unexpected events. As long as we are prepared at any time, some hegemonists who "preempt" will not dare to act rashly. Only in this way can lasting peace be maintained in the world, and a harmonious world be established. The third monograph, *Introduction to Radar Countermeasures in the New System*, mainly summarizes my 50-year radar countermeasure work and explores some radar countermeasure laws. Radar has a history of practical application for more than 70 years since it worked as military equipment in the 1930s. Radar has played a great role in World War II. After that, though no more world war happened, radar still played an important role in every local war. Radar is a clairvoyant eye, and it not detects targets only thousands of miles away but farther. With continuous development of electronic technology, many new radar systems have emerged, such as various systems in conventional pulse radar, pulse compression radar, pulse-Doppler radar, synthetic aperture radar, phased array radar, frequency-agile radar, and multiple input multiple output radar. Radar jamming comes into being with the emergence of radar and grows with the development of radar. That is to say, radar countermeasures were born since the day when radar was applied, and when radar develops a new system, jamming equipment against this new system also appears.

No radar cannot be jammed, and no jamming cannot be resisted. From the analysis of the laws of various radar systems, we can see that radar and various radio equipment have three common weaknesses:

(1) All kinds of signals working in the receiving passband of radar or radio equipment can be received, regardless of whether they are true or false and whether they are from the enemy or ourselves.
(2) No matter what signal processing method the radar adopts, as long as the jamming signal ratio reaches a certain threshold, it is impossible to extract useful information from the mixture of jamming and signal.
(3) No matter what kind of antenna the radar uses, it has sidelobes within 360°.

These three weaknesses exist because of the inherent characteristics of radar and radio equipment, so all kinds of radar system can be effectively jammed. However, if we want to achieve effective jamming, we must meet six necessary conditions:

(1) Aiming at the jamming frequency;
(2) The main lobe of the transmitter antenna of the jammer should aim at the jammed target;
(3) The jamming suppressing time should be generally continuous;
(4) Different jamming suppression coefficients are required for diverse radar systems;
(5) The spectrum width of the jamming signal should be larger than the bandwidth of the radar receiver;
(6) The polarization of the jamming signal should be consistent with that of the jammed signal.

Once meeting these conditions, one jammer can form effective suppressing jamming against radars with multiple systems in the same frequency band, which is called "one machine to multiple systems" jammer. If these six necessary conditions are not met, the jamming effect will deteriorate or be lost.

This book makes a quantitative analysis of the jamming effects of various radar systems, including conventional pulse radar, pulse compression radar, pulse-Doppler radar, synthetic aperture radar, phased array radar, and bistatic radar, etc., and the jamming effects of these radars installed on different platforms. The benchmark of jamming effect analysis is to evaluate the jamming effect of radars with different systems installed on various platforms, with the effective jamming areas and jamming exposure areas after jamming, or with the effective detection area decreasing percentage after jamming, or with the angle tracking error and distance tracking error's significant increases after jamming.

In this book, a unified jamming equation for various system radars is derived from the effective suppression jamming to different system radars. By using this suitable jamming equation, the required jamming equivalent power can be calculated on condition of deterministic effective jamming area when different system radars are installed on different platforms, and then the jamming program is simplified. In the unified equation of radar jamming, the effective jamming area is related to the following parameters:

- Radar radiation equivalent power;
- Transmission pulse width;
- Equivalent reflection area of the protected target;
- Distance between jammer and radar;
- Distance between radar and the protected target;
- Main/auxiliary gain ratio of radar antenna;
- Transmitter feeder loss;
- The transmission loss of signal and jamming signal in space;
- Equivalent radiation power of jammer;
- Feeder loss of jamming transmitter;
- Polarization loss of jammer;
- Ratio of spectrum width of jamming signal to passband width of radar receiver;
- Jamming blanket factor.

Therefore, there will be some errors when calculating jamming suppression and jamming zone, so the calculated value of effective jamming area exemplified in the book can only be used for reference. If we want to get more accurate data, we must combine it with the actual test.

This book tells the development of radar and radar countermeasures. It detailedly describes the countermeasures and jamming effects of new system radars, the calculation of the required jamming equivalent power, the calculation of jamming blanket factor required by radars of different systems, and the applicable formula of parameters.

This book is suitable for designers engaged in radar jammers, professionals in research and production of electronic warfare, teaching staff of electronic warfare specialty, military commanders, professional operators, and military and political leaders for reference. It can also help readers expand their knowledge and understand electronic warfare, information warfare, and future informatization warfare.

Chapters 1–7, 9–10 are written by the first author of this book, and the second author of this book writes Chap. 8 Broadband Antenna Technology. The third and second authors have compiled and corrected the charts, curves, and some contents in this book.

This book has received the guidance and support of experts including Researcher Gao xianwei, Deputy Director He tao, Deputy Chief Engineer Li yao, and Deputy Director Liu jiang, Zhang li, key laboratory of national defense electronic warfare. At the same time, we also received specific help and support from Mu hao, Liu yinghong, Feng mei, Zhang Zhongjie, and other comrades.

I would like to express my heartfelt thanks for this!

Chengdu, China

Xixiang Zhang
Kaiqi Xiao
Jie Gu

# Contents


**1 The Development of Radar and Radar Countermeasure** .......... 1
1.1 The Overview of Radar Development .......... 1
1.1.1 The Origin of Radar .......... 1
1.1.2 The Fundamental Principle of Radar .......... 2
1.1.3 The Fundamental Function of Radar .......... 3
1.1.4 The Basic Composition of Radar .......... 4
1.1.5 The Main Weakness of Radar .......... 5
1.1.6 The Frequency of Radar .......... 7
1.1.7 Typical Statistical Average of Radar Cross-Section (RCS) .......... 8
1.1.8 Radar Antenna .......... 12
1.2 Development of Radar Countermeasures .......... 14
1.2.1 Development of Radar Countermeasures Abroad .......... 15
1.2.2 Overview of the Development of Chniese Electronic Countermeasures .......... 16
1.2.3 Overview of the Development of Radar Countermeasure Technology .......... 17
1.2.4 Jamming Equation to Conventional Pulse Radar .......... 19
1.2.5 Jamming to Cone Scanning Radar .......... 21
1.2.6 Jamming to the Hidden Cone-Scanning Radar .......... 27
1.2.7 Jamming to Monopulse Tracking Radar .......... 28
1.2.8 Jamming to Frequency Agile Radar .......... 36
1.2.9 Countermeasure Against the "AGM-45 Shrike Anti-radiation Missile" .......... 40
1.3 Development of Electronic Warfare to Information Warfare .......... 47
1.3.1 Overview of the Battlefield of Informationized Operations .......... 48
1.3.2 The Concept of Information Warfare and Informationized Operations .......... 53
1.3.3 Use "Information Warfare" to Deal with "Informationized Operations" .......... 55

**2 Countermeasures Against Synthetic Aperture Radar** .......... 65
2.1 Operating Principle of Synthetic Aperture Radar .......... 67
2.1.1 Equation of Synthetic Aperture Radar .......... 67
2.1.2 Composition of Synthetic Aperture Radar .......... 71
2.1.3 Pulse Repetition Frequency Selection of Synthetic Aperture Radar .......... 73
2.2 Jamming Principle for Synthetic Aperture Radar .......... 75
2.2.1 Jamming Equation to SAR .......... 75
2.2.2 Jamming Equivalent Power Required for SAR .......... 77
2.3 Measurement and Analysis of SAR Jamming Suppression Coefficient .......... 78
2.3.1 Computer Simulation Measurement .......... 78
2.3.2 Indoors Simulation Test of Low Power Jamming Against Formal SAR .......... 80
2.3.3 Flight Test on the Jamming Suppression Coefficient for Airborne SAR .......... 81
2.3.4 Data Analysis of Airborne SAR Jamming Suppression Coefficient from Flight Test .......... 82
2.4 Analysis on Reflection Area of Point Target .......... 85
2.4.1 Mechanism of Radar Target High-Frequency Scattering .......... 85
2.4.2 Measurement on Equivalent Reflection Area of Complex Target's Scattering Center .......... 86
2.4.3 RCS of Regular Objects .......... 87
2.5 Timely Reconnaissance of Spaceborne SAR .......... 88
2.5.1 Necessity of Timely Reconnaissance for Spaceborne SAR .......... 88
2.5.2 Feasibility of Timely Reconnaissance for Spaceborne SAR .......... 90
2.5.3 Requirements for Timely SAR Reconnaissance Stations .......... 91
2.5.4 Composition of Timely Reconnaissance Stations for Spaceborne SAR .......... 98
2.6 Timely Suppression Jamming Against Spaceborne SAR .......... 100
2.6.1 Feasibility of Timely Suppression Jamming Against Spaceborne SAR .......... 100
2.6.2 Jamming Protection Zone of Spaceborne SAR from a Single Suppression Jamming Station .......... 102
2.6.3 Protection Zone of Space Borne SAR from Multiple Suppression Jamming Stations .......... 103
2.6.4 Composition of Suppression Stations for Spaceborne SAR .......... 106
2.7 Jamming Against the Radar Identification System .......... 114
2.8 Jamming Against Variable Polarization Height Measurement SAR .......... 116

2.8.1 Interferometric SAR Three-Dimensional Imaging Image ... 116
2.8.2 Principle of Random Polarization Jamming ... 117
2.9 Selection of Optimal Jamming Patterns for SAR ... 119

**3 Jamming Against Pulse Compression Radar** ... 123
3.1 Airborne Early Warning Pulse Compression Radar ... 123
3.2 Operating Principle of Airborne Early Warning Pulse Compression Radar ... 125
3.2.1 Operating Principle of Pulse Compression Radar ... 125
3.3 Jamming Principle for Pulse Compression Radar ... 133
3.3.1 Jamming Equation for Early Warning Aircraft Radar ... 133
3.3.2 The Value of Jamming Suppression Coefficient for Pulse Compression Radar ... 135
3.3.3 Analysis on Jamming Exposure Area of Early Warning Aircraft Radar ... 138
3.4 Test of Jamming Suppression Coefficient and Jamming Exposure Area of Pulse Compression Radar ... 142
3.4.1 Test on the Antenna Pattern and Polarization Characteristics of Radar ... 142
3.4.2 Test of Jamming Suppression Coefficient and Jamming Exposure Area of Pulse Compression Radar ... 144
3.4.3 Measurement and Calculation of Jamming Suppression Coefficient of Pulse Compression Radar ... 148
3.5 Jamming Suppression Zone of the Airborne Jammer Against Ground Pulse Compression Radar ... 150

**4 Jamming Techniques for Pulse-Doppler Radar** ... 155
4.1 Jamming Principle for Pulse-Doppler Radar ... 155
4.1.1 Operating Principle of Pulse-Doppler Radar ... 155
4.2 Analysis of Jamming Suppression Coefficient for Pulse Doppler Radar ... 166
4.2.1 Determination of Jamming Suppression Coefficient for Pulse-Doppler Radar ... 166
4.2.2 Calculation of Equivalent Jamming Power for Pulse-Doppler Radar ... 169
4.2.3 Example of Jamming Suppression Coefficient of Pulse-Doppler Radar ... 170
4.3 Jamming Exposure Area for Pulse-Doppler Radar ... 171

**5 Unified Jamming Equation and "One-to-Many Systems" Jammer** ... 173
5.1 Unified Jamming Equation of Radar Jammer ... 173

5.1.1 Calculation of Jamming Suppression Coefficient for Pulse Compression Radar ..... 176
5.1.2 Calculation of Jamming Suppression Coefficient of Pulse-Doppler Radar ..... 178
5.1.3 Calculation of Equivalent Power of Jamming Synthetic Aperture Radar ..... 180
5.2 "One-to-Many Systems" Radar Jammer ..... 182
5.2.1 Essential Elements of 'One-to-Many Systems' ..... 183
5.2.2 Composition of 'One-to-Many Systems' Jammer ..... 184
5.3 The Layout of Air-to-Space Integration of 'One-to-Many Systems' Jammer ..... 198

**6 Analysis on Configuration of Radar Jammer and Protected Target ..... 205**
6.1 Ground Jamming Angainstaircraft Radar to Protect Important Military Targets ..... 205
6.2 Jamming Exposure Area of Ground Jammer Against Early Warning Aircraft Radar ..... 213
6.2.1 Theoretical Calculation of Jamming Exposure Area for Early Warning Aircraft Radar ..... 213
6.2.2 The Measured Jamming Exposure Area for the Early Warning Aircraft Radar ..... 215
6.2.3 The Jamming Exposure Area Formed by Shipborne Jammer covering the fleet ..... 216
6.3 Analysis on the Jamming Exposure Area and Jamming Effective Area of Ground-to-Air Jammers Against Airborne SAR ..... 220
6.3.1 Analysis and Calculation on Jamming Effective Area for Airborne SAR ..... 220
6.3.2 Steps and Methods for Measuring the Jamming Effective Area to Airborne SAR ..... 226
6.4 Analysis and Calculation of Jamming Effective Area F Spaceborne SAR ..... 229

**7 Jamming Techniques for Bistatc Radar ..... 243**
7.1 Basic Working Principle of Bistatic Radar ..... 243
7.2 Jamming Equation and Jamming Exposure Area for Bistatic Radar ..... 245
7.2.1 Self-Defense Jamming to Bistatic Radar ..... 246
7.2.2 Exposure Area in Stand-Off Jamming to Bistatic Radar ..... 247
7.3 Analysis on Jamming Exposure Area for Bistatic Radar ..... 251
7.4 Selection of Beamwidth of the Jamming Antenna When Jamming Against Bistatic Radar ..... 251
7.4.1 Selection of Beamwidth of Antenna Azimuth ..... 251

7.4.2 Selection of Beamwidth of Jamming Transmitting Antenna Elevation ..... 252
7.5 Jamming Against Multi-Base Radar ..... 253
7.5.1 Cover Jamming for Multi-Base Radar ..... 253
7.5.2 Self-Defense Jamming to Multi-Base Radar ..... 254
7.6 Block Diagram of the Bistatic Radar Jammers' Compostion ..... 255
7.7 Jamming Against Bistatic Airborne SAR ..... 256

**8 Broadband Antenna Technology** ..... 261
8.1 Major Indicator of the Broadband Antenna ..... 262
8.2 The Technology of Multi-mode Antenna ..... 272
8.2.1 Small Aperture Broadband Adcock Antenna Array ..... 273
8.2.2 Four-Arm Spiral Antenna ..... 281
8.2.3 Multi-mode Log-Periodic Antenna Array ..... 289
8.3 The Technology of Multi-beam Antenna ..... 295
8.3.1 Bootlace Lens Multi-beamformer Bootlace ..... 296
8.3.2 Beamforming of Luneburg Lens ..... 304
8.3.3 Two-Dimensional Multi-beam Area Array Antenna ..... 309
8.3.4 Dielectric Lens Antenna with Matching Layer ..... 317
8.4 Broadband Phased Circular Array Antenna ..... 324
8.4.1 Broadband Phased Circular Array Antenna ..... 324
8.4.2 Broadband Phased Transmit Array Antenna ..... 331
8.4.3 Broadband Active Phased Array Antenna ..... 339
8.4.4 Broadband Planar Phased Array Antenna ..... 341
8.5 Broadband Array and Reflective Surface Composite Antenna ..... 348
8.6 Digital Beamforming Technology ..... 355
8.7 Broadband Antenna System ..... 359
8.7.1 Broadband Radome ..... 359
8.7.2 Broadband Polarization Cover ..... 368
8.7.3 Antenna Receiving and Transmitting Isolation Technology ..... 381
8.7.4 System Architecture of Broadband Antenna ..... 389

**9 Jamming Against Phased Array Radar** ..... 395
9.1 Overview of Phased Array Radar ..... 395
9.1.1 Phase Scanning ..... 396
9.1.2 Frequency Scanning ..... 397
9.1.3 Typical Phased Array Radar ..... 398
9.1.4 Planar Array Beam ..... 403
9.1.5 Sparse Array and Small Array Pattern ..... 405
9.1.6 Low Sidelobe Phased Array ..... 405
9.2 The Advantages of Phased Array Radar and Jamming Countermeasures ..... 410
9.2.1 The Advantages of Phased Array Radar ..... 411

9.2.2 Basic Jamming Countermeasures Against of Phased Array Radar ... 412
9.2.3 An Example of Jammer Designed for Intergrated Phased Array Radar ... 415
9.3 Analysis of Effective Area of Phased Array Radar Jamming ... 424
9.3.1 Jamming Exposure Area of Phased Array Radar ... 424
9.3.2 The Ratio of the Radar's Detection Area Under Jamming ... 426
9.4 The Analysis of the Jamming Effect on Phased Array SAR of Airborne (or Unmanned Aerial Vehicle) ... 428
9.4.1 Analysis of Jamming Effect on UAV-Borne SAR ... 428
9.4.2 Analysis of the Effective Area of Point Target Jamming in the Ground Attack of Missile-Borne SAR ... 435
9.5 Analysis of Jamming Effect on Spaceborne Phased Array SAR ... 439
9.5.1 The Jamming Equation of Spaceborne Phased Array SAR ... 439
9.5.2 Derivation of "Unified Jamming Equation" of Phased Array Synthetic Aperture Radar ... 443
9.5.3 Calculation of Equivalent Power of Jamming to Spaceborne Phased Array Synthetic Aperture Radar ... 445
9.5.4 Analysis of Jamming Effective Area Against Spaceborne Phased Array SAR ... 445

**10 New Concepts and Technologies of Electronic Warfare ... 453**
10.1 Combat Effect and Protection Technology of Electromagnetic Pulse ... 453
10.1.1 High Repetition Frequency Electromagnetic Pulse with No Carrier and Low Peak Power ... 454
10.1.2 Effect of Electromagnetic Pulse on Electronic Devices ... 456
10.1.3 Analysis of Combat Effect of Electromagnetic Pulse ... 458
10.1.4 Protection Technology for High Power Electromagnetic Pulse ... 461
10.2 Radar Countermeasures and High-Power Microwave Devices ... 465
10.2.1 Mechanism of High Power Microwave ... 465
10.2.2 The Power Required by Various Jammers and High-Power Microwave Weapons ... 467
10.2.3 Application Comparison of Solid-State Microwave Power Devices and Electric Vacuum Microwave Devices ... 468

10.3 Feasibility Analysis of Attacking Reconnaissance Satellites Using High-Power Microwave Weapons ..... 472
10.3.1 The Feasibility of Using High-Power Microwave Attacks on Passive Reconnaissance Satellites ..... 472
10.3.2 The Feasibility of Strong Microwave Jamming to Reconnaissance Satellites ..... 475
10.4 Application of UWB High PRF Jammer ..... 479
10.4.1 Demand Status ..... 480
10.4.2 Demand Analysis for High Repetition Frequency ..... 484
10.4.3 Pulse Width and Spectrum ..... 485
10.4.4 Jamming Effect on Meter Band Radar ..... 486
10.4.5 Application of UWB Jammer ..... 489
10.4.6 Test of UWB Jammer ..... 491
10.5 Integrated Electronic Warfare Technology Including Reconnaissance, Jamming, and Positioning ..... 494
10.5.1 Utilize the Integrated Technology of High-Power Continuous Wave Noise Jammer ..... 494
10.5.2 Using the Jammer's Angle Passive Tracking System to Realize the Integration of "Reconnaissance, Positioning, and Jamming" ..... 501
10.5.3 Ranging Error of Single Station Passive Positioning System Using Angle Passive Tracking System ..... 506
10.5.4 Single Station Passive Positioning Based on a Passive Angle Tracking System ..... 508
10.5.5 Typical Application of Single Station Passive Location Based on Angle Passive Tracking System ..... 509
10.6 Analysis of Jamming Effect of "Signal Hiding" Anti-jamming Radar ..... 511

**References** ..... 519

# Chapter 1
# The Development of Radar and Radar Countermeasure

## 1.1 The Overview of Radar Development

### *1.1.1 The Origin of Radar*

In 1885–1888, the German physicist Heinrich Hertz conducted his celebrated experiment that proved the existence of radio waves, which establish the core concept of radar. He verifies that the nature of radio waves being similar to visible light in their ability to be reflected from the metal object and refracted from the dielectric prism.

In the 1920s, when a ship passed through the receiver and transmitter located on both sides of the river, it accidentally observed a fluctuating signal on the receiver, which was called the continuous wave jamming system at that time.

In the 1930s, Radar has developed in at least eight countries—France, German, Italy, Japan, Netherlands, the Soviet Union, the United Kingdom, and the United States. They mostly used the frequency of 100–200 MHz. However, the United Kingdom was using 30 MHz, and Germany was using 600 MHz, which is the highest frequency before World War II. In 1936, the U.S. started the pulse radar development plan, by the time of Attack on Pearl Harbor, the 200 MHz's SCR-268 antiaircraft searchlight control radar was available for use.

Even though pulse radar developed later in the U.K. than in the U.S., The 30 MHz radar made great contributions in the Battle of Britain. By 1939, Britain established a 200 MHz airborne radar that could detect aircraft. With the invention of the high-power microwave magnetron control, the technology of the radar has achieved extremely significant development, and it has opened the door for radars operating at higher frequencies. Therefore, Radar soon developed in Germany, the Soviet Union, Italy, French, Japan, and other countries.

Since World War II, radar technology has developed rapidly. Some major technological breakthroughs have made radar performance more complete. For example, the successful development of moving target display technology and the high power amplifiers; the application of mono-pulse radar; technology of pulse compression enables people to get high energy from long pulse waveforms;



X. Zhang et al., *Theory to Countermeasures Against New Radars*,
https://doi.org/10.1007/978-981-16-6715-2_1

Airborne Synthetic Aperture Radar provides map-like high-resolution off-ground scene imaging; airborne radar can detect a target in severe clutter background by using Doppler processing; electronically scanned array radar provides fast scanning without mechanical movement of the antenna, and can also perform multi-target tracking; HF over-the-horizon radar expands the detection range of the radar by more than ten times. Radar has become an important tool for meteorologists. The rapid development of digital technology has made many theoretically digital signal processing and digital data processing possible. New radar technology and electromechanical integration strongly guarantee the continuous application of new radar applications.

Radar countermeasure is produced with the development of radar. Electronic warfare has been applied as early as World War II. In 1944, the Allied forces demonstrated the power of radar countermeasure from the landing operation on the Normandy. In this landing operation, there were more than 6000 ships and more than a million personnel. Since the allied forces adopted electronic countermeasures, the allied troops landed successfully with only six warships lost, making it a typical example of using electronic countermeasures to reduce casualties in World War II.

### *1.1.2 The Fundamental Principle of Radar*

Radar is mainly used to detect and locate objects with electromagnetic scatterings, such as aircraft, ships, space shuttles, vehicles, pedestrians, etc. High-precision direction finding, ranging, and all-weather working ability are the most important characteristics of radar. It allows radar work in the dark, moist, dense fog, rain, and snow.

Figure 1.1 shows the principle of radar ranging.

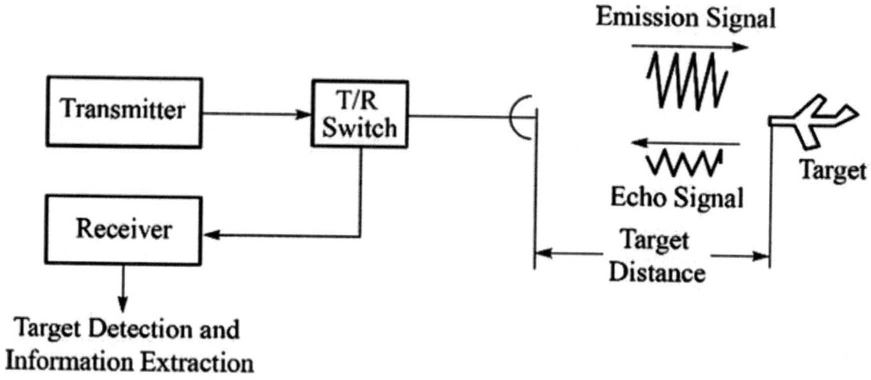


**Fig. 1.1** Fundamental principle of radar

The electromagnetic waves emitted by the radar propagate through the air at the speed of light, and it encounters a reflective object and returns to the radar antenna. The distance between the radar and the target is equal to the time the electromagnetic wave travels between the antenna and the reflective target divided by two and multiplied by the speed of light.

$$R = \frac{1}{2} CT \tag{1.1}$$

In the formula:

R Distance between radar and target
C Speed of light
T The time that electromagnetic waves travel back and forth between radar and target.

The direction-finding of the radar is relying on the directionality of the radar antenna to determine the direction of the target. It can use the directional characteristics of the radar antenna to measure the maximum echo direction, the azimuth, and elevation of the target to determine the three-dimensional coordinates of the target. The three-dimensional coordinates of the target will lead to the positioning of the target.

### 1.1.3 The Fundamental Function of Radar

Set radar's transmitted power as $P_t$, radar's antenna gain as $G_t$, equivalent reflective area of the target as $\sigma$, and the wavelength of radar as $\lambda$, radar signal power received by the target as $P_A$

$$P_A = \frac{P_t G_t}{4\pi R^2} A = \frac{P_t G_t}{4\pi R^2} \cdot \frac{\sigma \lambda^2}{4\pi} \tag{1.2}$$

Radar antenna receives the signal power reflected by the target as $P_r$

$$P_r = \frac{P_A}{4\pi R^2} \cdot G_r \tag{1.3}$$

radar's receiving antenna gain as $G_r$.

The receiving antenna and the transmitting antenna of the radar are the same, therefore $G_t = G_r$.

Plug (1.2) into (1.3)

$$P_r = \frac{PtG_t^2 \sigma \lambda^2}{(4\pi)^3 R^4} \tag{1.4}$$

$$R_{\max} = \left[ \frac{P_t G_t^2 \sigma \lambda^2}{(4\pi)^3 P_{r\,\min}} \right]^{1/4} \tag{1.5}$$

$R_{\max}$ Maximum detection distance of the radar
$P_{r\,\min}$ The smallest signal of useful information that received by the radar receiver.

### 1.1.4 The Basic Composition of Radar

The basic components of the radar include a transmitter, receiver, antenna, transceiver switch, terminal display, servo rotation, power supply, and other parts, as Fig. 1.2 shows.

Generally, there are two types of radar transmitters. The first type is the high-power main vibration source, which generates microwave pulse power in the order of megawatts. Most early radars used this type of transmission. Nowadays, the transmitter that radar used is the second type, that is, a single or multiple low-power continuous-wave oscillation sources are used, which have a relatively stable frequency and a long life, and then use pulse modulation to amplify the output through the pulse power amplifier. The pulse source is both the pulse modulation source of the transmitter and the synchronization pulse source of the display.

The pulse source is the reference pulse for radar ranging. The transceiver switch is an important device for the radar to transmit and receive a common antenna. When there is a high-power pulse signal input, it connects the transmitter to the antenna. When there is a weak signal input, it combines the antenna to the receiver, which makes the radar share antenna easily. A low-noise amplifier is usually for improving the sensitivity of the receiver so that the radar can increase the detection range. The typical radar receiver uses the superheterodyne receiving system. It uses the local oscillator and mixing to reduce the signal to the intermediate frequency, which makes the signal easy to amplify, and then uses the detector to extract the video pulse for the operator intuitively reads the coordinate position of the target. After the Second World War, although there was no World War III, local wars in the world continued.

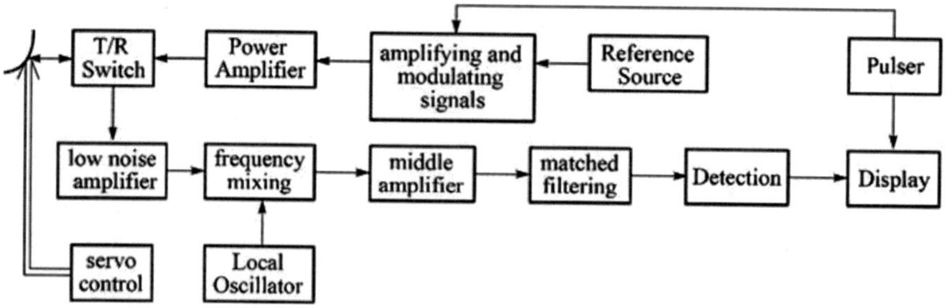


**Fig. 1.2** Block diagram of the basic composition of radar

Several regional wars that have a greater impact on the world, such as the Korean War, the Vietnam War, the four Middle East Wars, the Island War, the Gulf War, and the Iraq War. These wars have greatly promoted the development of radar and radar countermeasures.

At present, radar has been widely used to detect ground, sea, air, space, and underground targets. It has become an important part of air defense systems and offensive missiles and other combat weapons. In air defense, it has surveillance and weapon control functions. It can detect targets, target identification, target tracking, target assignment to weapon systems, target tracking weapon control, guide weapon interception, and assess the effectiveness of engagement. In the missile system, the radar is helpful for weapon guidance and detonation. The synthetic aperture high-resolution imaging radar has already been used to detect fixed targets and to move targets on the battlefield.

Radar is also widely used in civil applications, such as retesting, meteorological observations, planetary observations, and short-range underground detection, to guide navigation channels for maritime transportation, and to provide marine ice floe mapping.

Air traffic control has been used worldwide for flight control near airports. It involves aircraft safety and navigation so that it can avoid rain and danger zones measured by radar; Space vehicles have used radar for rendezvous and docking, as well as for lunar landings, and the establishment of astronomical units for accurate measurements; in other respects, radar can also be used for the detection of oil and gas, migration of insects and birds. Radar can also be used for oil, gas detection, insect, and bird migration. Some radars are small enough to fit in the palm of a person, while others are large enough to occupy several football fields. Their range of action is as close as almost touchable targets, as far as thousands of kilometers or to planets.

### *1.1.5 The Main Weakness of Radar*

Radar has a history of more than 70 years since its application. In these years, it can be said that it has developed rapidly and has become one of the main pillars of information warfare and one of the main components of long-range precision strike weapons.

Because radar is very important in information warfare, designers engaged in radar countermeasure must do everything possible to find its loopholes and weaknesses, and attack it to make it lose its effectiveness. In short, the radar summarizes the following vulnerabilities:

1. No matter if it is a useful signal or a harmful signal, as long as the frequency is within the passband of the radar receiver, it can enter the receiver. The harmful signal affects the radar's extraction of useful signals. The jammer is to use this principle weakness to send harmful signals into the radar receiver, making it

difficult to extract useful information correctly, thereby reducing or destroying the normal operation of the radar.

2. No matter what object, as long as it can reflect the radio signal, the radar can receive it. Some of these reflected objects are the target echoes that the radar needs to receive. Some reflective objects are the targets that the radar needs to find. Some reflected signals generated by the reflector harm the discovery of useful signals. Because of this feature of radar, jammers use reflectors to reflect a large amount of false information, causing difficulties in radar detection and affecting it to find useful target signals.
3. Unintentional jammings cause difficulties for the radar to extract useful signals Such as ground clutter, sea clutter, meteorology, and bird flocks. For this reason, the radar designer has spent a lot of energy and has taken various measures to eliminate these clutters. On the contrary, this clutter has helped the radar jammers.
4. Human-made jamming is to align the jamming frequency with the radar frequency accurately, and then modulate some signals that the radar is very unwilling to see so that the radar's useful signals are difficult to extract or lose. The radar can also extract some fake information that is difficult to identify. Even if the radar operator knows these jammings, it cannot be eliminated. Therefore, it can be said that the radar will be jammed with, which is an inevitable phenomenon.
5. The radar antenna is one of the important subsystems of the radar system. It can transmit useful signals as well as receive useful signals, but it inevitably receives jamming signals. In particular, the radar antenna inevitably generates sidelobes due to processing errors, feed blockage, and current distribution on the aperture. This provides favorable conditions for the shielded sidelobe jamming, making the shielded jamming and track deception jamming with opportunity.
6. No matter what kind of radar system, as long as it can extract useful information normally, there must be a signal-to-noise ratio requirement. This signal-to-noise ratio may be greater than one, less than one or equal to one. When the jamming signal makes its signal-to-noise ratio not meet this requirement, it cannot extract useful information, which means the radar has been jammed. Different radars have different required signal-to-noise ratio and jamming-to-signal ratio. This jamming signal ratio is called an jamming suppression coefficient in radar jamming, and it is one of the important basis for calculating radar jamming equivalent power.

Radar has a wide range of applications in civilian and military applications. Many civilian radars are also in the military range during wartime. It can be said there is no long-range precision strike without the support of radar in an information-based operation. However, the radar has its weaknesses, which allows jamming to be used. Therefore, in information-based operations, both radar and radar countermeasures are top priorities for development. They must have a fierce battle in informationized warfare. Which party gains the initiative in the radar countermeasure will also gain the initiative in the war.

Many war examples after World War II show that, such as the Vietnam War, the surface-to-air missile guidance radar has a hit probability of 14% when it is not disturbed, and its hit probability drops to 1.4% after being disturbed. In the first Middle East war, a new band of guidance radar appeared, which caused the loss of 74 Israeli aircraft in a few days of operation without jamming. During the Gulf War, the United States had early warning aircraft radar, combat aircraft, and electronic warfare jamming aircraft. Iraq has fighter aviation, surface-to-air missile forces, and other ground forces, but there are no jamming forces. When American planes entered Iraq, Iraqi ground anti-aircraft weapons were inaccurate, and the fighter aviation dared not take off so that US aviation troops won the Gulf War with zero losses. These warfare examples strongly show that the presence or absence of radar countermeasures has a huge impact on the outcome of wars and losses in battles.

### *1.1.6 The Frequency of Radar*

In World War II, the frequency of radar was mostly 100–200 MHz, and the highest frequency in Germany was 600 MHz. As Table 1.1 shows, the frequency of radar and other wireless civilian equipment has developed from extremely low frequency to sub-millimeter wave band.

The frequency range of the radar is particularly wide, ranging from 5 MH to 30 GHz or 30 MH to 300 GHz. It is called terahertz at 1000 GHz, it has many characteristics of different radio frequencies, and has strong penetration. Countries are currently studying their features and applications. There are many users in this broad electromagnetic spectrum, including communications, satellite navigation, radio and television, mobile communications, and radar.

To reduce the mutual influence, the International Radio Commission has assigned a frequency range to radar, which is very valuable for us to study radar countermeasures. Because radar is widely used in military applications, it is also possible for international powers to fail to comply with this rule. Therefore, this allocated dedicated radar frequency can be used as a reference or the preferred frequency for radar jamming. See Table 1.2 for standard radar frequencies.

**Table 1.1** Frequency range of radio and its typical application

| 3 kHz – 30 kHz | 30 kHz – 300 kHz | 300 kHz – 3 MHz | 3 MHz – 30 MHz | 30 MHz – 300 MHz | 300 MHz – 3GHz | 3GHz – 30 GHz | 30 GHz – 300 GHz | 1 000 GHz Terahertz |
|---|---|---|---|---|---|---|---|---|
| Very Low Frequency | Low Frequency | Intermediate Frequency | High Frequency | Very high frequency | ultrahigh frequency | ultra high frequency | extremely high frequency | |
| Global Communications | Remote communication broadcast | | | Relay Communication | Satellite Communication、Satellite Navigation | | | |
| submarine communication address detection | | | | Mobile Communication | | | Satellite Remote Sensing | |
| | | | | Radar frequency band | | | | |

**Table 1.2** The letter name of the radar standard frequency

| Band | Standard frequency range | Assigned radar frequency range |
|---|---|---|
| HF | 3–30 MHz | |
| VHF | 30–300 MHz | 138–144 MHz<br>216–225 MHz |
| UHF | 300–1000 MHz | 420–450 MHz<br>850–942 MHz |
| L | 1–2 GHz | 1215–1400 MHz |
| S | 2–4 GHz | 2300–2500 MHz<br>2700–3700 MHz |
| C | 4–8 GHz | 5250–5924 MHz |
| X | 8–12 GHz | 8500–10,680 MHz |
| Ku | 12–18 GHz | 13.4–14.0 GHz<br>15.7–17.7 GHz |
| K | 18–27 GHz | 24.05–24.25 GHz |
| Ka | 27–40 GHz | 33.4–36 GHz |
| V | 40–75 GHz | 59–64 GHz |
| W | 75–110 GHz | 76–81 GHz<br>92–100 GHz |
| mm | 110–300 GHz | 126–142 GHz<br>144–149 GHz<br>231–235 GHz<br>238–248 GHz |

It can be seen from (1.5) that the detection range of the radar is inversely proportional to the square of the operating frequency, and the target's reflection area is also closely related to the operating frequency of the radar. The relationship between radar and frequency and the relationship between Radar jamming and target reflection area are closely related.

### *1.1.7 Typical Statistical Average of Radar Cross-Section (RCS)*

Typical Statistical Average value of RCS for various type of aircraft is listed Table 1.3.

It can be seen from Table 1.4 that the cross-sectional reflection area of the aircraft has a close relationship with the frequency. For the F-117A stealth aircraft, its effective reflection area in the S-band is below 0.1 $m^2$, while in the VHF band, its reflection area is between 7 and 75 $m^2$. Therefore, the meter-band radar anti-stealth aircraft still has its advantages. On the other hand, it also tells us that we should pay enough attention to the jamming against meter-band radar.

**Table 1.3** Typical statistical average value of RCS for various types of aircraft ($\lambda = 5$ cm)

| Serial number | Model | Nose cone ± 45° RCS ($m^2$) | Positive side 90° ± 5° RCS ($m^2$) |
|---|---|---|---|
| 1 | Catalina B52 | 100 | 1000 |
| 2 | Fighter F-15 | 4 | 400 |
| 3 | Quasi-stealth fighter F-16S | 0.4 | 10 |
| 4 | Reconnaissance aircraft (reconnaissance series) | 0.2 | / |
| 5 | Stealth bomber B-2 | 0.1 | / |
| 6 | Stealth reconnaissance/strike fighter F-117A | 0.02 | 0.1 |
| 7 | Stealth drone CM-30, CM-44 | 0.001 | 0.1 |

**Table 1.4** Statistical average of two typical aircraft RCS (Nose Cone ± 45°)

| Aircraft | Band | | | | | | |
|---|---|---|---|---|---|---|---|
| | VHF ($m^2$) | UHF ($m^2$) | L ($m^2$) | S ($m^2$) | C ($m^2$) | X ($m^2$) | Ku ($m^2$) |
| F-16S | 6–40 | 4–6 | 0.4–1.2 | 0.4 | 0.4 | 0.4 | 0.4–0.8 |
| F-117A | 7–75 | 1–7 | 0.1–1 | 0.02–0.1 | 0.02 | 0.02 | 0.02–0.1 |

**Table 1.5** Reference value of RCS of head and the front side of AGM-88 HARM anti-radiation missile and AGM–65 Maverick

| Frequency band | Name | | | |
|---|---|---|---|---|
| | Anti-radiation missile RCS ($m^2$) | | Air-to-ground missile RCS ($m^2$) | |
| | HH polarization | VV polarization | HH polarization | VV polarization |
| S | / | / | 0.27/1.65 | 0.29/3.11 |
| X | 0.13/7.45 | 0.06/5.31 | 0.54/1.56 | 0.75/7.14 |
| C | 0.08/2.86 | 0.05/3.10 | 0.32/2.9 | 0.26/1.62 |
| Ku | 0.1/4.11 | 0.12/4.78 | 0.79/3.34 | 1.44/2.87 |

**Table 1.6** RCS valuation of various stealth aircraft and non-stealth aircraft

| Name | RCS/$m^2$ | Name | RCS/$m^2$ |
|---|---|---|---|
| CatalinaB-2 (ATB) | 0.1 | CatalinaB-52 | 100 |
| Stealth reconnaissance/strike fighter F-117A | 0.02 | Fighter F-15 | 4 |
| Fighter ATF (YF-22A) | 0.05 | Fighter F-1 6 | 4 |
| Stealth drone CM-3 0, CM-4 4 | 0.001 | Scout BQM-34A | 0.4 0.2 |
| Cruise missile (ACM) AGM-129 | 0.005 | Cruise Missile AGM-8 6B | 1 |
| Fighter F-1 6-S improved | 0.2–0.5 | | |

It can be seen from Table 1.5 that the reflection cross-sectional area of the target is not only closely related to the frequency, but also to the polarization. RCS valuation of various stealth aircraft and other targets are shown if Table 1.6 and 1.7.

Stealth aircraft is not related to frequency and transition, but also to flight attitude, as shown in Fig. 1.3.

RCS values of different landforms, different small buildings, and different cities are shown in Table 1.8, Table 1.9.

The average polarization RCS value per square meter unit area measured at different sweep angles (or elevation angles) at different wind speeds is shown in Fig. 1.10.

**Table 1.7** RCS valuation for other targets

| Name | RCS/m$^2$ | Name | Vertical RCS/m$^2$ | Horizontal RCS/m$^2$ |
|---|---|---|---|---|
| Automobile | 100 | Blue—winged locust | $10^{-3}$ | $10^{-4}$ |
| Cargo car | 200 | Alfalfa larva butterfly | $0.5 \times 10^{-4}$ | $0.2 \times 10^{-5}$ |
| Bicycle | 2 | Worker Bee | $10^{-4}$ | $0.3 \times 10^{-4}$ |
| Human | 1 | Big fly | $0.3 \times 10^{-4}$ | $0.2 \times 10^{-5}$ |
| Big bird | $10^{-2}$ | Green—necked flies | $0.2\,5 \times 10^{-4}$ | $10^{-5}$ |
| Medium bird | $10^{-3}$ | Agglomerate female beetle | $0.2 \times 10^{-5}$ | $10^{-4}$ |
| Big insect | $10^{-4}$ | Spider | $10^{-5}$ | $0.7 \times 10^{-5}$ |
| Small insect | $10^{-5}$ | | | |

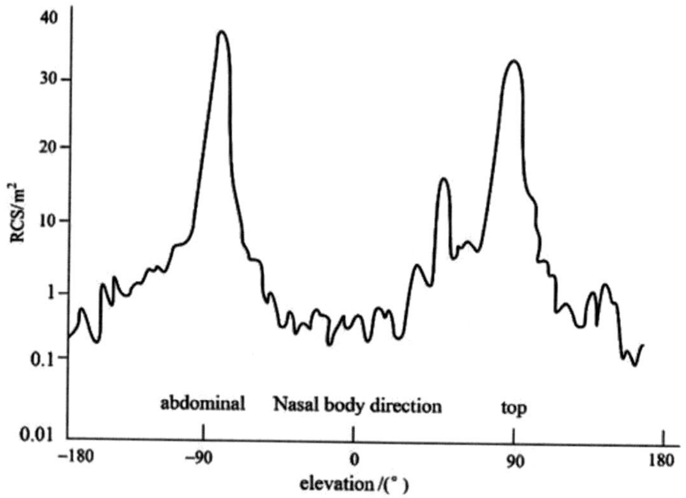


**Fig. 1.3** Variation curve of RCS measurement value of a typical stealth assault/reconnaissance aircraft with pitch attitude angle

**Table 1.8** Measured value of average RCS per square meter for different cross-sections of scattering terrain

| Frequency | Grazing angle | | | | | | | | |
|---|---|---|---|---|---|---|---|---|---|
| | Desert | | Cultivated land | | Outdoor wood | | Wooded hill | | |
| | 0°–1° | 10° | 0°–1° | 10° | 0°–1° | 10° | 0°–1° | 10° | |
| HF (32.8 MHz) | −6–55 | / | / | −75 to −65 | −60 | −65 to −55 | −50 to −30 | −60 to −45 | dB ($m^2$) |
| UHF (0.5 GHz) | −40 | / | −35 to −30 | −36 | / | −23 | −34 to −24 | −22 | dB ($m^2$) |
| S (3 GHz) | / | −25 | −25 to −35 | −21 | −23 | −23 | −47 to −32 | / | dB ($m^2$) |
| X (9.3 GHz) | −38 | −26 | −36 | −20 | −30 | −23 | −36 to −30 | −25 | dB ($m^2$) |
| Ka (35 GHz) | / | −22 | −20 | −20 | −25 | −19 | −17 | −19 | dB ($m^2$) |

**Table 1.9** The average RCS per square meter of small buildings and cities measured at different frequencies and ground angles

| Frequency | Grazing angle | | | |
|---|---|---|---|---|
| | Small building RCS ($dBm^2$) | | City RCS ($dBm^2$) | |
| | 0°–1° | 10° | 0°–1° | 10° |
| HF (32.8 MHz) | / | / | / | −10 |
| UHF (0.5 GHz) | / | −23 | −30 to −22 | −9 |
| S (3 GHz) | −35 | −23 | / | −15 |
| X (9.3 GHz) | −36 to −30 | −23 | −24 | −12 |

**Table 1.10** The average polarization RCS value per square meter unit area measured at different sweep angles (or elevation angles) at different wind speeds (sea surface wind speed 22 knots)

| Angle | | Grazing | | | | | |
|---|---|---|---|---|---|---|---|
| | | 0.3° ($dBm^2$) | 1° ($dBm^2$) | 3° ($dBm^2$) | 10° ($dBm^2$) | 30° ($dBm^2$) | 60° ($dBm^2$) |
| UHF (0.5 GHz) | VV | −75 | / | / | −25 | / | / |
| | HH | / | −65 | −55 | −46 | / | / |
| S (3 GHz) | VV | −50 | −38 | −35 | −28 | / | / |
| | HH | −44 | −42 | −37 | −38 | / | / |
| X (9.3 GHz) | VV | −39 | −33 | −31 | −26 | −20 | −10 |
| | HH | −39 | −33 | −32 | −31 | −24 | −12 |
| Ka (35 GHz) | VV | / | −31 | −30 | −26 | −20 | −4 |
| | HH | / | / | / | −27 | −20 | / |

The average polarization RCS value per square meter unit area measured at different sweep angles is listed in Table 1.10.

### *1.1.8 Radar Antenna*

Analyzing the radar antenna here, we do not understand how its mainlobe gain is obtained and how efficient it is. The focus is on the characteristics of the sidelobe studied, to what extent its lowest sidelobe can be reduced, and its impact on radar-shielded jamming.

$$\text{Radar directive gain } G_D = \text{Maximum radiation intensity} / \text{Average radiation intensity} \quad (1.6)$$

The average radiation intensity at the solid angle of the entire $4\pi$ solid radian is equal to the total power radiated by the antenna divided by $4\pi$. The directional gain of Eq. (1.6) can be written as:

$$G_D = 4\pi(\text{Maximum power radiated per unit solid angle}$$
$$/\text{Total power radiated by the antenna}) \tag{1.7}$$

$$G_D = \frac{4\pi P(\theta,\phi)_{\max}}{\iint P(\theta,\phi)\sin\theta d\theta d\phi} = 4\pi/B \tag{1.8}$$

B is Beam Area and is defined as:

$$B = \frac{\iint P(\theta,\phi)\sin\theta d\theta d\phi}{P(\theta,\phi)\max} \tag{1.9}$$

If $\theta_B$ and $\phi_B$ are half-power beamwidths (radians) in two orthogonal planes, beam plane B is approximately equal to $\theta_B\phi_B$. By Plug it in to (1.8):

$$G_D \cong \frac{4\pi}{\theta_B\phi_B} \tag{1.10}$$

An actual antenna is expressed in radians, and the antenna gain can be nearly written as:

$$G \cong \frac{26{,}000}{\theta_B\phi_B} \tag{1.11}$$

For example, if a needle beam $\theta_B$ and $\phi_B$ are both 1°, then G equals 44 dB.

The antenna pattern cannot be completely free of sidelobes. The sidelobes are radiated in the direction of 360°. Figure 1.4 is a radiation pattern of a particular parabolic antenna, which illustrates its main beam and sidelobes. Because it has sidelobe radiation, it can also be received in the sidelobe according to the antenna reciprocity characteristics. This allows jamming to be multiplied, so people implement jamming to the sidelobes to cover the target.

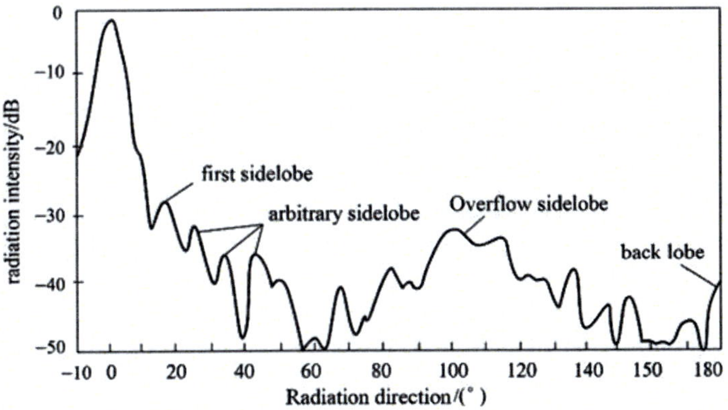


**Fig. 1.4** The radiation pattern of a particular parabolic antenna

The low sidelobe antenna pattern has other sidelobes besides the first sidelobe. The maximum side-lobe level of a traditional reflector antenna is approximately between −23 and −28 dB of the mainlobe peak. If the sidelobes are required to be between −35 and −40 dB, it is difficult to implement a conventional reflector antenna designed by a common method.

Generally low sidelobe antenna is −30 to −40 dB, ultra-low sidelobe antenna can be lower than −40 dB or −50 dB. Phased array radar, which performs sidelobe zeroing within 360°. That is, the sidelobe level in a certain direction is adjusted to the minimum. The sidelobe of the sidelobe zero adjustment point is less than −50 dB, and the number of zero adjustments is about 2–3. Sidelobe zeroing has little effect on the jamming array (or network).

The information age requires radar to play a role in ecological monitoring, environmental monitoring, resource exploration, transportation management safety guarantee, major natural disaster monitoring, ground observation, manned space flight, lunar exploration project, long-range precision strike, national defense security. Thereby promoting the development of radar in the following aspects:

1. Microwave imaging, high-resolution SAR, low-frequency SAR, ISAR, SAR positioning technology, SAR small satellite and constellation technology, multi-parameter synthesis technology, bistatic SAR technology, three-dimensional imaging
2. Space detection phased array radar technology, X-band active phased array technology, millimeter wave technology, laser meteorological radar technology
3. Air traffic control technology, S-mode secondary surveillance radar technology, new generation primary surveillance radar technology, precision tracking radar technology, small ground activity surveillance radar and its networking technology, and airport integrated radar technology
4. Ground-penetrating and safety radar detection technology, UWB high-power signal generation and transmission technology, UWB receiver technology, advanced signal processing technology, UWB radar design technology
5. Multifunctional integration technology, digital radar technology, and networked radar technology
6. Multi-sending and multi-sending branch networking

## 1.2 Development of Radar Countermeasures

Radar was produced on the eve of World War II, and developed during World War II, especially after World War II. Radar countermeasures are technologies developed with the advent of radar. Radar countering it is not like radar, it is designed according to its own needs and tactical requirements. Radar is developed according to its individual needs and tactical demands. Radar countermeasures are highly dependent. For a radar jammer, as long as the radar does not exist, the jammer will also be useless and will no longer exist.

### 1.2.1 Development of Radar Countermeasures Abroad

Foreign countermeasures formed different developments characterized by two camps after World War II. Their focus on developing electronic countermeasure equipment is also different:

Taking the former Soviet Union as an example, the characteristics of their development of electronic countermeasure equipment are mainly ground-to-air and ground-to-space jamming equipment, and subsequent development of airborne and shipborne electronic warfare equipment. However, the United States primarily focuses on the development of airborne, shipborne, and spaceborne electronic warfare equipment, and subsequent development of some ground electronic warfare equipment.

The Soviet Union ground-to-air electronic warfare equipment has developed four generations of products, and the first generation of ground-to-air jammers mainly based on СПБ-1 and СПБ-5; The second generation of Earth-to-air jamming system dominated by СПБ-13. The third generation of ground-to-air jamming systems represented by СБН-1, 2, 3; the fourth generation of jamming systems designated by improved СБН-1, 2, 3. The number of ground-to-air jamming systems is also astonishing. It is reported that there are more than 2000 sets of existing ground-to-air jammers in Russia.

The United States has more than 100 EA-6B jammers; hundreds of unmanned reconnaissance aircraft and jammers; more than 160 E-2C early warning aircraft; dozens of E-3A early warning aircraft; thousands of jamming aircraft, and hundreds of shipboard jammers.

These two major military powers have a different emphasis on the development of electronic warfare equipment, mainly from their different strategic positions. As in Russia today, its overall strategic position is strategic defense. Therefore, its electronic warfare equipment system is mostly ground-to-air and ground-to- space. But it also has certain strategic offensive capabilities, so it engages in some airborne electronic warfare equipment, shipborne electronic warfare equipment, and spaceborne electronic warfare equipment. The United States is now the only superpower in the world, and its strategic position is the position of strategic offensive. Therefore, they mainly developed airborne electronic warfare equipment, anti-radiation weapons, cruise missiles, spaceborne electronic warfare equipment, and global "network-centric warfare". Electronic defenses for them are secondary development. The United States now knows that it can only dare to attack other countries, and nobody in other countries dares to attack it first. If there is an attack, it can only be a terrorist attack. This is also determined by its current strategic position.

Judging from the overall situation, our country is still in a strategic defensive position. At present, although there are issues related to the Taiwan Strait and operations against Taiwan, that is also a strategic defense. Therefore, our electronic warfare equipment is mainly defensive and some offensive equipment.

### *1.2.2 Overview of the Development of Chniese Electronic Countermeasures*

China's electronic countermeasure was developed by relying on our wisdom and strength when the other party blocked the embargo. China's electronic countermeasures that can cope with the electronic attacks of modern powers, and can also have the electronic warfare offensive capability of local warfare.

It can be roughly divided into three development stages: primary stage, slow development stage, development stage:

1. Primary stage

   During the period of Anti-American Aid Korea and Anti-American Aid Vietnam, the enemy used not only heavy troops to attack North Korea, but also used active and passive jamming to jamming with our radar and communications.

   At that time, we relied on our strength to cope with these complex situations. When US-made U-2 and P2-V reconnaissance aircraft came to the mainland for harassment and surveillance activities, they were shot down by our brave air force. Among the wreckage of these aircraft are warning devices, answering jammers, etc. Although they blocked us, they sent us the actual reference equipment of the jammer. It is from these damaged reference devices that our country's electronic countermeasure equipment developers acquired knowledge of electronic warfare equipment, and gradually improved our level of electronic warfare equipment development.

2. Slow development stage

   Based on the basic theory learned and taking the wreckage of airborne jamming equipment along with Soviet ground-to-air jamming equipment as a reference, we gradually improved our independent research. During this period, we have successfully developed reverse-phase answering jammers, answering jammers to mono-pulse radars, anti-"100-tongue bird missile" radar add-ons, airborne warning equipment, vehicle-mounted ground-to-air jammers, vehicle-mounted reconnaissance equipment, etc. Although these devices cannot yet form the combat capability of electronic warfare, however, it has trained and cultivated a team of scientific research and teachers of electronic warfare equipment, which laid a good foundation for the subsequent independent development of Chinese electronic warfare equipment.

3. The relatively fast development period

   After the reform and opening-up, China's economic strength has continued to increase. The introduction of China's equipment has gradually increased, and it has also increased our research team and investment. Mainly after the Gulf War, we recognize the role of electronic warfare in war, and it is one of the key factors that determine the victory or defeat of the war. Besides, in 1994 and 1997, our country was twice harassed by unidentified air, making national leaders realize that electronic warfare is imminent. Electronic warfare is taking place every day. Therefore, we have strengthened the construction of electronic warfare development teams and electronic warfare combat forces, and at the same time, increased the investment quota of electronic warfare equipment. During this period, we independently developed jamming systems for early warning aircraft, airborne self-defense jammers, airborne jamming pods, airborne warning systems, photoelectric countermeasure systems, communication countermeasure systems, spaceborne reconnaissance equipment, shipborne reconnaissance, Jamming system, etc. China has enhanced the strength of electronic warfare, bringing the capability of electronic warfare to the advanced international level.

### 1.2.3 Overview of the Development of Radar Countermeasure Technology

After World War II, two camps in the world are actively developing new radars. The emergence of various fire control radars, long-range warning radar, mono-pulse radar, frequency agile conversion radar, intra-pulse multi-carrier frequency radar, phased array radar, pulse compression radar, Pulse-Doppler radar, and synthetic aperture radar has brought great pressure to the development of jamming radar and jammer technology. To be able to respond to the development of radar on time and effectively jamming with these new system radars, it takes a lot of human resources and material resources. Under the conditions of the other party's strict embargo and blockade, the technical personnel engaged in radar countermeasures in our country have paid a lot of energy to deal with these rapidly developing objects, as shown in Fig. 1.5.

Figure 1.5 is the timetable of radar and corresponding measures, showing the development process of radar and the time of successful jammer tests respectively. Generally speaking, radar jammers are produced with the new system radar and developed with the development of the new system radar. There must be a time difference, and this is the characteristic of Fig. 1.5.

There are several categories of radar. According to the purpose of radar, it can be divided into search warning radar, height measurement radar, three-coordinate radar, fire control radar, guidance radar, navigation radar, etc. According to the constitution of radar, it can be generally divided into continuous wave radar, conventional pulse radar, pulse compression radar, frequency conversion radar, Doppler radar, phased array radar, synthetic aperture radar, and inverse synthetic aperture radar. Some of the development of the radar system is improved and changed to overcome some of its shortcomings. Some are developed for anti-jamming, and some radars have developed a new system that overcomes certain shortcomings and increases anti-jamming capabilities. The development of the new radar system has also promoted the development of radar jamming technology, which is a pair of different parties. As long

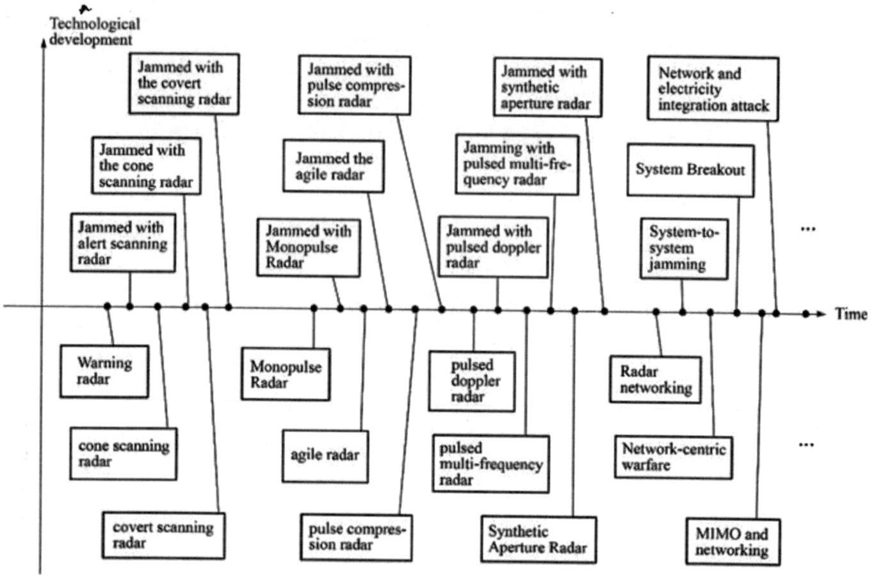


**Fig. 1.5** Schematic diagram of radar countermeasure and development

as there is war in the world, radar and radar countermeasure will develop mutually dependently and confront each other. Decades of experience in the development of radar countermeasure tells us that no radar cannot be disturbed, nor jamming that cannot be resisted. As long as you get to the bottom of each other, you can jam with their radars according to their characteristics and weaknesses. Similarly, knowing the other party's jamming radar can easily resist jamming. Therefore, in the radar countermeasure, both parties will pay great attention to confidentiality.

According to the carrier's platform classification, it can be divided into ground radar, shipborne radar, airborne radar, bomb-borne radar, spaceborne radar, vehicle-mounted radar, balloon-borne radar, etc. People describe radar as a clairvoyant, that is, when these platforms are equipped with radar, it seems like they have large eyes, so that these platforms can detect targets early and defend and attack in time. In terms of radar countermeasures, it is divided into self-defense jammers and cover jammers.

The self-defense jammer is to install the jammer on top of the shielded target. Covering jammers is to install jammers on carriers to cover other objectives. Whether it is a self-defense jammer or a cover jammer, it is also divided into ground countermeasure equipment, vehicle-mounted countermeasure equipment, shipborne countermeasure equipment, airborne countermeasure equipment, bomb-borne countermeasure equipment, and spaceborne countermeasure equipment. No matter what kind of classification, the most basic is that different radar systems need corresponding jamming methods and countermeasures.

### 1.2.4 Jamming Equation to Conventional Pulse Radar

Let us take the ground-to-air jamming as an example to illustrate the derivation process of its jamming equation.

The schematic diagram of ground-to-air jamming is shown in Fig. 1.6.

Establish a coordinate system based on the jamming station shown in Fig. 1.7.

A—the position of the jammer; B—the position of the radar; C—the position of the protected target; E—the projection point of the radar carrier on the ground; θ—Radar antenna sidelobe beam angle; $R_t$—the distance between radar and the protected target; $R_j$—the distance between radar and the jammer; d—the distance between jammers and protected target; H—the height of the carrier; E—the projection point of point B. β—the elevation angle of the jammer antenna.

$P_r$ is the echo strength of the signal reflected by the mainlobe of the radar antenna received

$$P_r = \frac{P_t G_t^2 \lambda 2\sigma L_d}{(4\pi)^3 R_t^4} \tag{1.12}$$

$P_n$ is the power of the jamming signal received by the radar antenna sidelobe

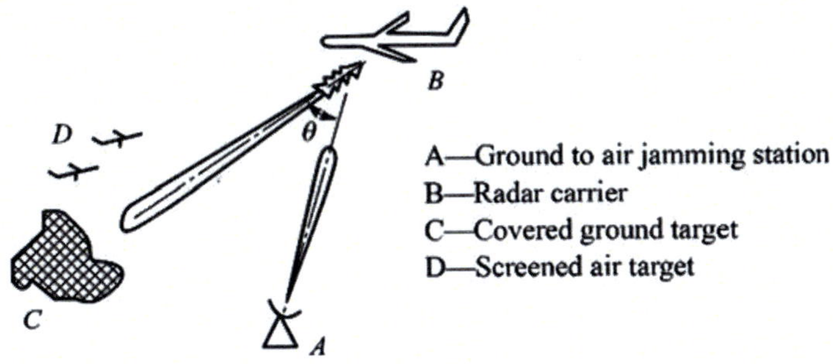


**Fig. 1.6** Schematic diagram of ground-to-air jamming

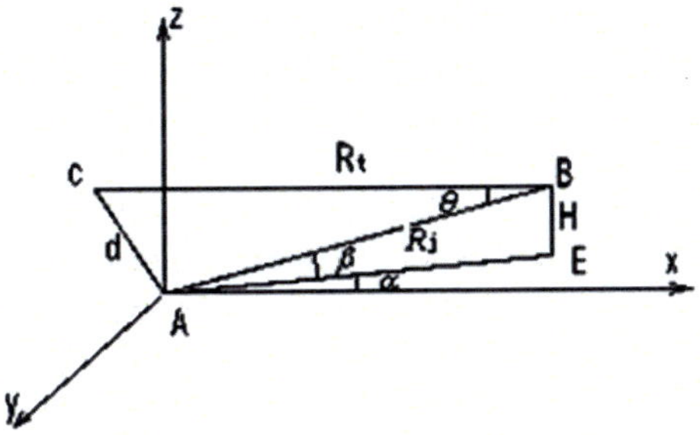


**Fig. 1.7** Coordinate system of ground-to-air jamming

$$P_n = \frac{P_j G_j K_h L_j}{4\pi R_j^2} \cdot \frac{\lambda^2}{4\pi} G_t(\theta) \tag{1.13}$$

According to the definition of jamming suppression coefficient:

$$K_j = \frac{P_n}{P_r} \tag{1.14}$$

Plug (1.12), (1.13) into (1.14): you can get the Power ratio of jamming signal and the echo signal entering radar receiver

$$K_j = \frac{4\pi \mathrm{P_j G_j R_t^4}}{P_t G_t \sigma R_j^2} \cdot \frac{Gt(\theta)}{Gt} \cdot \frac{K_h L_j K_f}{L_d} \tag{1.15}$$

$$P_j G_j = \frac{\mathrm{K_j P_t G_t} \sigma \mathrm{R_j^2}}{4\pi R_t^4} \cdot \frac{G_t}{G_t(\theta)} \cdot \frac{L_d}{K_h L_j K_f} \tag{1.16}$$

(1.15) or (1.16), collectively referred to as the radar jamming equation.

If it is self-defense jamming, the jammer is installed with the protected target, then:

$$P_j G_j = \frac{\mathrm{K_j P_t G_t} \sigma}{4\pi R_t^2} \cdot \frac{L_d}{K_h L_j K_f} \tag{1.17}$$

In the formula:

| | |
|---|---|
| $K_j$ | Jamming suppression coefficient. The minimum ratio of the power of the jamming signal to the power of the echo signal required at the input of the radar receiver when it can effectively jam with the radar. |
| $P_j G_j$ | The equivalent power radiated by the radar jammer. |
| $P_t G_t$ | Radar radiated equivalent power. |
| $\sigma$ | The equivalent area of the protection target in the irradiation area of the mainlobe of the radar antenna. |
| $R_j$ | Distance from the jammer to radar. |
| $R_t$ | Distance from the radar to the protected target. |
| $L_d$ | Radar transmitter 's feed loss coefficient. |
| $K_h$ | The loss factor of disturbance polarization mismatch. |
| $L_j$ | The loss coefficient of Disturbance feed. |
| $K_f$ | The ratio of radar receiver bandwidth to jamming spectrum width. |
| $G_t/G_t(\theta)$ | Radar antenna main/sub gain ratio in jammer direction. |
| $\sigma_i$ | Effective reflection area of terrain and surface features. As shown in Fig. 1.8. |

$$\sigma_i = r \Delta R_a \rho_i \tag{1.18}$$

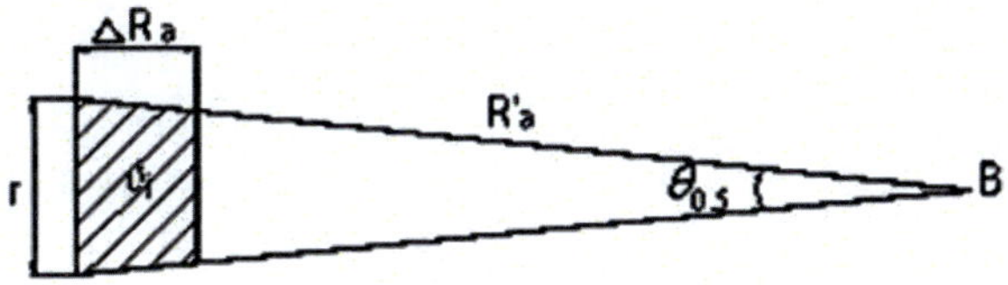


**Fig. 1.8** Schematic diagram of the equivalent reflection area of ground features

In formula, $r = 2R_a' \tan \frac{1}{2}\theta_{0.5}$; $\Delta R_a = \tau c$; $R_a' = R_a \sin \beta_1$.

Substitute the parameter into (1.17) formula:

$$\sigma_i = R_a \cdot \tau \cdot c \cdot \theta_{0.5} \cdot \rho_i \cdot \sin \beta_1 \tag{1.19}$$

$R_a$ The distance between the radar and the center of the protected target
$\tau$ Radar pulse width; C: Speed of light; $\theta_{0.5}$: Radar antenna horizontal beamwidth
$\beta$ Radar pitch angle to a ground target
$\rho_i$ Unit effective reflection coefficient.

It can be seen that the reflection coefficient of ground features is a complex function. It has different reflection coefficients depending on the geology. When the actual area of the target detected by the radar is greater than the irradiation area of the radar, the effective reflection area of the target should be analyzed using (1.19). If the actual area of the target (such as airplanes, tanks, small bridges, etc.) is smaller than the irradiation area of the radar, the equivalent reflection area of the target should be used for calculation. If the target needs to be divided into small units to analyze its equivalent reflection area, the equivalent reflection area in its resolution area will be analyzed in the synthetic aperture radar. The reflective area of the target is a specially developed subject. Especially with the emergence of stealth aircraft and absorbing materials, studying the equivalent reflective area of objects has its special application value and practical significance.

### *1.2.5 Jamming to Cone Scanning Radar*

After the Second World War, the main system adopted by the fire control radar was the cone scanning system. Its main working principle is to shift the maximum point of the antenna beam by an angle from the optical axis. Generally, the point of half power of the antenna beam coincides with the optical axis, and a small motor is used to drive the antenna feed to rotate to form a cone sweep as shown in Fig. 1.9.

AB is the optical axis of the antenna. BC is the maximum direction of the antenna beam. The maximum beam point of the antenna rotates in a dotted circle, and the frequency of rotation is the cone sweep frequency, which is generally tens of hertz (Fig. 1.10). If the target is in the C direction, the received pulse signal group has a low-frequency modulated signal, as shown in Fig. 1.11.

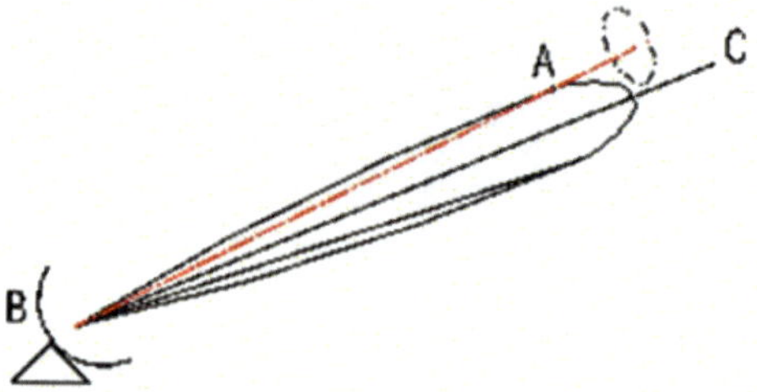


**Fig. 1.9** Schematic diagram of cone scanning antenna beam

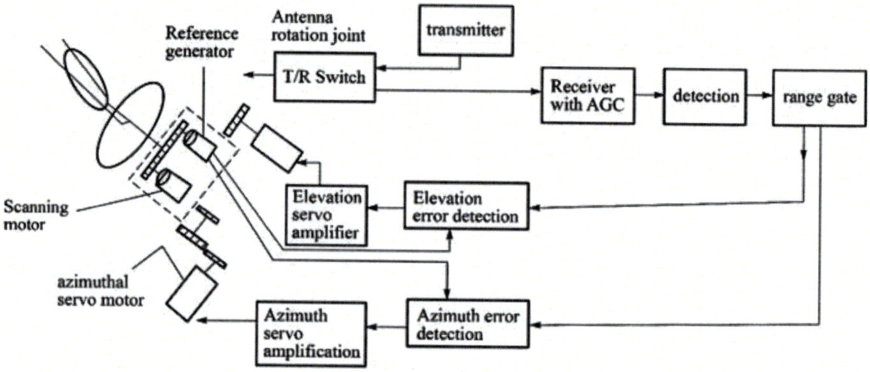


**Fig. 1.10** Schematic diagram of cone scanning antenna beam

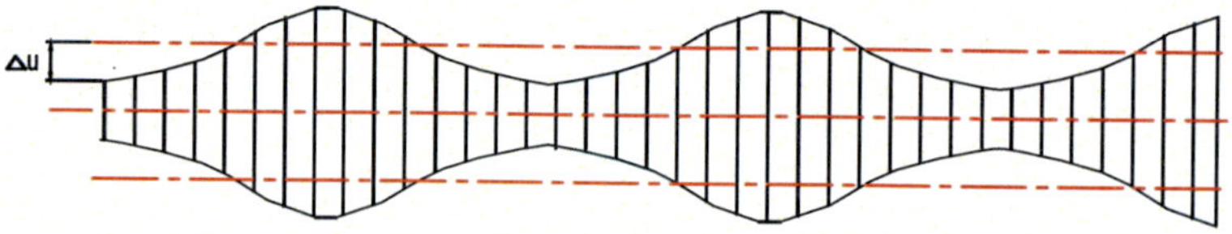


**Fig. 1.11** Schematic diagram of the radar amplitude modulation signal received on the target

When the radar is not disturbed, the target echo signal received by the radar is also modulated signal this way. The radar receiver amplifies this echo and then extracts the low-frequency signal after envelope detection. The frequency of this signal represents the cone sweep frequency. $\Delta u$ Modulation amplitude represents the size of the target polarization axis. The phase shift of the low-frequency signal represents the angle of the target of the optical axis. As long as the radar receiver can detect the presence of a low-frequency AM signal, the error signal is decomposed into azimuth error voltage or elevation error voltage to drive the azimuth motor and the elevation motor to rotate, so that the radar antenna rotates closer to the target. At this time, it rotates to the C direction, and the signal received in the C direction is a constant amplitude pulse train, as shown in Fig. 1.12.

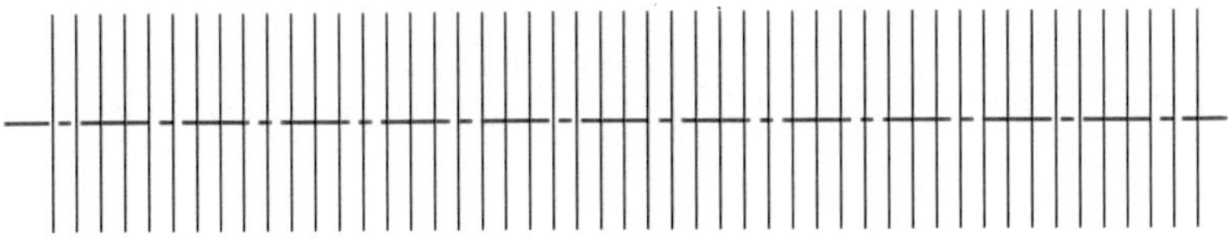

Fig. 1.12 Schematic diagram of the signal received on the target when the optical axis of the radar antenna is aligned with the target

At this time, the echo signal received on the radar antenna is also an equal amplitude burst. The radar receiver cannot solve the low-frequency modulation signal, indicating that the radar antenna optical axis is aligned with the target. When the target moves in a certain direction, another declination occurs between the optical axis of the radar antenna and the target. At this time, $\Delta u$ not equal to 0, and radar receiver can decode low-frequency modulated signals. This low-frequency modulated signal drives the radar's servo antenna to rotate in the direction of the target. When the target moves to that direction, the radar antenna will turn in that direction. This is the process of the cone scanning radar that automatically tracks the target.

#### 1.2.5.1 Phase-Reversal Answering Jamming

When installing a phase-reversal repeater jammer on the target machine, the radar cannot automatically track the target. The basic composition of phase-reversal answering jammer shows in Fig. 1.13.

The composition of the inverting jammer includes the receiving and transmitting antennas, low power amplification, low-frequency envelope-demodulation, inverting amplifier, square wave modulation amplifier, last-stage pulse power amplifier, etc. In Fig. 1.13, the output waveforms of points a, b, c, and d are shown in Fig. 1.14.

Figure 1.14 point a is the burst waveform of the received radar radiation.

Figure 1.14 point b is the amplified output waveform of low-frequency envelope-demodulation.

Figure 1.14 point c is the output waveform after inverting amplification.

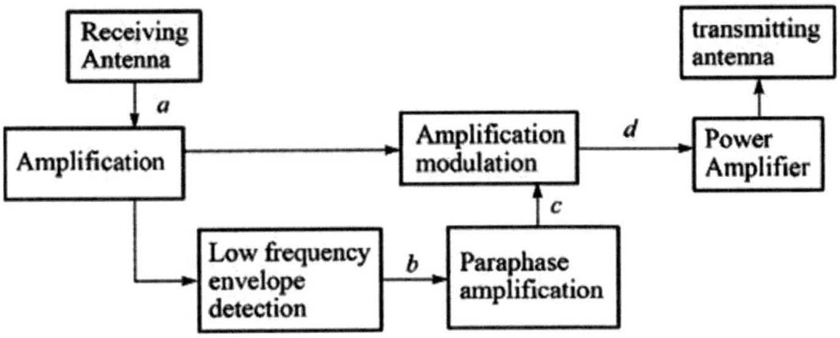


Fig. 1.13 Basic composition of phase-reversal answering jammer

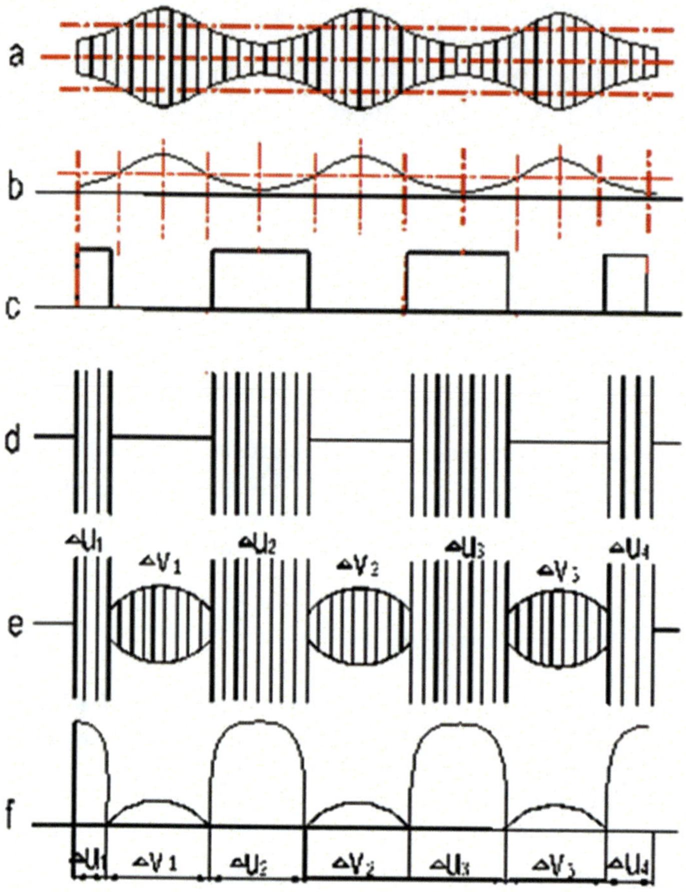


**Fig. 1.14** Schematic diagram of the jamming emission waveform, the echo received by the radar, and the mixed waveform of the jamming signal

Figure 1.14 point d is the waveform of amplified modulation output. The jamming antenna radiates this waveform after the power amplifier.

In Fig. 1.14, a, b, c, d are the waveforms of points a, b, c, d in Fig. 1.13. The waveform at point e in Fig. 1.14 is the mixed waveform of the jamming and echo signals received by the radar receiving antenna. $\Delta u_{1,2,3,4}$ are the phase-reversal answering jamming signal. $\Delta V_{1,2,3}$ are target echo signals. $\Delta u_{1,2,3,4}$ drive the motor to rotate away from the target. Due to $\Delta u_i > \Delta V_i$, after the radar is disturbed, the radar antenna is turned away from the target so that the radar cannot automatically track the target.

#### 1.2.5.2 Distance Deception and Phase-Reversal Answering Jamming

The reversing phase jammer shown in Fig. 1.13 needs to answer the jamming signal pulse power is several times larger than the echo signal to have a good jamming-effect. If the answering-type inverted jamming is combined with the distance spoofing jamming, then the jamming pulse power only requires 1: 0.8 to play a good jamming-effect. The composition of distance spoofing plus phase reversal jammer is shown in Fig. 1.15.

In Fig. 1.15, to generate range spoofing, the RF pulse of the radar signal must be widened. The RF broadening circuit used in the past was mainly composed of a coupler, RF delay line, and amplifier. Point a in Fig. 1.15 is the narrow pulse of the received radar signal, with a typical value of 0.5 μs. There is a delay of about 0.1 μs after the delay line. The delay line has a loss, so use the amplifier to cancel the delayed loss and lose to b. The first pulse is widened by 0.1 μs, and the signal of b is coupled to the input of the delay line through the coupler. After the delay line is delayed by another 0.1 μs, the pulse is extended by 0.2 μs in the second cycle. Forty cycles in this way, the pulse is widened by 4 μs. The microwave signals at points a and b are shown in Fig. 1.16.

In Fig. 1.16, c is the moving waveform generated by the moving pulse. Its first pulse has a delay time, which is about 0.1 μs. To the nth pulse, delay $\Delta\tau n$, generally take 4–6 μs. d is the jamming signal transmitted by answering and distance deception. After receiving this jamming signal, the radar first automatically tracks the wave gate and automatically moves backward with the deceptive signal. When the range gate shift (radar pulse width), all the pulse signals output from the distance tracking circuit are jamming signals. These jamming signals are controlled by an e-point square wave, which just causes the low-frequency signal detected by the radar to deviate from the original, useful signal by 180°. This signal causes the radar motor to drive the radar antenna to rotate away from the target. Because its useful signal is not output in the

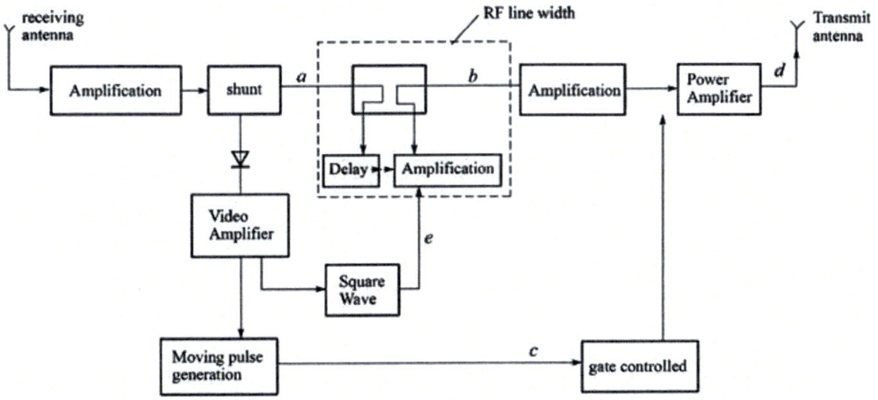


Fig. 1.15 Block schematic of the composition of distance spoofing plus phase reversal answering jammer

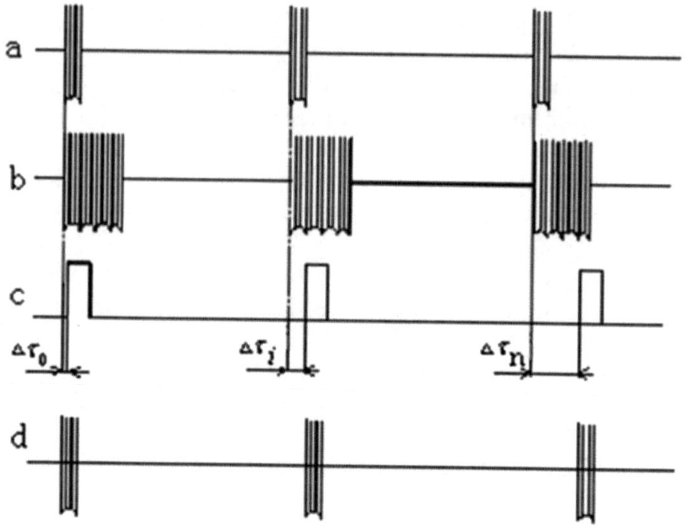


**Fig. 1.16** Schematic diagram of the waveform generated at each point

radar angle tracking branch and the jamming signal has output, the jamming signal is smaller than the echo and can also cause the radar to lose the target. Range spoofing plus inverted answering jamming, the waveform relationship generated in the radar is shown in Fig. 1.17.

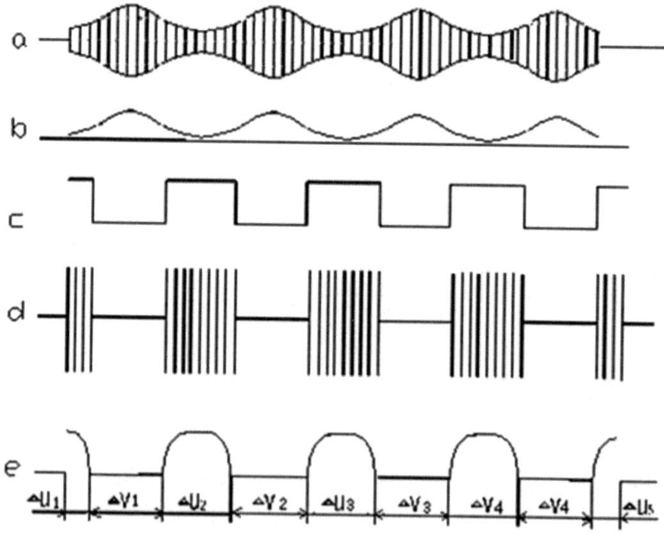


**Fig. 1.17** Waveform diagram of main contacts of distance spoofing plus phase reversal answering jammer

In Fig. 1.17, a and b are the normal target echoes received when the radar is not disturbed. Point e is the waveform at point e in Fig. 1.15. Point d is the jamming waveform output at point d in Fig. 1.15. Point f is the input waveform of the radar angle tracking branch.

$\Delta V_{1,2,3\ldots}$ are the voltage that causes the drive motor to rotate in the target direction. Because the radar's automatic range gate is disturbed by deception and has no output, that is $\Delta V_i = 0$, the drive motor cannot rotate the radar antenna in the direction of the target. However, $\Delta u_{1,2,3\ldots}$ have voltage output, which makes the drive motor rotate away from the target, and the radar loses the target. As long as the jamming signal can move the range gate to follow the jamming signal, the radar will lose the target in the azimuth, and the jamming/signal ratio is generally 0.8.

In the early 1960s, the distance spoofing plus answering -type inverted jammer was installed on the P2-V of the American-made Chiang Kai-shek airplane. Our fighter's radar is a cone-scanning fire control radar. When the driver searches for the target with the radar and turns the radar into the aiming state, the radar loses the target. The fighter pilot relied on bravery and shot down the P2-V aircraft at close range. It was only after analyzing the P2-V wreckage that this distance deception plus inverted answering jammer appear.

Since the cone-scanning tracking radar is very susceptible to jamming, this radar was retired from service in the mid-1960s. Therefore, the inverted answering jamming will lose its existence value. Among the anti-jamming technicians, an improved scheme for the cone-shaped radar is proposed, namely the hidden cone-scanning radar.

### *1.2.6 Jamming to the Hidden Cone-Scanning Radar*

The original cone scanning radar is also called exposed cone scanning. The reason is that it can install a receiver on the target to receive the frequency of its conical-scan low-frequency signal. The hidden cone-scanning radar emits a constant amplitude pulse burst. Added cone scanning in the receiving system, the original circuit can still be used after receiving the cone scanning.

This improvement prevents the jammer from receiving the low-frequency signal of the cone sweep, and thus it is impossible to carry out the phase reversal jamming. This improvement prevents the jammer from receiving the low-frequency signal of the cone sweep, and thus it is impossible to carry out the phase reversal jamming.

After the prototype of the hidden cone-scanning radar was made, the superior asked us to find a way to jamming with it. Based on the characteristics of the hidden cone-scanning radar, we also made a prototype of the jamming to the hidden cone-scanning radar. Both parties agreed on the time and place of the countermeasure test and conducted back-to-back tests. First, let the radar work, track the target normally, then the jammer starts. After a few seconds, the radar antenna swings at a large angle and then loses its target. The results of repeated experiments are the same.

Finally, we tell the radar researchers about the method of disturbing the hidden cone-scanning radar. They think that this method can indeed jamming with the hidden cone-scanning radar. Since then, the hidden cone-scanning radar has been retired without production equipment. Speaking of which, everyone also wants to ask questions, what method do you use to jamming with the hidden cone-scanning radar? It is also very simple, that is, a low-frequency sweep circuit is added based on the original inverted answering jammer. The low-frequency sweep frequency is 20–80 Hz, and the frequency is scanned at a frequency of 3–5 Hz/s. When the low-frequency sweep frequency sweeps close to the radar's concealed cone sweep frequency, the radar antenna begins to swing a little. At this time, the receiver on the target receives the amplitude modulated signal of the radar burst. Detect this amplitude modulation signal and control the low-frequency sweep source to stop sweeping. The low-frequency signal is fixed at $f_a$, and the radar's hidden cone sweep frequency is $f_a'$. If $f_a'$ and $f_a$ are close to each other, the radar angle tracking system is equivalent to being jammed with by the inverted response. The antenna wobbles and then loses track of the target. The fire control radar abandoned the cone scanning system and the concealed cone scanning system and turned to the development of a monopulse tracking radar.

### *1.2.7 Jamming to Monopulse Tracking Radar*

The jamming to the monopulse tracking radar is very different from that of the cone scanning tracking radar:

1. Monopulse radar extracts tracking error information in one pulse, while cone scanning tracking radar requires 6–8 pulses to obtain angular tracking error information.
2. The monopulse radar angle tracking receiver uses multi-channel reception. Generally, there are two azimuth and two elevation receivers. The conical tracking receiver is a single-channel receiver.
3. The tracking error of monopulse tracking radar can be very small, up to a few micrometers. The tracking error of conical scanning tracking radar is generally in mils.
4. The anti-jamming capability of monopulse tracking radar is stronger than that of conical scanning tracking radar. For these reasons, the monopulse tracking radar has been vigorously developed, and the cone scanning radar was finally eliminated. The composition of monopulse radar is shown in Fig. 1.18.

In the mid-to-late 1960s, we researched jamming monopulse radar. At the time, the first contact with monopulse radar was to learn by doing. We listened to the technicians who engaged in monopulse radar, said monopulse radar could not jamming. There is indeed such a situation, and the use of an inverted answering jammer does not affect it. Phase-inverted jamming has become its beacon tracking and is very stable. However, these phenomena do not mean that there is no other way to jamming with

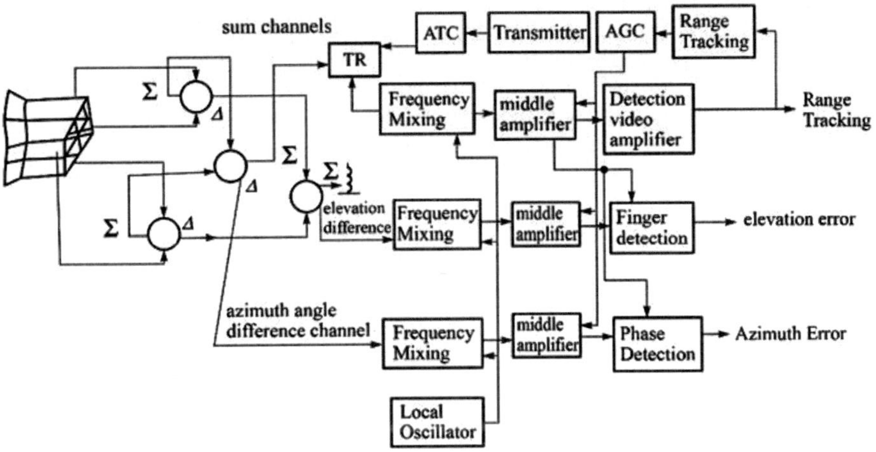


**Fig. 1.18** Block schematic of the composition of monopulse radar

monopulse radar. Later, based on learning the working principle of monopulse radar. Identified its weak link and jammed with it. We also received very good jamming effects.

### 1.2.7.1 Ground Jamming Test to Monopulse Radar

The jamming test to monopulse radar is shown in Fig. 1.19.

The jamming test prototype is composed of a signal source, a noise source, an amplifier, a moving pulse generator, and a transceiver antenna. A corner reflector (about 10 $m^2$) is placed next to the jamming transmitting antenna.

Test methods and steps:

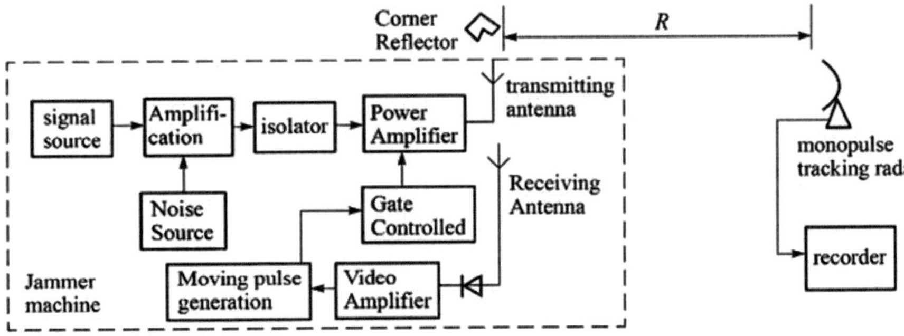


**Fig. 1.19** Composition of monopulse tracking radar jamming test

The first step: Turn on the radar. Make sure radar working normally, which it can track the angled reflector stably;
The second step: Adjust the frequency of the signal source to be the same as the radar frequency;
The third step: Turn on the jammer, and the antenna will start to swing or lose the target within about 2 s.

The recorder needs to be turned on all the time while doing experiments. The distance automatic tracking voltage, azimuth angle automatic tracking error voltage, and elevation angle automatic tracking error voltage are recorded.

The results of the test are the same after repeating the test several times.

The distance tracking error voltage, azimuth tracking error voltage and elevation tracking error voltage of the monopulse radar recorded on the recorder are shown in Fig. 1.20a, b.

In Fig. 1.20, $U_R$ is the monopulse radar distance automatic-tracking voltage.

$U_\beta$ is the automatic tracking voltage of mono-pulse radar elevation.

$U_\alpha$ is the automatic tracking voltage of mono-pulse radar 's azimuth.

The error voltage curves of $U_R$, $U_\beta$, $U_\alpha$ recorded from the recorder are consistent with the intuitive view of the antenna's swing and the target's loss after the radar is disturbed.

#### 1.2.7.2 Air Jamming Test on Monopulse Radar

The composition of air jamming test on monopulse radar is shown in Fig. 1.21.

Install the jamming test prototype on a transport plane, and the monopulse tracking radar on the fighter. First, let the monopulse radar track the target stably, and then notify the jammer to turn on. After the jammer is turned on, the radar antenna swings or loses its target. Three experiments were done, all of which were the same.

The automatic pulse recording display on the image display of the monopulse radar is consistent with the swing of the radar antenna after jamming, as shown in Fig. 1.22.

In Fig. 1.22, represents the direction of the muzzle. represents the image of the target being attacked. In this dotted circle, it indicates the allowed aiming error circle. If the target image is within this circle, firing cannons may hit the target. If the target image runs out of the circle, it means that the muzzle is not aimed at the target.

Figure 1.22 is the target image taken by a fast camera at four frames per second. From pictures 1 to 5 in Fig. 1.22, the target image has reached the edge of the dotted circle. The sixth picture has run out of the circle, and the eighth picture has run out of the circle too far, which again shows that the single pulse tracking radar can be use tojamming. However, at present, there are many articles on the jamming to monopulse radar, each with its own argument. Some analysis conclusions are still very different. Regardless of the analysis, it should be based on experiments.

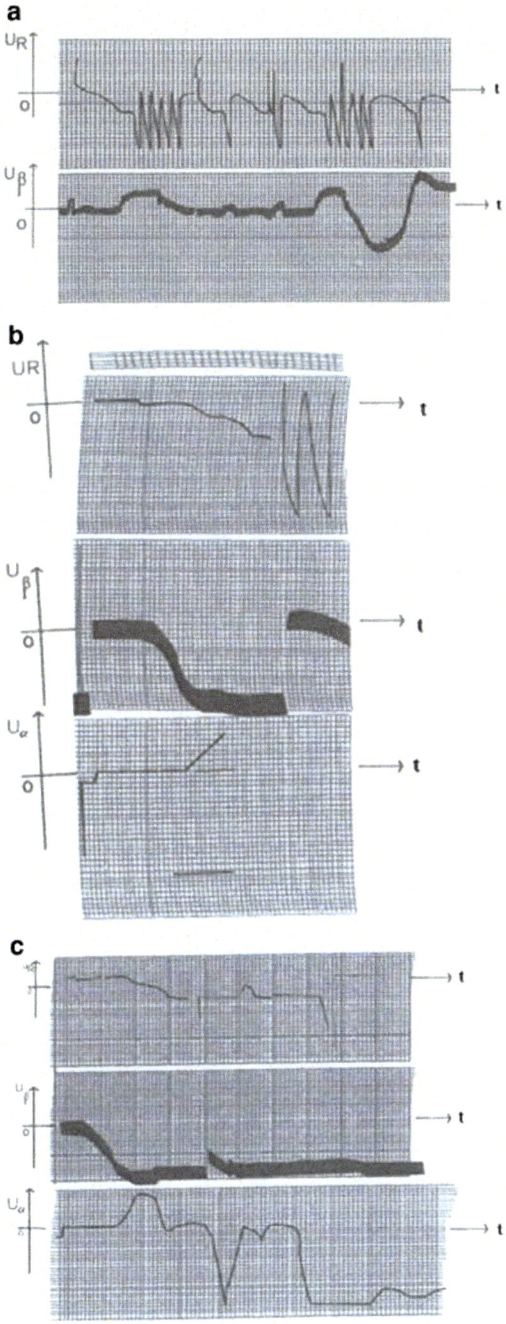


**Fig. 1.20** **a** The curve of $U_R$ and $U_\beta$ when the S/I ratio (voltage value) is 1.3, **b** $U_j/U_r = 2.5$ is the monopulse radar distance automatic-tracking voltage, **c** The curve of UR, Uβ, and Uα when the S/I voltage ratio greater than 2.5 times

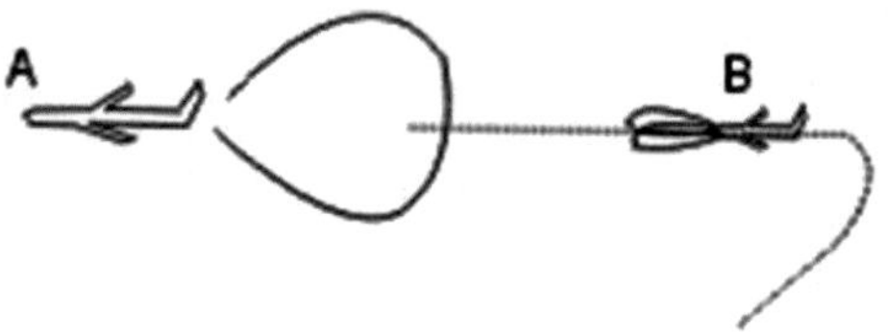


**Fig. 1.21** Air jamming test against monopulse radar. **A** Self-defense jamming carrier **B** Monopulse radar carrier

#### 1.2.7.3 Principle Analysis of Monopulse Tracking Radar Jamming

##### Broadband Aiming Continuous-Wave Noise Jamming

We have tried this jamming method many times, and the jamming effect is significant. How do we explain this principle of jamming? Originally, the monopulse radar can track the noise source from the basic principle. In the actual jamming test, it failed to track the noise source.

The reason we analyzed is: There are two assumptions for monopulse radar to start tracking: (a) The two channels of the tracking receiver system are balanced. (b) The bandwidth of the tracking receiver is infinitely wide, which is $B\tau > 1$ (B is the receiver bandwidth and $\tau$ pulse signal width).

The two channels of the tracking receiver are not completely balanced, and the receiver bandwidth is also limited.

We can think of the noise source as a random narrow pulse. For example, the receiver bandwidth is 2 MHz, and the random noise is a narrow pulse with a pulse width of 0.05 Ω. So, $B\tau = 2\times 10^6 \times 0.05\times 10^{-6} = 0.1 < 1$. Because the assumption of monopulse radar is not satisfied, it cannot track the noise source.

##### Range Spoofing Plus Noise Impulse Jamming

This is the experiment we did on the ground to jamming with monopulse radar. The test data shows that the jamming noise pulse voltage/signal echo pulse voltage is greater than 2.5 times can effectively jam with the monopulse radar, causing the target to be lost within two or three seconds after jamming. When the range spoofing jamming and range wave gate all track the noise pulse and then suddenly remove the jamming, the radar automatically turns to the angle search after losing information in the distance. This is to use distance spoofing to achieve the purpose of jamming in angle. The principle and method of range spoofing jamming to monopulse radar is the same as that to conical tracking radar.

##### Jamming of Monopulse Radar with Phase Wavefront Distortion

The jamming of monopulse radar with phase wavefront distortion is shown in Fig. 1.23.

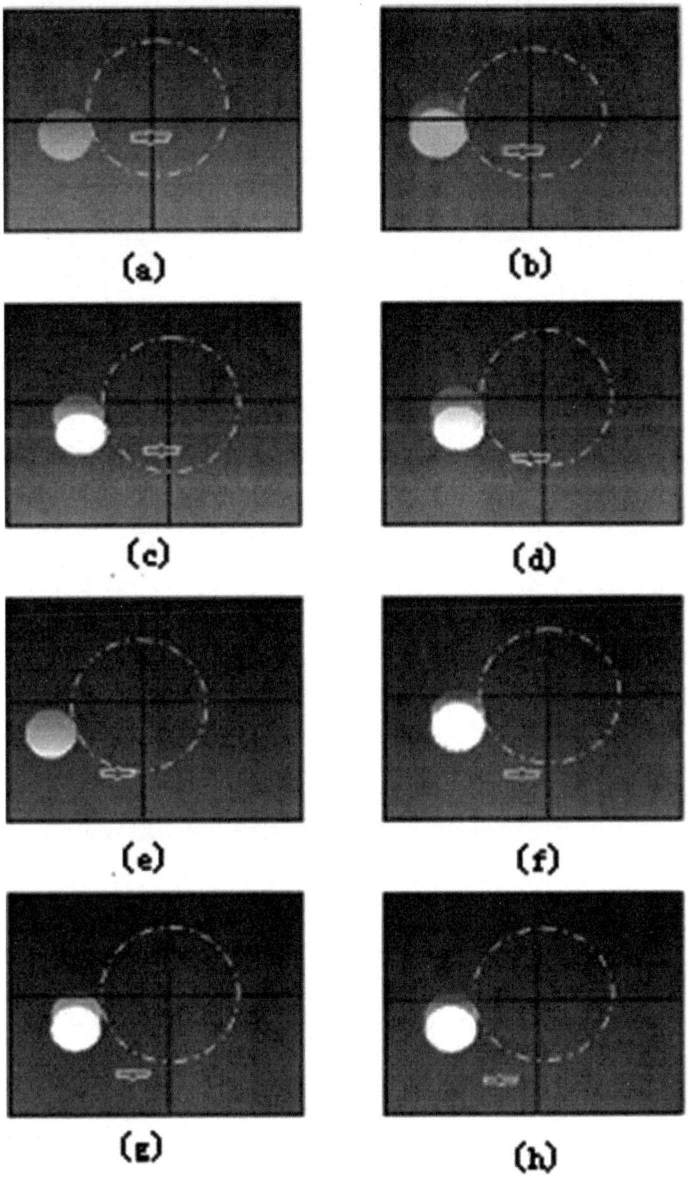

Fig. 1.22 Jamming monopulse tracking radar records in the air

**Fig. 1.23** The relative position between radar and target

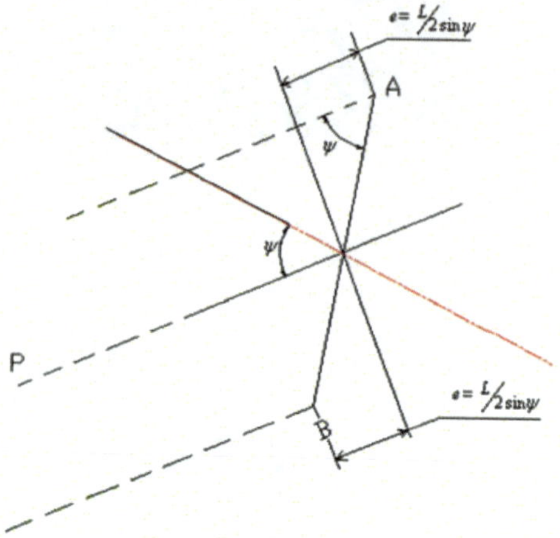


The two points A and B are the positions of the jamming antennas installed on the wings of the protected aircraft. Point P is the location of the radar. Point A is the jamming antenna 1. Point B is the jamming antenna 2. The distance between $\overline{AB}$ is L. If AB is perpendicular to PO, the signals radiated by A and B to P have no path difference. If it is tilted by an angle $\psi$, a wave path difference, that is, a phase inconsistency occurs. This error $\theta'$ is defined as:

$$\theta' = \left(\frac{L\cos\psi}{2}\right)\left(\frac{1-a^2}{1+a^2+2a\cos\psi+L\sin\psi}\right) \tag{1.20}$$

In the formula:

$\theta'$ The radar orientation error caused by the wavefront distortion generated by the AB point source.
a The amplitude ratio of the two-point source signal.
$\psi$ The angle between the connection of two-point sources and the vertical line of the radar.

According to the formula (1.20), the relationship curve between phase difference and phase $\psi$ can be obtained Fig. 1.24. At a = 0.966, the tracking error produced by $\psi = 180°$ is 29 times of L. For example, the two jamming antennas are installed on the wing of the protected aircraft. The two ends are 20 m apart. The ratio of the energy emitted by the two antennas is 0.966. The phase difference generated on the mouth of the radar antenna is 180°. Then the tracking deviation to the radar is 20 × 29 = 580 m. It can be seen from this example that the principle of phase wavefront distortion is used to jamming with monopulse radar.

At both ends of the wing of the protected aircraft, the jamming antenna is added, and an jamming emission source is shared. The amplitude and phase of the two outputs adjusted so that the output amplitudes of the two antennas are nearly equal

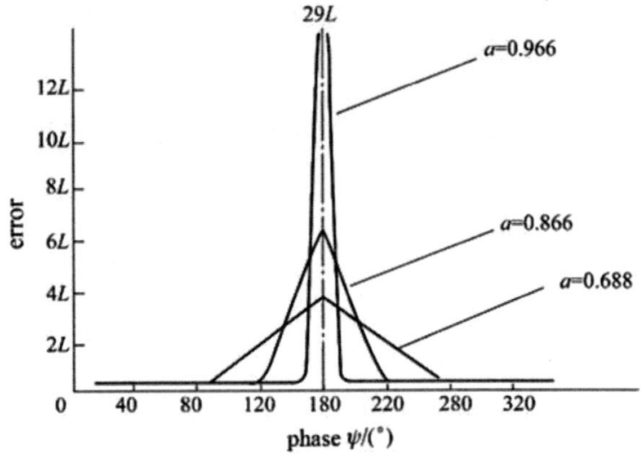


**Fig. 1.24** The relationship between jamming error of two-point sources and phase $\psi$

but not completely equal. The phase of the two channels is close to 180°. The jamming of two signals in space distorts the wavefront of the radiation source. When the radar receives this signal, the beam has deviated. Therefore, the deviation of angle tracking is introduced, which increases the angle tracking error of monopulse radar, making it unable to meet the requirements of aiming and shooting. This kind of jamming is completely established in principle. However, it is difficult to achieve this jamming. Because it is not easy to adjust the amplitude of the two channels to nearly equal, and it is not easy to adjust the phase of the two channels to nearly 180°. The required jamming power is also relatively large. Although the theory is established, the actual implementation is too difficult.

Frequency Detuning Jamming to Monopulse Radar

The center frequency of the echo signal carrier frequency of the monopulse radar is the center frequency of the passband of its receiver. If the center frequency of the jamming signal is at the edge of the passband of its receiver, plus $B\tau < 1$, according to the principle analysis and experimental results of monopulse radar, the polarity of monopulse radar error does not change with the position of the tracking antenna. And there is positive and negative error pulse output. Therefore, the oscillation of the tracking antenna causes the angle tracking error of the monopulse radar to exceed the tactical index allowable value. The theory of monopulse radar jamming is still not perfect. We still need more practice, from practice to recognition, and then repeated many times. We must constantly sum up experience and extract regular things from it. Then rise to theory to further guide effective jamming to monopulse radar.

### *1.2.8 Jamming to Frequency Agile Radar*

The frequency agile radar is produced specifically to resist the jamming of targeting noise. In the previous frequency aiming jamming machine, the general aiming frequency-time: In the 1950s it was in the order of seconds; In the 1970s it was in the order of milliseconds; In the 1980s it was in the order of microseconds; In the 1990s it was in the order of tens of nanoseconds.

In the 1970s, Since the aiming time needs to be in the order of milliseconds, the frequency agile radar changes its frequency every time a pulse is sent. The aiming frequency jammer was unable to keep up with its frequency change. So for a while, the frequency agile radar has played its anti-jamming role. In the 1980s, with the successful development of the instantaneous frequency measurement receiver, combined with the immediate frequency guidance, it can shorten the aiming time to the order of microseconds.

#### 1.2.8.1 Ground Shield Jamming

In the ground shield jammer, the jammer can be placed in front of the protected target. The receiver in the jammer receives the radar signal first, and then reflects the signal after the target receive the signal, as shown in Fig. 1.25.

In the formula:

A—The center of the protected target. r—The radius of the protected target. B—the position of the jammer. d—The distance from the jamming station to the protected target. θ—The invasion angle of the enemy aircraft.

According to the triangle theorem in Fig. 1.25.

$$\mathrm{r} + \mathrm{a} = \mathrm{d}\cos\theta \tag{1.21}$$

$$\mathrm{a} = \mathrm{d}\cos\theta - \mathrm{r}$$

Let $\tau$ be the delay time required for frequency measurement and aiming, $\nu$ be the speed of light

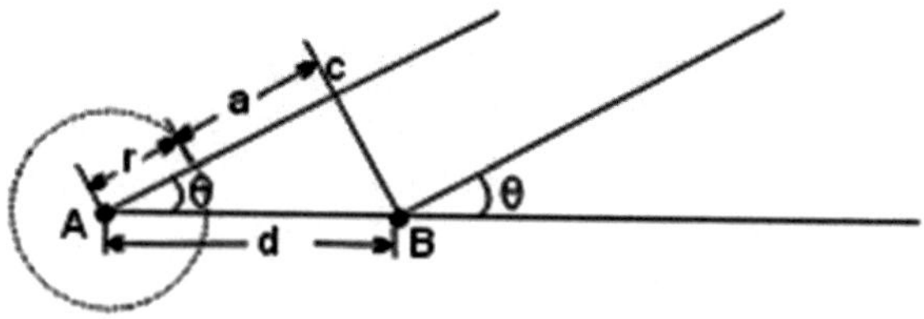


**Fig. 1.25** Schematic diagram of the configuration of the ground shield jammer and the protected target

**Table 1.11** Relationship between the jamming parameter and the delay time

| θ° | | 0 | 30 | 40 | 50 | 60 | 70 | 75 | 78 |
|---|---|---|---|---|---|---|---|---|---|
| $\tau_{\mu s}$ | d = 5 | 26.6 | 22 | 18.6 | 14 | 10 | 4.66 | 3.26 | 2.6 |
| | d = 3 | 13 | 10.6 | 8 | 6.1 | 3.3 | 0.18 | / | / |

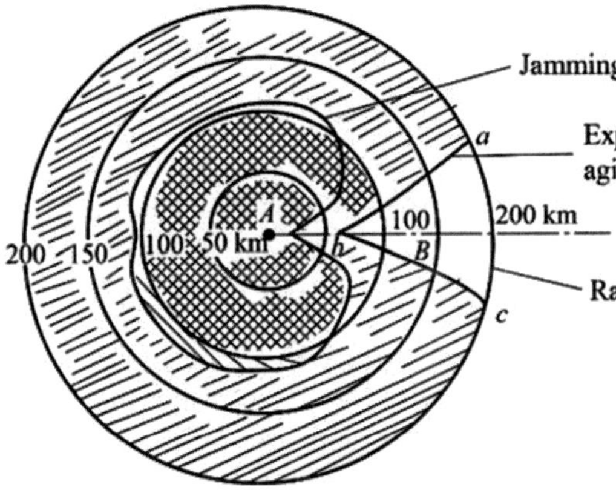


**Fig. 1.26** Schematic diagram of the jamming exposure area to conventional pulse radar and broadband random frequency agile radar. Point A is the position of radar. Point B is the position of the jammer

$$\tau = \frac{a}{\nu/2} = \frac{d\cos\theta - r}{1.5 \times 10^8}$$

If d = 3 or 5 km, r = 1 km, we get Table 1.11 after plug it in to (1.21).

According to Table 1.11, if the jammer is placed 3 km in front of the protected target and the sighting frequency is less than 3.3, within the range of θ is ±60°, it can effectively jam with the frequency agile radar. In other words, this kind of jamming is the same as jamming conventional pulse radar. As long as the jamming configuration meets the requirements of Fig. 1.26, the jamming effect is still very good.

#### 1.2.8.2 Spot Jamming for Frequency Agile Radar from Self-screening Jammer

The frequency agile radar also has different types of frequency-agile conversion, such as magnetron frequency agile conversion, fixed point frequency agile conversion, fixed multi-point frequency agile conversion, digital encoding frequency agile conversion, etc. For different frequency agile radar, there can be different jamming methods.

### Jamming to ESC Magnetron Frequency Agile Radar

The frequency agile of the magnetron has certain frequency agile rules. The frequency conversion law of magnetron frequency agile conversion is generally intensive at both ends of the frequency band, and relatively rare in the middle of the frequency band. Using frequency spectrum measurement, you can find out its frequency variation law and make frequency prediction. The frequency prediction method is generally based on the law of magnetron frequency agile conversion. List a mathematical expression equation. After receiving fourth, fifth pulses of the agile signal, it can be predicted at which frequency the sixth pulse should jump. After receiving the sixth pulse, it can be predicted that the seventh pulse should jump to that frequency, and so on. The frequency of agile radar conversion is precisely within the spectrum of the jamming signal. Its jamming effect is close to that of jamming conventional pulse radar.

Another jamming method for frequency agile magnetron: since the operating frequency point of the agile magnetron mostly works at both ends of the frequency band, the jamming spectrum of the jammer is pre-adjusted at both ends of the jammed radar frequency band, and the middle frequency band of the frequency band is jammed by fast sweep. Using this method to perform self-defense jamming on the magnetron frequency agile conversion radar can cause 85% of the pulses to be hindered. Therefore, its jamming effect can also be close to that of conventional pulse radar.

### Jamming to Fixed-Point Frequency Agile Radar

This frequency agile radar has some fixed frequency points, and random frequency hops on these frequency points. For example, if frequency points are f1, f2, f3, f4 ... f8, the radar will randomly jump at these eight frequency points. There is a frequency measurement memory device in the jammer. As long as it is measured once, these eight frequency points can be recorded. It transmits the jamming signals of these 8 frequency points, plus noise modulation, and no matter which frequency point the frequency agile radar jumps can be jammed with by noise suppression. At this time, the equivalent power of jamming at each frequency point is reduced by a factor of eight, which increases the jamming exposure distance by 1.7 times. For example, when the range of conventional pulse radar jamming exposure is 1 km, after changing it to frequency agile radar, its jamming exposure distance becomes 1.7 km, which still has a good jamming-effect.

### Jamming to Frequency Agile Radar with Fixed Multi-point Frequency Hopping

The fixed frequency hopping points n generally does not exceed 20 frequency points. For example, in the X frequency band, its operating frequency range is 500 MHz, and the number of frequency hopping points is 20. For this kind of frequency agile radar, the frequency storage memory can still be used. Jammer simultaneously transmits jamming signals at 20 frequency points, so that the radar will be subject to noise suppression jamming no matter which frequency point it jumps. This jamming increases the exposure distance by a factor of $(n)^{1/4}$. For example, when n = 20, the jamming exposure distance rises by 2.1 times. When there are more frequency points, the multi-frequency modulation loss of the jamming transmitter increases, and the jamming exposure distance is more than $(n)^{1/4}$ times.

#### Jamming to Frequency Agile Radar with Random Digital Code Frequency Hopping

For the jamming of this frequency agile radar, broadband noise barrage jamming is generally used. This jamming can increase the multiple of the jamming exposure distance $K_x$ (Empirical formula).

$$K_x = \frac{K_f \cdot \Delta F_a}{K_a \cdot \Delta f} \quad \text{Empirical formula} \tag{1.22}$$

In the formula:

- $K_f$ The ratio of receiver bandwidth to jamming spectrum width when conventional pulse radar jamming, which is $K_f = \frac{\Delta f}{\Delta F}$.
- $\Delta f$ The radar receiver bandwidth.
- $\Delta F$ The jamming spectrum width when jamming with conventional pulse radar.
- $\Delta F_a$ Jamming spectrum width when barrage jamming.
- $K_a$ The Coefficient of noise spectrum quality degradation in wide spectrum Jamming.

Substitute $K_f$ into (1.22) formula, $K_x$ becomes (1.23) formula:

$$K_x = \frac{\frac{\Delta f}{\Delta F} \cdot \Delta F_a}{K_a} = \frac{\Delta F_a}{K_a \cdot \Delta F} \tag{1.23}$$

For example, $\Delta F_a = 500$ MHz, $\Delta F = 5$ MHz, $K_a = 0.5$, after plug it in to (1.23), we get $K_x = 200$.

Then the Jamming exposure distance is increased by $(K_x)^{1/4}$ times, which is 3.8 times.

For broadband jamming, the Jamming exposure distance $R_{\min x}$ is defined as

$$R_{\min x} = \left[\frac{\Delta F_a}{K_a \Delta F}\right]^{1/4} \cdot R_{\min x} \tag{1.24}$$

In the formula:

$R_{\min x}$ The jamming exposure distance when jamming with conventional pulse radar.

Broadband barrage jamming has relatively little effect on self-defense jamming. However, the jamming exposure area has a great impact on the forecast radar. For example, the exposed area formed by jamming of jammer to certain radar is shown in Fig. 1.26. Because the radar uses random broadband frequency hopping anti-jamming, and the jamming uses broadband barrage jamming (other parameters remain unchanged). The two jamming exposure areas formed by these two types of jamming are shown in Fig. 1.26.

From Fig. 1.26, we can see that when the radar is not disturbed, its detection area is a circle with a radius of 200 km. If it is a conventional pulse radar, when the aiming continuous wave noise jamming with it, its effective detection area (jamming exposure area) is within ▩. Under the condition that the radar radiated equivalent power is the same, it is changed into a broadband agile radar. However, when the equivalent radiated power is the same, the jammer changes into broadband barrage jamming (Kx = 200), so that the effective detection area of the radar becomes the ▨ area. It increases the jamming exposure area greatly, making the effective area of jamming into a, b, c triangle area. It can be seen that the anti-jamming effect of the broadband random frequency agility radar is still very significant. To counter the anti-jamming measures of random broadband agile frequency conversion, we can increase the equivalent jamming power of the jammer or increase the number of jamming transmitters. Making it into narrow-band jamming or continuous with each other in the jamming band can effectively deal with random broadband frequency agile radar.

### 1.2.9 Countermeasure Against the "AGM-45 Shrike Anti-radiation Missile"

On the battlefield in Vietnam, the United States first applied the "AGM-45 Shrike anti-radiation missile". The "AGM-45 Shrike anti-radiation missile" is along the main beam of the radar antenna, and the signal radiated by the radar guides the missile to the radar itself. So it can only attack the fire control radar and the guidance radar, but not the alert search radar. The "AGM-45 Shrike anti-radiation missile" has a high probability of hitting in the target shooting test of the cooperative target. However, when used in the actual battlefield in Vietnam, its hit probability is not very high. According to statistics, the first attack against "AGM-45 Shrike anti-radiation missile" using fire control radar, 15 were launched, and two hits, with a hitting probability of 13%. Although only two hits were made, it had a great impact on the radar operator's state of mind. It caused fear for the operators when using radar to aim at the target.

In fact, shooting with or without radar has a great influence on the probability of a hit. According to statistics from World War II, when firing without fire control radar, an average of 3000 shells was required to shoot down an aircraft. While using fire control radar to aim and shoot, it takes an average of 83 bullets to shoot down an airplane, which increases the hit rate of anti-aircraft guns by 36 times. This is a very important data. Therefore, it is necessary to increase the method of countering the "AGM-45 Shrike anti-radiation missile" in the fire control radar.

There are many measures from both tactical and technical to counter the "AGM-45 Shrike anti-radiation missile". The general idea is: When it is found that the "AGM-45 Shrike anti-radiation missile" has come over, then take an alarm on the radar, or deflect the antenna or turn off the transmitter, so that the missile loses its tracking source, thereby increasing its angular tracking error and reducing its probability of hits.

#### 1.2.9.1 The Composition of Add-On of Anti-"AGM-45 Shrike Anti-radiation Missile" Radar

As shown in Fig. 1.27, attach add-on to the radar circuit. When the "AGM-45 Shrike anti-radiation missile" reaches a certain distance, the circuit automatically alarms or automatically cuts off the high-voltage power supply of the radar transmitter. As a result, the "AGM-45 Shrike anti-radiation missile" loses its tracking source and generates a large tracking error, reducing the probability of hits. As long as the "AGM-45 Shrike anti-radiation missile" reaches a certain distance, the operator

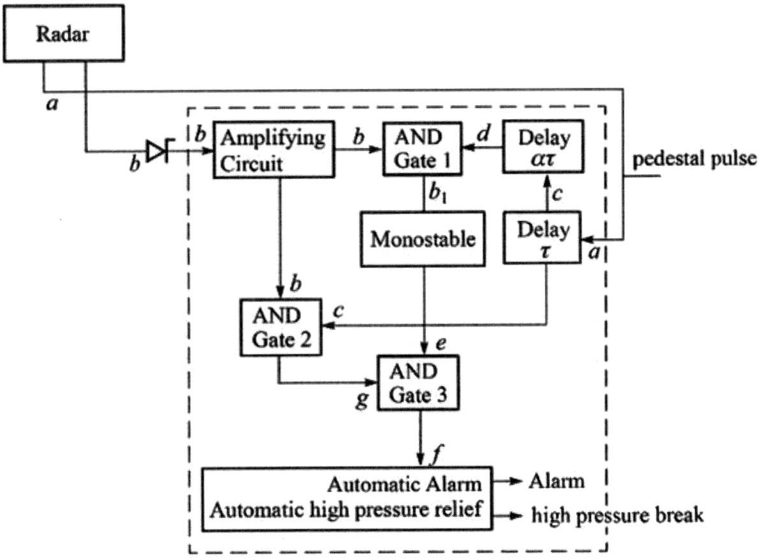


**Fig. 1.27** Schematic diagram of the composition of add-on of anti-"AGM-45 Shrike anti-radiation missile" radar

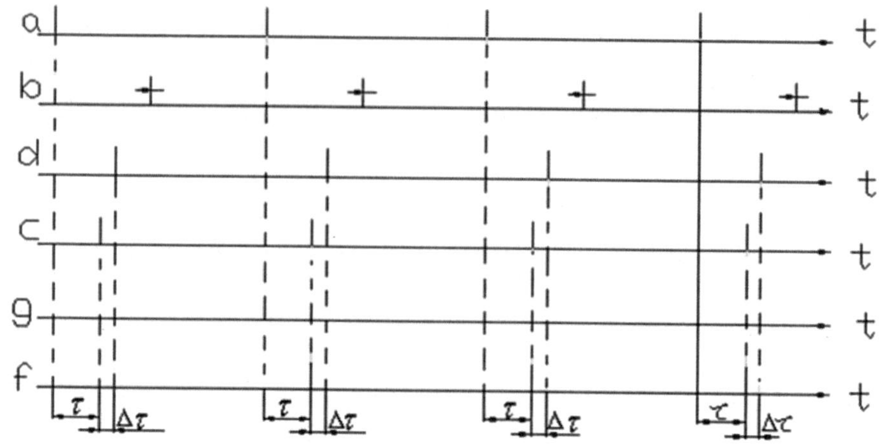


**Fig. 1.28** Schematic diagram of the time relationship of the main waveforms of the composition of add-on of anti-"AGM-45 Shrike anti-radiation missile" radar

can know that he can also adopt multiple countermeasures to ensure the safety of equipment and personnel.

The waveform diagram of each calibration point in Fig. 1.27 is shown in Fig. 1.28. Point a is the reference pulse of the radar. Point b is the reflected pulse of the target. Point c is the pulse after the reference pulse delay $\tau$. Point d is the pulse after the reference pulse delay time $\tau + \Delta\tau$. Point e is the Monostable output square wave. Point g is the pulse output by AND gate 2. b1 is the output waveform of AND gate 1.

Suppose that the speed of the "AGM-45 Shrike anti-radiation missile" is 500 m/s. if the radar finds that the distance of the "AGM-45 Shrike anti-radiation missile" is 8 km, so that the $\tau$ delay time is 26.7 μs, and $\Delta\tau$ is 1.7 μs. When the target is 8 km away, AND gate 1 has no output. When the target is 8 km, there is an output from AND gate 1, b1 has a signal, and the monostable circuit is triggered to generate a square wave that lasts for 1 s. The target continues to fly forward, and when it flies for one second, it just coincides with the monostable square wave. Then the output signal g of AND gate 2 and the monostable square wave are combined, and the f waveform is output. If the target is slow (300 m/s), it takes 1.7 s to fly to AND gate 3 and $\alpha$. At this time, the monostable square wave has disappeared, so AND gate 2 has no output. Only when the target speed is greater than or equal to the set speed of 500 m/s, the AND gate 3 has output. Only when the target speed is greater than or equal to the set speed of 500 m/s, the AND gate 3 has output, and the "automatic alarm" and "automatically disconnect the high voltage of the transmitter" control circuits will work. Ensure the credibility of the "add-on" work. The amount of delay $\Delta\tau$ is the speed of the selected target, the working time. The delay time $\tau$ is to determine at what distance the target alarms. Both $\tau$ and $\Delta\tau$ can have a certain adjustment range.

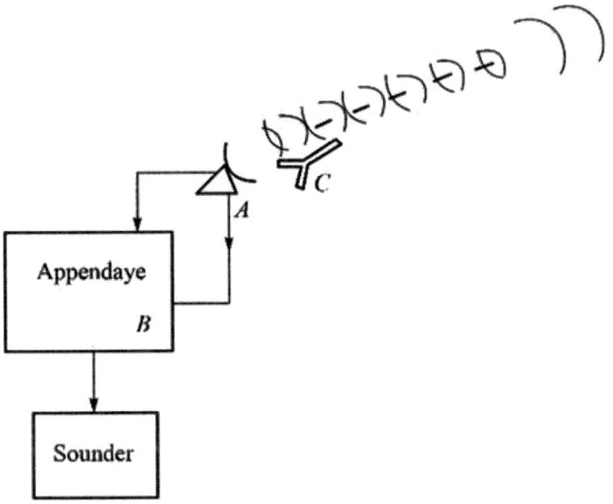


**Fig. 1.29** Schematic diagram of the effect experiment of the add-on

#### 1.2.9.2 The Test of Add-On of Anti-"AGM-45 Shrike Anti-radiation Missile" Radar

After the radar add-on is made, we need to verify that it can alarm or automatically disconnect the high voltage of the transmitter.

The composition of the experiment is shown in Fig. 1.29.

Point A is the experimental fire control radar. Point B is the add-on of anti-"AGM-45 Shrike anti-radiation missile" radar. Point C is a 130 cannon's shell. The experiment was conducted at the XXX base. The distance that the fire control radar can automatically track the 130 cannon's shell is about 8 km. Adjust $\tau$ the $\Delta\tau$ and delay time of the add-on so that the radar can issue an alarm or power off at about 7 km after tracking the shell. Because the "AGM-45 Shrike anti-radiation missile" attacked from the far away to the radar, and the test shells were fired from the radar to far. Therefore, the add-on of "d" and "c" was replaced during the test. Several rounds of shells were tested to confirm that the radar automatically tracked the shells when the shells reached 2 km. "add-on" automatically alarms when the shell reaches 7 km.

#### 1.2.9.3 The Effect of 'Add-On'

According to information, the probability of a hit by the United States' "AGM-45 Shrike anti-radiation missile" in actual combat is 13%. After taking countermeasures against it, its hit probability drops to 3–6%. In principle and practical simulation experiments, the "add-on" can play the role of "anti-AGM-45 Shrike anti-radiation missile". It can be deduced that as long as it is an anti-radiation missile that enters from the main beam of the antenna, this kind of "add-on" can automatically alarm or automatically cut off the high voltage. Provide an early warning signal to the operator so that the operator can take all countermeasures to fight against the AGM-45 Shrike anti-radiation missile.

For example, the distance of "radar add-on" pre-warned to the "AGM-45 Shrike anti-radiation missile" can be calculated according to the basic formula of the radar equation:

$$R = \left[ \frac{P_t G_t^2 \sigma \lambda^2 L_d}{(4\pi)^3 P_{\min}} \right]^{1/4} \tag{1.25}$$

In the formula:

$P_t$ The pulse power of the radar transmitter, which is 200 kW;
$G_t$ The radar antenna gain, which is 30 dB;
$\sigma$ The effective reflection surface of the AGM-45 Shrike missile, which is 0.1 $m^2$;
$\lambda$ The radar operating wavelength, which is 0.1 m;
$L_d$ The radar feeder loss, which takes 0.4;
$P_{\min}$ The actual sensitivity of the radar, which take 10–12 W;

Plug it in (1.25).

We get: $R = \left[ \frac{200 \times 10^3 \times 10^6 \times 0.1 \times 10^{-2} \times 0.4}{(4\pi)^3 \times 10^{-12}} \right]^{1/4} = 14\,\text{km}$.

Fire control radar can warn the AGM-45 Shrike missile at a distance of 14 km. If in alarm, the add-on can automatically cut off the high radar voltage. At this moment, the AGM-45 Shrike missile will lose its passive tracking source and hit the radar target with the deviation of inertial guidance. If the error angle of inertial guidance can be 0°–2°, the error distance from 10 km into inertia is 0–340 m. If the missile's blast radius is 10 m, the average hit probability of the AGM-45 Shrike missile is reduced to 3.4%.

It can be seen from this that although the United States used the "AGM-45 Shrike anti-radiation missile" in the anti-American aid to Vietnam. We have adopted the "radar add-on" against the "AGM-45 Shrike anti-radiation missile" so that it does not pose a serious threat to our army. Later, anti-radiation missiles were improved. It does not enter from the mainlobe of the antenna, but the sidelobe of the antenna. Therefore, the "Radar add-on" also lost its application value. However, currently phased array electrical scanning radars are gradually emerging. Even if the anti-radiation missile enters from the sidelobe, it can be detected by phased array electrical scanning. "Radar add-on" may still play its role after proper improvement.

#### 1.2.9.4 The Countermeasure of AGM Anti-radiation Missiles

Following the AGM-45 Shrike anti-radiation missile, the United States has developed the four generations of AGM anti-radiation missile series. It is attacked from the sidelobe of the radar antenna, aimed at entering from the sidelobe. We must develop new countermeasures equipment and adopt new countermeasures for this kind of

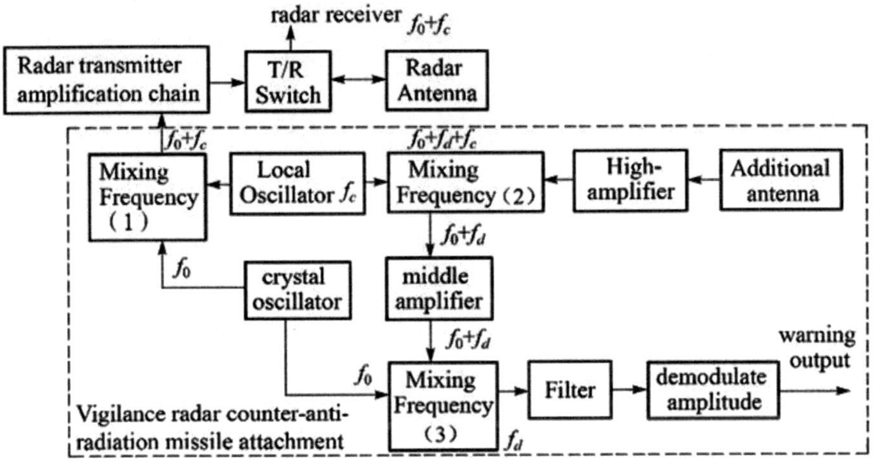


**Fig. 1.30** Alert radar against anti-radiation missile add-on

missiles. All new equipment and new measures of electronic countermeasures are generated according to the specific application characteristics of the other party's electronic equipment. As long as there is war, electronic warfare will continue to develop.

For example, when the anti-radiation missile attacks the warning radar, the anti-radiation missile can only enter from the sidelobe of the radar antenna because the warning radar antenna keeps rotating to search for targets. At this time, the receiving sensitivity of anti-radiation missiles is required to be improved. For the warning radar, it has to resist the entry of anti-radiation missiles from the sidelobes of the antenna. A set of "additional devices" should also be added to extract the useful information of anti-radiation missiles, and then warn the approach of anti-radiation missiles.

As shown in Fig. 1.30, the signal frequency of the warning radar radiation is $f_c + f_0$. The frequency of the echo signal received by the add-on is $f_c + f_0 + f_d$. After the mixer (2), the output frequency is $f_0 + f_d$. After the middle amplifier, the output frequency is $f_d$. The anti-radiation missile is coming in the radial direction of the radar, $f_d = \frac{2v_d}{\lambda}$. $v_d$ is the speed of the anti-radiation missile. $\lambda$ is the radar operating wavelength. $v_d$ is the passband width of the 200–1000 m/s filter, with a width of 4–20 kHz. As long as there is a signal input within the filter band, the filter will output and issue an alarm. In this "add-on," the anti-radiation missile enters from the sidelobe of the antenna of the radar. In this "add-on," the anti-radiation missile enters from the sidelobe of the antenna of the radar. The "add-on" specifically increases the gain $G_r$ of a rapidly rotating receiving antenna, then the approach warning distance $R(\theta)$ of the "add-on" to the anti-radiation missile is defined as:

$$R(\theta) = \left[ \frac{P_t G_t(\theta) \sigma \lambda^2 L_d G_r}{(4\pi)^3 P_{\min}} \right]^{1/4} \tag{1.26}$$

Table 1.12 Relationship between the warning distance and the angle

| $\theta$ (°) | 0 | 10 | 20 | 30 | 60 | 90 | 180 |
|---|---|---|---|---|---|---|---|
| $G_t(\theta)$ (dB) | 40 | 12 | 6 | 0 | −6 | −6 | −6 |
| $R(\theta)$ (km) | 90 | 17.6 | 12.6 | 9 | 6.4 | 6.4 | 6.4 |

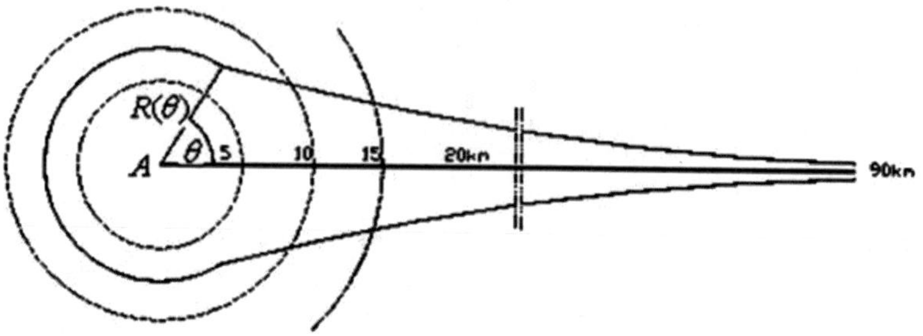


Fig. 1.31 Schematic diagram of the warning distance of the warning radar in the range of 360°

In the formula:

$P_t$ The output power of the radar transmitter.
$G_t(\theta)$ The gain of the anti-radiation missile aligned with the sidelobe of the radar antenna.
$\sigma$ The equivalent reflection area of the anti-radiation missile.
$\lambda$ The radar operating wavelength.
$L_d$ The feeder loss.
$G_r$ The additional antenna gain.
$P_{\min}$ The actual working sensitivity of the receiver.

If $P_t$ is 1 MW, substitute the parameter into (1.26), and the calculated $G_t(\theta)$ data is shown in Table 1.12.

Drawing Fig. 1.31, according to Table 1.12, the warning distance of the warning radar is showing in the range of 0°–360°.

If the radar receives the warning signal, it will cut off the high voltage, and the anti-radiation missile will turn into inertial guidance. In this case, the average hit probability of anti-radiation missiles is much reduced.

If the anti-radiation missile radar add-on is not added, the hit probability is more than 13%. It can be seen that after installed the add-on on the anti-radiation radar, the hit probability of the anti-radiation missile is greatly reduced.

#### 1.2.9.5 Measures of Phased Array Radar Against AGM Series Anti-radiation Missile

Phased array radar has a short search time. In general, the period of searching for certain airspace is estimated by 500 ms. Therefore, it can use the antenna mainlobe

Table 1.13 Relationship between the FPS-115 radar signal power and the distance

| R (km) | 1 | 2 | 3 | 5 | 10 |
|---|---|---|---|---|---|
| W (J) | $2 \times 10^{-3}$ | $0.5 \times 10^{-3}$ | $2.1 \times 10^{-4}$ | $0.8 \times 10^{-4}$ | $2 \times 10^{-5}$ |

to find anti-radiation missiles. For example, the FPS-115 warning radar has a peak power of 600 kW, and antenna gains of 40 dB and an average transmit power of 140 kW. It found that the distance of the target of 0.1 $m^2$ is calculated according to (1.5) formula, which can reach more than 1000 km. If the anti-radiation missile attacks it, the anti-radiation seeker receives the energy of the radar signal at 5–10 km:

$$W = P_n \tau = \frac{P_t G_t}{4\pi R^2} \cdot \frac{\lambda^2}{4\pi} \cdot G_a \cdot \tau \tag{1.27}$$

If $P_t = 600\,\text{kW}, \quad G_t = 40\,\text{dB}; \quad \lambda = 0.1\,\text{m}; \quad Ga = 10; \quad \tau = 5\,\text{ms}$.

When R is 1, 2, 3, 5, and 10 km, the data in Table 1.13 can be calculated.

General receiver components (such as amplifiers, detectors, etc.) can be damaged at $10^{-4}$ to $10^{-6}$ J. Therefore, the high-power phased-array warning tracking radar is not afraid of the anti-radiation missile's attack. The reasons are the following: (1) it can launch high-pulse energy to find anti-radiation missiles at tens of kilometers away; (2) it can use its high power characteristics to damage the anti-radiation seeker for several kilometers and make it lose the ability that can attack radiation source. For targets to be attacked by anti-radiation missiles, specific targets should be analyzed to achieve better results. In addition, radar can also increase anti-radiation decoys against anti-radiation missiles.

## 1.3 Development of Electronic Warfare to Information Warfare

Changes in weapons are the basis for changes in combat tactics. Changes in weapons are the basis for changes in combat tactics. It is an alternating process. When the two armies fought in the cold weapon era, they often faced each other. You can scold at a short distance. Then the cavalry charged, and the infantry stepped forward. In the era of hot weapons, the two sides are far away from each other in the trenches. The two sides also built strong fortifications and concealed each other. When attacking an enemy, they often bombard each other's position with a cannon. Then bombed with planes, and then the tanks fire to cover infantry stealthily approaching the enemy. In the information age, the two sides are far apart. Both sides have no idea where the other's personnel are concealed and where the target of the attack is transferred.

In all kinds of wars, there are generally four steps when attacking the opponent's target: The first step is reconnaissance, which is looking for the target of the attack; the second step is information transmission, which is passing the information of the discovered target to the command post; the third step is sorting and decision-making of the transmitted information; The fourth steps are to execution and attack the target. In the era of cold weapons or the era of mechanized warfare, these four steps took longer. Some are tens of hours or tens of days or even months. For information warfare, the time required for these four steps is as short as tens or dozens of minutes. Because information warfare generally uses electronic reconnaissance, which is radio network transmission and computer processing decisions. Weapons controlled with electronic equipment at long distances, and then carry out precise strikes. So it takes a shorter time, which is unmatched by previous wars.

The most notable feature of informationized operations is to strike the command organ first, then the periphery and the grassroots. This kind of warfare is called "Decapitation Strike" by Americans. For example, in the Iraq War, it was quickly discovered where Saddam was hiding, and the missiles came immediately after it. This is an example of "Decapitation Strike".

### *1.3.1 Overview of the Battlefield of Informationized Operations*

In Fig. 1.32, when the long-range precision strike does not encounter electronic countermeasures, its long-range precision strike can get good results. If it encounters a strong electronic countermeasure (as shown in Fig. 1.33), its long-range precision strike may not necessarily be effective.

Overview of electronic warfare in the informationized combat area. It includes navigation satellites, communication satellites, reconnaissance satellites, early warning aircraft, jamming aircraft, combat aircraft, drones, airborne electronic platforms, ground-based jammers, airborne jamming aircraft, spaceborne jamming satellites, bomber jammers, ground-based missile defense systems, and missile launch systems. As shown in Fig. 1.33.

The ground (sea), air, and space integrated electronic warfare are divided into two combat groups, A and B. From the perspective of combat strength, Party B is larger than Party A, which is also called asymmetric strategy. Party B is on the offensive, and Party A is on the defensive. Which party can win the war? There are many reasons: one is to look at people's hearts; two is to look at morale; three is to look at economic strength; four is to look at commander strategies; five is to look at military strength; This article does not analyze the other factors of the victory or defeat of the war, only the position and role of electronic warfare in the information warfare.

There are three notable features of informationized operations, one is fast, the second is to hit the head office first, and then the grassroots, and the third is a long-range precision strike.

Let's talk about 'fast.' There are always four steps in the fighting. This is true no matter in the era of cold weapons, the age of hot weapons, or the era of information weapons. These four steps are reconnaissance, information transmission, information

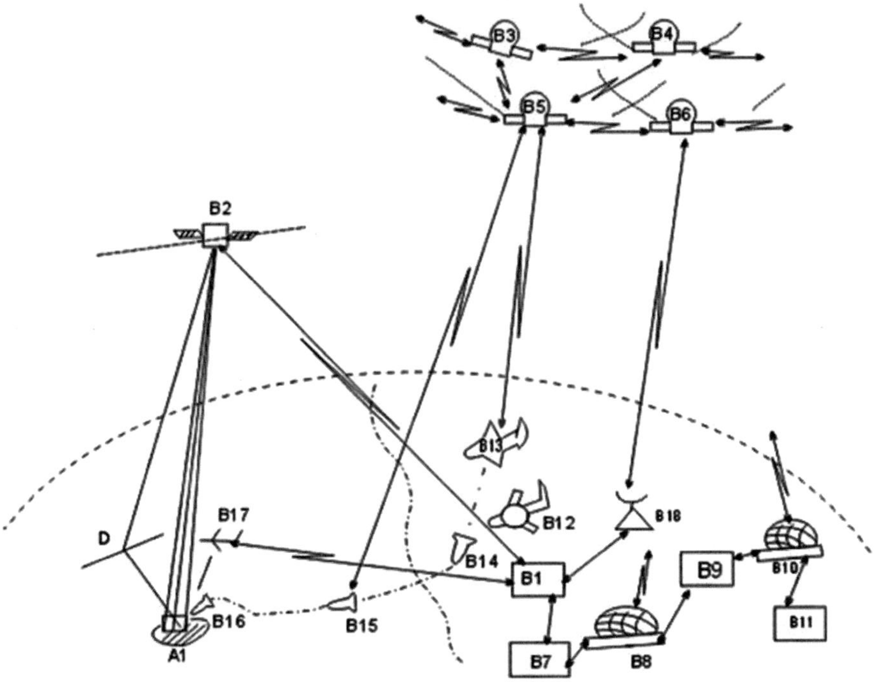


**Fig. 1.32** Schematic diagram of the composition of long-range precision strike

sorting decision, and execution. The completion of these four steps takes several months or tens of days in the era of cold weapons. In the information age, the completion of these four steps only takes tens or ten minutes to reach the office of the head office.

Countermeasure of "four steps" in informationized operations.

From the perspective of electronic warfare, the "four steps" of informationized operations can be dealt with. Electronic equipment has its advantages. It can quickly and accurately transfer data to each other. It can also quickly and accurately aim at the target and control the weapon launch. But it also has a weak side. This is because the radio equipment is susceptible to jamming and disturbance, which prevents it from performing its original tasks correctly. In the "Four Steps," their working process is serial, as long as one step is jammed and disturbed, the entire task fails.

Another feature is the precise strike from a distance. First, obtain the image data of the bombed target from the reconnaissance satellite. After airborne reconnaissance and ground verification, the ground control station sends accurate data to the GPS navigation satellite. The bomber obtains accurate data from the GPS navigation satellites and can launch cruise missiles thousands of miles away from the bombed target. After the bombing is completed, the bomber returns. The cruise missile flies to the target by itself and reaches the target to start terminal guidance, such as a laser-guided smart bomb.

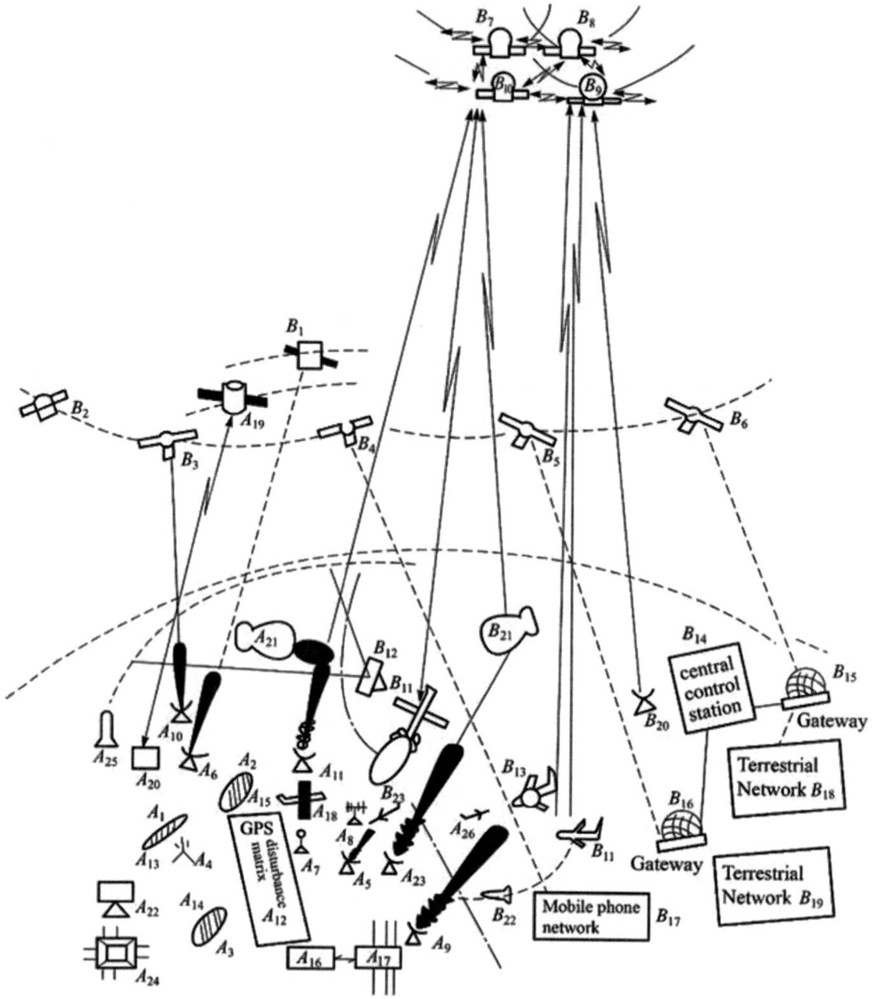


**Fig. 1.33** Schematic diagram of the integrated electronic warfare offense and defense posture in the future regional information operations

If the attacked party does not have such advanced weapons, it has very advanced jamming equipment and effectively jamming with navigation, guidance, and reconnaissance, and making the attacking party invisible, unable to drop, and unable to target. It can also invalidate the attacker's long-range precision strike into close-range combat. This can be used to defeat opponents with superior equipment.

In Fig. 1.33, B is the attacker, and A is the defender. Offense and defense are relative in combat. There is defense in attack and attack in defense.

B1 Low-orbit reconnaissance imaging satellite, which may be jammed by the ground-based jammer of Party A;

B2, 3, 4, 5, 6 Spaceborne communication network, which may be subject to ground-based and space-based jamming;

B7, 8, 9, 10 GPS navigation satellites, which may be affected by ground-based jamming.

B11 Jamming aircraft jamming with the ground radar of Party A;

B12 Ground anti-missile system, timely discovery, tracking or interception of missiles in the take-off phase of Party A.

B13 Early warning aircraft, search and track the target of party A, guide the attack, and may be jammed by the other party;

B14 Gateway central control station, which can be networked with ground gateways B15, 16, ground network B18, 19, or may be subject to network countermeasure;

B20 Ground to GPS data transmission operation station;

B21 Bombing aircraft, which can accurately strike the target according to the accurate data obtained by the GPS constellation, launching a smart bomb (B22) which can directly receive the precise command data of GPS to hit the target;

B23 After a high-altitude drone finds a target roughly in a low-orbit satellite, it can fly directly over the target for precise detection. The data accurately detected by the drone can be transmitted to B20, which can transmit it to the GPS constellation. Then the GPS constellation was passed to the bomber to launch a smart bomb to hit the target accurately. The whole process took about 10–20 min.

A1, 2, 3 Protected important military targets on the ground, such as command posts, military ports, missile positions, etc.;

A4 Anti-aircraft gun placed next to the protected target;

A5 Surface-to-air missiles cooperate with anti-aircraft guns to attack enemy aircraft and missiles close to the target;

A6 Ground-based jammers can jam with spaceborne low-orbit radars, as well as airborne radars and unmanned aerial radars. They should be integrated air-to-air jammers;

A7 Ground communication equipment;

A8 Ground radar equipment;

A9 Ground-based radar jammer system for early warning aircraft can also jamming with spaceborne radar in the same frequency band;

A10 Jammer to the communication network;

A11 Jammer to GPS constellation;

A12 For GPS array jamming systems on cruise missiles, it can use dozens of low-power GPS jammers to form an array, which can cause cruise missiles to yaw or land due to jamming in an area of 100,000 km$^2$;

A13, 14, 15 Comprehensive jamming to the missile end guidance radar and end guidance for infrared and laser guidance systems;

A16 Battlefield electronic warfare command center;

A17 Electronic warfare data transmission area;

A18 Electronic jamming aircraft;

| | |
|---|---|
| A19 | Electronic reconnaissance satellite; |
| A20 | Satellite ground control station; |
| A21 | Airborne jamming platform; |
| A22 | Ultra-long-range radar station and radar networking system; |
| A23 | Ground-based radar jammer against enemy airborne platform; |
| A24 | Ground Network Center; |
| A25 | Long-range ballistic missile launch base; |
| A26 | Reconnaissance drone. |

Informationized warfare has such characteristics: it is an integrated operation of the earth (sea), air, and space. It first obtains images or data of the attacked target from the reconnaissance satellite. After the airborne survey and ground verification, the ground operation station sends accurate data to the GPS navigation satellite mount for viewing. The bomber obtains accurate data from the GPS navigation satellite mount and can launch cruise missiles thousands of miles away from the target. After being dropped, the bomber can return home, and the cruise missile can navigate to the target by itself. When reaching near the attacked target, the missile final guidance is turned on, and the ultimate direction will directly hit the missile. During the Kosovo War, the Chinese embassy in Yugoslavia was bombed, which is the method used. Its attack precision is very high. The bomb hit from the roof to the basement explosion. In an asymmetric strategy, this kind of thing happens when one side has very advanced weapons and advanced military electronic equipment guidance.

As shown in Fig. 1.32.

| | |
|---|---|
| B1 | Command center; |
| B2 | Reconnaissance satellite; |
| B3, 4, 5, 6 | GPS navigation satellite; |
| B7, 8, 9, 10, 11 | Earth-to-satellite data transmission network and gateway; |
| B12 | Early warning aircraft; |
| B13 | Long-range bomber; |
| B14, 15, 16 | Cruise missile (or smart bomb); |
| B17 | Unmanned reconnaissance aircraft; |
| B18 | Ground to GPS console; |
| A1 | The target being attacked. |

In an asymmetric strategy, when Party A does not have any countermeasure capabilities, the B2 reconnaissance satellite finds the attacked target A1, and B2 transmits the information to the B1 command center and directly directs the B17 unmanned reconnaissance aircraft to perform actual calibration and the accurate data transmitted B1. B1 sends accurate data to the GPS satellite navigation seat through the control station B18, and the GPS satellite navigation directly transmits the accurate data to the bomber. The bomber launches a cruise missile or smart bomb, and then direct navigation of the GPS satellite mount directs the cruise missile to the vicinity of the attacked target. The missile is controlled by the terminal guidance on the missile to hit the target A1 directly. In the absence of electronic countermeasures, long-range precision strikes are still accurate. In the Gulf War and the Kosovo War, the Americans were able to win the attack quickly with little loss because it did not encounter a strong electronic countermeasure. If Party A has a fairly strong electronic countermeasure capability, as shown in Fig. 1.33, to win the war, it will cost a lot. The final victory also depends on many other factors.

### *1.3.2 The Concept of Information Warfare and Informationized Operations*

Informationized combat mainly expresses a combat process and combat steps and the division of combat phases.

The four steps and the meaning of the steps in informationized operations:

1. Reconnaissance and surveillance: The content of reconnaissance and surveillance is very wide. There are reconnaissance and surveillance during wartime, and reconnaissance and surveillance in peacetime. There are also many reconnaissance platforms, including ground vehicle, shipborne, airborne, unmanned aerial vehicle, high altitude airborne, and spaceborne reconnaissance equipment. There are also optical photography, infrared, passive radar reconnaissance, active radar imaging reconnaissance, communications reconnaissance, etc. Satellite reconnaissance is the current and future main platform in this reconnaissance. It can detect 99% of the targets in the detection area, and it is not affected by climate, day and night, national borders, etc. Some developed countries have listed spaceborne reconnaissance and surveillance as the most important means. The importance of satellite reconnaissance has been shown in the Gulf War, Iraq War, and the Afghanistan War.
2. Information transmission: Taking satellite-based reconnaissance as an example, it mainly transmits the detected information to the command center by means of radio communication and data link.
3. Signal sorting identification and decision-making: The information passed through signal sorting processing, license, and decision-making. And it was then distributed to the command post and commanders at all levels through the network.
4. Execution: Execution can be divided into two situations. In peacetime, store the sorted information in the database. In times of war, it is possible to execute an attack order. It is possible to use non-contact long-range precision strikes. For example, in the war between Iraq and Afghanistan, from reconnaissance to signals, and transmitted to the command post. It only takes more than ten minutes for the signal sorting processing decision to execute the attack command. For the anti-missile system, from the time of reconnaissance to when the other party has launched the ballistic missile information, then to the time to issue the command to discover the incoming missile is shown in Fig. 1.34.

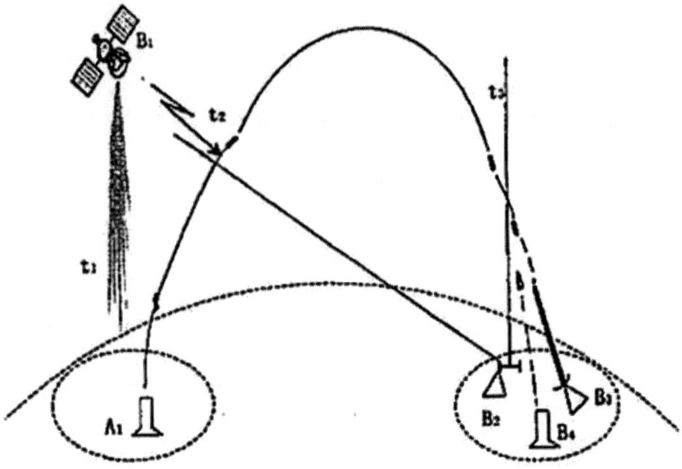


**Fig. 1.34** Schematic diagram of the time required for the anti-missile system to find the target and issue the attack command

In Fig. 1.34:

A The missile launcher.
B1 The defender's early warning satellite.
B2 The defender's warning radar.
B3 The defender's tracking radar.
B4 The defender's kinetic weapon launcher.
t1 The time of early warning satellite discovery target.
t2 The moment when the warning radar finds the target.
t3 The moment when the tracking radar keeps up with the target.

Assume the time required for the missile to fly from t1 to t2 is 10 min, and the flight time from t2 to t3 is 8 min. Then the missile defense system, from the information from the reconnaissance satellite to the launch of the attacking missile, the total time of information transmission, signal sorting processing, identification decision, issuing instructions, and the start of attack is less than 20 min.

The main contents of information warfare include electronic warfare, network offensive, and defensive, and psychological warfare. Electronic warfare includes electronic offensive, electronic defense, and electronic reconnaissance. Therefore, information warfare includes all the contents of electronic warfare during the information era. It is also an extension of all the content of electronic warfare that we usually conduct. The range of information warfare is shown in Table 1.14.

**Table 1.14** The basic content of information warfare

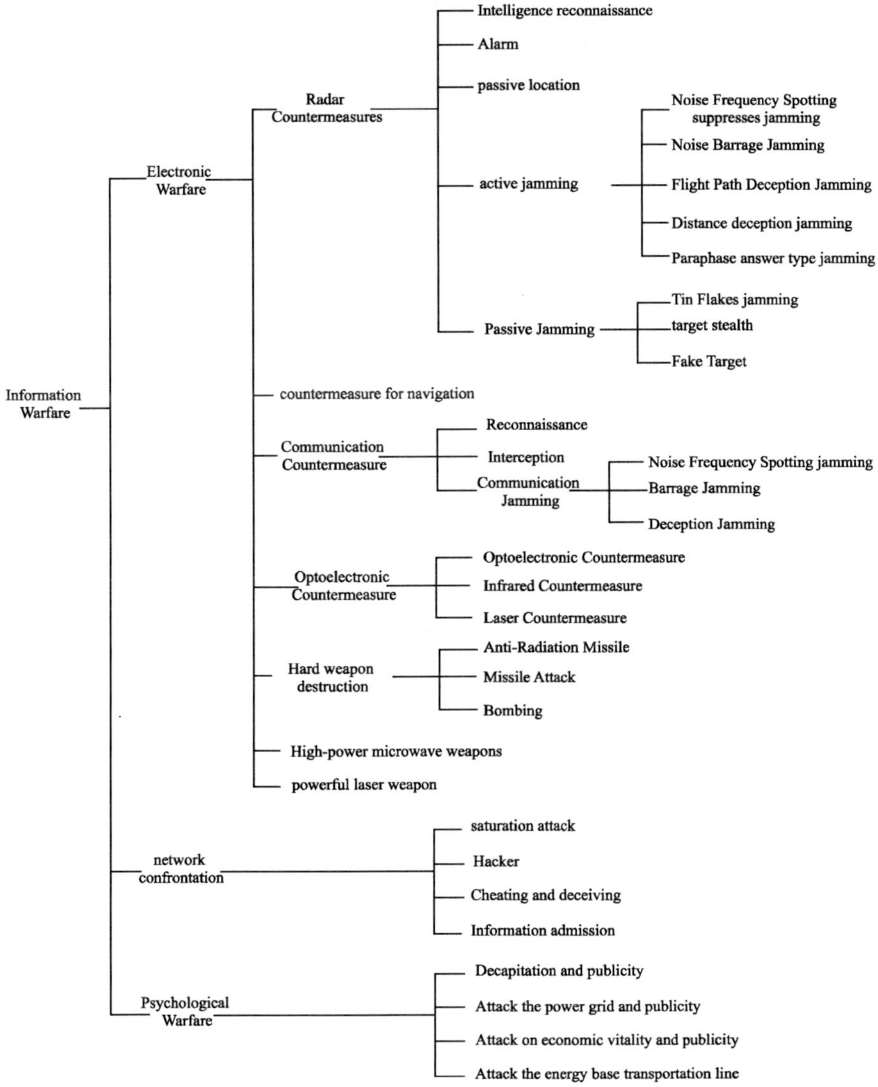


## *1.3.3 Use "Information Warfare" to Deal with "Informationized Operations"*

Generally speaking, information warfare is the most economical and effective means to deal with informationized operations. Take the Gulf War as an example to illustrate this issue. In the six months before the start of the Gulf War, the United States used more than 100 military satellites to layout Iraq's military. It was clear that the troops were transferred, their geographic location, and their commanding leaders were reconnaissance. Even unmanned reconnaissance aircraft and high-altitude reconnaissance aircraft have been used for verification. At the beginning of the war, the US military first used stealth aircraft to attack the key targets. During the penetration, the US military used the EA-6B jamming aircraft to suppress jamming against the Iraqi warning radar strongly. There are also E-3A and E-2C early warning aircraft that search for targets and command in the air, preventing Iraqi aircraft from taking off. Before the Iraqi aircraft reached the operational height, it was discovered by the US military's early-warning aircraft, and it commanded the air fighter to shoot it down.

In the Gulf War, the US military's mode of warfare used the command mode of information operations. However, Iraq adopts the mechanized combat command method. Therefore, in the Gulf War, the US military achieved overall victory with very little loss. If the Iraqi authorities have long been aware of dealing with informationized operations, they must use "information warfare" to deal with them. Then the measure they can take is to spend 10% of the funds for the purchase of more than 300 combat aircraft to buy electronic warfare equipment. 10% of the funds spent on purchasing SCUD missiles are used to buy electronic warfare equipment. When the United States mobilizes a large number of military satellites to investigate Iraq's military layout and military facilities, it should take action. Use passive jamming to cover up real targets and create false targets. Use active jamming to jamming with the satellite's reconnaissance of key Iraqi departments. After this step is completed, the US military cannot accurately grasp the layout of the Iraqi army. The US military may have two results: One is that they have not clarified the layout of the Iraqi military and dared not rush into the attack or postpone the attack. The other is offensive. Although they can win, they have to pay a higher price and get a lesson. Dare not to invade Iraq any more easily. U.S. military will not dare to invade Iraq any more easily.

We can use the Gulf War and the Iraq War, to sum up, many experiences and lessons in dealing with "informationized operations". Under the current circumstances, how to deal with economically powerful and militarily powerful countries like the United States that have the capability of "informationized operations" in economically weak countries. Using information warfare to deal with informationized operations, from the current situation, still, proceed from the four steps of informationized operations.

1. Countermeasure with reconnaissance and surveillance

Reconnaissance and surveillance are the eyes and ears of information operations. There are basically three types of countermeasures against it: suppression jamming, deception jamming, and camouflage. Because reconnaissance is comprehensive, jamming also needs to be comprehensive. Reconnaissance and surveillance are some of the main contents of electronic warfare. If it can be disabled, the next three steps will not be completed. Therefore, reconnaissance and surveillance is an important step. Therefore, reconnaissance and surveillance is an important step. It is also going on in peacetime, forming the fact that "electronic warfare is fighting every day". The E-P3 incident in Hainan is an example that was publicly announced. Major military powers use airborne platforms and satellite platforms loaded with various types of electronic reconnaissance equipment to detect military intelligence of other countries' electronic equipment parameters, military layout, and performance of newly developed products. Usually, it stores these reconnaissance parameters in the database for use in wartime. The United States has more than 200 military reconnaissance satellites. For some developing countries, the use of camouflage measures is one of the effective measures. For example, the missile launcher is covered with color and microwave-absorbing materials; some fake launchers are made; aircraft and tanks are placed in appropriate positions to disguise and deceive, etc.

Active suppression jamming can also be used. In information warfare, the two sides are competing for space more intensely. Taking the initiative in space is equivalent to taking the initiative in the war, as shown in Fig. 1.35. Space includes synchronous satellites, low-orbit reconnaissance satellites, early-warning reconnaissance aircraft, reconnaissance satellites, and early-warning aircraft. Although they have their characteristics and advantages, they also expose some of their shortcomings. For example, the time that the surveillance satellite monitors important ground targets are too short. According to actual combat needs, a near platform is likely to appear shortly, with an altitude of 20–100 km and a flight speed of Ma3–Ma8. This solves the purpose of expanding both the surveillance area and continuous surveillance.

C1, 2, 3—One-to-many integrated jammers. A1, 2, 3—The ground point target protection jammers of Party A. M1—The local combat control area of Party B. M2—The local combat control area of Party A. B8—The airfield of Party B. A5—The airfield of Party A. B9—The combat command center of Party B. A6—The combat command center of Party A. B10—The satellite communications launch site on the ground of the other party. B3—The communications satellite of Party B. B2—The ground satellite communication receiving station of Party B. B5—The near warning platform of Party B. B7—The low-orbit spaceborne SAR of Party B. B6—The other party's early warning aircraft.

Figure 1.36 is a stereo image of a target that can be detected by multi-sector reconnaissance.

Figure 1.37 shows that full-time carrier-based jammers, cover fleet formation, and carrier formation are available in the ocean.

With these ground (sea) air-to-space jammers, the other party's space, and air superiority cannot function properly, making it far away. No matter in peacetime or wartime, they dare not act rashly without detecting information. Without information during wartime, accurate long-range strikes cannot be achieved. Make the long-range precision strike into close combat. The two armies meet the brave to win. In close combat, the strengths of developing countries' military forces can be fully utilized. Therefore, during the informationized period, the weaker party must prepare for ground-to-air (air) and air-to-air electronic warfare. In local wars, do a good job of jamming. This is a force multiplier that spends less money, defeats the enemy, and give full play to one's own advantages.

In Fig. 1.37, B1, 2, 3 are multi-satellite imaging systems; B4 is the opponent's airborne platform detection system; B5 is the airborne early warning aircraft system;

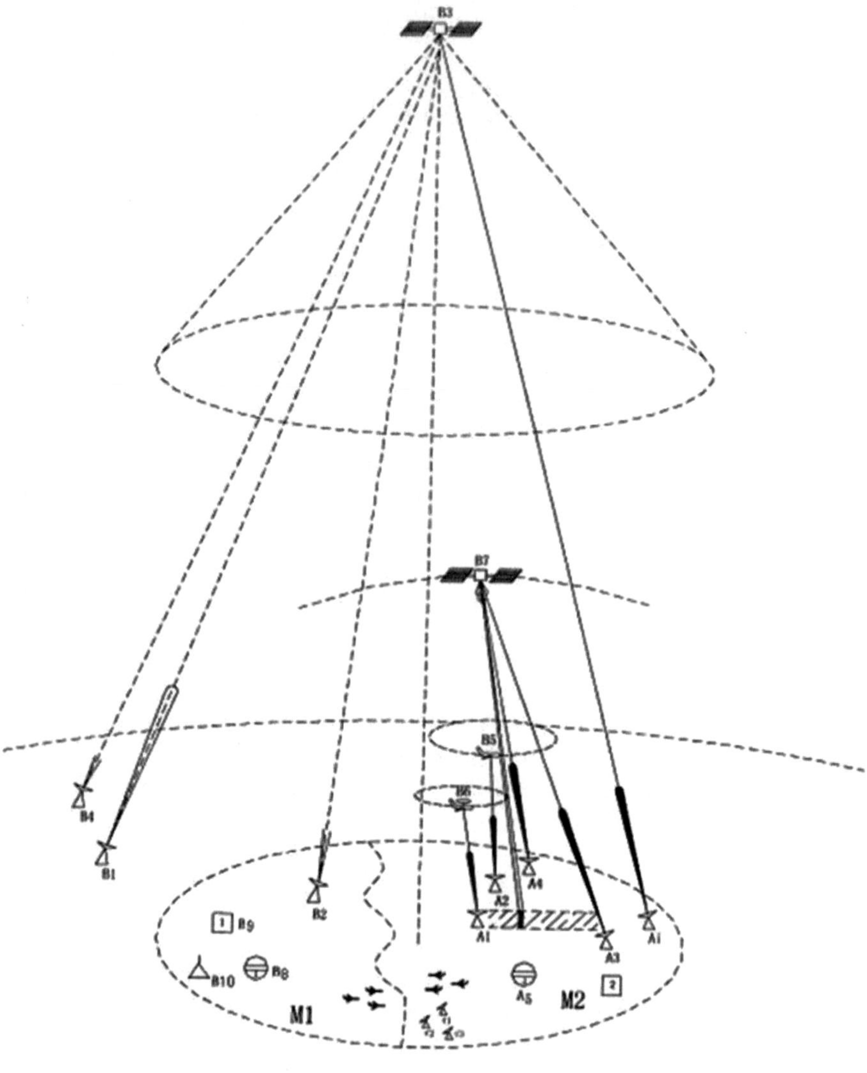


**Fig. 1.35** Schematic diagram of the 'One-to-many systems' jamming system

M1 is the B-side control area; M2 is the A-side control area; A1, 2, 3, 4, 5 are "one-to-many and air-to-air" integrated jammer system.

In Fig. 1.38, A1, 2, 3 are full-time carrier-based jamming ships, cover fleets or aircraft carrier formations; The configuration distance between A1 and A2 is about 100–200 km; The configuration distance between A2 and A3 is about 100–200 km.

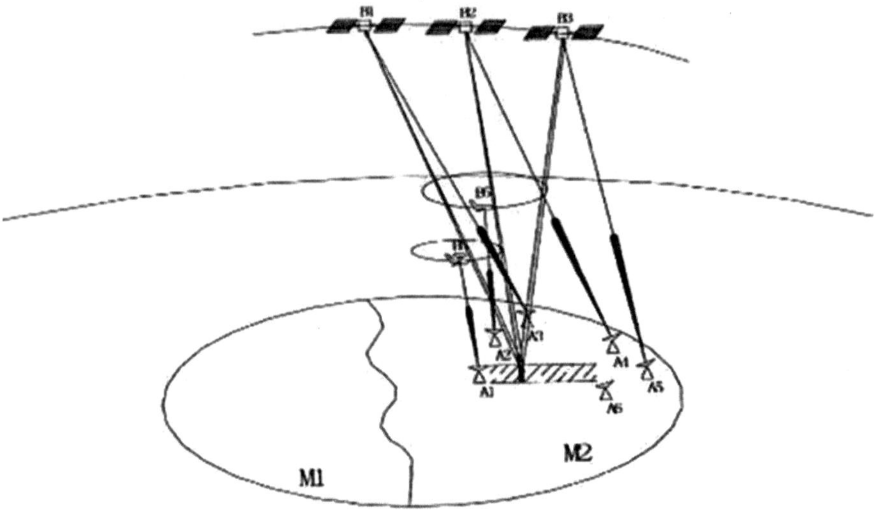


**Fig. 1.36** Schematic diagram of all in one system of 'one-to-many systems', multi-satellite, and multi-machine

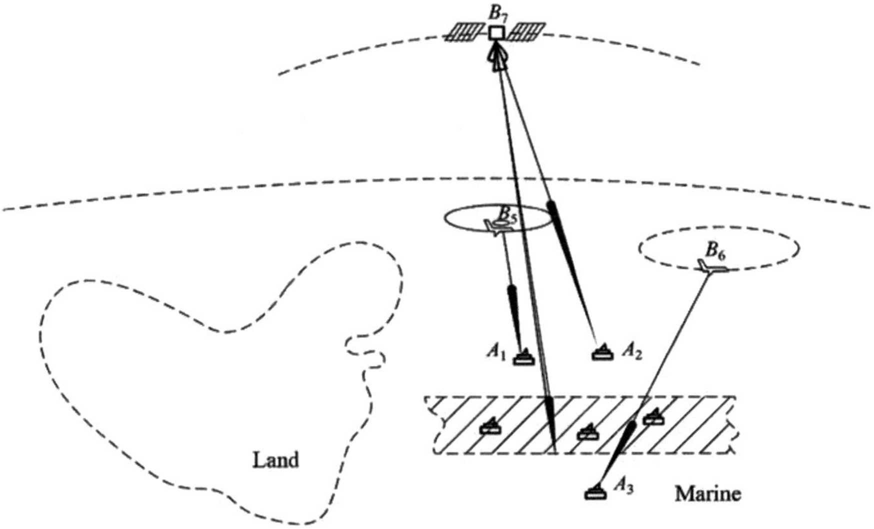


**Fig. 1.37** Schematic diagram of the shipborne 'to-air,' 'to-space,' and 'one to many system' jamming

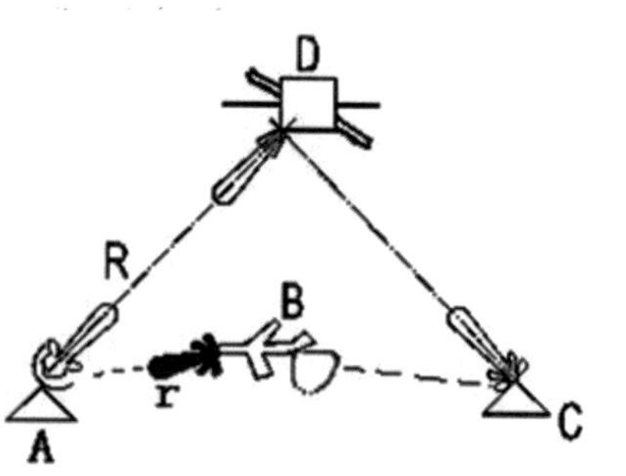


**Fig. 1.38** A schematic diagram of spaceborne reconnaissance equipment performing digital transmission to a ground receiving station, and using airborne jamming equipment to jamming with a ground digital transmitting receiving station

2. Countermeasure with Signal Transmission

Information transmission is through the communication and data link to transmit the detected information to the command center. For the jamming to the ground receiving station, its working frequency can be found from the signal transmitted by the satellite. Using this frequency plus some jamming signals for modulation, the satellite ground receiving station receives the jamming information. Because the jamming is stronger than its useful signal, its ground receiving station cannot obtain correct data and images. If these can cause it to be disturbed and disturbed, the "four steps" will fail.

In the "four steps," the jamming to the digital transmission system is a relatively difficult step, which has many tactical uncertainties. For example, the airborne jammer jamming with the satellite-to-ground digital transmission system, as shown in Fig. 1.38.

A is the ground receiving station of Party B. B is the airborne jammer on Party A. C is the ground reconnaissance frequency guidance equipment of Party A. D is the satellite-borne reconnaissance data transmission equipment of Party B.

Station A: The power of the digital signal received by the main antenna lobe is

$$P_A = \frac{P_D G_D}{4\pi R^2} \cdot \frac{\lambda^2}{4\pi} \cdot G_A \tag{1.28}$$

Station B: The jamming power to the sidelobe of Station A is

$$P_A(\theta) = \frac{P_j G_j}{4\pi r^2} \cdot \frac{\lambda^2}{4\pi} \cdot G_A(\theta) \tag{1.29}$$

Station C: Receive the downlink signal from the onboard equipment, aim at the frequency, and forward it to Station B (or B station directly amplifies/modulates/forwards).

Substitute (1.28) and (1.29) into (1.30).

According to the definition of the suppression coefficient:

$$K_j = \frac{P_A(\theta)}{P_A} = \frac{\frac{P_j G_j}{4\pi r^2} \cdot \frac{\lambda^2}{4\pi} G_A(\theta)}{\frac{P_D G_D}{4\pi R^2} \cdot \frac{\lambda^2}{4\pi} \cdot G_A}$$

After simplification:

$$K_j = \frac{P_j G_j R^2}{P_D G_D r^2} \cdot \frac{G_A(\theta)}{G_A} \tag{1.30}$$

In the formula:

| | |
|---|---|
| $P_j G_j$ | Equivalent radiated power of airborne jammer; |
| $P_D G_D$ | Equivalent radiated power of spaceborne digital transmission equipment; |
| R | The distance from the ground reconnaissance station to the satellite; |
| r | The distance from the ground reconnaissance station to the jamming aircraft; |
| $G_A(\theta)/G_A$ | Auxiliary/mainlobe ratio of ground reconnaissance station antenna. |

The equivalent jamming power required to jamming with the aircraft is

$$P_j G_j = \frac{K_j P_D G_D r^2}{R^2} \cdot \frac{G_A}{G_A(\theta)} \tag{1.31}$$

When $K_j = 0.5$; $r = 100\,\text{km}$; $R = 700\,\text{km}$; $\frac{G_A}{G_A(\theta)} = 10^3$; $P_D G_D = 10 \times 10^3$. We require $P_j G_j = \frac{0.5 \times 10^4 \times 10^4}{49 \times 10^4} \times 10^3 = \frac{0.5 \times 10^7}{49} \cong 100\,\text{kW}$.

It is still easy to build this jammer. But under the following conditions, the difficulty of jamming is increased:

a. If the location of the reconnaissance station is uncertain, the antenna of the airborne jammer would misalign the reconnaissance station, and it would become a sidelobe to sidelobe jamming. In this case, it is necessary to increase the jamming power to obtain the desired jamming effect, and this will become even more difficult.
b. If a circle of absorption material is added around the antenna of the ground reconnaissance station, the gain of the remote sidelobe (sidelobe with a large angle θ) of the antenna beam can be reduced by 15–20 dB ($G_A/G_A(\theta)$ increase). This will cause increasing the jamming equivalent power, and this will also become even more difficult.

In short, there are many uncertain factors in the jamming of the digital transmission system, and there are also many uncertain factors in the jamming effect. Therefore, in the four steps of informationized operations, it is better to choose the most critical and surest section to achieve effective jamming (reconnaissance, data transmission, sorting process execution). Among the four steps of information-based operations, "reconnaissance" and "execution" are the most critical, and the jamming is almost assured. Therefore, the main energy and input of electronic warfare should be spent on "reconnaissance" and "execution".

3. Countermeasure with Signal Sorting and Decision-Making of Transmitted Information

This step mainly needs to use the network. It is necessary to use network countermeasures or strong microwave attacks to saturate its network, chaotic programs, or damage to some devices, so that it cannot process the correct information and data, resulting in decision-making errors. If this step is successful, the "four steps" will also fail.

4. Execution

During the war, it can carry out long-range precision strikes, cruise missile attacks, anti-radiation missile attacks, or send bombers to attack. These cruise missiles or anti-radiation missiles are also weapons controlled by electronic equipment. As long as they jamming with electronic equipment, these weapons will not function properly. This means that the last step of the "four steps" failed. Informationized warfare has played a large role in the Gulf War, the Afghanistan War, and the Iraq War. Its "four steps" did not encounter a large countermeasure, so the missile hit probability is high. Some typical examples of warfare, coupled with American propaganda, have caused some fear of "informationized operations". In fact, if you know their laws and master the countermeasures against them, the long-range precision strikes supported by informationized operations will also become "not far" and "inaccurate", so there is nothing to fear. This is the role of psychological warfare in information warfare. The United States has publicized the power of information warfare and caused fear. It makes people afraid that the United States will precisely strike from a long distance, so as to achieve the purpose of surrendering without fighting.

We advertised that information warfare is not so terrible. As long as you have mastered the law of "four steps" in information operations, you will have countermeasures adapted to each of them. Turn non-contact long-range precision strikes into contacted close-range inaccurate strikes. In this way, the United States will lose the advantages of its global offensive strategy, and the world may become peaceful, stable, and peaceful.

To do a better job on effective countermeasures against information-based operations, we should also do a good job in the transmission of land, sea, air, and space integrated information grid networks. This allows combatants to share information, improve combat efficiency, and shorten combat time. In summary, informationized operations are a powerful pillar of "non-contact long-range precision strike". However, information operations have many of their advantages. However, it also has inherent weaknesses, which is that it is susceptible to various disturbances and jammings. In the four steps of information warfare, each step can be disturbed. As long as one of the four steps is disturbed, it will affect its information combat results. Making non-contact long-range precision strikes will result in "hits not too long nor accurate". As long as we can concentrate our strength, carefully study the "Four Steps of Informationized operations", study every detail of the entire process of information warfare, and find its weak links and countermeasures. It is possible to destroy its information warfare composed of huge amounts of money with little cost. This is one of the best strategies for countries in the world to fight back against power.

Of the four steps in this informationized operation, active reconnaissance is the easiest to achieve, and the effects of jamming are clearly and surely. It is also the most critical step in informationized operations. Therefore, the focus of jamming in electronic warfare should be on the radar and various reconnaissance sensors. The data transmission system is a step with many uncertain factors. Of the four steps in this informationized operation, active reconnaissance is the easiest to achieve, and the effects of jamming are clearly and surely. It is also the most critical step in information operations. Therefore, the focus of jamming in electronic warfare should be on the radar and various reconnaissance sensors. The data transmission system is a step with many uncertain factors of jamming-effect. When manpower and material resources are lacking, this part can be avoided. "Execution" is a very important step in informationized operations during the war. Therefore, we should also work hard to jamming with it.

Jamming to synthetic aperture radar, jamming to pulse compression radar, jamming to Pulse-Doppler radar, jamming to phased array radar, etc., all belong to "reconnaissance" before jamming. These countermeasures are specifically discussed in the following chapters.

# Chapter 2
# Countermeasures Against Synthetic Aperture Radar

Since the concept of imaging radar was proposed by an American scientist in 1951, the United States first developed the AN/APQ-97 airborne Synthetic Aperture Radar (SAR). After the 1970s, SAR extended from airborne to spaceborne, marking a new era of space operations for reconnaissance and remote sensing of the earth.

Since 1978, the United States has launched five types of spaceborne SAR. In 1978, it launched the SEASAT-A spaceborne SAR, which operated in space for 100 days; In 1981, the model SIR SAR was carried on the space shuttle; In 1988, 1991, and 1997, LAROSSE Lacrosse star-borne SAR monitors every corner of the world every day. Lacrosse spaceborne SAR was launched three times, with an effective operating period of 7 years. In the 1991 Gulf War, he directly participated in military reconnaissance and target instructions for Iraq, and discovered tanks buried in the desert. In the Kosovo War, it provided information for the United States to attack military targets and provided evidence for the assessment of the effectiveness of the strike, as shown in Fig. 2.1. In the Afghan war, it was an expert in space reconnaissance, making contributions to the fight against Al-Qaeda. In the 2003 war against Iraq, it was also a pioneer in reconnaissance.

The arrow on the left of Fig. 2.1 refers to the image before the attack of the Ministry of Internal Affairs building, and the arrow on the right of Fig. 2.1 refers to the image after the attack of the Ministry of Internal Affairs building. The comparison between the two images shows the attack on the Ministry of Internal Affairs building.

The success of spaceborne SAR in obtaining information arouses people's special attention to it. From the perspective of the profession of the electronic countermeasure, the responsibilities of technicians who engaged in electronic countermeasures field should focus on the most effective electronic equipment.

In 2002, the United States launched the spaceborne SAR model RADAR1, with a 7-year cycle of operation and a resolution of 1 m × 1 m. From 2003 to 2004, the United States launched the Discove-rer-2 demonstration satellite, which is a distributed multi-satellite SAR that can turn the irradiation area of the multi-satellite SAR into a synthetic area. It has a strong three-dimensional sense for reconnaissance targets, as shown in Fig. 2.2.



X. Zhang et al., *Theory to Countermeasures Against New Radars*,
https://doi.org/10.1007/978-981-16-6715-2_2

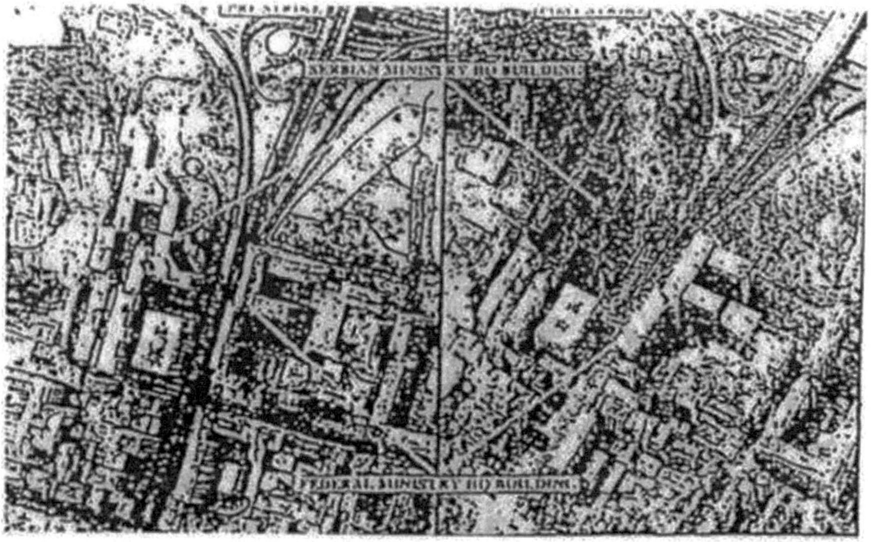

Fig. 2.1 Spaceborne SAR (image of attacking the Yugoslav Ministry of Internal Affairs building)

Fig. 2.2 Schematic diagram of multi-satellite distributed SAR imaging

| | |
|---|---|
| 用户保密通信 | User confidential communication |
| 基于多任务的可重组的系统 | Reconfigurable system based on multitasking |

The figures, missile launchers, airplanes, tanks, etc. in the picture have a strong sense of three-dimensional. In addition to the wide range of military applications,

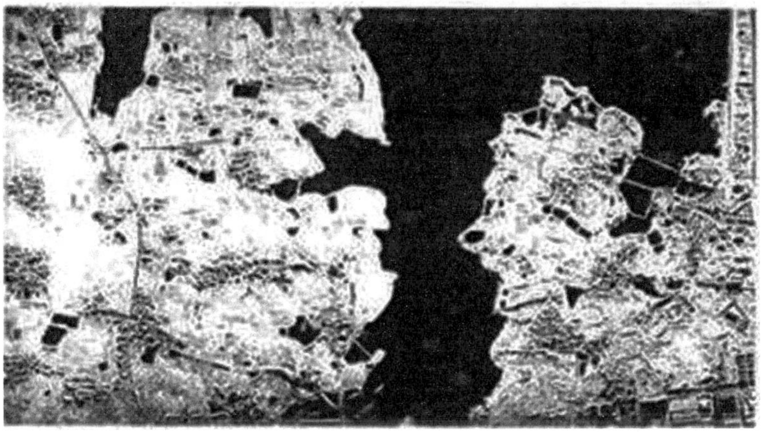

**Fig. 2.3** Images of peaceful use of SAR

spaceborne SAR also has many uses in civil applications, such as map surveying, flood control, marine census, etc. (as shown in Fig. 2.3).

In addition to the various SARs launched by the United States, Russia, Canada, Europe, Japan, Germany have also launched spaceborne SAR, shown in Table 2.1.

The military application of spaceborne SAR is becoming more and more popular, and it has been continuously improved and developed. This has caused technicians and military commanders engaged in electronic warfare to pay special attention to spaceborne SAR and should come up with effective countermeasures to blind these "FAST". The President of the United States threatened, "Whoever jamming with my satellite is equivalent to declaring war to the United States," which makes no sense that only allows their satellites to arbitrarily reconnaissance the terrain and military facilities of other countries, and not allow others to implement anti-reconnaissance measures.

## 2.1 Operating Principle of Synthetic Aperture Radar

### *2.1.1 Equation of Synthetic Aperture Radar*

The operating principle of synthetic aperture radar is similar to conventional pulse radar, except some measures are taken in range resolution and azimuth resolution, which increase the two-dimensional resolution to the order of meters or less than 1 m, as shown in Fig. 2.4.

To improve the resolution in the distance, the principle of pulse compression is adopted, emitting a wide pulse to increase the detection range of the radar and then compressing the received echo into a narrow pulse. Radar range resolution $\Delta R$,

**Table 2.1** Spaceborne SAR launched by Russia, Canada, Europe, Japan, Germany, etc.

| Index | Satellite model | Country | Launch date and life | Altitude (km) | Working frequency (GHz) | The bandwidth of surveying and mapping (km) | Resolution (m) | Type and size of antenna (m) |
|---|---|---|---|---|---|---|---|---|
| 1 | SEASAT-A | The United States | 1978<br>100 days | 800 | 1.275 | 800 | 25 × 25 | Microstrip surface array 10.7 × 2.16 |
| 2 | SIR-A space shuttle | The United States | 1981<br>2.4 days | 259 | 1.278 | 50 | 40 × 40 | Microstrip surface array 9.4 × 2.2 |
| 3 | ERS-1 | The Europe | 1991<br>2000 days | 785 | 5.25 | 100 | 25 × 13.5 | Slots array 10 × 1.0 |
| 4 | ENVTSAT-1 | The Europe | 2002<br>5 years | 800 | 5.331 | 5–100 | 30 × 30 | Active phased array 10 × 1.3 |
| 5 | RADAR1 | The United States | 2002<br>7 years | 703 | X Waveband | 4 × 4 | 1 × 1 | parabolic reflector 9 m |
| 6 | RADARSAT-1 | Canada | 1995<br>5 years | 798 | 5.3 | 50–500 | 8–100 × 8–100 | Crack waveguide surface phased array 15 × 1.5 |
| 7 | RADARSAT-2 | Canada | 2003<br>7 years | 798 | 5.405 | 10–500 | 3–100 × 3–100 | Active phased array 15 × 1.37 |
| 8 | JERS-1 | Japan | 1992<br>6 years | 568 | 1.275 | 75 | 18 × 18 | 12 × 2.2 |
| 9 | ALOS-1 | Japan | 2003<br>3–5 years | 700 | 1.27 | 250–300 | 10 × 10 | Active phased array 3.5 × 8.9 |
| 10 | RESURS-DK | Russia | 2000 | | | | 3 × 3 | |

(continued)

**Table 2.1** (continued)

| Index | Satellite model | Country | Launch date and life | Altitude (km) | Working frequency (GHz) | The bandwidth of surveying and mapping (km) | Resolution (m) | Type and size of antenna (m) |
|---|---|---|---|---|---|---|---|---|
| 11 | ALMAZ-1 | Russia | 1991<br>2 years | 300 | 3.0 | 25–45 | 10–15 | Slots array 15 × 1.5 |
| 12 | LACROSSE | The United States | 1988/91/97<br>8 years | 680 | L Waveband<br>X Waveband | | 1–3 | D18m paraboloid and phased array |
| 13 | SAR-Lupe | Germany | 2005 | | X Waveband | | 0.3 | |
| 14 | Terra SAR-L<br>Terra SAR-X | Germany | 2004<br>100 | 500<br>600 | L Waveband<br>X Waveband | 10 × 10<br>40 × 60 | | L Waveband 12.5<br>X Waveband 6.3 |
| 15 | Discoverer-2 | The United States | 2003–2004<br>Demonstration distributed satellite | 770 | | | 1–0.3 | Phased array 5 × 8 |

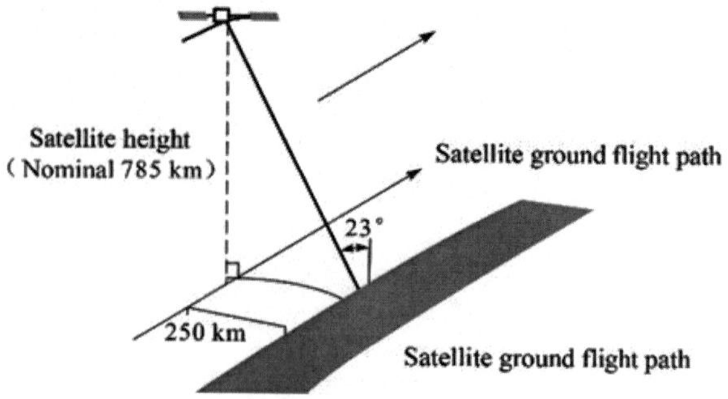


Fig. 2.4 Schematic diagram of the work of spaceborne SAR

$$\Delta R = \frac{1}{2} C \Delta \tau \quad (2.1)$$

After pulse compression, the radar has a gain factor $\eta_{\gamma}$

$$\eta_{\gamma} = \frac{\tau}{\Delta \tau} \quad (2.2)$$

In order to improve the resolution in azimuth, the signal in a beam is compressed, which is equivalent to the effect of a very narrow beam antenna, as shown in Fig. 2.5.

The carrier moves in a straight line. The radar antenna is installed on the side of the carrier. The radar antenna does not rotate. When the carrier moves, the radar antenna moves with it. If the azimuth wave beamwidth of the radar antenna is $\theta_{0.5}$

$$R_T = R \cdot \theta_{0.5} = \frac{h}{\cos \beta} \cdot \theta_{0.5} \quad (2.3)$$

The time $T_{\gamma}$ for the carrier to pass $R_T$ is

$$T_{\gamma} = \frac{R_T}{V} = \frac{\theta_{0.5} \cdot h}{V \cdot \cos \beta} \quad (2.4)$$

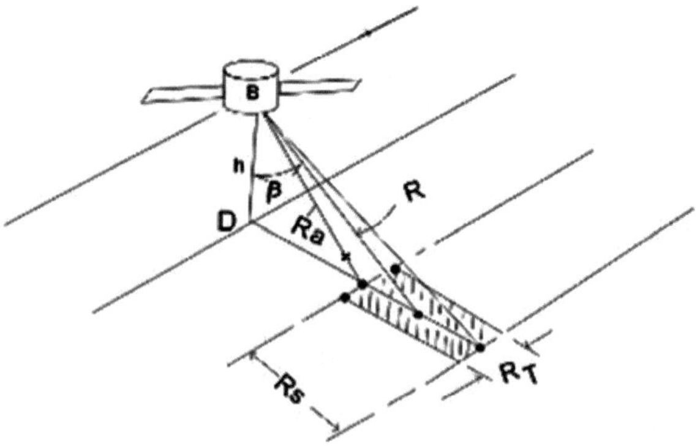


Fig. 2.5 Schematic diagram of a method of SAR to improve azimuth resolution

The gain factor for azimuth compression $\eta_b$

$$\eta_b = f_\gamma \cdot T_\gamma = \frac{f_\gamma \cdot \theta_{0.5} \cdot h}{V \cdot \cos\beta} \tag{2.5}$$

The signal echo power of the radar-based on radar equation can be written as

$$P_\gamma = \frac{P_t \cdot G_t^2 \cdot \lambda^2 \cdot \sigma \cdot L_d}{(4\pi)^3 \cdot R^4} \tag{2.6}$$

Assuming that the radar system is linearly amplified, the relevant receive gain factor is $\eta_\gamma \cdot \eta_b$.

When the signal power at the output of the radar is $S_\gamma$, then

$$S_\gamma = P_\gamma \cdot \eta_\gamma \cdot \eta_b \tag{2.7}$$

$$S_\gamma = \frac{P_t \cdot G_t^2 \cdot \lambda^2 \cdot \sigma \cdot L_d}{(4\pi)^3 \cdot R^4} \cdot \frac{\tau}{\Delta\tau} \cdot \frac{f_\gamma \cdot \theta_{0.5} \cdot h}{V \cdot \cos\beta} \tag{2.8}$$

In the formula:

$P_t$ Radar transmitter's output power;
$G_t$ Radar antenna gain;
$\lambda$ Radar working wavelength;
$L_d$ Radar transmission feed loss and atmospheric loss;
$\tau$ Radar launch pulse width;
$\Delta\tau$ The pulse width or the reciprocal of the radar receiver bandwidth after the radar is compressed;
$f_\gamma$ Radar pulse repetition frequency;
$\theta_{0.5}$ Radar antenna beamwidth;
$\beta$ Radar beam elevation;
R The distance between the radar and the target (beam center point);
h Radar platform height;
V Radar platform speed.

### 2.1.2 *Composition of Synthetic Aperture Radar*

Synthetic aperture radar consists of transceiver antennas, transceiver switches, high-power amplifier transmitters, receivers, frequency stabilization and local oscillator, basic pulse emitter, range pulse compression processing and azimuth compression processing, medium and low voltage power supplies, servo drive control extensions and subsystems.

Block diagram of composition principle of Synthetic aperture radar, shown in Fig. 2.6.

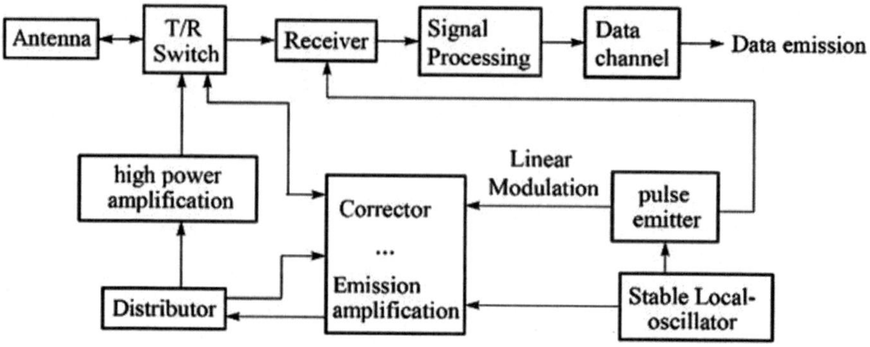


**Fig. 2.6** Block diagram of the SAR system

There are many types of SAR antennas. Phased Array Antenna, slot antenna, and parabolic antenna are generally used, depending on the use and performance of SAR.

The local oscillator of the receiver is the same source as the transmitted signal. Hence, the intermediate frequency it outputs is very stable, and it can also output the Doppler frequency of the echo stably.

Transmitter/receiver functions:

(1) Generate a linear frequency modulation signal with appropriate frequency;
(2) Amplify the frequency-stabilized pulse to the required power level;
(3) Coherent demodulation of the echo signal received by the antenna.

The signal generator: It includes a constant temperature crystal oscillator and a frequency synthesizer. This oscillator produces a very stable signal, which is the time standard and phase reference of the entire SAR system. In order to achieve high-quality SAR images, the short-term frequency stability can reach $0.5 \times 10^{-9}$/ms, then it sends $f_0$ to the frequency synthesizer, and the frequency synthesizer generates 4 $f_0$, 6 $f_0$, 16 $f_0$, and 208 $f_0$. 208 $f_0$ is the transmitting frequency. 208 $f_0$ and 16 $f_0$ are used for the receiving unit. 6 $f_0$ signals are sent to the signal processing subsystem, and the video signal is sampled by A/D conversion.

Receiver and controller: The receiver is composed of a switch, a low-noise amplifier, and a receiving unit. It should also prevent the leaked transmit pulse from burning the first-stage low-noise amplifier. The received signal is added to the low-noise amplifier through the transceiver switch, which has high requirements for the SAR transmitter and receiver compared to the conventional pulse radar, as shown in Table 2.2 the performance table of the transmitter/receiver.

After understanding the characteristics of SAR signals, it is convenient for reconnaissance analysis of SAR signals. The stability of the SAR signal requires SAR to be a high indicator and represents a new level of the radar in the new system. From the perspective of reconnaissance jamming, it is also a weak point, for example, its carrier frequency is very stable, the change does not exceed 0.6 ten thousandths, and the frequency cannot be hopped during the time the carrier flies a beam distance (in seconds). This is very beneficial to SAR detection and jamming.

**Table 2.2** Performance of transmitter/receiver

| Parameter | Indicator | Measured value |
|---|---|---|
| Center frequency | $fc \pm 0.0015\%$ | Meet the target |
| Frequency stability (short time) | $\leq 2.6 \times 10^{-9}/5\,\text{ms}$ | $0.31 \times 10^{-9}/5\,\text{ms}$ |
| Frequency stability (long time) | $\leq 5 \times 10^{-6}/2\,\text{ms}$ | $\leq 0.03 \times 10^{-6}/2\,\text{ms}$ |
| Linearly frequency modulated pulse width | $\tau_{-0}^{+0.5}\ \mu\text{s}$ | |
| Linearly frequency modulated bandwidth | $B \pm 0.06\%$ | |
| Repetition frequency | 1505.8, 1530.1, 1555.2, 1581.1, 1606.0 ± 1.3 Hz | |
| Pulse amplitude smoothness | ≤2 dB | |
| Pulse phase fluctuation | ≤8° | |
| Smallest receiver input | ≤−92 dBm | |
| Receiver gain control | 70–92 dB | |
| Receiver output | 0 dBm | I: −0.5 ± 0.6 dBm<br>Q: −0.8 ± 0.6 dBm |

### *2.1.3 Pulse Repetition Frequency Selection of Synthetic Aperture Radar*

The repetition frequency of spaceborne SAR must be properly selected. If the repetition frequency is too high, the range ambiguity will appear; and if the repetition frequency is too low, the average power of the radar will decrease, affecting the detection range of the radar. The typical value of general SAR repetition frequency selection is 1680 Hz. For the repetition frequency, it needs to receive the radar echo back to the satellite before transmitting the next pulse.

The method of adjustment the repetition frequency is that the useful echo signal falls into the 10th and 11th transmission pulses and so on, as shown in Fig. 2.7. Between the two pulses, there is noise and Echo, N represents noise; E represents echo.

I represent the replica signal; T represents the transmitted pulse; E represents the echo signal; N represents the system noise.

The repetition period of SAR is determined by the width of the SAR detection strip. It can be seen from Fig. 2.5 of the detection strip width $R_S$ and Repetition cycle $T_Y$, their relationship must satisfy

$$R_S \leq \frac{T_Y \cdot C}{2}$$

which is

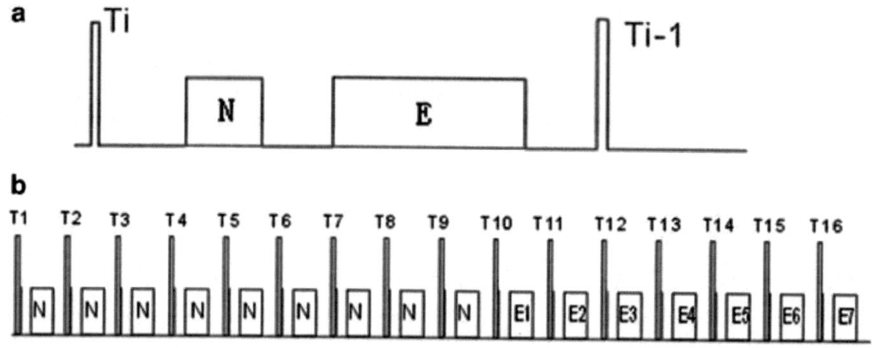


**Fig. 2.7 a, b** Schematic diagram of the SAR pulse sequence

$$T_\gamma = \frac{2R_{S\max}}{C} \tag{2.9}$$

$R_{S\max}$ is the maximum width of the SAR detection strip.
C is the speed of light ($3 \times 10^8$/s).

For example, when the maximum width of the SAR detection strip is 100 km, $T_\gamma$ equals 0.666 ms.

But $f_\gamma = \frac{1}{T_\gamma} = 1501.5\,\text{Hz}$.

This repetition frequency needs to be adjusted. Depending on the height of SAR, the pulse at the edge of n pulse launch bars is $T_{\gamma a} \cdot n$. (n is a positive integer)

$$R_a = \frac{h}{\cos(\beta - \beta_{0.5})} = \frac{T_{\gamma a} \cdot C \cdot n}{2} \quad T_{\gamma a} = \frac{2 \cdot h}{n \cdot C \cdot \cos(\beta - \beta_{0.5})} \tag{2.10}$$

If h = 700 km, $\beta - \beta_{0.5} = 40°, 60°$, the value of n and $T_{\gamma a}$, shown in Table 2.3.

The principle of choosing n:

(1) The closet to $T_\gamma$
(2) The n value is slightly larger than $T_\gamma$.

**Table 2.3** Relationship between the spaceborne and the elevation angle

| n / $T_{\gamma a}$ (ms) / $\beta - \beta_{0.5}$ | 4 | 5 | 6 | 7 | 8 | 9 | 10 | 11 |
|---|---|---|---|---|---|---|---|---|
| 40° | 1.522 | 1.218 | 1.015 | 0.87 | 0.761 | 0.676 | 0.609 | 0.553 |
| fr/Hz | 657 | 821 | 985.2 | 1149.4 | 1314.1 | 1479.3 | 1642 | 1808.3 |
| 60° | 2.332 | 1.866 | 1.555 | 1.332 | 1.166 | 1.036 | 0.933 | 0.848 |
| fr/Hz | 429.1 | 535.9 | 643.1 | 750.7 | 857.6 | 965.2 | 1071.8 | 1179.2 |

If these principles are not followed, the average power will be reduced, or the distance will be blurred. According to these two principles, when h is 700 km, $\beta - \beta_{0.5} = 40°$ and n = 9, $T_\gamma = 0.676$ ms. But for $T_\gamma = 0.666$ ms, the repetition frequency of radar needs to be 1479.3 Hz.

When $\beta - \beta_{0.5} = 60°$ and n = 14, $T_{\gamma a} = 0.7176$ ms; when n = 15, $T_{\gamma a} = 0.622$ ms. So, n = 14.

This data tells us that the repetition frequency of the spaceborne SAR is selected according to the height of the carrier platform and the elevation angle position of the radar antenna, and it does not change arbitrarily. As long as the repetition frequency cannot be changed arbitrarily, it has certain advantages to achieve deception jamming.

## 2.2 Jamming Principle for Synthetic Aperture Radar

### *2.2.1 Jamming Equation to SAR*

The power of the jamming signal received by the SAR antenna is $N_j$

$$N_j = \frac{P_j \cdot G_j}{4 \cdot \pi \cdot R_j^2} \cdot \frac{\lambda^2}{4 \cdot \pi} \cdot G_t(\theta) \cdot L_j \cdot \gamma_j \tag{2.11}$$

In the formula:

$P_j\, G_j$ is jammer equivalent radiated power;
$\lambda$ is jammer working wavelength;
$R_j$ is distance between jammer and radar;
$G_t(\theta)$ is the sidelobe gain of the jammer launch antenna aligned with the radar antenna;
$L_j$ is cable loss of jammer and atmospheric loss;
$\gamma_j$ is polarization loss of jammer.

If the radar receiving system also linearly amplifies the jamming signal, the jamming signal at the output of the radar is

$$N_{j\gamma} = MN_j \tag{2.12}$$

According to the definition of jamming suppression coefficient, at the minimum effective jamming distance, the ratio of the power of jamming signal to the power of the radar echo signal at the output of the radar receiving antenna is defined as the equivalent jamming suppression coefficient of the jammer, denoted by $K_j$.

The jamming signal entering the radar receiver is also limited by the passband of the radar receiver. No matter how wide the jamming signal spectrum is, it can only enter the spectrum signal with the same bandwidth as the receiver, that is $\Delta f_j$. And $K_f = \Delta f_\gamma / \Delta f_j$. So,

$$K_j = \frac{\Delta f_j \cdot N_j / M}{\Delta f_r \cdot S_r / M} = \frac{N_j}{S_r \cdot K_f} \tag{2.13}$$

The coefficient of suppression of SAR jamming is expressed by $K_{j\Sigma}$. Substitute (2.11) and (2.8) into (2.13), then get

$$\begin{aligned} K_{j\Sigma} &= \frac{\frac{P_j \cdot G_j}{4\pi R_j^2} \cdot \frac{\lambda^2}{4\pi} \cdot G_t(\theta) \cdot L_j \cdot K_f \cdot r_j}{\frac{P_t \cdot G_t^2 \cdot \lambda^2 \cdot \sigma_i \cdot L_d}{(4\pi)^3 \cdot R_{\min}^4} \cdot \frac{\tau}{\Delta\tau} \cdot \frac{f_r \cdot \theta_{0.5} \cdot h}{V \cdot \cos\beta}} \\ &= \frac{4\pi P_j \cdot G_j \cdot R_{\min}^4}{\overline{P}_t \cdot G_t \cdot \sigma_i \cdot R_j^2} \cdot \frac{V \cdot \cos\beta}{B_r \cdot \theta_{0.5} \cdot h} \cdot \frac{r_j \cdot L_j \cdot K_f}{L_d} \cdot \frac{G_t(\theta)}{G_t} \end{aligned} \tag{2.14}$$

In the formula:

$P_j G_j$ is the equivalent power of the jammer;
$\overline{P}_t$ is the average power of the radar transmitter ($\overline{P}_t = P_t f_\gamma \tau$);
$G_t$ is the radar antenna gain;
$\sigma_i$ is the effective reflection area in the radar resolution area;
V is the speed of radar carrier movement;
$\beta$ is the elevation position of the radar antenna;
h is the height of the radar carrier;
$B_r$ is the bandwidth of the radar received;
$\theta_{0.5}$ is the radar azimuth beam width;
$L_j$ is jamming cable loss and atmospheric attenuation loss;
$\gamma_j$ is the polarization loss of jammer;
$K_f$ is Ratio of radar receiver bandwidth to jamming spectrum width;
$L_d$ is radar cable loss and atmospheric loss;
$G_t(\theta)/G_t$ is the ratio of main lobe and side lobe on the direction of the radar antenna side lobe to the jammer;
$R_{\min}$ is the minimum jamming distance to the radar.

Formula (2.14) shows that the effective jamming suppression coefficient for SAR is related not only to the equivalent power of the radar and jammer, but also to various factors such as the moving speed, altitude, and attitude of the moving carrier.

To simplify the jamming equation, we can write (2.8) as

$$S_\gamma = \frac{P_t \cdot G_t^2 \cdot \lambda^2 \cdot \sigma_i \cdot L_d}{(4\pi)^3 \cdot R^4} \cdot \eta_a \cdot \eta_b \cdot M \tag{2.15}$$

Substitute (2.15) and (2.11) into (2.13), we get

$$K_j = \frac{N_j}{S_r \cdot K_f} = \frac{\frac{P_j \cdot G_j}{4\pi R_j^2} \cdot \frac{\lambda^2}{4\pi} \cdot G_t(\theta) \cdot L_j \cdot \gamma_j \cdot \Delta f_j}{\frac{P_t \cdot G_t^2 \cdot \lambda^2 \cdot \sigma_i \cdot L_d}{(4\pi)^3 \cdot R^4} \cdot \eta_a \cdot \eta_b \cdot \Delta f_r}$$

$$= \frac{4\pi P_j \cdot G_j \cdot R_{\min}^4}{P_t \cdot G_t \cdot \sigma_i \cdot R_j^2} \cdot \frac{r_j \cdot L_j \cdot K_f}{L_d} \cdot \frac{G_t(\theta)}{G_t} \cdot \frac{1}{\eta_a \cdot \eta_b} \tag{2.16}$$

For synthetic aperture radar, it requires the largest jamming suppression coefficient

$$K_{j\Sigma\max} = K_j \cdot \eta_a \cdot \eta_b \tag{2.17}$$

Substitute (2.16) into (2.17), get

$$K_{j\Sigma\max} = \frac{4\pi P_j \cdot G_j \cdot R_{\min}^4}{P_t \cdot G_t \cdot \sigma_i \cdot R_j^2} \cdot \frac{r_j \cdot L_j \cdot K_f}{L_d} \cdot \frac{G_t(\theta)}{G_t} \tag{2.18}$$

## 2.2.2 *Jamming Equivalent Power Required for SAR*

Transform (2.18) into

$$P_j \cdot G_j = \frac{K_{j\Sigma\max} P_t \cdot G_t \cdot \sigma_i \cdot R_j^2}{4\pi \cdot R_{\min}^4} \cdot \frac{L_d}{r_j \cdot L_j \cdot K_f} \cdot \frac{G_t}{G_t(\theta)} \tag{2.19}$$

In the process of deriving the maximum jamming suppression coefficient ($K_{j\Sigma\max}$), it is assumed that the SAR is receiving, and the amplifying system is linear. In fact, it is not completely linear. There is an error between theory and practice. For the actual $K_{j\Sigma} = \left(K_{j\Sigma\max}\right)^{\rho}$, $\rho$ is 0.8 generally. Therefore, we get:

$$P_j \cdot G_j = \frac{K_{j\Sigma} P_t \cdot G_t \cdot \sigma_i \cdot R_j^2}{4\pi \cdot R_{\min}^4} \cdot \frac{L_d}{r_j \cdot L_j \cdot K_f} \cdot \frac{G_t}{G_t(\theta)} \tag{2.20}$$

The formula (2.19) is similar to our traditional jamming equation. The difference lies in the different values and concepts. The other parameters in the equation are exactly the same as those in the traditional jamming equation. In the traditional concept, $\sigma_i$ is the effective reflection area of the target in the radar beam illuminated area. However, in SAR jamming, $\sigma_i$ is the effective reflection area in the SAR equivalent resolution area. The difference between these two areas makes an effective reflection area very different. For example, the Nanjing Yangtze River Bridge in a radar beam irradiation area has an equivalent reflection area of about 50,000 $m^2$ and in a radar resolution area of "1-m × 1 m", which its an effective reflection area is a few square meters. This result shows that in order to protect the Yangtze River Bridge from the other party's radar, no matter what radar system the other party uses, such as general pulse radar, pulse compression radar, or synthetic aperture radar, the equivalent jamming power required for them is the same. In other words, the two-dimensional compression processing of the synthetic aperture radar only brings great benefits to its radar to improve the received signal-to-noise ratio. For jamming, the jamming power required by it does not change much.

This conclusion is not difficult to understand from a physical concept. For example, the receiver noise power at the input of the SAR receiver is −120 dBW. For the point target signal, after two-dimensional compression processing, it is improved by 50 dB. When the point target signal is −170 dBW, it can just extract the point target signal. If an additional noise signal power is −117 dBW, then the SAR cannot extract the signal from the noise. For a conventional radar, it can extract the signal with a signal-to-noise ratio of 1 dB, while the receiver noise power is still −120 dBW, and additional noise power is still −117 dB. At this time, conventional radar cannot extract useful signals. The signal-to-noise ratio required by these two radars is very different, but the noise of jamming needed power is the same.

## 2.3 Measurement and Analysis of SAR Jamming Suppression Coefficient

### *2.3.1 Computer Simulation Measurement*

The composition of the test is shown in Fig. 2.8.

The SAR echo signal is the recorded SAR echo signal. The jamming signal is a self-made continuous-wave noise source.

Measurement methods and steps:

(1) When no jamming is added, use a small power meter to measure the power of the SAR echo signal as $P_\gamma$;
(2) Remove the small power meter and add the SAR signal to the image display after signal processing. The observed image is shown in Fig. 2.9.

    Replace the small power meter with a spectrum analyzer and measure the spectrum width of the SAR signal as $B_\gamma$;
(3) Adjust the spectrum width of the noise source to make it equal to $B_\gamma$, and then remove the spectrum analyzer;
(4) Add the SAR signal and the jamming signal at the same time, and change the attenuator to increase the jamming signal gradually. From the display, you

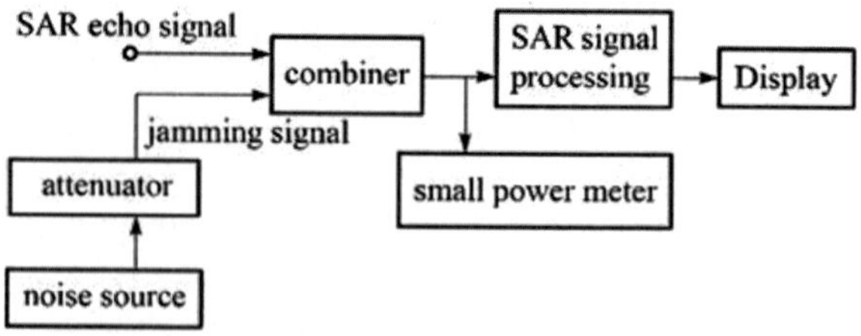


**Fig. 2.8** Block diagram of computer simulation jamming test

**Fig. 2.9** Images of simulation test SAR echo (without jamming) (the picture is a seaside salt field, clearly visible)

can see that the image is gradually blurred. When the signal is strengthened again, the image in Fig. 2.10 appears, and it is considered that the jamming completely destroys the image.

Remove the SAR signal, measure the noise power at this time as $P_j$, and compare $P_j/P_y$ as 16 dB. This kind of signal is the signal of the surface target, and the effective jamming suppression system is 16 dB;

(5) Disconnect the jamming, remove the SAR echo signal, and replace it with a self-made point target echo signal. Use a small power meter to measure the power as $P'_\gamma$, and the image appears on the image display, see Fig. 2.11;
(6) Add the point target signal and the noise signal at the same time, change the attenuator to increase the jamming gradually. From Fig. 2.11a–d, the point target has not been completely suppressed, but because the power of the jamming source is not large enough. The test is done so far. Use a small power meter to measure the jamming signal power, and compare $P'_j/P'_\gamma$ as 40 dB.

It is concluded that noise suppression jamming is performed on the target signal at the SAR point. The jamming/signal suppression ratio is 40 dB, and the signal has not been suppressed.

It can be seen from the experiment that the jamming suppression coefficients required for the surface target signal and the point target signal of SAR differ greatly.

**Fig. 2.10** SAR signal image after effective jamming

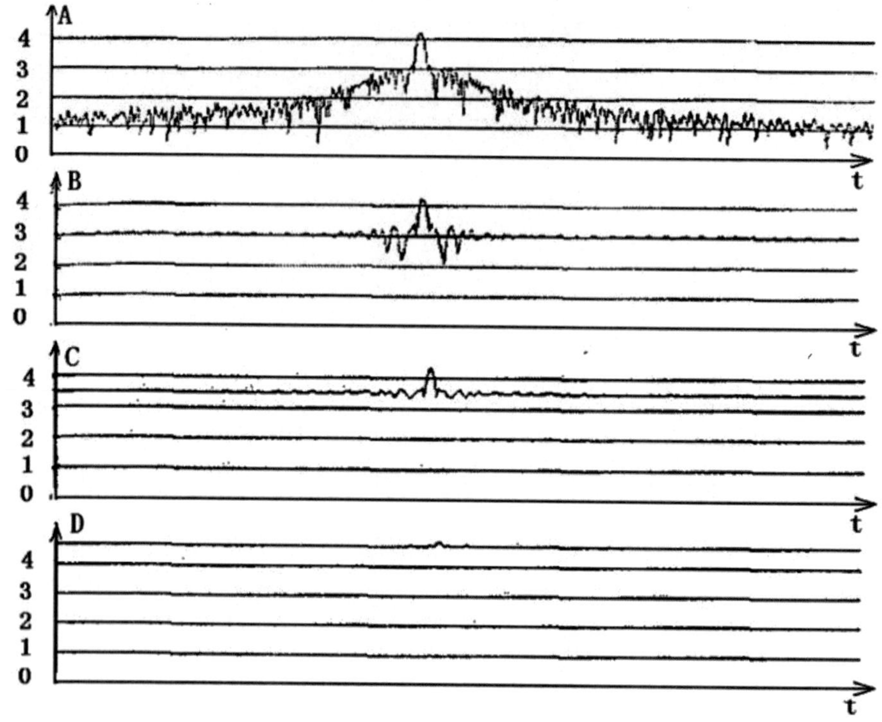


**Fig. 2.11** Jamming suppression image of SAR point target echo

## 2.3.2 *Indoors Simulation Test of Low Power Jamming Against Formal SAR*

The test composition is shown in Fig. 2.12.

Test steps and methods:

(1) Adjust the output of SAR point frequency signal and normally display on display;

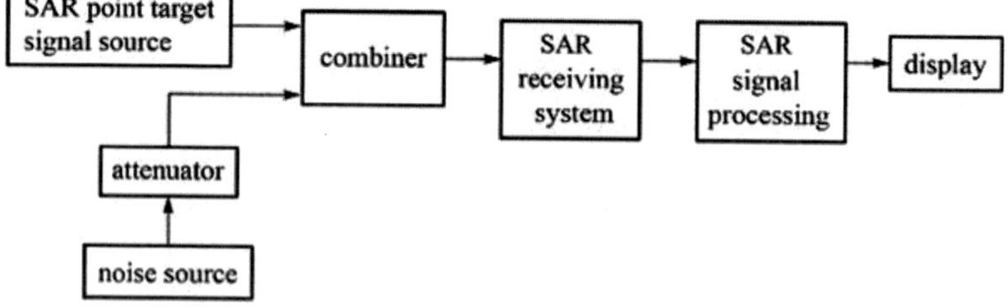


**Fig. 2.12** Block diagram of SAR indoor jamming test

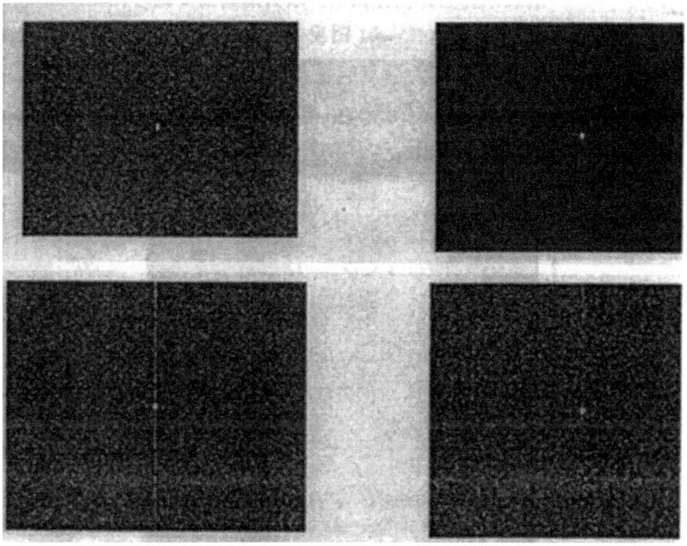

Fig. 2.13 Test image of indoor low-power jamming to SAR

(2) When no jamming is added, the image in Fig. 2.13a appears on display. Add the jamming and gradually increase the jamming, when the jamming suppression ratio reaches 25 dB, Fig. 2.13d appears, and the signal is still not suppressed;

(3) Reintensify the jamming signal until the jamming suppression coefficient is 60 dB, and the point target echo will be completely suppressed, that is, there will be a noise on display, and the point target signal can no longer be found.

### 2.3.3 Flight Test on the Jamming Suppression Coefficient for Airborne SAR

The layout of the target covered by the jamming shown in Fig. 2.14.

The interval between the nine corner reflectors is 30 m, and the interval between the two rows is 50 m. The distance between the jamming car and the corner reflector

Fig. 2.14 Partial schematic diagram of target detected by airborne SAR

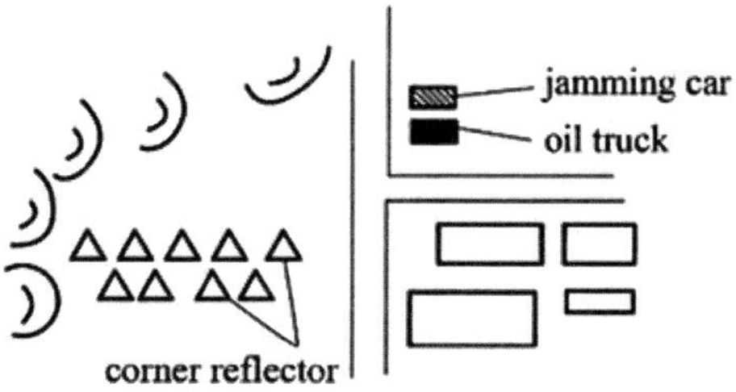

Fig. 2.15 Chart of jamming to airborne SAR

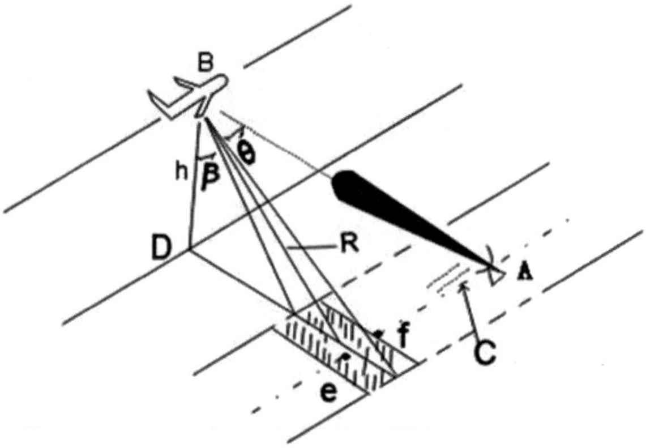


is about 500 m. There are two types of equivalent reflection areas of corner reflectors, namely 10 and 20 $m^2$. The relationship between the flight path, jammer, and detection target is shown in Fig. 2.15.

A is the jammer; B is the SAR carrier; C is the detection target; D is the projection point of B; h is the target height; is the SAR irradiation width of azimuth beam.

Process of test flight:

The airborne SAR enters the predetermined route, keeping h and $\beta$ unchanged. The jammer automatically tracks the airborne SAR. The angle 6 changes from large to 0, and then becomes 6.

When no jamming is added, the imaging map of the airborne SAR to the detection target is shown in Fig. 2.16 (on the left side of the detection target is a county).

After the jamming is added, the airborne SAR will fly according to the predetermined route. When flying angle changes from 6° to 0°, the detection target will be suppressed by the jamming signal. Repeat the flight multiple times to adjust the output power of the jamming transmitter. When the transmitter power is adjusted to 0.3 W, in the SAR mainlobe, the detection target echo is just suppressed, as shown in Fig. 2.17b.

All ground features in the radar's mainlobe, including mountains, cities, and detected reflector echoes, are all suppressed by jamming. When the jamming is adjusted to 0.3 W, all the echoes of the detected reference are suppressed by the jamming as shown in Fig. 2.17b.

From the three images in Fig. 2.17, it is confirmed that $P_j = 0.3$ W is the critical power of effective suppression of jamming.

### 2.3.4 *Data Analysis of Airborne SAR Jamming Suppression Coefficient from Flight Test*

Known parameters of the tested airborne SAR:

**Fig. 2.16** Target image detected by airborne SAR without jamming

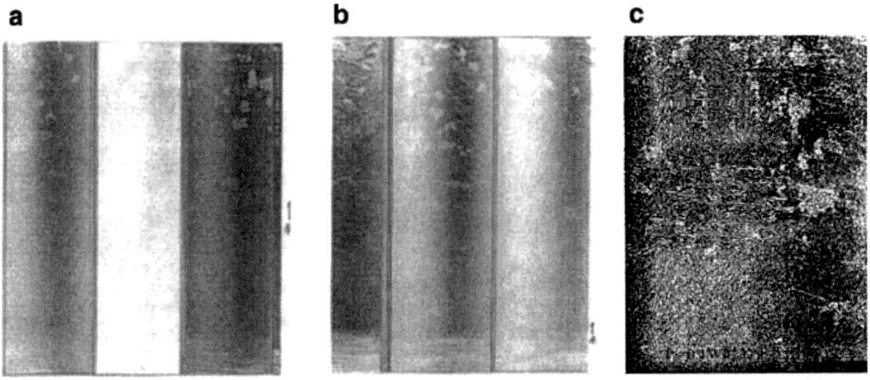

**Fig. 2.17** **a** The jamming image of airborne SAR with jamming power of 1 W. **b** The jamming image of airborne SAR with jamming power of 0.3 W-h. **c** The jamming image of airborne SAR with jamming power of 0.1 W-h

The peak power is 300 W; the antenna gain is 17 dB; the azimuth beamwidth is 8°; the transmission pulse width is 33 μs; the compressed pulse width is 16 ns; the flying height is 7200 m; the mapping distance is 20 km; and the flying speed is 150 m/s.

Parameter of jammer:

The antenna gain is 24 dB; the jamming emission output power is 0.3 W (experimental test data). It jamming with the SAR mainlobe, and the effective reflection area of the detected point target is 20 m$^2$.

(1) The jamming suppression coefficient required for theoretical calculation:

Distance gain: $\eta_a = \tau/\Delta\tau = \frac{33\times10^{-6}}{16\times10^{-9}} = 33\,\text{dB}$.

Position gain: $\eta_b = f_r \cdot T_s = f_r \theta_{0.5} \cdot R/V = 600\times8^\circ\times20\times10^3/150 = 41\,\text{dB}$.

The maximum jamming suppression factor required is $\eta_a \cdot \eta_b = 41 + 33 = 74\,\text{dB}$.

Considering the nonlinear loss of the amplifying system, the actual jamming suppression coefficient required is:

$$K_{j\Sigma}(K_{j\Sigma\max})^{0.8} = 59.2\,\text{dB}$$

(2) The measured jamming suppression coefficient $K_{j\Sigma0}$:

$$K_{j\Sigma0} = \frac{4\pi P_j \cdot R_{\min}^4}{\overline{P}_t \cdot G_t \cdot \sigma_i \cdot R_j^2} \cdot \frac{r_j \cdot L_j \cdot K_f}{L_d} \cdot \frac{G_t(\theta)}{G_t}$$

In the formula $R_j = R_{\min}$, $\frac{G_t(\theta)}{G_t} = 1$, $\frac{r_j \cdot L_j \cdot K_f}{L_d} \approx 1$, $P_j = 0.3\text{W}$, $G_j = 24\,\text{dB}$.

Then $K_{j\Sigma0} = \frac{4\pi P_j \cdot G_j \cdot R_{\min}^2}{\overline{P}_t \cdot G_t \cdot \sigma_i} = \frac{4\pi\times48\times(20\times10^3)^2}{15\times10^3\times20} = 59\,\text{dB}$.

It is proved that the theoretical calculation result is basically the same as the jamming suppression coefficient measured by the test flight.

Tested with three methods (for point targets), the jamming suppression coefficient measured for SAR jamming is basically the same.

In fact, there is a fourth test method for the jamming suppression coefficient required by jamming SAR, using Fig. 2.12. Directly measure the improvement of the received signal/noise ratio of SAR. If the measured signal/noise ratio is −60 dB, the jamming suppression coefficient for it is also 60 dB. This is the basic rule obtained after three tests.

From (2.19), we can see that there are two parameters in the jamming equation to SAR, one is the SAR jamming suppression coefficient, the other is the effective reflection area of the point target. Here is some analysis of the values.

## 2.4 Analysis on Reflection Area of Point Target

### *2.4.1 Mechanism of Radar Target High-Frequency Scattering*

When the size of the object is much larger than the wavelength of the electromagnetic wave, the interaction between the electromagnetic wave and the object has a "local" characteristic and is closely related to the shape of the object. This is the local characteristic of high-frequency scattering.

The radar cross-section has seven scattering mechanisms (RCS). These mechanisms include:

(1) Mirror Reflection;
(2) Surface discontinuity scattering, such as edges, corners, and tips;
(3) Scattering of the discontinuity of the derivative of the side surface;
(4) Diffraction of creeping waves or shadow boundaries;
(5) Scattering of the traveling wave;
(6) Scattering in concave areas, such as cavity, dihedral angle and trihedral angle;
(7) Multiple scattering.

Large scatters are often composed of many scattering centers, which are classified according to the range of RCS:

(1) The strongest scattering source is generated by plate, bright lines on cylinders, cavity, etc. These scatterings constitute the mainlobe of the scattering diagram;
(2) The contribution of geometric optical reflection and edge diffraction field to the composition of scattering is small;
(3) Spherical waves of scattering from the edges, vertices, and end of thin wires contribute the least to scattering.

For example, Fig. 2.18 is the distribution map of the scattering center of a certain type of aircraft. These scattering center areas are the equivalent reflection area of the SAR point target.

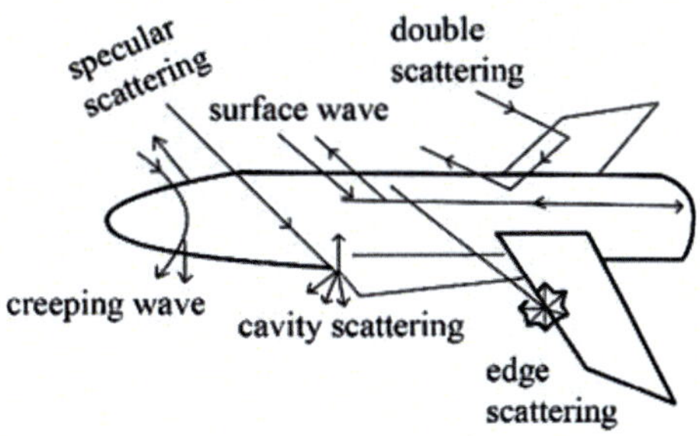


**Fig. 2.18** Effect of scattering center on RCS

In the total scattering field, resonance scattering tends to be the main influence of RCS at certain angles. Electrical Resonant scattered waves are equivalent to coherent phase jamming.

(4) For the RCS of low-resolution radar, its mathematical expression is

$$\sigma = \lim 4\pi \cdot R^2 \left| \sum_{K=1}^{n} \frac{1}{\sqrt{4\pi}R} \cdot \sqrt{\sigma_K} \cdot e^{j(4\pi d_K/\lambda)} \right|^2$$
$$= \left| \sum_{K=1}^{n} \sqrt{\sigma_K} \cdot e^{j(4\pi d_K/\lambda)} \right|^2.$$

$\sigma_K$ is the kth single scatterer RCS. If it comes from some point within a target at a distance R, the radar range is $d_K$ for each scatterer pair.

(5) RCS of high-resolution radar.

High-resolution radar can distinguish the various scatterers of the target. To the extent that the target scatterer can be resolved, its fluctuations are reduced.

The way to improve the range resolution is to use narrow pulses. Range resolution $\Delta \gamma = \frac{\Delta \tau \cdot C}{2}$ If the distance resolution is required to be 1 m, it can be deduced that the pulse width should be 6.7 ns, which requires the spectrum width of the radar emission to be 150 MHz. The principle of the synthetic aperture can improve the azimuth resolution of the radar. Radars with high-resolution range, azimuth, and speed are synthetic aperture (SAR) and inverse synthetic aperture (ISAR) radars.

## 2.4.2 *Measurement on Equivalent Reflection Area of Complex Target's Scattering Center*

Testing of RCS waves at each scattering center point of aircraft as shown in Fig. 2.19.

Figure 2.19a is a side view of an aircraft foam stand, and Fig. 2.19b is a top view of the aircraft.

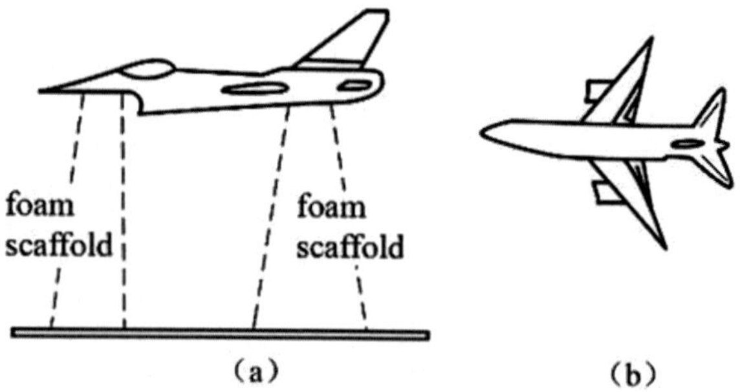


**Fig. 2.19** Schematic diagram of the RCS wave test of each scattering center point of the aircraft

**Table 2.4** Correlation of contribution of each scatterer

| Order number | Aircraft typical scattering points | Range (m) | Relative distance (m) |
|---|---|---|---|
| 1 | Nose cone | 0.5 | 3.28 |
| 2 | Leading-edge of sit wheel cover | 3 | 4.3 |
| 3 | Engine import | 5 | 5.33 |
| 4 | Rear edge of the sit wheel cover | 2 | 6.08 |
| 5 | Leading-edge of the wing root | 4 | 6.67 |
| 6 | Rear edge of the wing root | 2 | 8.4 |
| 7 | Vertically stabilized wing root 1 | 1.5 | 8.8 |
| 8 | Vertically stabilized wing root 2 | 0.5 | 9.9 |
| 9 | Top of vertical stabilizer | 0.5 | 10.6 |
| 10 | Leading-edge of horizontal wing | 1.5 | 9.2 |
| 11 | Rear edge of horizontal wing | 1.0 | 10.2 |
| 12 | Interaction of turntable | 3.0 | 11.7 |

Test method:

(1) Transmit a short pulse (less than 5 ns), then separate the target from the echo
(2) Use fast Fourier transform to transform the received echo into a time-domain signal
(3) Set the signal outside the predetermined window to zero, and use the wave gate to filter out the interactive echo
(4) Invert the selection echo to the frequency domain.

Correlation of contribution of each scatterer and $\sqrt{\sigma}$:

$$\sqrt{\sigma} = \sum_{K=1}^{12} \sqrt{\sigma_K} \cdot e^{j2\gamma_K} \tag{2.22}$$

$\sigma_K$ is the kth scatterer RCS:

It is the relative distance of the scatterer on the line of sight of radar.

The measured data $\sqrt{\sigma}$ of 12 scatterers are shown in Table 2.4.

### *2.4.3 RCS of Regular Objects*

(1) A metal cylinder has a diameter of 7.3 cm and a length of 37 cm. The RCS it contributes is shown in Fig. 2.20.
(2) A metal cannonball body whose RCS changes with frequency is shown in Fig. 2.21 (horizontal polarization).
(3) The maximum RCS of different reflectors is shown in Table 2.5.

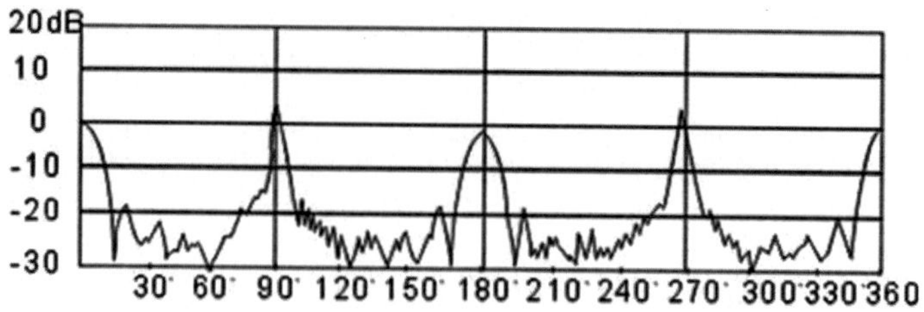


**Fig. 2.20** Simplified data graph of the metal cylinder

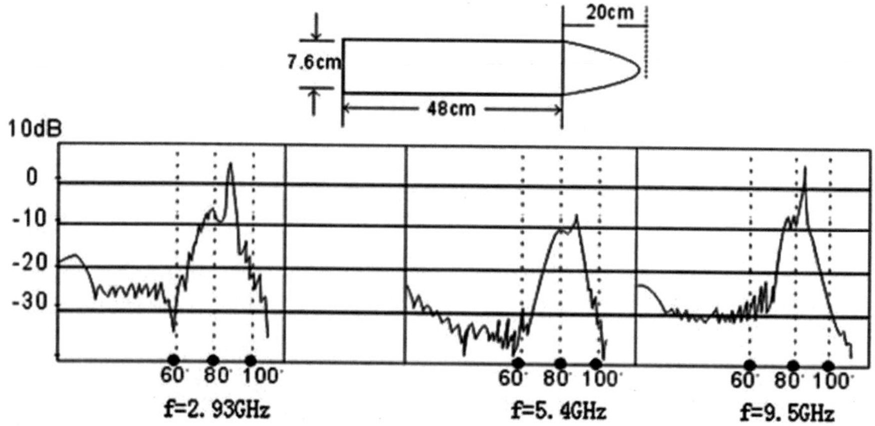


**Fig. 2.21** Change chart of RCS due to frequency change of shell-shaped objects

## 2.5 Timely Reconnaissance of Spaceborne SAR

### *2.5.1 Necessity of Timely Reconnaissance for Spaceborne SAR*

From the post-World War II period to the 1980s, U.S. reconnaissance and surveillance to other countries has been mainly based on airborne reconnaissance aircraft, high-altitude reconnaissance aircraft, and unmanned reconnaissance aircraft. The airborne reconnaissance platform is susceptible to the shadowing of ground objects, and can generally detect 27% of the signals in the detection area. The spaceborne reconnaissance is generally not affected by the shadow of the ground or little affected, and it can detect 99% of the targets in the reconnaissance area. Also, it is not restricted by airspace and climate. So, spaceborne SAR is widely used and has broad development prospects.

In order to conduct stereo imaging reconnaissance of geomorphic features and military targets, distributed multi-satellite SAR reconnaissance has been developed to obtain satisfied imaging results, as shown in Fig. 2.2.

**Table 2.5** Maximum RCS of different reflectors is shown in Table

| Reflector shape | Reflector name | Maximum RCS expression |
|---|---|---|
| 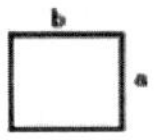 | Square tablet | $\sigma = \frac{4\pi \cdot a^2 \cdot b^2}{\lambda^2}$ |
|  | Cylinder | $\sigma = \frac{2\pi \cdot a \cdot b^2}{\lambda}$ |
|  | Sphere | $\sigma = \pi \cdot a^2$ |
|  | Dihedral angle | $\sigma = \frac{8\pi \cdot a^2 \cdot b^2}{\lambda^2}$ |
|  | Trihedral angle | $\sigma = \frac{4\pi \cdot a^4}{3\lambda^2}$ |
| 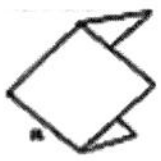 | Square face triangular face | $\sigma = \frac{12\pi \cdot a^4}{\lambda^2}$ |
|  | Arc face triangular face | $\sigma = \frac{0.507\pi^3 \cdot a^4}{\lambda^2}$ |
| 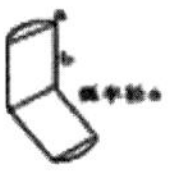 | Reflector | $\sigma = \frac{2\pi \cdot a \cdot b^2}{\lambda \cdot \cos^3(\phi)}$ |

From the 1990s onwards, the U.S. reconnaissance and surveillance of other countries would gradually evolve from airborne reconnaissance to spaceborne surveillance. Therefore, the anti-reconnaissance and anti-surveillance of the country would also be changed from anti-airborne reconnaissance and surveillance to anti-satellite reconnaissance and surveillance, which is an inevitable trend of military technology development.

Spaceborne reconnaissance equipment includes optical reconnaissance, infrared reconnaissance, television reconnaissance, and radar imaging reconnaissance. Among these types of reconnaissance methods, optical, infrared, and television reconnaissance are susceptible to weather or other shelters, making it impossible to be an all-weather reconnaissance device. Only the radar imaging reconnaissance equipment is not affected by the weather or some shelter. It can be called all-weather reconnaissance equipment. Spaceborne SAR provides both imaging and all-weather reconnaissance. It can be said to be a pair of "space eyes" for reconnaissance and monitoring the world. Now the United States has mastered it, posing a threat to all countries in the world. Other countries must take countermeasures so that the United States can't use spaceborne SAR to reconcile and monitor other countries in peacetime, nor can it detect the military deployment of the invaded country during wartime.

The development of timely reconnaissance for spaceborne SAR is the prelude to the development of jamming to spaceborne SAR. If it is impossible to detect spaceborne SAR promptly, it is impossible to timely jamming to spaceborne SAR. There are tens of thousands of multiple space targets in space currently. Some of them are abandoned satellites; Some are carrier rocket debris. There are only about 800 satellites in operation in space, including about 200 military satellites and around ten spaceborne SAR. These spaceborne SARs can detect and monitor every corner of the world and can monitor the military activities and military deployments in all countries. If you want to jamming with it on time, you must first identify spaceborne SAR from these thousands of satellites to jamming with them effectively.

### *2.5.2 Feasibility of Timely Reconnaissance for Spaceborne SAR*

Timely reconnaissance of spaceborne SAR is very necessary, but there are still many difficulties in achieving it.

(1) It is more difficult to find the spaceborne SAR.

Spaceborne SAR is rarely powered on, and it only starts when it is flying near the airspace of the area where it needs reconnaissance. The boot time is usually a few minutes. If there is no fast search reconnaissance system, it is difficult to seize the opportunity of its short boot. The remote warning radar can also find some carriers in space, but it is difficult to identify if it is the object we need to detect and jammed with.

(2) The antenna beam of the spaceborne SAR is narrow, and the sidelobe level is very low. Its antenna does not search. It just moves with the carrier. Normally, reconnaissance can only detect its sidelobe, which requires that the sensitivity of the reconnaissance receiver is relatively high, which is two to four orders of magnitude higher than that of a conventional reconnaissance receiver.

(3) The signal spectrum of SAR is relatively wide, generally ranging from tens of megahertz to hundreds of megahertz. Besides, the sensitivity of the receiver is required to be relatively high, which makes the development of reconnaissance receivers more difficult.

(4) The equivalent jamming power of the jammer is required to be relatively large, which requires increasing the gain of the jamming transmitting antenna and the power of the jamming transmitter to compensate for these requirements.

(5) The jamming transmits power increases, the receiver sensitivity increases, and the system's isolation of transmission and reception requirements increase.

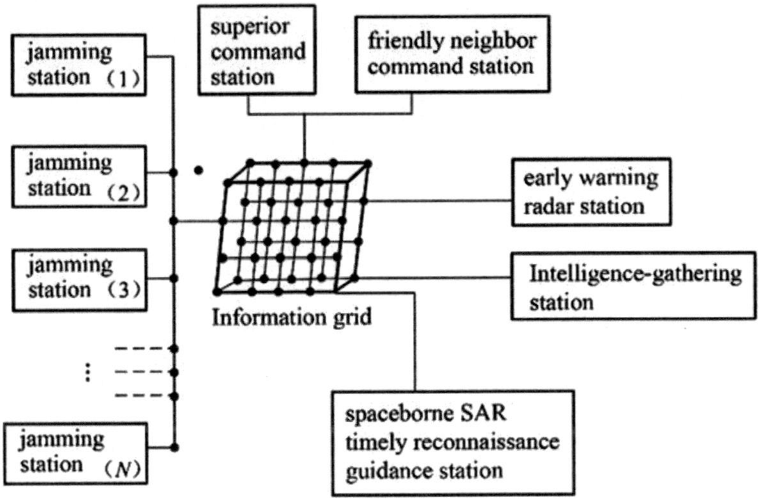


**Fig. 2.22** Schematic diagram of the composition of spaceborne SAR timely reconnaissance guidance jamming (region) system

The isolation of transmission and reception of the jamming system is one of the key problems in the jammer system. In the reconnaissance and jamming system for space-borne SAR, the isolation of transmission and reception has been added new difficulties.

Despite these difficulties, according to our analysis and research, the actual test of SAR proves that it can effectively jam with SAR and can also be detected in time. The spaceborne SAR jamming system consists of multiple jamming stations, timely reconnaissance and guidance stations, target warning radar stations, intelligence collection, and data distribution systems, see Fig. 2.22.

## 2.5.3 Requirements for Timely SAR Reconnaissance Stations

(1) Identify the investigated object.
(2) Determine the frequency band of reconnaissance.
(3) The time for timely guidance is $t_S$. It should ensure that the station can 100% intercept of the spaceborne SAR signal within time $t_S$ (assuming that SAR is turned on within the detection area of this station).
(4) The maximum distance $R_{\max}$ of the station to detect spaceborne SAR:

$$R_{\max} = \frac{h}{\cos \beta_{\max}}. \tag{2.23}$$

(5) Elevation range for SAR detection: 5°–85°.

(6) Azimuth range of SAR detection: 0°–360°.
(7) Requirements for equivalent sensitivity of spaceborne SAR reconnaissance receiving system.

$$S_j = \frac{P_t \cdot G_t(\theta)}{4\pi \cdot R_j^2} \cdot \frac{\lambda^2}{4\pi} \cdot G_j \tag{2.24}$$

As shown in Fig. 2.23: A is the reconnaissance station; B is the spaceborne SAR; D is the satellite projection point; $\beta$ is the elevation; $\theta$ is the angle between the detection beam and reconnaissance; R is the SAR detection distance; C is the center of the detection point.

According to Fig. 2.23.

$$R = \frac{h}{\cos\beta}, \quad d^2 = R^2 + R_j^2 - 2R \cdot R_j \cdot \cos(\theta)$$

Can be written as

$$R_j^2 - 2R \cdot R_j \cdot \cos(\theta) + R^2 - d^2 = 0$$

According to the solution of the quadratic equation

$$R_j = \frac{2R \cdot \cos(\theta) + \left[4(R \cdot \cos(\theta))^2 - 4\left(R^2 - d^2\right)\right]^{1/2}}{2}, R = \frac{h}{\cos(\beta)} \tag{2.25}$$

Then

$$R_j = \frac{2h\frac{\cos(\theta)}{\cos(\beta)} + \left[4\left(h \cdot \frac{\cos(\theta)}{\cos(\beta)}\right)^2 - 4\left(\frac{h}{\cos(\beta)}\right)^2 + 4d^2\right]^{1/2}}{2} \tag{2.26}$$

Substitute formula (2.26) into formula (2.24)

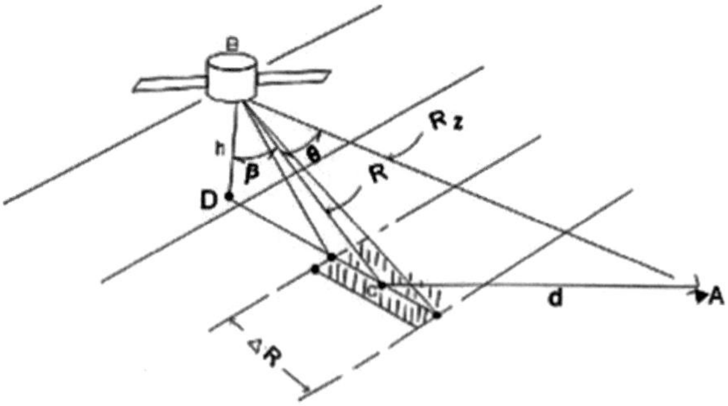

**Fig. 2.23** Schematic diagram of spaceborne SAR reconnaissance

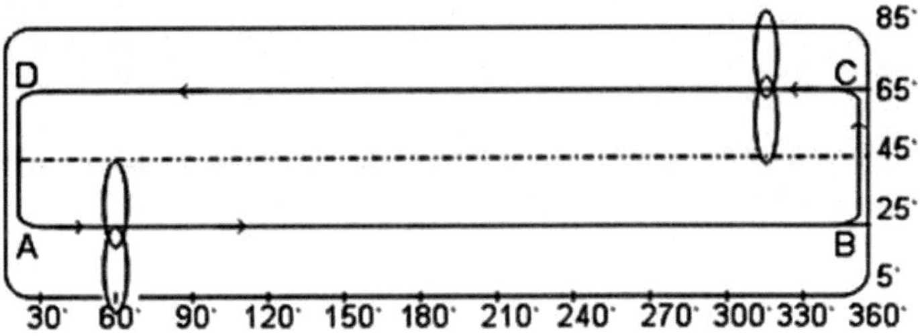


**Fig. 2.24** Schematic diagram of the search trajectory of the search antenna

$$S_j = \frac{P_t \cdot G_t(\alpha,\beta)\cdot\lambda^2\cdot G_{ja}}{4\pi^2\left\{2h\frac{\cos(\theta)}{\cos(\beta)} + \left[4\left(h\cdot\frac{\cos(\theta)}{\cos(\beta)}\right)^2 - 4\left(\frac{h}{\cos(\beta)}\right)^2 + 4d^2\right]^{1/2}\right\}^2} \tag{2.27}$$

The sensitivity of the spaceborne SAR receiver is related to various factors such as the peak power of the spaceborne SAR transmitter, the antenna sidelobe gain, the elevation of the antenna, the height of the satellite, the azimuth of the radar and the reconnaissance station.

(8) Requirements for the antenna beamwidth of the reconnaissance station.

In order to meet the one-time reliable search, take the width of elevation as $\beta_{0.5}$; take the width of azimuth beam as $\theta_{0.5}$; let the range of elevation search from 5° to 85°; let the range of azimuth search from 0° to 360°; Antenna search uses quadrilateral search. Shown as Fig. 2.24.

The elevation of the search antenna is set at 25°, and the azimuth angle is turned from 0° to 360°. When turning to 360°, the azimuth stops, and the elevation turns upward. When turning to 65°, the elevation stops, and the azimuth angle turns from 360 ° to 0 °. When turning to 0°, the elevation turns downward from 65°. The elevation stops when turning to 25°, and the azimuth angle turns from 0° to 360°. As long as the elevation angle is 5°–85°, the distance is in the detected range. The direction is from 0° to 360°, and then from 360° to 0°, the SAR signal should be intercepted.

The crossbeam is used for elevation, and the beamwidth on each side is 20°. The beamwidth $\theta_{0.5}$ of the azimuth angle should be determined according to the search speed $V_t$ of the antenna, the speed of the frequency search, and the requirements for reliable search.

a. Searching by machine.

If the search antenna sweeps a beam for time $T_s$ then

$$T_S = \frac{\theta_{0.5}}{V_t} \tag{2.28}$$

The time $\Delta T_S$ required for the frequency synthesizer to sweep through a frequency band and ensure that more than three pulses are received within time $\Delta T_S$, then $\Delta T_S = 3T_\gamma$.

$T_\gamma$ is the repetition period of the radar pulse. When the antenna illuminates a beam, ensuring the frequency can search a frequency band $\Delta F = f_1 - f_2$. The time required to search this band is $T_S = n \cdot T_S = 3n \cdot T_\gamma$.

We can get $T_S = 3n \cdot T_\gamma = \frac{\theta_{0.5}}{V_t}$

$$V_t = \frac{\theta_{0.5}}{3n} \cdot f_\gamma \tag{2.29}$$

In the formula:

$\theta_{0.5}$ is the azimuth beamwidth of the search antenna, assuming it is 2°.
$f_\gamma$ is the radar pulse repetition frequency, assuming it is 1400 Hz;
n is 20.

So

$$V_t \le \frac{2 \times 1400}{3 \times 20} = 46.6^\circ/\text{s}.$$

The time required for the entire search is $t_S$, so:

$$t_S = \frac{2 \times 360^\circ}{V_t} + \frac{2 \times 40^\circ}{V_\beta} \tag{2.30}$$

In the formula:

360° The range of direction search
40° The range of elevation search
$V_t$ The speed of azimuth search, choosing to be 46°/s
$V_\beta$ The speed of elevation search, choosing to be 20°/s.

Substitute into formula (2.29), we get

$$t_S = \frac{720}{46} + \frac{80}{20} = 156 + 4 = 196\,\text{s}$$

That is, a period of reliable search for spaceborne SAR (including reliable frequency search) takes 20 s. According to the tactical requirements, the search time of the signal expected to be as fast as possible. Because the spaceborne SAR flight time is about 60 s in the ground-to-star effective jamming area. If it adopts some tactical anti-jamming measures, the boot time maybe 20–30 s. So, the search guidance time is preferably less than 2 s.

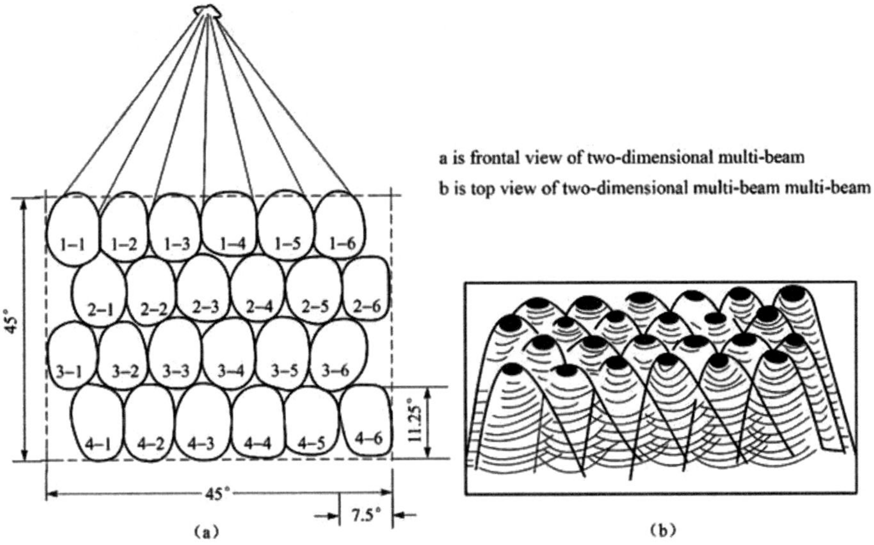


**Fig. 2.25** Schematic diagram of two-dimensional multi-beam reconnaissance antenna beam

b. Using two-dimensional multi-beam search antenna

If it is known that the two-dimensional multi-beam array antenna has 24 beams, as shown in Fig. 2.25. Each beam has an azimuth of 7.5° and an elevation of 11.25°.

There are six segments in the azimuth, and each double width is 7.5°. There are 4 segments in the elevation direction, and each segment width is 11.25°. A combination of two-dimensional multi-string is 24 segments, forming 45° × 45° reconnaissance area. If the detection area is 45° × 45°, it requires detection time $T_S = 3n \cdot T_\gamma = 3 \times 20 \times 071 = 428\,\text{ms}$.

If the azimuth rotates at a speed of 100°/s and turns through a 360° × 45° airspace, the time required is about 4 s. If there is a target guidance radar guiding, the search time can be less than 1 s. It can be seen that different search antenna systems and guidance methods can be selected according to different requirements to ensure timely interception of spaceborne SAR signals.

(9) The selection of the instantaneous bandwidth and sensitivity of the receiver.

In principle, the instantaneous bandwidth of the receiver should be equal to or slightly larger than the spectrum width of the radar signal. However, if the receiver sensitivity is critical, the receiver bandwidth can be divided into two types, one is the narrow frequency band, which is used to search and track the receiver. First, ensure the sensitivity and track the angle after intercepting the signal. The other is a wideband receiver in which frequency bandwidth is greater than or equal to the radar signal spectrum width. Since the reconnaissance antenna continuously tracks the spaceborne SAR, the wideband receiver can perform delay accumulation to improve the receiver's signal/noise ratio.

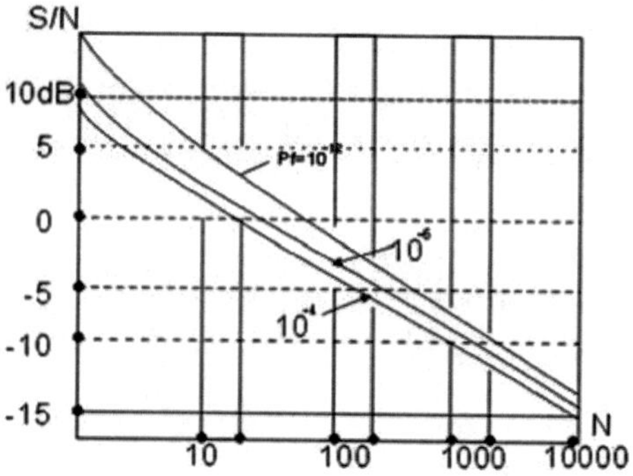


**Fig. 2.26** Relationship between pulse number accumulation and S/N

The narrowband receiver tracking system has been used to track the SAR signal. In the broadband receiving branch, the probability of each discovery is $P_{Di}$, and the probability of not finding is $P_{Mi}$, so $P_{Mi} = (1 - P_{Di})$.

The cumulative discovery probability is

$$P_D = 1 - P_M = 1 - \sum_{i=1}^{n} (1 - P_{Di}) \tag{2.31}$$

The greater n is, the greater the chance of finding SAR is. The relationship curve of S/N is shown in Fig. 2.26.

It is found that the probability $P_D = 90\%$, the signal has fluctuations. When the false alarm probability is $P_f = 10^{-6}$ and n = 10, the S/N is 3 dB; when n = 100, S/N is -4 dB; when n = 1000, S/N is −8 dB. This shows that the effect of accumulation on improving the receiver's signal/noise ratio. The accumulation mentioned here refers to delaying the received pulse n in turn by an appropriate time because the repetition frequency of SAR is stable within a certain time range. The pulse received first has more delay, the pulse received later has less delay, and the last pulse has no delay. Add them at the same moment. Since the signal is regular, and the noise is irregular, random, and chaotic, so the result of the cross-period addition improves the signal/noise ratio.

As can be seen from Fig. 2.26, when n is three pulse accumulation, S/N needs 8 dB; when n is ten pulse accumulation, S/N needs 3 dB; when n is 20 pulse accumulation, S/N needs 1 dB. In the broadband branch, if the receiver sensitivity is not enough, delay accumulation measures can be used to improve the signal/noise ratio of the receiver.

(10) The capability of measuring the trajectory of spaceborne SAR.

Assuming that the spaceborne SAR is a low-orbit satellite, the curvature of the earth can be ignored. The coordinate relationship between the spaceborne SAR and the reconnaissance station is shown in Fig. 2.27.

Let the satellite move from point B to point D, the time required is $t_0$;

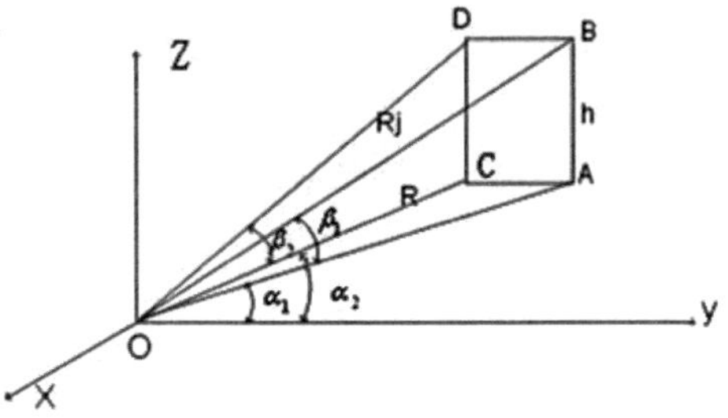


**Fig. 2.27** Schematic diagram of the trajectory of the reconnaissance station's measurement of spaceborne SAR

Point B is the position of the target at time $t_1$;
Point D is the target reaching point D at time $t_2$ after $t_0$ time;
Point A is the projection point of the target at point B;
Point C is the projection point of the target at point D;
Point 0 is the location of the reconnaissance station;
$\alpha_1$ is the azimuth angle of the target at the projection point A;
$\alpha_2$ is the azimuth angle of the target at the projection point C;
$\beta_1$ is the elevation of the target at point B;
$\beta_2$ is the elevation of the target at point D;
h is the height of the target.

At time $t_1$, the target is at point B, the projection point A, the corresponding azimuth angle is $\alpha_1$, and elevation is $\beta_1$; At time $t_2$, the target is at point D, the projection point C, the corresponding azimuth angle is $\alpha_1$, and elevation is $\beta_2$. The satellite carrier flight speed is $V_t$. According to Fig. 2.27, we can get

$$\overline{DB} = \overline{AC} = (t_1 - t_2)V_t = t_0 \cdot V_t$$

$$\overline{OC} = R. \tag{2.32}$$

Solve the simultaneous equations

$$\overline{CA}^2 = \overline{OC}^2 + \overline{OA}^2 - 2\overline{OA} \cdot \overline{OC} \cdot \cos(\alpha_2 - \alpha_1) \tag{2.33}$$

$$\tan\beta_1 = \frac{h}{\overline{OA}} \tag{2.33}$$

$$\tan\beta_2 = \frac{h}{\overline{OC}} \tag{2.34}$$

Get

$$R = \frac{V_t \cdot t_0}{\left[1 + \left(\frac{\tan\beta_2}{\tan\beta_1}\right)^2 - 2\frac{\tan\beta_2}{\tan\beta_1} \cdot \cos(\alpha_2 - \alpha_1)\right]^{1/2}} \quad R_j^2 = R^2 + h^2 \tag{2.35}$$

So

$$R_j = \left[ \frac{(V_t \cdot t_0)^2}{1 + \left(\frac{\tan\beta_2}{\tan\beta_1}\right)^2 - 2\frac{\tan\beta_2}{\tan\beta_1} \cdot \cos(\alpha_2 - \alpha_1)} + h^2 \right]^{1/2} \tag{2.36}$$

After the satellite's trajectory is determined, $V_t$ is a known number;
Satellite height h is also a known number;
$t_0$ is the interval from $t_1$ to $t_2$ or $t_2$ to $t_3$ ..., $\beta_{1,2}, \alpha_{1,2}$ or $\beta_{2,3}, \alpha_{2,3}$ can be measured timely. Therefore, $R_j$ is a number that can be determined at time $t_2$ and time $t_3$.... According to $R_j$ at different moments, the reconnaissance station can draw the trajectory of the satellite to the reconnaissance station.

If the spaceborne SAR adopts orbit transferring and speed changing, the measurement will produce a large error. However, its orbit transferring and speed changing generally occur in a relatively short period, so the track change and speed change measures are not often adopted. Otherwise, its imaging is not good. Therefore, the trajectory measured by the reconnaissance station during the stable operation of the spaceborne SAR is relatively accurate.

(11) The ability to measure and displaying parameters of spaceborne SAR.
(12) It can be connected with the "information grid network" of the jamming system to provide the reconnaissance parameters required by the system.
(13) Search and reconnaissance system can switch to passive tracking after found targets automatically.
(14) It equipped with carrying and power supply.

## 2.5.4 *Composition of Timely Reconnaissance Stations for Spaceborne SAR*

### 2.5.4.1 Composition of Timely Reconnaissance Station

See Fig. 2.28.

Timely reconnaissance stations for spaceborne SAR include antenna feeder sub-system, low-noise amplification sub-system, sum-difference formation network, frequency synthesizer, frequency mixer, detection, and video amplifier, error detection, servo amplification and driving, signal sorting processing, display, Radar parameter measurement, spectrum measurement, and other sub-systems.

In order to satisfy the simultaneous search in the air and frequency domains, satisfy the allowed reliable search time, automatically record the frequency and angle data after searching for SAR signals, and automatically switch to angle tracking and frequency tracking, the selection and design of the principal block diagram must be considered comprehensive.

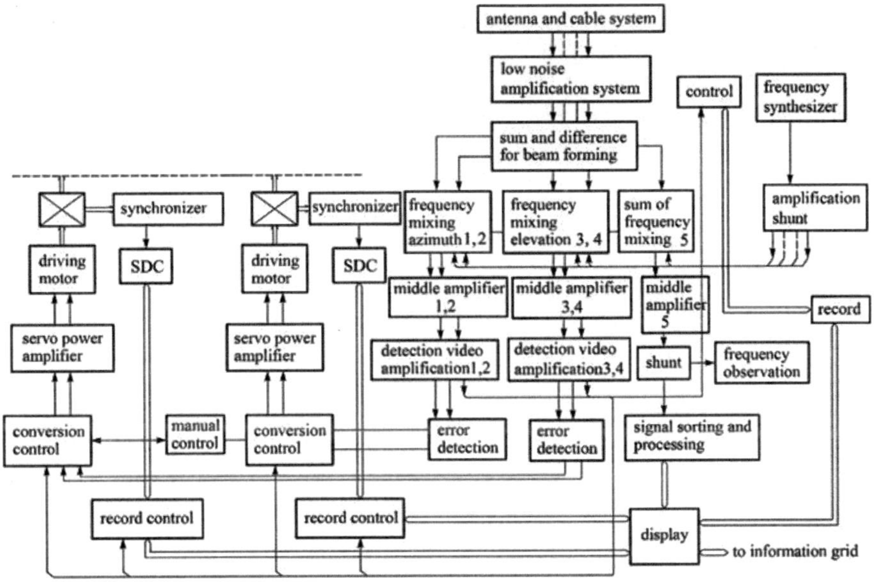


**Fig. 2.28** Block diagram of the composition of timely reconnaissance station of a spaceborne

#### 2.5.4.2 Description of Timely Reconnaissance of Spaceborne SAR

The reconnaissance antennas are searched in the shape of line or square, or use multi-beam receiving antenna array to wait for reception in local airspace. No matter what kind of the antenna search method is, it must meet the frequency of searching for a frequency band (such as L, S, C or X frequency band) within the time of sweeping an antenna search area, and then divide this frequency band into n segments for frequency search. For example, the time for the antenna to sweep a beam of the target is 0.1 s, and a frequency band is divided into 20 (n) frequency bands. The receiving time of each frequency band is better than 4.95 ms. During this period, if the radar repetition frequency is 1400 Hz, you can receive six pulses reliably, and meet the requirement of 3 reliable received pulses.

When the receiver reliably receives more than three pulses, the frequency band is automatically locked in the $n_i$ frequency band, and $\alpha_i$, $\beta_i$ are automatically recorded. Because the mechanical search angle of the antenna is relatively large, when a stop pulse is given suddenly, it may not stop at $\alpha_i$, $\beta_i$ but rush out of an angle due to inertia. Therefore, some ancillary circuits are needed to make the antenna automatically return to a, b, and then enter the automatic tracking state. At this time, we believe that the spaceborne SAR has been completely intercepted. During this intercepted time, you can perform parameter measurement, intra-pulse feature measurement, spectrum measurement on the signal, and you can also guide the jammer to jamming with it effectively. At the same time, the data can be transmitted to the superior command post, friendly neighbor command post, jammer command system through the "information grid network". The data obtained by other stations can also be transferred to this reconnaissance station in time to realize the interconnection, intercommunication, and resource sharing of each station, which will greatly improve the combat capability of the system.

## 2.6 Timely Suppression Jamming Against Spaceborne SAR

### 2.6.1 *Feasibility of Timely Suppression Jamming Against Spaceborne SAR*

Jamming equation for spaceborne SAR:

$$P_j G_j = \frac{K_{j\Sigma} \cdot P_t \cdot G_t \cdot \sigma_i \cdot R_j^2}{4\pi \cdot R_{\min}^4} \cdot \frac{L_d}{K_f \cdot L_j \cdot r_j} \cdot \frac{G_t}{G_t(\theta)}$$

In this equation, it is the same as the jamming to the conventional radar, but two parameters are quite different from the conventional radar, $K_{j\Sigma}$ and $\sigma_i$. The jamming suppression coefficient $K_{j\Sigma}$ is much larger than that of conventional radar, and the maximum is 4–5 orders of magnitude. The effective reflection area of $\sigma_i$ target is much smaller than the equivalent reflection area of the conventional radar beam irradiation area if it is calculated as a point target, and the minimum can be as small as 4–5 orders of magnitude. We analyzed the radars of many different systems. We found out that the production, $K_{j\Sigma}\sigma_i$, of jamming suppression coefficient of synthetic aperture radar on point target and equivalent reflection area of point target is very similar to $K_{j\Sigma 0}$ of other systems. It can be seen that the synthetic aperture radar system itself does not have much advantage against continuous wave noise. Instead, through two-dimensional compression, the resolution of azimuth and distance is improved, so that it can image and improve the signal-to-noise ratio of the radar itself, enabling it can extract useful information of point targets on the order of −60 dB. But this does not mean that much jamming equivalent power is required. But this does not mean that much jamming equivalent power is needed.

For example, a spaceborne SAR works in the L-band, with an altitude of 700 km, an elevation of 25°, a maximum elevation of 60°, a peak power of 5 kW, a pulse width of 35 us, a compressed pulse width of 15 ns, and a flight speed of 7.3 km/s and azimuth beamwidths are 2°, and the elevation beamwidth is 3°, as shown in Fig. 2.29.

It can be seen from Fig. 2.29,

$$R_a = h \cdot \tan\left(\beta_{\min} - \frac{1}{2}\beta_{0.5}\right) = 700 \times 10^3 \times 0.43 = 304\,\text{km}$$

$$R_{\min} = \frac{h}{\cos(\beta_{\min} - \beta_{0.5})} = \frac{700 \times 10^3}{\cos 23.5^\circ} = 769\,\text{km}$$

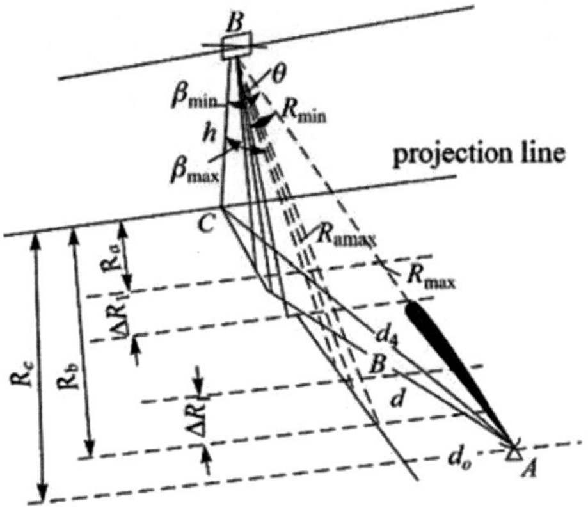


**Fig. 2.29** Schematic diagram of single station's suppressing jamming to spaceborne SAR

$$R_b = h \cdot \tan\left(\beta_{\max} + \frac{1}{2}\beta_{0.5}\right) = 700 \times \tan 61.5^\circ = 1289\,\text{km}$$

$$R_{\max} = \frac{h}{\cos 61.5^\circ} = 1489\,\text{km}$$

$$R_c = R_a + R_b = 304 + 1289 = 1593\,\text{km}$$

$$R_{\max}^{'2} = h^2 + R_c^2$$

$$R_{\max}' = \left(h^2 + R_c^2\right)^{1/2} = 1324\,\text{K}$$

$$R_{j\max} \cong \left(R_{\max}^{'2} + d_0^2\right)^{1/2} = 1339\,\text{km}$$

$$n_a = \frac{\tau}{\Delta\tau} = \frac{35 \times 10^{-6}}{15 \times 10^{-9}} = 33.6\,\text{dB}$$

$$n_b = f_\gamma \cdot \theta_{0.5} \cdot R_{\min}/V_t = \frac{1400 \times 2 \times 0.017 \times 769}{7.3} = 45.5\,\text{dB}$$

$$K_{j\Sigma} = (n_a \cdot n_b)^{0.8} = 63.3\,\text{dB}$$

$\sigma_i$ take the point target value, generally 5 $m^2$.

Substitute parameters into (2.19).

$$P_j G_j = \frac{K_{j\Sigma} \cdot P_t \cdot G_t \sigma_i \cdot R_{j\max}^2}{4\pi \cdot R_{\min}^4} \cdot \frac{L_d}{K_f \cdot L_j \cdot r_j} \cdot \frac{G_t}{G_t(\theta)}$$

$$= \frac{2 \times 10^6 \times 10^8 \times 5 \times (133)^2 \times 10^{12} \times 1 \times 10^4}{4\pi \times (0769)^4 \times 10^{24}}$$

$$= \frac{1.76 \times 10^{31}}{4\pi \times 0.35 \times 10^{24}} = 0.39 \times 10^7 = 66\,\text{dBW}$$

The existing technical basis can achieve an equivalent jamming power of 66 dBW.

### 2.6.2 *Jamming Protection Zone of Spaceborne SAR from a Single Suppression Jamming Station*

(1) Jamming protection zone for spaceborne SAR in the linear region.

As can be seen from Fig. 2.29, if the satellite's trajectory passes straight through a certain area, the effective jamming area is shown in Fig. 2.30. The angle can be obtained

$$d_1^2 = R_{j\max}^2 - 2R_{j\max} \cdot R_{\max} \cdot \cos\theta_1 + R_{\max}^2$$

or $d_2^2 = R_j^2 - 2R_j \cdot R \cdot \cos\theta_2 + R_{\min}^2$

$$\cos\theta_i = \frac{R_j^2 + R_{\min}^2 - d_i^2}{2R_j \cdot R_{\min}}$$

The closer the point A of the jamming station is to the satellite projection line, the greater the value of $\theta$.

If the jamming station is located on the middle line of $\Delta R_1$, $d_0 = 200\,\text{km}$.

$R_j \cong \left(d_0^2 + R_{\min}^2\right)^{1/2} = \left(769^2 + 200^2\right)^{1/2} = 794\,\text{km}$, so that

$$\theta = \cos^{-1}\left(\frac{R_j^2 + R_{\min}^2 - d_0^2}{2R_j \cdot R_{\min}}\right) = \cos^{-1}\left(\frac{794^2 + 769^2 - 200^2}{2 \times 794 \times 769}\right)$$
$$= \cos^{-1} 0.96 = 16.2°.$$

$d_0 = 200\,\text{km}$, $R_a = 304\,\text{km}$, $R_b = 1289\,\text{km}$. After the spaceborne SAR is disturbed, the ground object cannot be imaged within $R_b \times 2d_0$ area ($1289 \times 400 = 515,600\,\text{km}^2$).

(2) The jamming protection zone when the satellite-borne SAR track shifts are shown in Fig. 2.31.

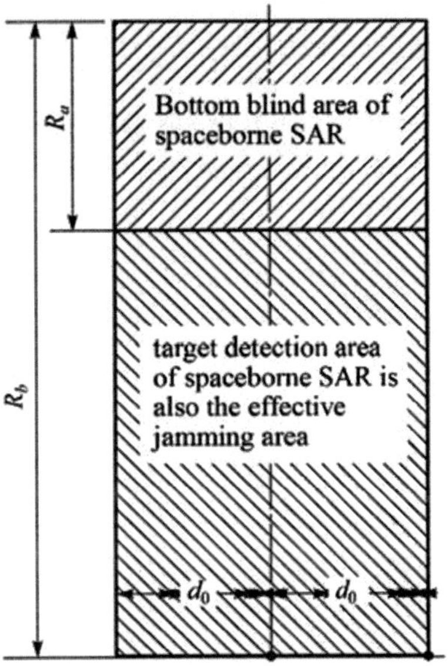


**Fig. 2.30** Schematic diagram of suppressed jamming area of the current trajectory of spaceborne SAR

H = 700 km, $\beta = 25°$, d = 320 km, Inside the circle is the effective jamming area. Outside the circle is the jamming invalid area. Point A is the position of the jammer.

### *2.6.3 Protection Zone of Space Borne SAR from Multiple Suppression Jamming Stations*

The protection of spaceborne SAR jamming stations requires the establishment of multiple jammers, mainly for three reasons:

1. In the future, many spaceborne SARs will use phased array antennas. In addition to the rapid electrical scanning, the advantages can also change the antenna beam zero. So $G_t(\theta)$ is actually $G_t(\alpha, \beta)$. It can adjust the sidelobe to the minimum in the direction of the jammer up to the order of −60 dB or −70 dB. We chose $G_t/G_t(\theta) = 40$ dB in Sect. 2.6.1, which is not enough, and the equivalent jamming power is still 2–3 orders of magnitude worse. Therefore, multiple jamming stations are required to jamming with it in multiple directions (because its zero-setting direction is generally within three). If we want to use four jamming stations to jamming with it simultaneously, its antenna beam zeroing will not work.
2. In the future, spaceborne SAR will develop to multi-satellite distributed reconnaissance to make the targets detected by spaceborne SAR with higher resolution and more stereoscopic, as shown in Fig. 2.32. The distance between

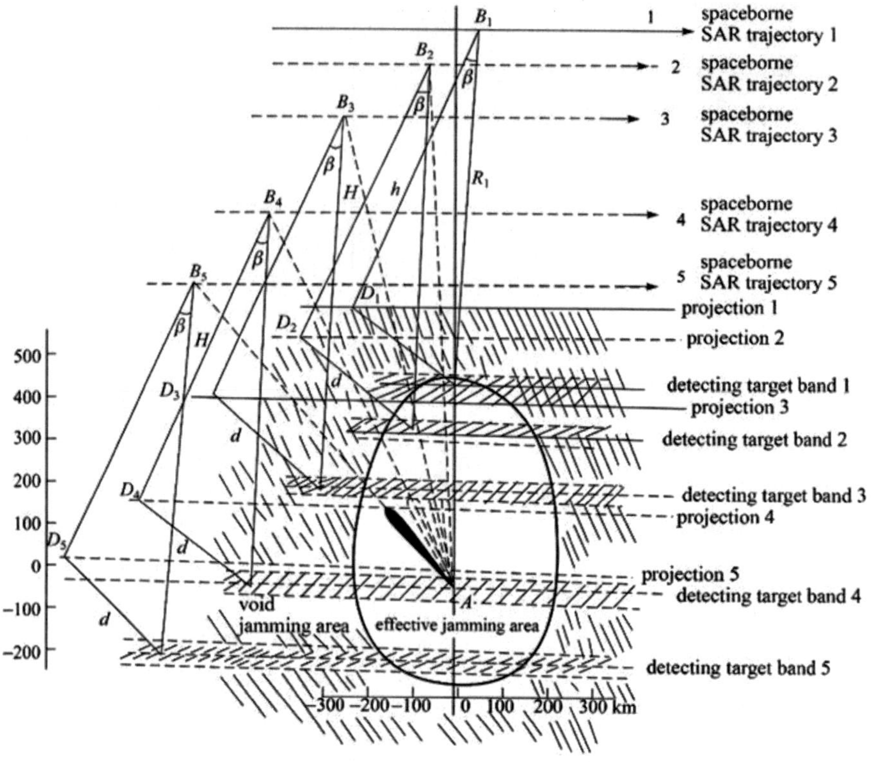


**Fig. 2.31** When spaceborne SAR's track shift and h, β, d are unchanged, the schematic diagram of jamming effective area

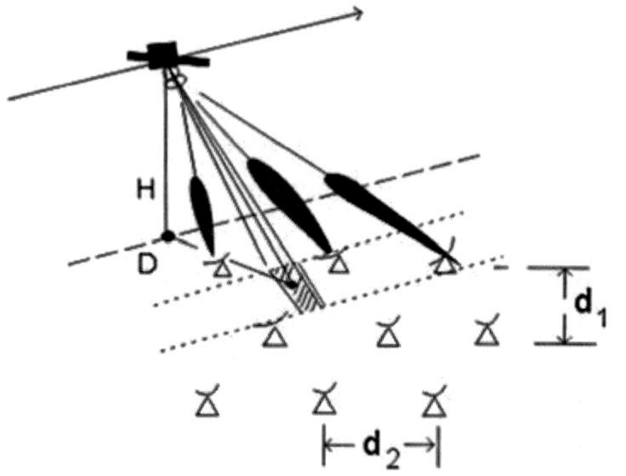


**Fig. 2.32** Schematic diagram of the multi-jamming station to single satellite SAR (phased array antenna)

multi-satellite SAR satellites is generally a few kilometers to tens of kilometers. There can also be some differences in frequency. It also requires multiple jamming stations to jamming with multi-satellite SAR effectively.

3. In an emergency, it requires 24 h of continuous duty to monitor and jamming. Therefore, the jamming station needs to be backed up, and there should be multiple stations. The composition of the jamming system for spaceborne SAR is shown in Fig. 2.32.

   a. The spaceborne SAR jamming to the phased array antenna is shown in Fig. 2.32.

      The $d_2$ distance in the figure is generally 150–200 km.

      The range of its protected area is affected by the working range and altitude of the satellite-borne SAR elevation angle. Generally, it should be between 100,000 and 500,000 $km^2$. If the SAR elevation angle working range is 20°–60°, it will reach 3 million $km^2$.

   b. The jamming to distributed multi-satellite SAR is shown in Fig. 2.33.

      A1, 2, 3 … are the jamming station. B1, 2, 3 … are distributed spaceborne SAR. C is a distributed multi-satellite SAR reconnaissance station. D is the data transmission and information grid network. E is a remote warning radar station.

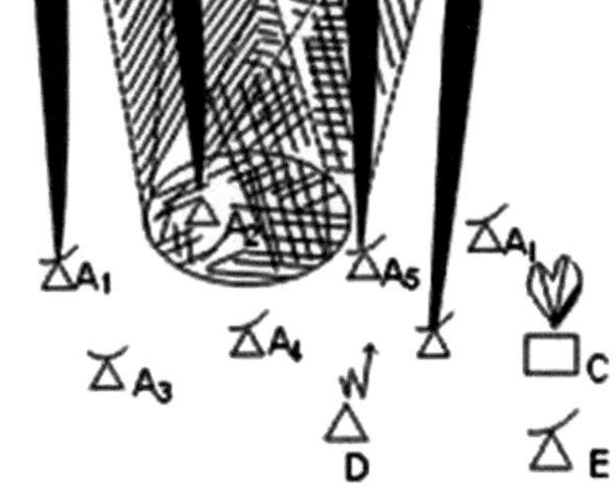


**Fig. 2.33** Schematic diagram of jamming from multiple jamming stations to jamming with multi-satellite stations

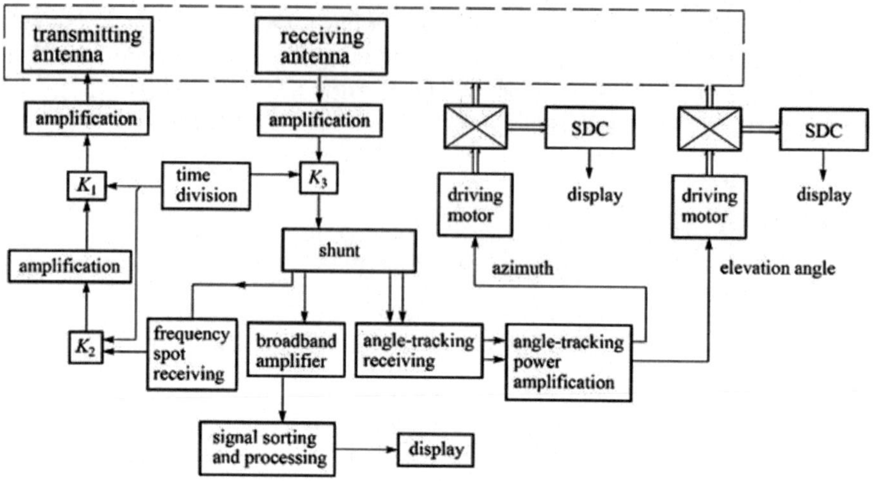


**Fig. 2.34** Block diagram of the composition of spaceborne SAR jamming station

## 2.6.4 *Composition of Suppression Stations for Spaceborne SAR*

### 2.6.4.1 Block Diagram of the Composition of Spaceborne SAR Jamming Station

The block diagram of the composition of spaceborne SAR jamming stations is shown in Fig. 2.34. Its basic composition includes: transmitting antenna, receiving antenna, jamming transmitter subsystem, aiming receiver subsystem, the wideband signal receiving subsystem, angle tracking receiver subsystem, display sub-system, sending and receiving time interval sub-system, power supply system, etc.

From the principle block diagram, it is not much different from ordinary continuous-wave noise jammers, but their parameter requirements are very different. The isolation of the transceiver antenna is one of the key parameters. Originally in conventional jammers, this is a key technology that is difficult to solve. In jammers to spaceborne SAR, its receiver sensitivity is 2–3 orders of magnitude higher than conventional receivers, which means that its transceiver isolation should also be increased by 2–3 orders of magnitude.

### 2.6.4.2 Sensitivity of Jammer Receiving System

According to Fig. 2.23, the receiver's sensitivity Eq. (2.27) of the jammer is introduced

$$S_j = \frac{P_t \cdot G_t(\alpha, \beta) \cdot \lambda^2 \cdot G_{ja}}{4\pi^2 \left\{ 2h \frac{\cos\theta}{\cos\beta} + \left[ 4\left(h \frac{\cos\theta}{\cos\beta}\right)^2 - 4\left(\frac{h}{\cos\beta}\right)^2 + 4d^2 \right]^{1/2} \right\}^2} \tag{2.37}$$

Taking a spaceborne SAR as an example: let it work in the L-band, h is 700 km, the maximum elevation is 60°, the peak power is 5 KW, and the flight speed is 7.3 km/s, $\theta = 60°$, $G_t(\alpha, \beta) = 03$, $G_{ja} = 25\,\text{dB}$.

We can get $\cos 16° = 0.96$; $\cos 60° = 0.5$.

Substitute parameters into formula (2.26).

$$S_j = \frac{5 \times 10^3 \times 0.3 \times 0.24^2 \times 300}{4\pi^2 \left\{ 2.68 \times 10^6 + \left[ 7.2 \times 10^{12} - 3.92 \times 10^{12} + 0.64 \times 10^{12} \right]^{1/2} \right\}}$$

$$= \frac{0.258 \times 10^5}{4\pi^2 \left\{ 2.68 \times 10^6 + 1.97 \times 10^6 \right\}^2} = \frac{0.258 \times 10^5}{2.93 \times 10^{14}} = -101\,\text{dBW}$$

#### 2.6.4.3 Transceiver Isolation of Jammer

According to the definition of degree of transceiver isolation:

$$D = \frac{\sum_{i=1}^{n} P_{ji}}{S_j} \cdot (S/N) \tag{2.38}$$

In the formula:

| | |
|---|---|
| D | The degree of transceiver isolation. |
| $\sum_{i=1}^{n} P_{ji}$ | The sum of various noises entering the receiver. |
| $S_j$ | The minimum signal received by jammer's receiver. |
| S/N | The signal/noise ratio for jammer's receiver to work normally. |

The noise entering the receiver includes internal jamming of the device, thermal noise of the receiver itself, jamming noise reflected by ground objects, and noise of the transceiver antenna entering the receiver from the sidelobe. Among these four noise sources, because the elevation of the jamming antenna to the spaceborne SAR is relatively high, the jamming noise reflected by the ground objects can be ignored. The internal noise of the receiver has been made it ineffective when selecting the receiving sensitivity. The internal jamming of the equipment has been considered to adopt measures so that it will not affect the normal operation of the receiver. Only the noise entering from the sidelobe of the transceiver antenna is the main factor affecting self-sending and self-receiving, then:

$$\sum_{i=1}^{n} P_{ji} \cong \frac{P_{tus} \cdot G_t(\alpha, \beta)}{4\pi \cdot r^2} \cdot \frac{\lambda^2}{4\pi} \cdot G_j(\alpha, \beta) \tag{2.39}$$

Substitute formula (2.38) into formula (2.36)

$$D = \frac{P_{tus} \cdot G_t(\alpha, \beta) \cdot \lambda^2 \cdot G_j(\alpha_a, \beta_a) \cdot (S/N)}{(4\pi \cdot r)^2 \cdot S_{j\,\min}} \tag{2.40}$$

In the formula,

$P_{\text{tus}}$ is the power of jamming transmitter or background noise of jamming transmitter.
$G_t(\alpha, \beta)$ is sidelobe gain from the jamming antenna to the direction of the receiving antenna.
$G_j(\alpha_a, \beta_a)$ is Sidelobe gain for the jammer's receiving antenna to the direction of transmitting antenna;
$\lambda$ is the working wavelength.
r is the distance between Jamming transmitting antenna and Jamming receiving antenna.
$S_{j\,\min}$ is the minimum signal received by jammer's receiving antenna from spaceborne SAR.
S/N is the signal/noise ratio for jammer's receiver to work normally.

As shown in Fig. 2.35.

(1) Simultaneously sending and receiving.

Assume the power of transmitter is 5 KW, the power of background noise is 10 mW; $G_t(\alpha, \beta)$ is 5; $G_j(\alpha_a, \beta_a)$ is 5; S/N = 10; $\lambda = 0.24$ m; $r = 50$ m; $S_{j\,\min}$ = −101 dBW.

Substitute parameters into formula (2.39), we can get

$$D_1 = \frac{5 \times 10^3 \times 5 \times 0.24^2 \times 5 \times 10}{(126 \times 50)^2 \times 0.8 \times 10^{-10}} = 93.5\,\text{dB}$$

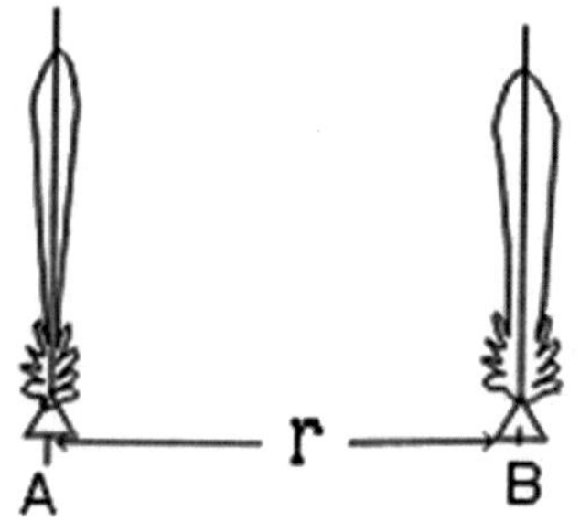


**Fig. 2.35** Working diagram of transceiver antenna

If $r = 5$ m, the degree of transceiver isolation is 113.5 dB.

(2) If the time separation system is adopted, that is, when the transmitter is turned off, the receiver does not receive the signal, and when receiving, the transmitter stops transmitting. At this time, only the background noise of the transmitter is 10 mW.

Substitute parameters into formula (2.39), then

$$D_2 = \frac{10^{-2} \times 5 \times 0.24^2 \times 5 \times 10}{(126 \times 50)^2 \times 0.8 \times 10^{-10}} = \frac{0.14}{3.12 \times 10^{-5}} = 0.44 \times 10^4 = 36.5\,\text{dB}$$

If $r$ changes to 5 m, the degree of transceiver isolation needs to be 56.5 dB.

If $r$ changes to 1 m, the degree of transceiver isolation needs to be 76.5 dB.

(3) The transceiver antenna platform is installed separately, adopting the time separation system

The required transceiver isolation in this state is related to the antenna's rotation angle $\theta$.

$$D(\theta) = \frac{P_{jus} \cdot G_t(\theta) \cdot \lambda^2 \cdot G_j(\theta)(S/N)}{(4\pi\lambda)^2 \cdot S_{j\min}} \tag{2.41}$$

As shown in Fig. 2.36

In order to effectively receive and jamming with the target, the receiving and transmitting antennas must be parallel. Therefore, transceiver isolation is a function of the antenna rotation angle $\theta$, selected $r$ as 20 m. See Table 2.6 for the $D(\theta)$ data.

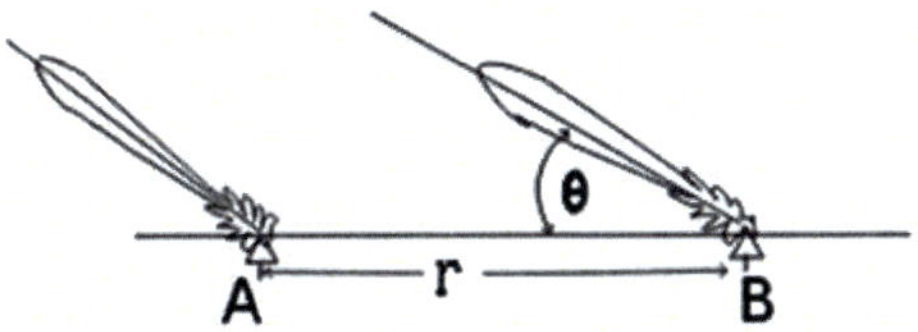


**Fig. 2.36** Schematic diagram of the degree of isolation at different positions of the transceiver antenna

**Table 2.6** Antenna sidelobe gain for various angles when r=20m

| $\theta$ (°) | 0 | 30 | 60 | 90 | 120 | 150 | 180 | 210 | 240 | 270 |
|---|---|---|---|---|---|---|---|---|---|---|
| $G_j(\theta)$ | 40 | 8 | 7 | 5 | 7 | 8 | 10 | 8 | 7 | 5 |
| $G_t(\theta)$ | 10 | 8 | 7 | 5 | 7 | 8 | 30 | 8 | 7 | 5 |
| $D(\theta)$ | 79 | 45 | 43 | 39 | 43 | 45 | 60 | 45 | 43 | 39 |

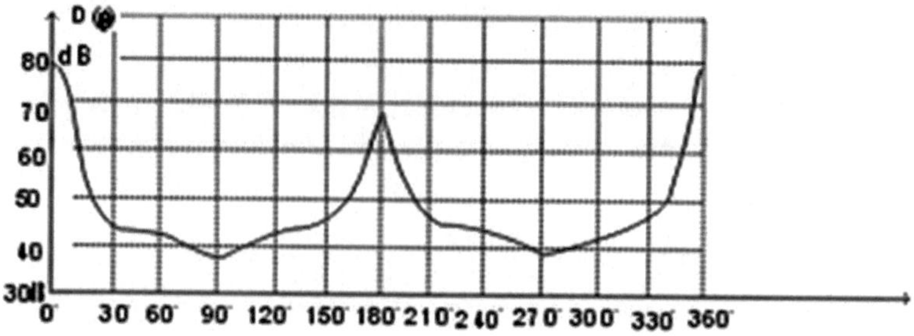


**Fig. 2.37** Relationship between $D(\theta)$ and $\theta$

Draw the curve of $D(\theta)$ and $\theta$ according to Table 2.6, as shown in Fig. 2.37.

It can be seen from Fig. 2.37, if the transceiver isolation can reach more than 79 dB, the jammer can normally work in the range of 0°–360°.

If the transceiver isolation is only 69 dB, the jammer will not work properly within ±15°

If the transceiver isolation only reaches 60 dB, the jammer will not work properly in the range of 0° ± 15° and 180° ± 10°.

In addition, when the transceiver is installed on two platforms, it also brings difficulties in the calibration of the parallelism of the azimuth axis and the pitch axis. The angle tracking system will also increase the tracking error, which limits the increase of the jamming antenna gain.

(4) The time-separation system is adopted for the transceiver antenna installed on one platform.

As shown in Fig. 2.38, the two antennas A and B are installed on a turntable, allowing r to be 3–5 m and selecting 4 m for analysis. The adjustment of the parallelism of the photoelectric axis of two antennas can be kept unchanged, and an absorption isolation layer can be added between the A and B antennas to improve the isolation of the transmitting and receiving antennas, as shown in Fig. 2.38.

**Fig. 2.38** Situation that the transceiver antenna is installed on one platform

A is the transmitting antenna, and B is the receiving antenna. These two are 4 m apart. An absorption isolation layer is added between A and B to improve the isolation between the transmitting and receiving antennas.

Substitute parameters into formula (2.39).

$P_{j/k}$ is 10 mW, $G_t(\theta)$ and $G_j(\theta)$ changes to 0 dB because of the addition of isolation layer, S/N = 10, r = 4 m, $S_{j\,\min} = 0.8 \times 10^{-10}$

We can get

$$D = \frac{10^{-2} \times 1 \times 0.24^2 \times 1 \times 10}{(126 \times 4)^2 \times 0.8 \times 10^{-10}} = 44.5\,\text{dB}.$$

When the transceiver antenna is installed on a turntable, measures to solve the transceiver isolation:

a. Improve the performance of the absorbing material between A and B so that the sidelobe gain of the two antennas on the 90° direction is as low as possible.
b. The background noise of the transmitter power amplifier should be considered in the design of the tube so that the output of the background noise is as low as possible. At present, the background noise of general high-power traveling wave tubes is less than 10 mW, and some can be less than 1 mW.
c. After the background noise of the tube is determined, the control of the high-voltage power supply of the tube can also be used to make it further reduce the background noise of the tube during the receiving time.
d. Improve the design of the transceiver antenna so that its sidelobe gain around 90° is very low.
e. Change the position of A and B antennas so that the sidelobe gain at 90° is as low as possible.

After taking various measures, under normal circumstances, the transceiver antenna can be installed on a rotating platform to meet the requirements of transceiver isolation.

The transceiver antenna installed on a turntable has obvious advantages: the equipment is compact; the position selection is easier; after the parallelism of the two antenna axes is set, there is no need to readjust when changing the position; there is no tracking error of the servo system, etc. Therefore, the transceiver antennas of most of our common high-power jammers are installed on a turntable.

#### 2.6.4.4 Distributed Jamming to Spaceborne SAR

The multi-site array jamming was introduced earlier. This section describes distributed jamming and the main differences between them:

1. Array jamming, which uses high-power jammer to jamming with the sidelobe of SAR. The protection area of an jamming station is over 100,000 km$^2$. The

distributed jammer is mainly focused on jamming with the mainlobe of the SAR antenna. It has a small jammer cover area and a large number of jammers.

2. The layout of jammer equipment is more complicated. Jamming with SAR in the frequency domain, space domain, and time domain. The jamming used emits a narrow antenna beam and requires two-dimensional angle tracking. The distributed jamming equipment is relatively simple, only need to aim in the frequency domain, and use a wide beam antenna in the air domain. Therefore, the angle does not need two-dimensional tracking.
3. Number of jammers required for distributed jamming

   a. For SAR flying horizontally.

      The area covered by a single high-power jammer 400 × 900 = 360,000 $km^2$. The number of jammers is needed to use distributed jamming against in the range of 360,000 $km^2$:

      Let the azimuth beamwidth of the SAR be 0.8°. The elevation beamwidth is 2°; R = 1000 km. Then the azimuth spacing $R_a = 0.8 \times 0.017 \times 1000 = 13.6$ km of distributed jammers and the elevation spacing $R_b = 2° \times 0.017 \times 1000 = 34$ km.

      The array of distributed jammers is shown in Fig. 2.39.

      In the horizontal direction, 400/13.4 = 30 units. In the vertical direction, 900/34 = 26 units. According to Fig. 2.39, the total number of jammers M = 780.

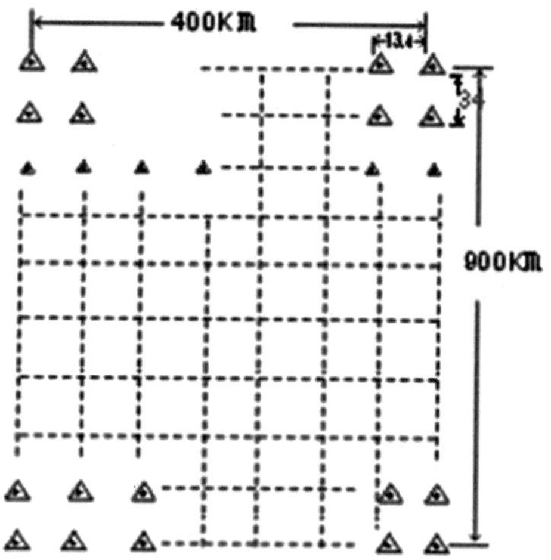


**Fig. 2.39** Schematic diagram of the layout of jammer array in distributed jamming

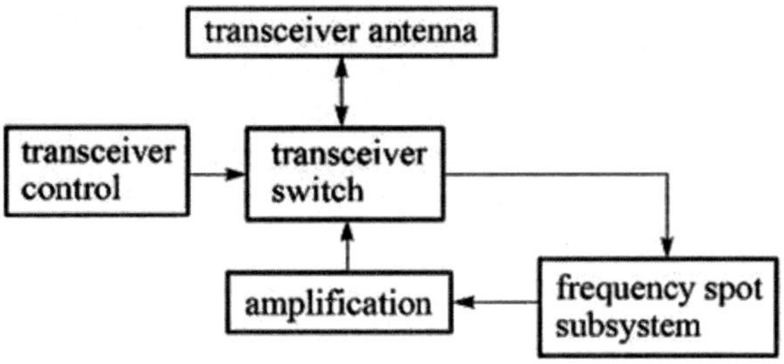


**Fig. 2.40** Composition of the distributed jammer

b. Composition of the distributed jammer.

The composition of the distributed jammer is shown in Fig. 2.40, including the common antenna for transceiver, transceiver switch, frequency-receiving system, transmission amplification subsystem, transceiver switch control, and its power supply, etc.

The transceiver antenna must be a wide beam. It can jam with SAR when SAR flies over the entire protected area.

c. Equivalent jamming power of the distributed jammer.

The equivalent jamming power of the distributed jammer can use (2.20), which is

$$P_j G_j = \frac{K_{j\Sigma} \cdot P_t \cdot G_t \cdot \sigma_i \cdot R_j^2}{4\pi R_{\min}^4} \cdot \frac{K_f \cdot L_d}{L_j \cdot \gamma_j} \cdot \frac{G_t}{G_t(\theta)}.$$

In this formula, we only change

$$G_t / G_t(\theta) = 1.$$

In Sect. 2.6.1, the equivalent power of the SAR side-lobe jammer is $G_t/G_t(\theta) = 40$ dB, and its equivalent power is 66 dB. For a distributed jammer that jamming with the mainlobe, its equivalent power is 66–40 = 26 dBW. To jamming with a specific area, the antenna gain of the distributed jammer can be 5 dB, then the power of the jamming transmitter is (26–5) = 21 dBW = 130 W.

#### 2.6.4.5 Passive Deception Jamming to Spaceborne SAR

Passive jamming plays an important role in the countermeasure against radar camouflage. It can create echoes of bridges, change river and bays, and can still passively deceive the spaceborne SAR. Absorb with absorbing material at the strong reflection point to weakening the reflected echo. Add passive jammings such as reflectors at

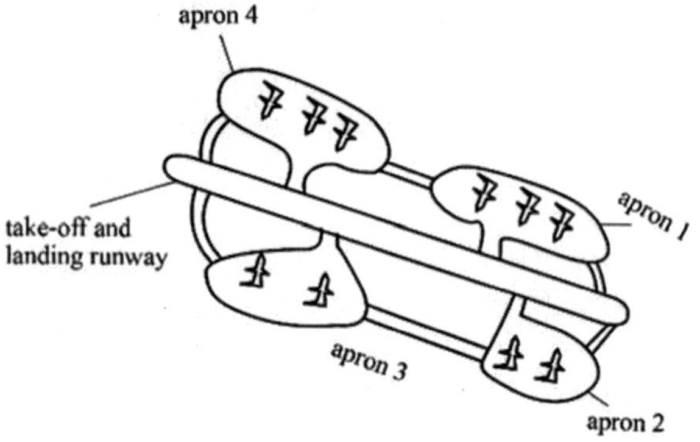


**Fig. 2.41** Schematic diagram of passive camouflage radar

the weak reflection point to change the SAR imaging map, so that it cannot get the real image.

The picture of passive jamming is shown in Fig. 2.41.

The aprons 1, 2, and 3 are real airplanes covered with absorbing materials so that the spaceborne SAR cannot detect the airplanes. The apron 4 is a fake airplane, made of wood, with gold foil or metal powder.

## 2.7 Jamming Against the Radar Identification System

The target is not moving, and the radar is moving. This imaging radar is called a synthetic aperture radar (SAR). When the target moves, the radar antenna beam does not move, and there is relative motion between these two. This imaging is called inverse synthetic aperture radar (ISAR). The basic imaging principle is the same, but the application occasions are different. For example, SAR is used for the reconnaissance of mapping, flood relief, and ground military facilities. At the same time, ISAR is suitable for the identification of multiple warheads of missiles and the identification of small moving targets (Fig. 2.42).

A is the missile launch site; B is the area under attack; C is the ISAR; D is the kinetic energy weapon; E is the long-range warning radar.

The missile's trajectory is determined after the pusher comes off. The long-range warning radar can measure the missile's trajectory and predict the missile's trajectory for ISAR. ISAR can wait in the airspace. When the warhead passes its beam, it can image the distribution of multiple warheads. Each warhead is equivalent to a point target.

We still take a radar as an example: it works in the L-band; $P_t G_t$ is 80 dBW; pulse width is 35 μs; pulse width after compression is 5 ns; elevation beamwidth is 2°; the speed of the missile's descent is 10 km/s; the effective reflection area of the warhead is 0.1 m$^2$; $f_\gamma$ is 1400 Hz; $R = 1000$ km.

We can get

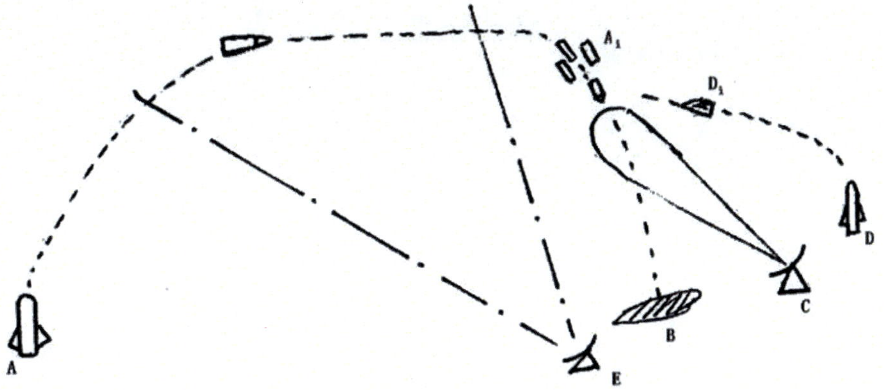


**Fig. 2.42** Schematic diagram for identifying multiple warheads of missiles

$$\eta_a = \frac{\tau}{\Delta\tau} = \frac{35 \times 10^{-6}}{5 \times 10^{-9}} = 37\,\text{dB}$$

$$\eta_b = \frac{f_\gamma \cdot \theta_{0.5} \cdot R}{V} = \frac{1400 \times 2 \times 0.017 \times 1000}{10} = 47{,}600 = 46.7\,\text{dB}$$

$$K_{j\Sigma} = (\eta_a \cdot \eta_b)^{0.8} = 66.9\,\text{dB}$$

$$G_t/G_t(\theta) = 1$$

According to formula (2.20).

$$P_j G_j = \frac{K_{j\Sigma} \cdot P_t \cdot G_t \cdot \sigma_i \cdot R_j^2}{4\pi R^4} \cdot \frac{L_d}{L_j \cdot \gamma_j \cdot K_f} \cdot \frac{G_t}{G_t(\theta)}$$
$$= \frac{4.8 \times 10^6 \times 10^8 \times 0.1 \times 10^{12}}{12.6 \times 10^{24}} \times \frac{1}{0.5} \times 1 = \frac{0.96 \times 10^{26}}{12.6 \times 10^{24}} = 7.6$$

It can be seen that to protect multiple warheads, one or two warheads can be equipped with a 3-W jamming transmitter, and the antenna gain is 5 dB. (Assume that multiple warheads are distributed in a sphere with a radius of 1 km.) It is possible to detect ISAR at a distance of 1000 km, and cannot distinguish the distribution of multiple warheads. It can be said that jamming against ISAR can be achieved.

## 2.8 Jamming Against Variable Polarization Height Measurement SAR

When the radar wave illuminates the target, some information of the target itself is modulated on the scattered wave. Generally, the target has the characteristics of improving the polarization mode of the radar detection signal. Experiments prove that the scattering characteristics describing the target's echo energy (target effective scattering cross-section RCS) not only depend on the relationship between the target's morphological position and surface vegetation but also have a close relationship with the polarization mode of the detection signal. Therefore, when the radar extracts the information of target scattered echo, not only the echo amplitude, phase, Doppler frequency shift, and waveform must be extracted, but also the radar polarization effect must be considered. The use of radar polarization information is an important aspect of radar technology development.

At present, the radar polarization theory system has been formed, and the corresponding polarization measurement technology is also tending to improve. Radar uses polarization information for measurement or anti-jamming. Therefore, it is necessary to take effective countermeasures against radar polarization.

### 2.8.1 *Interferometric SAR Three-Dimensional Imaging Image*

Interferometric SAR three-dimensional imaging is based on the two-dimensional SAR imaging, using the principle of the jamming of the time phase of the echo signal. After the phase is deburred, the height information of the target is extracted. This is an effective SAR three-dimensional imaging technology.

In the principle of three-dimensional imaging, it is assumed that the trajectories of the two satellites are parallel; the scene does not change within the time interval between the two observations; the atmospheric propagation does not change, and the reflection characteristics of the two are the same.

The phase $\phi$ is

$$\phi \cong \frac{4\pi}{\lambda}\left(B_{//} + B_{\perp}\frac{h}{\gamma \cdot \sin\theta}\right) \tag{2.42}$$

In the formula, $B_{//}$ and $B_{\perp}$ represent the projected components of the baseline parallel to the slant distance and the vertical slant distance, reflecting the relationship between terrain height and jamming phase; h is the target's height; $\gamma$ is the slant distance between the target and the radar; $\theta$ is the angle between the slant distance and the vertical height of the carrier, as shown in Fig. 2.43.

It can be shown from formula (2.30) that the echo behavior of the SAR signal includes the target's height information.

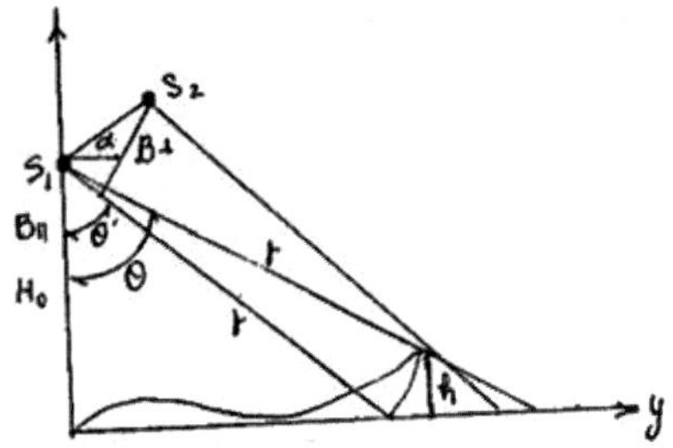

Fig. 2.43 Geometric diagram of ISAR

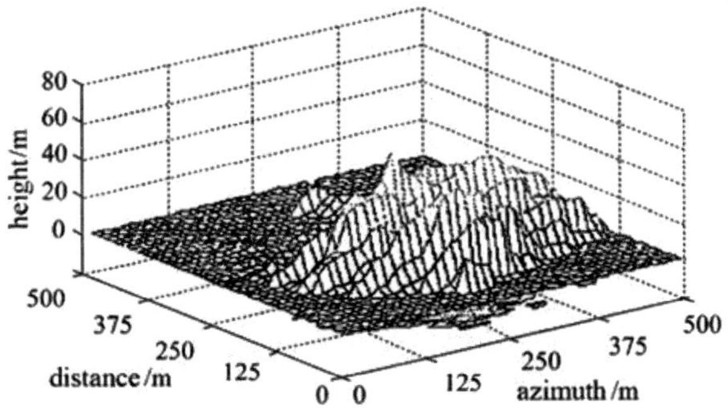


Fig. 2.44 3D relief map of Luzhou Island (SAR) in Hong Kong

An example of polarized SAR three-dimensional imaging is the terrain height skeleton map of polarized three-dimensional imaging of Luzhou Island in Hong Kong. The maximum height of the three-dimensional imaging of Luzhou Island is 66 m, and the actual maximum height is 69 m.

The sea surrounds Luzhou Island, and the islands rise out of the sea. The three-dimensional imaging relief map of Luzhou Island (SAR) in Hong Kong is shown in Fig. 2.44.

SAR can image in three dimensions by using polarization. For the personnel engaged in electronic countermeasures, it is necessary to use jamming polarization to destroy the radar, use polarization anti-jamming, and use polarization imaging to measure height. For the personnel engaged in electronic countermeasures, it is necessary to use jamming polarization to destroy the radar, use polarization anti-jamming, and use polarization imaging to measure height.

## 2.8.2 Principle of Random Polarization Jamming

Using a random polarization feed, one channel is input from the vertical channel, and the other channel is input from the horizontal channel. The two signals are from the same source.

The signal of one channel is $P_x = A \cdot \sin \omega t$, the other one is $P_y = B \cdot \cos \omega t$. Observing with an oscilloscope, add a horizontal deflector on one channel, and a

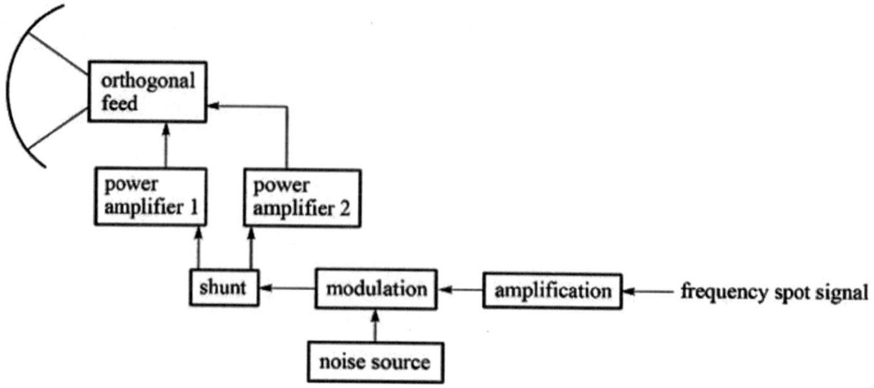


**Fig. 2.45** Schematic diagram of random polarization generation

vertical deflector on the other one. If A = B, it is an ellipse on display. If one channel is set to zero, a line will appear. If these two signals are added to the orthogonal polarization feed, the change of A and B may be circular polarization, elliptical polarization, linear polarization, etc. The input signal is modulated by noise, so the output signal of the orthogonal polarization feed is randomly polarized, as shown in Fig. 2.45.

Because it is noise-modulated, its polarization may change many times within 1 μs. It can turn a phase-modulated jamming signal into an amplitude-modulated noise signal, which increases $K_f$ (radar bandwidth and jamming spectrum width) from 0.3 to 1 and can increase the power density of the jamming signal. It is beneficial for jamming. For radars that use variable polarization anti-jamming or use polarized three-dimensional imaging, after this signal enters the radar, it not only effectively jamming in amplitude information, phase information, or polarization information, but also causes SAR three-dimensional imaging to be difficult due to the randomly polarized jamming.

In a word, spaceborne SAR can be jammed with. It is possible to use existing jamming technology, existing devices, and processing facilities to develop jammers that are effective for existing SAR.

Several difficulties in combating spaceborne SAR:

(1) It is difficult to detect it in time. Spaceborne SAR is rarely powered on, only when it reaches the airspace where it needs reconnaissance. The boot time is generally tens of seconds or 1–2 min.
(2) Its antenna beam is narrow, the sidelobe is low, and only its sidelobe can be collected at ordinary times, which requires a relatively high receiver sensitivity.
(3) The signal spectrum is relatively wide, generally tens of megahertz to hundreds of megahertz, which requires high sensitivity and the wide instantaneous band of the receiver. There are still many difficulties in the development of this type of receiver.

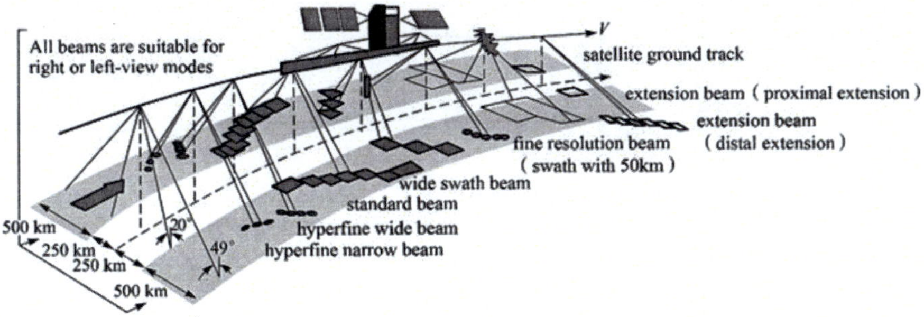


**Fig. 2.46** Schematic diagram of the imaging area of various working modes of Radarsat-2

(4) It requires the equivalent jamming power to be large, the beam of the transmitting antenna to be narrow, and angle tracking need to two-dimensional tracking with high accuracy.
(5) High standard transceiver isolation is required, and solutions must be proposed in the top-level design of the jammer. The jamming suppression coefficient required for SAR jamming can be normalized to a uniform value $K_d$, that is, its pulse compression coefficient D is raised to the 0.8th power, $K_d = D^{0.8}$.
(6) Spaceborne SAR has multiple imaging modes, as shown in Fig. 2.46. It can be imaged on the left and right of the satellite's track, and the beamwidth and elevation can be searched and extended. Also, the probability of detecting and jamming with its antenna mainlobe is very low.

## 2.9 Selection of Optimal Jamming Patterns for SAR

Synthetic aperture radar (SAR) is a new system radar, which has three characteristics compared with general radar:

(1) The working process of SAR, the antenna is fixed an angle, not the search state.
(2) In the process of target imaging, the distance between SAR and target is constant, and the imaging band of SAR is formed with the movement of SAR carrier.
(3) It can image the detected target. According to these characteristics of SAR, its anti-jamming ability is stronger than the general radar, it is easy to resist some external jamming pulses.

For example: SAR to 'multi-pulse jamming', 'pulse answer jamming', 'image deception jamming'… , are easy to resist, as shown in Fig. 2.47.

$B$ is SAR carrier, $D$ is SAR pojection point, $C$ is SAR direction finding center, and $H$ is SAR height.

$R$ is the distance from SAR to direction finding center,$d_i$ is the distance between direction finding center line and projection line, and $\beta_i$ is the included Angle between $R$ and $H$ (generally 20°–70°).

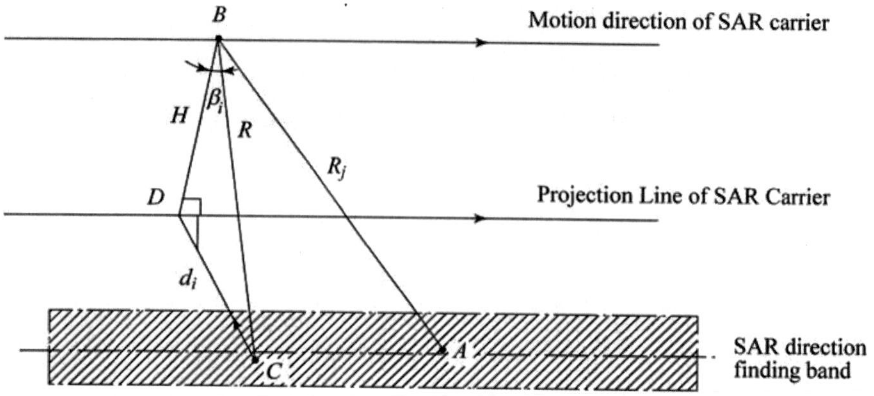


**Fig. 2.47** SAR direction finding diagram

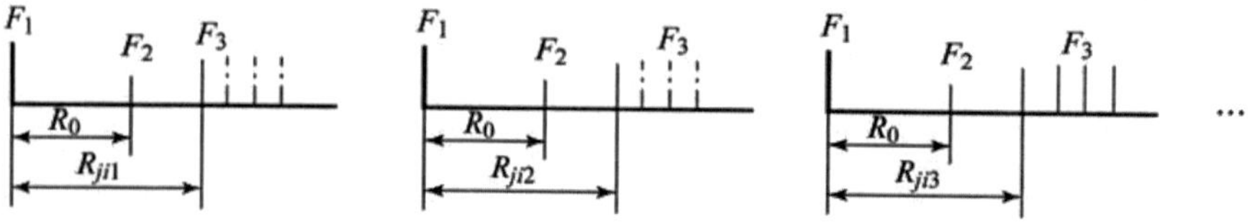


**Fig. 2.48** Time relation among transmitting pulse, echo pulse and jamming pulse

$d_i = Ht_{in}\beta_i$, when the SAR carrier moves, the SAR detects the image in the direction finding band. If $\beta_i$ changes, another image in the direction finding band is detected.

According to the process of SAR image measurement, the relationship between transmitting pulse and echo pulse and jamming pulse can be plotted, as shown in Fig. 2.48.

When $\beta_i$ is $\beta_0$, the distance between the SAR and the direction finding center $R_0$ is constant, while $R_j$ changes as the SAR moves. In general $R_j > R_0$, $R_j = R_0$ is only when the main lobe of the SAR antenna is aligned with the main lobe of the jammer antenna.

If a gate-gate pulse is added to the input end of the SAR receiver, only the SAR echo pulse can be output, and other pulses different from the gate-gate pulse cannot be output, which makes most of the jamming pulses cannot enter the signal processing circuit. Therefore, pulse jamming has poor effect on the SAR jamming. This is the main reason for the poor jamming effect of various pulse spoofing jamming on SAR.

According to the working principle of SAR, what kind of jamming has the best effect on SAR jamming is "continuous wave noise aiming frequency suppression jamming", which has the best effect on SAR jamming, as shown in Fig. 2.49.

The power required for jamming equivalence of SAR can also be calculated by "unified equation of radar jamming". As long as the input end of the SAR receiver meets the requirements of the jamming suppression coefficient $K_j$ and $\Delta F > \Delta f$

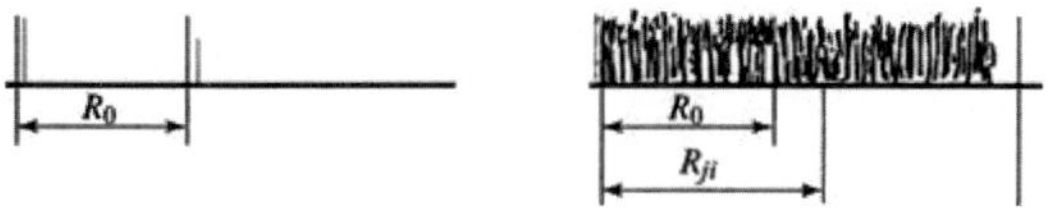


**Fig. 2.49** Schematic diagram of "continuous wave noise suppression interference". **a** When interference does not exist. **b** When interference exists

($\Delta F$ is the spectral width of the jamming signal and $\Delta f$ is the bandwidth of the radar receiver), the SAR will not be able to image normally.

The anti-jamming ability of SAR has been tested in a variety of jamming styles. Such as "multi-pulse spoofing", "answer pulse jamming", "narrow spectrum noise jamming", "continuous wave noise aiming frequency suppression jamming"···. Other jamming can be eliminated, only "continuous wave noise aim bandwidth spectrum suppression jamming" has not been eliminated. The experimental result is consistent with the theoretical analysis result.

# Chapter 3
# Jamming Against Pulse Compression Radar

The generation of pulse compression radar was not developed for anti-jamming, but designed according to the radar's own needs. According to the demand for the warning radar, it is required to be able to find the target at a farther distance. However, the maximum detection distance of the radar is proportional to the fourth power of the radar's pulse energy. The pulse energy of the radar is equal to $P\tau$. P is the peak power of the radar pulse; $\tau$ is the radar pulse width. Due to the limitation of the peak power output of the current electric vacuum device, it has reached the megawatt level, but it is much harder to do further. Therefore, the problem of increasing the detection distance is solved by increasing the pulse width and increasing the pulse energy. The other kind of device is a solid device. Its peak power cannot be made very large, but its pulse width can be made wide, which gives a new emission source for the development of pulse compression radar. After the pulse width is widened, the problem of increasing the detection distance is solved. However, the range resolution is proportional to the pulse width—the wider the pulse width, the lower the range resolution. From the perspective of range resolution, it hoped that the pulse width is narrow. The radar designers used the method of widening the transmitted pulse to solve the problem of increasing the detection distance and narrowed the received echo signal pulse to solve the problem of range resolution. In this way, a new body radar, pulse compression radar, is produced.

## 3.1 Airborne Early Warning Pulse Compression Radar

The pulse compression radar is mainly used for long-range warning and guidance. The airborne long-range warning radar is the radar on the Airborne Early Warning. To find long-range targets, many of them use a pulse compression radar system. In order to detect low-altitude moving targets, it also uses the Pulse-Doppler system. Pulse compression radar is also widely used in ground long-range warning radars. This chapter mainly describes the jamming technology, the effective jamming area,



X. Zhang et al., *Theory to Countermeasures Against New Radars*,
https://doi.org/10.1007/978-981-16-6715-2_3

and jamming exposure area formed by the jamming to the early warning aircraft pulse compression radar and ground warning pulse compression radar.

The rapid development of modern military electronic technology and weapon systems has caused profound changes in the scale, methods, and means of modern warfare. Modern warfare has developed into comprehensive warfare integrated with land, sea, air, space, and electronics. Advanced weapons are controlled by advanced electronic equipment. If it loses control of advanced electronic equipment, it will also lose its superiority. Non-contact wars will also become contact wars if they lose control of advanced electronic equipment. One of the important functions of information warfare/electronic warfare is to make some advanced weapons out of control and turn non-contact wars into contact wars. In a contacted war, each country can play its special skills. Relying on short soldiers and bravery to destroy the other side, the advantage of a country with advanced weapons cannot be fully played. Still, countries with less developed weapons can give full play to the advantage of bravery and good combat. In modern warfare, airborne early warning is specialized in performing electronic security, air command, control, and communication in the air, which is equivalent to a small $C^3I$ system. In the Gulf War, early warning aircraft played a leading role in air operations and air management. Therefore, the development of a comprehensive jamming system for Airborne Early Warning is great significance in the future modern warfare.

The comprehensive jamming to the airborne early warning is technically feasible, which will be further analyzed in the relevant parts below.

Jamming against early warning aircraft is also economically very cost-effective. For example, buying an early warning aircraft costs $200 million or $300 million, and producing an jamming system for an early warning aircraft costs only $20 million. That is to say, only one-tenth of the money of the early warning aircraft can make the early warning aircraft fail to play its role.

Foreign airborne early warning aircraft were developed from the Vietnam War in the 1960s. At that time, the U.S. military wanted to find Vietnamese aircraft farther away. While the U.S. military's shipborne radar stopped at sea, it was affected by the curvature of the earth so that finding long-range targets was restricted. Therefore, the US military wanted to raise the radar to observe targets in a wide range. By installing a warning radar on an aircraft, long-range targets can be found. This is the reason for the development of early warning aircraft.

There are numbers of technical problems for installing a warning radar on the aircraft, such as the installation of the antenna, the reduction of the volume and weight, and the suppression of the ground echo, etc. These problems all need to be properly handled to make the early warning aircraft radar function normally.

After several improvements, the airborne early warning has a good downsight function, which can detect incoming targets in low-altitude in time, provide long enough early warning time, and prepare the combat platform to make countermeasures against attacking targets in advance. Besides, the early warning aircraft detection has a wide coverage that can deal with 600 targets at the same time. Therefore, airborne early warning is a very important means of combat in modern integrated electronic warfare. Due to the continuous improvement of the air defense system on the ground and the sea, the large amount of equipment of precision-guided missile weapon and aviation missile has forced combat aircraft to penetrate the submarine at low altitude. Ground radars and shipborne warning radars generally have large low-altitude dead zones, making it difficult to find high-speed low-altitude or sea-skimming flying targets in time. Early warning aircraft can detect intrusive low-altitude targets early and provide sufficient early warning time.

At present, there are more than 50 manufacturers in more than 20 countries that have developed variety of early warning aircraft. The types and radars of these early warning aircraft are shown in Table 3.1.

The early warning aircraft that have been tested in actual combat are the US E-2C, E-3A, and E-8A. These early warning aircraft have played well in the Gulf War, Kosovo War, Afghanistan War, and the US-British War against Iraq. More than 160 E-2C early warning aircraft were produced in the United States, sold to Japan, Singapore, India and other countries, and also sold to Taiwan (remodeled to E-2T). E-3A have been produced about 30 but did not sell to other countries. More than ten E-8A have been produced, and they are still being improved after actual combat.

In Table 3.1, the parameters of the early warning aircraft radars of the United States and the former Soviet Union are mainly included, and the early warning aircraft radars of countries such as the United Kingdom and Israel are not included in the table.

## 3.2 Operating Principle of Airborne Early Warning Pulse Compression Radar

The radar of early warning aircraft mainly uses two systems; one is the pulse compression system; the other is the Pulse-Doppler system.

### *3.2.1 Operating Principle of Pulse Compression Radar*

In order to obtain a long-range warning, the radar needs a transmitter with high peak power. However, after the high-power transmitter reaches the peak power of megawatts, it is more difficult to achieve further improvement. The maximum detection distance of the radar is proportional to the 1/4th power of the pulse energy of the radar transmitter, and the pulse energy is equal to the peak power times the pulse width. When the peak power of the radar transmitter reaches a certain value, widening the transmission pulse width is also a means to increase the detection range of the radar. However, after increasing the radar transmission pulse width, the radar's range resolution decreases because the radar's range resolution is proportional to 1/2 pulse width. In order to increase the range of the radar, the pulse width should be increased; in order to improve the range resolution of the radar, the pulse width

**Table 3.1** parameters of the early warning aircraft radars from different manufactuer

| Radar naming | AN/ASP-20 | AN/ASP-70 | AN/ASP-138 | AN/APY | Jack | Pillar | Moss | Spacemaster | AN/APG-66 | AN/TPS-63 | AN/DPS-5 |
|---|---|---|---|---|---|---|---|---|---|---|---|
| Manufacturer | General Electric Company | General Electric Company | General Electric Company | Westinghouse | Soviet Union | Soviet Union | Soviet Union | | Westinghouse | Westinghouse | American radio |
| Installed models | Shekton | | E-2C | 707–320 | Figure-126 | IL-76 | Figure-126 | Defender | Airship | Airship | Airship |
| Frequency (GHz) | 2.9 | 0.34 | UHF | S band | | | L band | 9.7 | X band | L band | |
| Installation location | Bottom | | Bottom | Bottom | | | Bottom | Head | Bottom | Bottom | Bottom |
| Antenna type | Paraboloid | | Endfire antenna array | Planar waveguide antenna array | | | | Paraboloid | Paraboloid | Paraboloid | |
| Antenna size | 2.3 × 1 | | 7.4 × 0.77 | 7.6 × 1.5 | | | 11 (round) | 1.9 × 0.9 | 1 × 1.3 | 4 × 7.6 | 3.7 × 7.3 |
| Antenna gain | | 21 dB | 21.5 dB | 40 dB | | | | 38 dB | | 34.5 dB | 37 dB |
| Average transmit power | 450 W | 3.6 kW | 3.8 kW | 10 kW | | | | 550 W | | 1.5 kW | 600 W |
| Pulse width (μs) | 2 | 6 | 13 | 46 | | | | 17.8 | | | 2 |
| Pulse repetition frequency (Hz) | 300 | 300 | 300 | 250, 300 | | | | Low, medium | | 375 | 250 |
| Antenna speed | 6 rpm | 6 rpm | 6 rpm | 6 rpm | | | | 20 rpm | | | |
| Clutter processing | Digital moving target display | 30 dB cancellation | Pulse suppression | Pulse pressure Pulse Doppler | | | Moving target display | Pulse Doppler | | 60 dB cancellation | Moving target display |

(continued)

**Table 3.1** (continued)

| Radar naming | AN/ASP-20 | AN/ASP-70 | AN/ASP-138 | AN/APY | Jack | Pillar | Moss | Spacemaster | AN/APG-66 | AN/TPS-63 | AN/DPS-5 |
|---|---|---|---|---|---|---|---|---|---|---|---|
| Target | Air and sea | Air and sea | Air, sea and land | Air and sea | | | | Air and sea | Air | Air | |

should be reduced. This is a pair of contradictions. To solve this contradiction, the radar designers came up with a clever solution, that is, the pulse compression radar system. Pulse compression radar launches wide pulses to increase the radar's detection range. After receiving the echo signals, the received echo signals are processed into narrow pulses, which improves the radar's range resolution. Therefore, this kind of radar is widely used.

#### 3.2.1.1 The Relationship Between Radar's Detection Distance and Pulse Width

Maximum effective detection distance $R_{max}$ of radar

$$R_{max} = R'_{max} \cdot g(\theta)^{1/4} \cdot e^{-a \cdot R_{max}} \tag{3.1}$$

In the formula:

$\theta$ The elevation of target

$g(\theta)$ The function of the ground effect on the maximum distance, which has a greater impact on the meter-band radar, and little effect on the centimeter-wave radar. $\alpha$ is the absorption coefficient in the atmosphere. If $R_{max}$ is in meters, then $\alpha$ is equal to the absorption coefficient in decibels times 0.115. The value of $\alpha$ in the centimeter waveband cannot be ignored. It must be solved graphically or by other convenient methods to solve Eq. (3.1) to find $R_{max}$.

For receivers with very little noise, cosmic jamming, industrial jamming, and atmospheric jamming can reduce the maximum effective detection distance of the radar in free space to 50%. $R_{max}$ is the maximum effective distance of the radar in free space. In free space, it is assumed that the radar has no external jamming and no maximum effective distance for atmospheric absorption.

Under this circumstance,

$$R'_{max} = \left( \frac{G_t \cdot G_\gamma \cdot A_e \cdot E_t \cdot \lambda^2}{(4\pi)^3 \cdot E_{min}} \right)^{1/4} \tag{3.2}$$

$$E_{min} = a \cdot \overline{NF} \cdot K \cdot T \text{ (J)} \tag{3.3}$$

In the formula:

a Pulse visibility factor of a certain radar indicator;

K Boltzmann's constant, $1.38 \times 10^{-23}$ J/1°;

T the absolute temperature, and the unit is Kelvin (K);

$\overline{NF}$ the noise factor of receivers. When the ideal receiver is matched with the antenna, $\overline{NF}$ equals 2.

After derivation, in the absence of external jamming and atmospheric absorption, the general formula for the maximum detection distance for large fighters or light bombers can be written as $R_I$

$$R_I = 791.8\left[\frac{E_t \cdot L_t \cdot L_\gamma \cdot G'_\gamma \cdot G'_t}{\frac{K_1 \cdot \tau \cdot B}{4} \cdot \left(1 + \frac{K_2}{B\tau}\right)^2 \cdot f^2 \cdot \overline{NF}}\right]^{1/4} \cdot F^{1/12} \tag{3.4}$$

$$E_t = P_t \cdot \tau$$

In the formula:

| | |
|---|---|
| $E_t$ | The pulse energy (J) of transmitter output. |
| $P_t$ | The peak power of the transmitted pulse. |
| $\overline{NF}$ | The noise figure of the receiver's input. |
| $G'_t, G'_\gamma$ | The transmitting antenna and receiving antenna half-power point gain. |
| $L_t, L_\gamma$ | The loss factor of feed and switch of receiving antenna and transmitting antenna. |
| $\tau$ | The effective pulse width (s). |
| B | The effective width of the receiver (Hz). |
| F | The pulse repetition frequency. |
| f | The radar working carrier frequency (MHz). |
| $K_1$ | Coefficient related to the indicator type, antenna rotation speed, beamwidth, scan type, etc. |
| $K_2$ | Coefficient related to the pulse shape. |

Equation (3.4) includes all factors that the radar designers can control that affect the maximum range of the radar.

The pulse energy $E_t$ should be as large as possible. Increasing the transmission pulse width is a means to increase the effective detection distance. Pulse compression radar is to increase the transmission pulse width to increase the effective detection distance of the radar.

Pulse width $\tau$ and receiver bandwidth B, in the formula, only appear in the form of $\tau \cdot B$. When $\tau \cdot B = 1$, the effective radar distance is the largest (Generally, when designing radar, it takes 1.3). When all other factors in the above formula have been determined, the maximum radar detection range decreases with increasing frequency. Therefore, the carrier frequency of the long-range warning radar is generally low. Most of them are P and L bands.

According to the analysis of Formula (3.4), when the other parameters are determined, increasing the transmitted pulse width can increase the radar range.

#### 3.2.1.2 Pulse Compression Signal Waveform

Requirements for signal waveform:

1. It has enough energy to meet the requirement of detection distance;

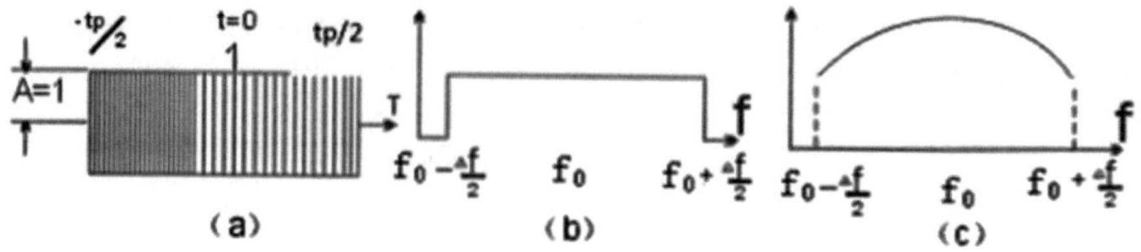


**Fig. 3.1** Chirp diagram

2. It has sufficient ability to distinguish the target.
3. For the unnecessary echo, it has sufficient suppression ability.

Pulse compression radar has a wider pulse width. It has enough energy to be linearly modulated (or other coded) in the pulse carrier frequency. So, the operating frequency varies. The signal has a wide frequency spectrum, which provides conditions for fine distance measurement. When receiving, the matched filter is used to process the echo signal, and the wide pulse is compressed into a narrow pulse.

Take linear frequency modulation as an example to analyze pulse compression. The wide-pulse chirp signal is designed according to the principle revealed by the ambiguity function, which is a widely used signal form in radar. It has both a large time width and a large frequency bandwidth, so it has good range resolution and radial velocity resolution.

The basic concept of the chirp signal is shown in Fig. 3.1.

Fig. 3.1a is a diagram of the chirp signal, which is a signal by a rectangular pulse. Its width changes linearly within the pulse of the carrier frequency, starting at a high frequency and then decreasing linearly (it can also start at a low frequency and then increase linearly).

$\omega_0 = 2\pi \cdot f_0$ represents the unmodulated carrier frequency. The frequency change law of the chirp signal can be written as

$$\omega = \omega_0 + \mu \cdot f, \quad t \leq t_p/2 \tag{3.5}$$

where, $\mu$ is the frequency modulation index.

The law of phase change is Formula (3.6)

$$\theta(t) = \int \omega dt = \omega_0 \cdot t + \frac{1}{2}\mu \cdot t^2 \tag{3.6}$$

Figure 3.1b shows the amplitude characteristics of the spectrum.

Figure 3.1c is the phase characteristic of the spectrum.

$\Delta f$ indicates the difference between the high and low end of the signal frequency.

After receiving the chirp signal, the receiver needs to be processed with a matched filter to compress the wide pulse of FM into a narrow pulse, shown in Fig. 3.2.

If the filter is lossless, the signal input energy will be completely output. When the power of the input signal is $P_i$ and the power of the output signal is $P_0$, then:

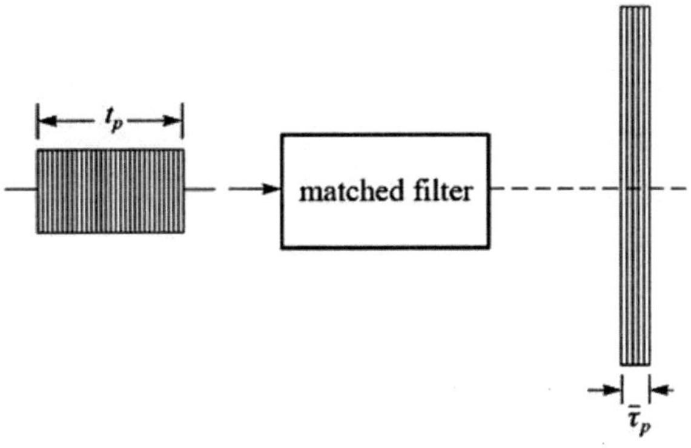


**Fig. 3.2** Compression effect of matched filter

$$P_i \cdot t_p = P_0 \cdot \tau_p \tag{3.7}$$

$$P_0 = P_i \cdot \frac{t_p}{\tau_p} \tag{3.8}$$

$t_p/\tau_p$ is called the pulse compression ratio. In general chirp radars, the compression ratio is generally 30 to several hundred.

To ensure the compression effect, the matched filter must have different delay characteristics for different frequencies. The above chirp signal has a high starting frequency and a low frequency at the end. Therefore, the filter has a longer delay for high frequencies and a shorter delay for low frequencies, as shown in Fig. 3.3.

The difference between the delay at the high-frequency end and the low-frequency end is $\Delta t$

$$\Delta t \approx t_p$$

characteristics of filter phase $\theta(\omega)$

$$\theta(\omega) = \int t_d d\omega = k(\omega - \omega_2)^2 + b \cdot \omega + c \tag{3.9}$$

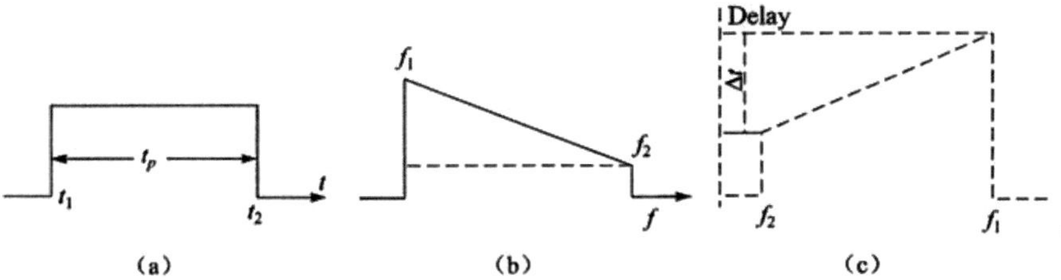


**Fig. 3.3** LFM signal parameters and filter characteristics. **a** Pulse envelope; **b** frequency variation; **c** filter delay characteristics

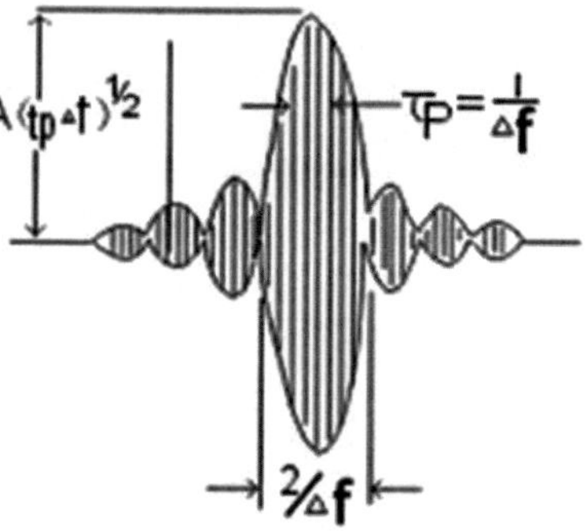


**Fig. 3.4** Actual waveform after compression

where c is the integral constant. Carefully select the frequency modulation index $\mu$ and the delay coefficient k to make the filter meet the match. The output compression waveform is shown in Fig. 3.4.

#### 3.2.1.3 Generation of Chirp Signal

There are two basic methods for generating a chirp signal. One is the active method, which uses a voltage-controlled oscillator to generate frequency-modulated waves, and the control voltage changes according to the required frequency-regulated law. The other is the passive method, which uses an extended filter to generate a chirp signal. Currently, this method is mainly used, as shown in Fig. 3.5.

Suppose the frequency spectrum of the excitation pulse $S_i(t)$ is $S_i(\omega)$, and the transfer function of the extended filter is $H(\omega)$. After the reception, the transfer function of the compression filter will be $H^*(\omega)$, and the output waveform of the transmitter is

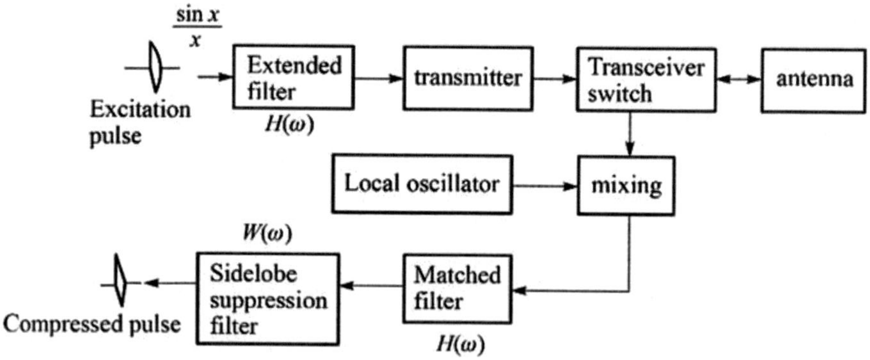


**Fig. 3.5** LFM signal transmission and reception system using conjugate filter

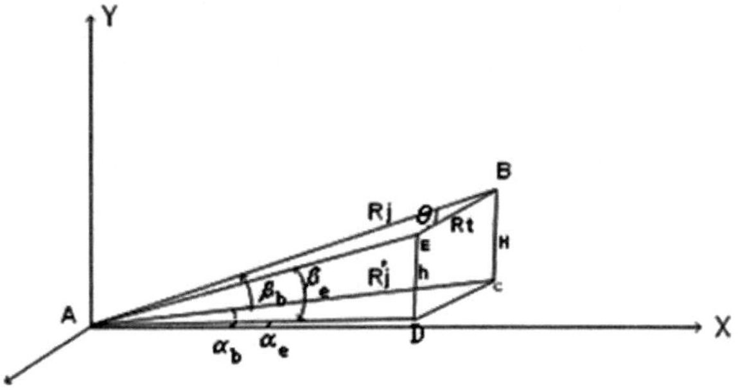


**Fig. 3.6** Relative position map of jammer and early warning aircraft

$$S_i(t) = \frac{1}{2\pi}\int_{-\infty}^{\infty} S_i(\omega)\cdot H(\omega)\cdot e^{j\omega t}d\omega \tag{3.10}$$

The selection of excitation pulse should make the signal after expansion meets the requirements of chirp. The more suitable form is

$$S_i(t) = \frac{\sin\frac{\mu\cdot t_p\cdot t}{2}}{\frac{\mu\cdot t_p\cdot t}{2}}\cdot\cos\omega_0 t \tag{3.11}$$

The compressed output signal from the receiver is

$$g(t) = \frac{1}{2\pi}\int_{-\infty}^{\infty} S_i(\omega)\cdot H(\omega)\cdot H^*(\omega)\cdot e^{j\omega t}d\omega = \frac{1}{2\pi}\int_{-\infty}^{\infty} S_i(\omega)\cdot |H(\omega)|^2 e^{j\omega t}d\omega \tag{3.12}$$

$\sin x/x$ shaped excitation pulse can use a narrow pulse, which its frequency band is greater than $\Delta f$. Add this pulse to a wideband $\Delta f$, the frequency characteristic would show on the rectangular filter. The phase characteristic of the filter is linear. After the pulse passes through this type of filter, the desired waveform will be obtained (Fig. 3.6).

## 3.3 Jamming Principle for Pulse Compression Radar

### *3.3.1 Jamming Equation for Early Warning Aircraft Radar*

Set point A as the position of the jammer; set point B as the position of the early warning aircraft; set point C as the projection point of point B; set point E as the position of the covered target; the power of radar signal received by the target is $P_1$;

the height of the early warning aircraft is H; the height of the covered target is h; the distance of the covered target from the early warning aircraft is $R_t$.

According to the basic formula of electronic countermeasures, the following formula can be written:

$$P_1 = \frac{P_t \cdot G_t \cdot \sigma}{4\pi R_t^2} \tag{3.13}$$

$$P_\gamma = \frac{P_1}{4\pi R_t^2} \cdot A = \frac{P_1}{4\pi R_t^2} \cdot \frac{\lambda^2}{4\pi} \cdot G_t \tag{3.14}$$

Substituting Eq. (3.13) into Eq. (3.14) is

$$P_\gamma = \frac{P_t \cdot G_t^2 \cdot \lambda^2 \cdot \sigma \cdot L_t}{(4\pi)^3 R_t^4} \tag{3.15}$$

Power of jammer radiation received by radar is $P_N$

$$P_N = \frac{P_j \cdot G_j}{4\pi R_j^2} \cdot \frac{\lambda^2}{4\pi} \cdot G_t(\theta) \cdot K_f \cdot \gamma_j \cdot L_j \tag{3.16}$$

The ratio of the power of jamming signal received by the radar to the power of the echo signal received by the radar at the input of the radar receiver is expressed as $K_{jt}$

$$K_{jt} = \frac{P_N}{P_\gamma} \tag{3.17}$$

Substitute (3.15) and (3.16) into (3.17)

$$K_{jt} = \frac{4\pi P_j \cdot G_j \cdot R_t^4}{P_t \cdot G_t \cdot \sigma \cdot R_j^2} \cdot \frac{K_f \cdot L_j \cdot \gamma_j}{L_t} \cdot \frac{G_t(\theta)}{G_t} \tag{3.18}$$

In the formula:

| | |
|---|---|
| $R_t$ | The distance between the covered target and the radar. |
| $R_j$ | The distance between jammer and the radar. |
| $P_jG_j$ | The equivalent power of the jammer. |
| $P_tG_t$ | The equivalent power of radar. |
| $L_j$ | The feeder loss of jammers and atmospheric loss in space. |
| $\gamma_j$ | The polarization mismatch loss of jammer. |
| $L_t$ | The feeder loss of radar and atmospheric loss in space. |
| $K_f$ | The ratio of radar receiver bandwidth to jamming spectrum width. |
| $G_t(\theta)$ | The gain of the sidelobe of the radar antenna aimed by the jammer. |
| $\sigma$ | The effective reelection area of the covered target. |

According to Formula (3.18), the jamming equivalent power required by the jamming early warning aircraft radar is

$$P_j \cdot G_j = \frac{K_{jt} \cdot P_t \cdot G_t \cdot \sigma \cdot R_j^2}{4\pi \cdot R_t^4} \cdot \frac{L_t}{K_f \cdot L_j \cdot \gamma_j} \cdot \frac{G_t}{G_t(\theta)} \tag{3.19}$$

In the Formula (3.19), $R_t$, $P_tG_t$,$G_t/G_t(\theta)$, $\sigma$, $R_j$ are all definite data. Only the values of $K_{jt}$ are based on different radar systems. Therefore, it is necessary to determine the value of $K_{jt}$ for jamming design on the pulse compression radar and calculating its equivalent power.

### *3.3.2 The Value of Jamming Suppression Coefficient for Pulse Compression Radar*

Pulse compression radar widens the transmitted pulse, which increases the maximum detection range of the radar.

1. The first algorithm:

Assume that the radar's gain factor is $K_{dl}$

$$K_{dl} = \left(\frac{R_{\max(2)}}{R_{\max(1)}}\right)^4$$

$R_{\max(1)}$ is the maximum detection distance with conventional radar pulse width.
$R_{\max(2)}$ is the maximum detection distance when the radar transmission pulse width is widened.

The radar gain factor $K_{dl}$ is multiple that the jamming suppression factor should increase. The maximum detection distance $R_{\max}$ of the radar can be written as $R_I$ Eq. (3.4) when atmospheric absorption and ground echo jamming are not considered.

$$R_I = 791.8\left(\frac{E_t \cdot L_t \cdot L_\gamma G_t' \cdot G_\gamma'}{f^2 \cdot \overline{NF} \cdot \frac{K_1 \cdot \tau \cdot B}{4} \cdot \left(1+\frac{K_2}{\tau \cdot B}\right)^2}\right)^{1/4} \cdot (F_1)^{1/12} \tag{3.20}$$

If a conventional radar $E_t$ emits pulse energy $E_t = P_t\tau_1$.
In the formula:

$\overline{NF}$ The noise figure of receiver;
$G_t'$, $G_\gamma'$ The gain of transmitting and receiving antenna;
$L_t$, $L_\gamma$ The feeder loss of transmission and receiving.
$\tau_1$ The pulse-width of transmission (S).
$B_1$ The bandwidth of receiver (Hz);

$F_1$ The pulse repetition frequency.
f The radar operating frequency (megahertz);
$K_1$ Coefficient related to indicator type, antenna rotation speed, beamwidth, etc.;
$K_2$ Coefficient related to the pulse shape (rectangular pulse is 1).

Substitute this constant into Eq. (3.20)

$$R_{I(1)} = 796.4\left(\frac{P_{t1}\cdot\tau_1\cdot L_t\cdot L_\gamma\cdot G_t\cdot G_\gamma}{f^2\cdot\overline{NF}}\right)^{1/4}\cdot\left(\frac{K_1\cdot\tau_1\cdot B_1}{4}\cdot\left(1+\frac{K_2}{\tau_1\cdot B_1}\right)^2\right)^{-1/4}\cdot(F_1)^{1/12}.$$

If the peak power of the transmitter of this radar is $P_{t2}$; the transmit pulse width is $\tau_2$; the receiver's bandwidth is $B_2$; the pulse width after compression is $\tau_3$; $K_2$ becomes $K_2'$, other parameters remain unchanged, substituting Eq. (3.4), it can be written as:

$$R_{I(2)} = 796.4\left(\frac{P_{t2}\cdot\tau_2\cdot L_t\cdot L_\gamma\cdot G_t\cdot G_\gamma}{f^2\cdot\overline{NF}}\right)^{1/4}\cdot\left(\frac{K_1\cdot\tau_3\cdot B_2}{4}\cdot\left(1+\frac{K_2'}{\tau_3\cdot B_2}\right)^2\right)^{-1/4}\cdot(F_2)^{1/12}$$

Substitute into Eq. (3.20)

$$K_{d1} = \left\{\frac{\left(\frac{P_{t2}\cdot\tau_2\cdot L_t\cdot L_\gamma\cdot G_t\cdot G_\gamma}{f^2\cdot\overline{NF}}\right)^{1/4}\cdot\left(\frac{K_1\cdot\tau_3\cdot B_2}{4}\cdot\left(1+\frac{K_2'}{\tau_3\cdot B_2}\right)^2\right)^{-1/4}\cdot(F_2)^{1/12}}{\left(\frac{P_{t1}\cdot\tau_1\cdot L_t\cdot L_\gamma\cdot G_t\cdot G_\gamma}{f^2\cdot\overline{NF}}\right)^{1/4}\cdot\left(\frac{K_1\cdot\tau_1\cdot B_1}{4}\cdot\left(1+\frac{K_2}{\tau_1\cdot B_1}\right)^2\right)^{-1/4}\cdot(F_1)^{1/12}}\right\}^4$$

After simplification

$$K_{d1} = \frac{\frac{P_{t2}\cdot\tau_2}{\tau_3\cdot B_2\cdot\left(1+\frac{K_2'}{\tau_3\cdot B_2}\right)^2}P_{t1}\cdot\tau_1}{\frac{1}{\tau_1\cdot B_1\cdot\left(1+\frac{K_2}{\tau_1\cdot B_1}\right)^2}}\cdot\left(\frac{F_2}{F_1}\right)^{1/3} \tag{3.21}$$

If $K_0$ is a coefficient that causes the change of the rectangular parameter after the width of the radar transmission pulse becomes wider.

$$\text{Take } K_0 \cong \frac{\left(1+\frac{K_2'}{1.3}\right)^2}{\left(1+\frac{K_2}{1.3}\right)^2} = \left(\frac{1.3+K_2'}{1.3+K_2}\right)^2$$

$$\text{Then}\quad K_{d1} = \frac{P_{t2}}{P_{t1}}\cdot\frac{\tau_2}{\tau_1\cdot K_0}\cdot\left(\frac{F_2}{F_1}\right)^{1/3} \tag{3.22}$$

If $P_{t1} = P_{t2}$; $F_1 = F_2$

$$K_{d1} = \frac{\tau_2}{\tau_1 \cdot K_0} = \frac{\frac{\tau_2}{\tau_3}}{\frac{\tau_1}{\tau_3} \cdot K_0} = \frac{D}{D_0 \cdot K_0} \tag{3.23}$$

In the formula:

$\tau_2$ the pulse width of the pulse compression radar;
$\tau_3$ the pulse width after pulse compression radar receiving compression;
$\tau_1$ the pulse width of the conventional pulse radar;
D the pulse compression coefficient of the pulse compression radar;
$D_0$ the coefficient of variation of the signal-to-noise ratio after the conventional pulse radar becomes a pulse compression radar.

For example, when a radar is not changed to a pulse compression radar, its pulse width is 2; after changing to a pulse compression radar, the transmission pulse width is 40. After receiving compression processing, the pulse width is 0.3. The rectangular coefficient of the transmitted pulse is 1.2 when the pulse pressure radar is not changed. After changing to the pulse pressure radar, the rectangular pulse coefficient may become worse (take 1.8). When the pulse pressure radar is unchanged, the peak power is 1 MW. The peak power after changing to pulse pressure is 0.8 MW, and other parameters remain unchanged.

Substitute into Formula (3.22), then $K_0 = 1.537$

$$D_0 = \frac{2\,\mu\text{s}}{0.3\,\mu\text{s}} = 6.7$$

$$D = \frac{40\,\mu\text{s}}{0.3\,\mu\text{s}} = 133.3$$

$$K_{d1} = \frac{D}{D_0 \cdot K_0} = 10.4$$

2. The second algorithm

Other parameters are unknown, and it is known that the average power of the radar changes.

$$B_1\tau_1 \approx 1;\ B_2\tau_2 \approx 1$$

$$E_{t1} = \overline{P_{t1}}/F_1;\ E_{t2} = \overline{P_{t2}}/F_2$$

Substitute parameters into Formula (3.4)

$$R_{I(1)} = 796\left(\frac{\frac{\overline{P_{t1}}F_1}{\cdot} L_t \cdot L_\gamma \cdot G_t \cdot G_\gamma}{f^2 \cdot \overline{NF}}\right)^{1/4} \cdot \left(\frac{K_1 \cdot \tau_1 \cdot B_1}{4} \cdot \left(1 + \frac{K_2'}{\tau_1 \cdot B_1}\right)^2\right)^{-1/4} \cdot (F_1)^{1/12}$$

$$R_{I(2)} = 796\left(\frac{\frac{\overline{P_{t2}}}{F_2}\cdot L_t\cdot L_\gamma\cdot G_t\cdot G_\gamma}{f^2\cdot\overline{NF}}\right)^{1/4}\cdot\left(\frac{K_1\cdot\tau_2\cdot B_2}{4}\cdot\left(1+\frac{K_2'}{\tau_2\cdot B_2}\right)^2\right)^{-1/4}\cdot(F_2)^{1/12}$$

$$K_{d2} = \left(\frac{R_{I(2)}}{R_{I(1)}}\right)^4 = \frac{\overline{P_{t2}}}{\overline{P_{t1}}}\cdot\left(\frac{F_1}{F_2}\right)^{1/3} \tag{3.24}$$

For example: when a radar does not change to the pulse pressure radar, its average power is 0.9 kW. After changing to a pulse pressure radar, its average power is 10 kW, and other parameters remain unchanged. Then the radar's gain coefficient $K_{d2}$ is

$$K_{d2} = \frac{\overline{P_{t2}}}{\overline{P_{t1}}} = \frac{10}{0.9} = 11.1$$

3. The third algorithm

The radar parameters are unknown. It is only known that when the pulse pressure radar is unchanged, the discovery distance for a 2-m$^2$ RCS target is 210 km. After changing to a pulse pressure radar, the discovery distance for a 2-m$^2$ target is 370 km. The radar gain coefficient $K_{d3}$ at this time is

$$K_{d3} = \left(\frac{370}{210}\right)^4 = 9.6$$

4. The fourth algorithm

Some parameters of the radar are unknown. Only the receiver 's signal/noise ratio improvement value after compression processing is generally known as −10 to −15 dB. So, the gain factor of the pulse compression radar can be taken as 10 to 15 dB.

The value of the jamming suppression coefficient of pulse compression radar is generally about 15 dB.

### 3.3.3 *Analysis on Jamming Exposure Area of Early Warning Aircraft Radar*

Assuming that the early warning aircraft radar is a pulse compression radar. Its equivalent radiated power is $P_tG_t = 96\,\text{dBW}$, $\sigma = 3.5\,\text{m}^2$.

The sidelobe characteristics of the antenna are shown in Fig. 3.7.

It is required that within the range of ±20°, the jamming exposure distance to the covered target is less than 50 km, and the distance between the jammer and the radar is 300 km. Find the required equivalent power of the jammer.

According to the jamming equation (3.19)

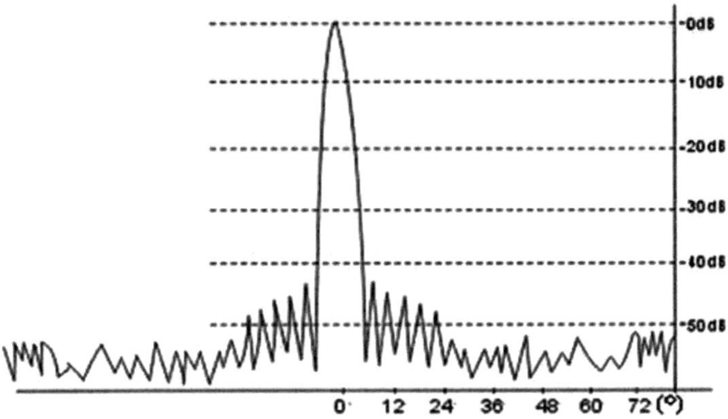


**Fig. 3.7** (10 cm) typical antenna lobe of early warning aircraft radar

$$P_j G_j = \frac{K_{jt} \cdot P_t \cdot G_t \cdot \sigma \cdot R_j^2}{4\pi R_{\min}^4} \cdot \frac{L_t}{L_j \cdot \gamma_j \cdot K_f} \cdot \frac{G_t}{G_t(\theta)}$$

Take $K_{jt} = 10$; $P_t G_t = 4 \times 10^9$; $\sigma = 3.5\,\text{m}^2$; $R_j = 300\,\text{km}$; $L_t/L_j \cdot \gamma_j \cdot K_f = 2$; $G_t/G_t(\theta) = 4 \times 10^4$; $R_{\min} = 50\,\text{km}$.

Substitute into Eq. (3.19)

$$P_j G_j = \frac{10 \times 4 \times 10^9 \times 9 \times 10^{10} \times 2 \times 4 \times 10^4}{12.6 \times 0.5^4 \times 10^{20}} = 67\,\text{dBW}$$

Jamming exposure distance and jamming exposure area

Jamming exposure distance means that when the radar is jammed with, its detection distance should be shortened. In the case of jamming, the maximum distance that the radar can find the target, also known as the minimum effective distance of jamming, can be described by Eq. (3.25):

$$R_{\min} = \left( \frac{K_{jt} \cdot P_t \cdot G_t \cdot \sigma \cdot R_j^2}{4\pi P_j \cdot G_j} \cdot \frac{L_t}{L_j \cdot \gamma_j \cdot K_f} \cdot \frac{G_t}{G_t(\theta)} \right)^{1/4} \tag{3.25}$$

The jamming exposure area refers to the minimum area generated by different $\theta$ corresponding to different $R_{\min}$ where the jammer generates different jamming exposure distances when it jamming with the radar from different sidelobe angles of the radar.

For example: known $P_t G_t = 96\,\text{dBW}$; $P_j G_j = 67\,\text{dBW}$; $K_{jt} = 10\,\text{dB}$; $\sigma = 3.5\,\text{m}^2$; $L_t/L_j \cdot \gamma_j \cdot K_f = 2$, find the jamming exposure areas when $R_j$ equals to 200 km, 300 km, 400 km.

Substitute the parameters into the Formula (3.25) to get the data in Table 3.2.

According to Table 3.2, draw the jamming exposure area shown in Fig. 3.8.

**Table 3.2** jamming exposure area with different θ and Rj

| $\theta(0)$ | $G_t(\theta)/G_t$ (dB) | $R_j = 200\,\text{km}$ (km) | $R_j = 300\,\text{km}$ (km) | $R_j = 400\,\text{km}$ (km) |
|---|---|---|---|---|
| 1 | 0 | 3.6 | 4.3 | 5 |
| 5 | −10 | 6.5 | 6.7 | 9.1 |
| 8.5 | −20 | 11.5 | 13.8 | 16.1 |
| 13 | −40 | 36 | 43.2 | 50.4 |
| 18 | −56 | 90.4 | 108 | 126 |
| 27 | −46 | 50 | 60 | 70 |
| 45 | −55 | 84.2 | 100.8 | 117.6 |
| 90 | −55 | 84.2 | 100.8 | 117.6 |
| 180 | −52 | 70.9 | 85.1 | 99.3 |

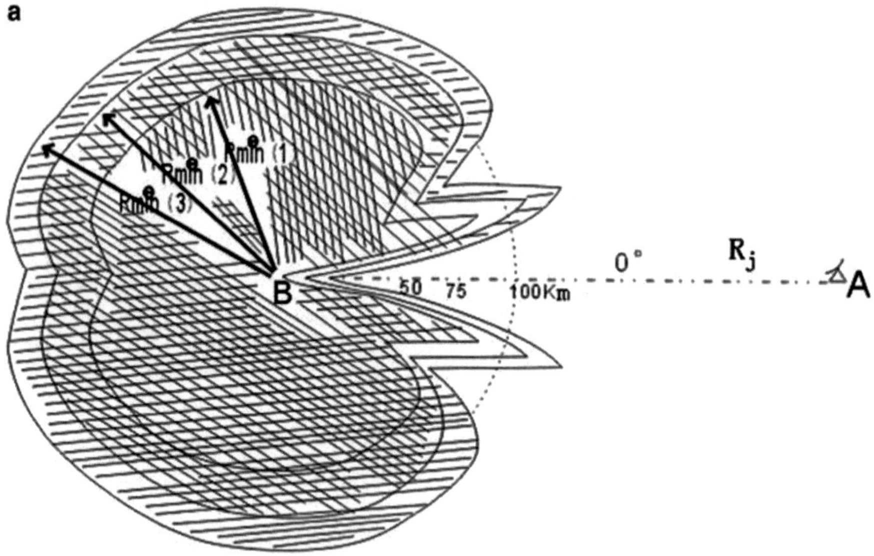


**Fig. 3.8** **a** Schematic diagram of the area exposed to the early warning aircraft radar jamming. Point A is the location of the jammer. Point B is the location of the early warning aircraft radar. $R_{\min(1)}(\theta)$ is the jamming exposure area when $R_j = 200\,\text{km}$. $R_{\min(2)}(\theta)$ is the jamming exposure area when $R_j = 300\,\text{km}$. $R_{\min(3)}(\theta)$ is the jamming exposure area when $R_j = 400\,\text{km}$. **b** Schematic diagram of the jamming exposure area when the radar jamming distance of the early warning aircraft is different. A is the position of the jammer; B1 is the jamming exposure area when the early warning aircraft radar is 200 km away from the jammer; B2 is the jamming exposure area when the early warning aircraft radar is 400 km away from the jammer; C1, C2, and C3 are fighter aircraft attacking the early warning aircraft; **c** Schematic diagram of the jamming exposure area formed when three jammers jamming at the same time. A1, A2, A3 are the position of the jammer; B is the position of the early warning aircraft; C1, C2, C3 are fighter aircraft

**b**

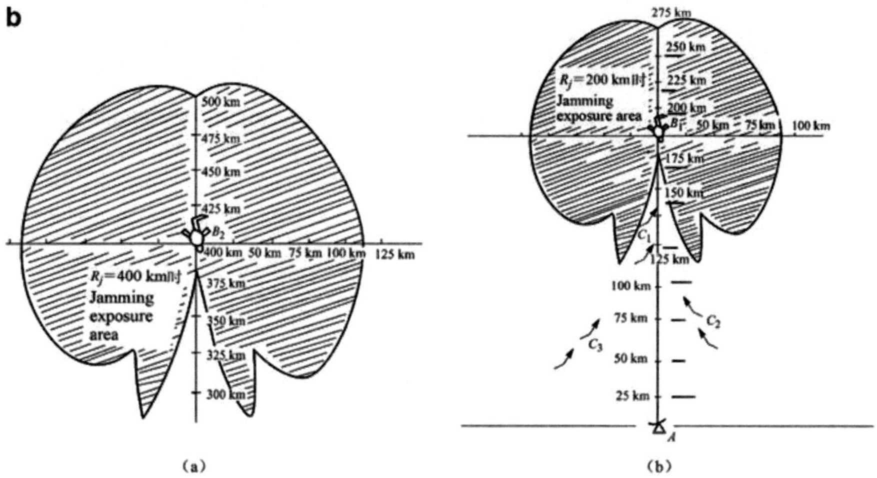


**Fig. 3.8** (continued)

**c**

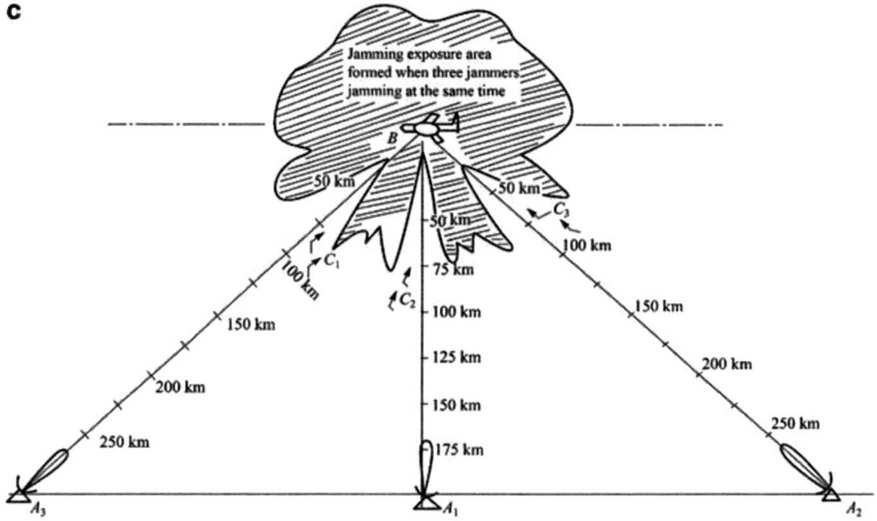


**Fig. 3.8** (continued)

The connection between radar and jammer is the datum line.

It can be seen from Fig. 3.8a that when the early warning aircraft radar is not jammed with, its detection range is 460 km. If it is jammed with, its average detection distance is within 100 km, and its effective detection area is reduced by more than 90%.

The detection distance of the early warning aircraft radar when it does not jamming is 470 km. If it is jammed, its average detection range is reduced to less than 100 km, and within the range of $0° \pm 15°$, and its detection range is reduced to less than 50 km. Therefore, the target of the covered attacking early warning aircraft is best to enter from the $0° \pm 10°$ direction. In this way, the distance between the attacked target and the early warning aircraft radar detection can be minimized, which is conducive to attack it.

## 3.4 Test of Jamming Suppression Coefficient and Jamming Exposure Area of Pulse Compression Radar

### *3.4.1 Test on the Antenna Pattern and Polarization Characteristics of Radar*

1. The test composition is shown in Fig. 3.9.
2. Steps and methods of test

    (a) Adjust the equipment under test and test equipment according to the composition of the test (the test antenna is a wide beam antenna);
    (b) Turn on radar and test equipment;
    (c) Adjust the mainlobe of the radar antenna to the mainlobe of the test antenna to maximize the test indication (record the test power value). At this time, the azimuth and elevation angle to the radar is 0°;
    (d) The radar azimuth remains unchanged, and a set of data is measured by changing the elevation angle of the radar antenna;
    (e) Change the radar azimuth angle to $\alpha_1$, and then change the elevation angle to $\beta_{0,1,2,3}$, and measure the values of horizontal polarization and vertical polarization reception separately;
    (f) Change the radar azimuth angle to $\alpha_2$, and then change the elevation angle to $\beta_{0,1,2,3}$, and measure the values of horizontal polarization and vertical polarization reception separately;

By analogy, get the data in Table 3.3.

**Fig. 3.9** Composition of radar antenna pattern and polarization characteristics test. A is the display system of receiving test; B is the radar station under test; $R_j$ is the distance between the radar and the test antenna

**Table 3.3** antenna gain with different azimuth, elevation and polarization

| Azimuth (°) | Elevation (°) | Horizontal polarization value (dB) | Vertical polarization value (dB) |
|---|---|---|---|
| 0 | 0 | −1.1 | −29.6 |
| 0 | 1 | −10.5 | −40.86 |
| 0 | 2 | −23.0 | −50.3 |
| 0 | 2.6 | −21.3 | −49.7 |
| 0 | 3 | −19 | −47.3 |
| 10.3 | 0 | −44 | −47 |
| 10.3 | 1 | −44 | −46 |
| 10.3 | 4 | −48 | −49 |
| 15 | 0 | −46.2 | −40 |
| 15 | 1 | −45.6 | −44.7 |
| 15 | 2 | −53.4 | −45.2 |
| 15 | 3 | −49.5 | −44.9 |
| 45 | 0 | −54.2 | / |
| 45 | 1 | −53.8 | / |
| 45 | 2 | −54.7 | / |
| 45 | 3 | −54.0 | / |
| 90 | / | / | / |
| 180 | / | / | / |

| Azimuth (°) | Elevation angle (°) | Horizontal polarization value (dB) | Vertical polarization value (dB) |
|---|---|---|---|
| 0 | 0 | −1.1 | −29.6 |
| 0 | 1 | −10.5 | −40.86 |
| 0 | 2 | −23.0 | −50.3 |
| 0 | 2.6 | −21.3 | −49.7 |
| 0 | 3 | −19 | −47.3 |
| 10.3 | 0 | −44 | −47 |
| 10.3 | 1 | −44 | −46 |
| 10.3 | 4 | −48 | −49 |
| 15 | 0 | −46.2 | −40 |
| 15 | 1 | −45.6 | −44.7 |
| 15 | 2 | −53.4 | −45.2 |
| 15 | 3 | −49.5 | −44.9 |
| 45 | 0 | −54.2 | / |
| 45 | 1 | −53.8 | / |
| 45 | 2 | −54.7 | / |
| 45 | 3 | −54.0 | / |
| 90 | / | / | / |

(continued)

Table 3.3 (continued)

| Azimuth (°) | Elevation angle (°) | Horizontal polarization value (dB) | Vertical polarization value (dB) |
|---|---|---|---|
| 180 | / | / | / |

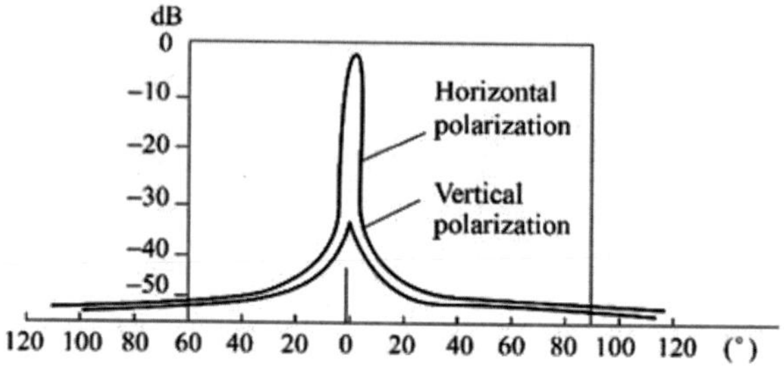


**Fig. 3.10** Sidelobe characteristics and polarization characteristics of narrow beam antennas

Draw Fig. 3.10, according to the data in Table 3.3. This curve not only reflects the main and sidelobe characteristics of the antenna but also reflects the polarization characteristics of the antenna. In the mainlobe area of the antenna, the vertical polarization and the horizontal polarization of the antenna are very different. Still, in the sidelobe area of the antenna, there is very little difference between the vertical polarization and the horizontal polarization.

When the radar is horizontally polarized, and the receiving antenna is horizontally polarized, the received signal is strong. When the receiving antenna is vertically polarized, the signal received in the mainlobe region is about 28 dB different. Outside region $\pm 10°$ of the sidelobe, regardless of whether the receiving antenna is horizontally or vertically polarized, the received signal strength is very close. Therefore, for jammers that jamming with the sidelobes of the radar antenna, the requirements of polarization of the jamming antenna are not strict.

### 3.4.2 *Test of Jamming Suppression Coefficient and Jamming Exposure Area of Pulse Compression Radar*

1. The composition of the test

It includes the radar participating in the test, the cooperative target aircraft, the jammer being tested, the radar monitoring the target's track, and the test command center, as shown in Fig. 3.11.

Route 1 is direction 0°, and the cooperation target enters along with line A and B;

Route 2 enters in the direction of $\theta_1 = 15°$;

Route 3 enters in the direction of $\theta_2 = 45°$.

The cooperation target enters from 60 km away from point B, and the effective reflection area of the cooperation target is 3.5 $m^2$.

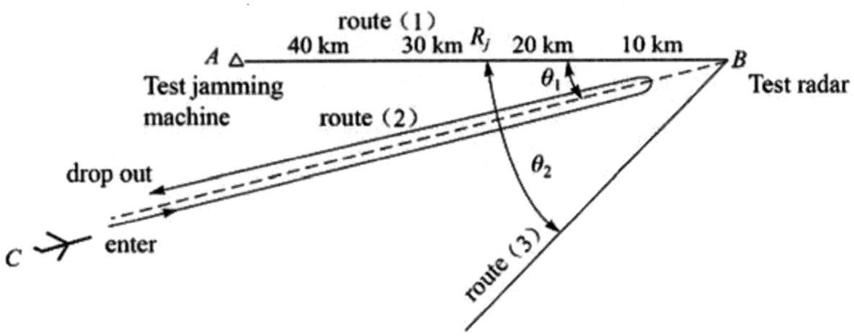


**Fig. 3.11** Schematic diagram of jamming suppression coefficient flight test. A is the jammer under test; B is the radar participating in the test; C is the cooperative test target; $R_j$ is the distance between the radar and the jammer

2. Test procedures and methods

   (a) Debugging equipment
   (b) The cooperation target entered the route from 60 km away, and the radars participating in the test began to work. At the first-time enter without jamming, the radar can normally detect cooperation targets.
   (c) When the cooperation target enters for the second time, the jammer is turned on, and the radar cannot detect the target. When the target reaches a certain distance, the radar starts to find the target. If the radar can find the target in for successive two sweeps, it is found to be effective, and the distance to the target is recorded as $R_{\min(1)}$. The target continues to fly forward, and the radar can detect that the target returns after reach the predetermined distance and exits the route. Make three jamming against the entries on each route, and find the root mean square value of the three discovered distances as the jamming exposure distance.
   (d) Perform the test on the third route according to the above test, and measure the minimum jamming exposure distance when entering from the third route.
   (e) When entering from the first route, its jamming exposure distance is less than the minimum detection distance of the radar. Therefore, the minimum jamming exposure distance of the radar cannot be measured.

The jamming exposure distance measured by the test is shown in Table 3.4.

During the test, the radar works at a single carrier frequency. Single carrier frequency is the way to detect the maximum distance of the target. The jamming adopts the method of continuous-wave noise jamming.

3. The measured data of the jamming exposure area and the jamming exposure area.

The measured jamming exposure area drawn from the data in Table 3.4 is shown in Fig. 3.12.

**Table 3.4** jamming exposure area with different flight path

| Effect of jamming | Flightpath (0) | $G_t(\theta)/G_t$ (dB) | Measured value of burn-through distance $R_{\min}$ (km) | Elevation of burn-through point (0) | RMS of burn-through distance r $R_{\min(1)}$ (km) | $R_j = 200\,\text{km}$ Estimated value $R_{\min(2)}$ (km) | $R_j = 400\,\text{km}$ Estimated value $R_{\min(3)}$ (km) |
|---|---|---|---|---|---|---|---|
| Tertiary jamming | 15 | −43 | 18 | 3.08 | 22.1 | 46.4 | 66.3 |
| | 15 | −44 | 23 | 2.36 | | | |
| | 15 | −44 | 22.1 | 2.52 | | | |
| Tertiary jamming | 45 | −53 | 34.4 | 1.48 | 34.6 | 72.7 | 104.1 |
| | 45 | −53 | 32.5 | 1.78 | | | |
| | 45 | −53 | 37 | 1.44 | | | |
| Tertiary jamming | 0 | | Less than 12 | | 12 | 25.2 | 36 |

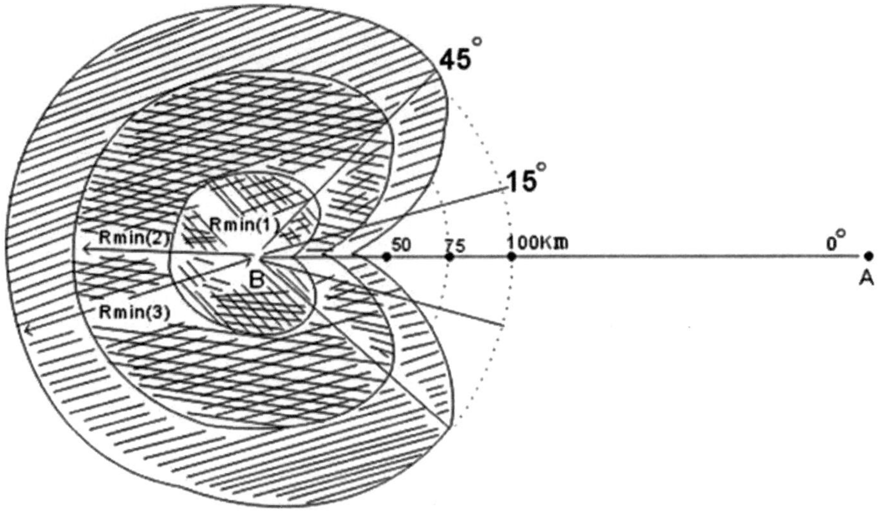


**Fig. 3.12** Test the jamming exposure area of the pulse compression radar jamming test flight and calculate the jamming exposure area according to the actual measurement

$R_{\min(1)}$ is the measured jamming exposure area;

$R_{\min(2)}$ is the estimated value when the measured exposure area changes $R_j$ to 200 km;

$R_{\min(3)}$ is the estimated value when the measured exposure area changes $R_j$ to 400 km.

Parameters of the test radar: $P_tG_t = 4.4 \times 10^9$ W; $\sigma = 3.5\,\text{m}^2$; $R_j = 45$ km; $P_jG_j = 7 \times 10^6$ W; $G_t/G_t(\theta)$, $R_{\min}$ as shown in Table 3.4.

Compare Figs. 3.8 and 3.12:

In Fig. 3.8, the jamming exposure area is calculated by applying Formula (3.25) according to the parameters of radar and jammer. Due to the different values of the antenna pattern, there is a protruded point of the jamming exposure distance at point $\pm 18°$. The deviation of the jamming exposure area for $R_j = 200$ km and $R_j = 400$ km are larger (100.8 and 117.6 km).

In Fig. 3.12, using the same radar parameters and jammer parameters for the actual test get the jamming exposure area. The actual test flight distance $R_j$ is 45 km. $R_{\min(1)}$ is its jamming exposure area. Use $R_{\min(1)}$ to estimate that when $R_j = 200$ km, we get $R_{\min(2)}$ jamming exposure area; Use $R_{\min(1)}$ to estimate that when $R_j = 200$ km, we get $R_{\min(3)}$ jamming exposure area.

Comparing Fig. 3.8 with Fig. 3.13, the measured jamming exposure area is slightly smaller than the theoretically calculated jamming exposure area, indicating that the theoretically calculated value is completely credible.

Point A is the position of the jammer; point $B_1$ is the position of the radar during the test; $B_2$ is the jamming exposure area estimated based on the experiment when

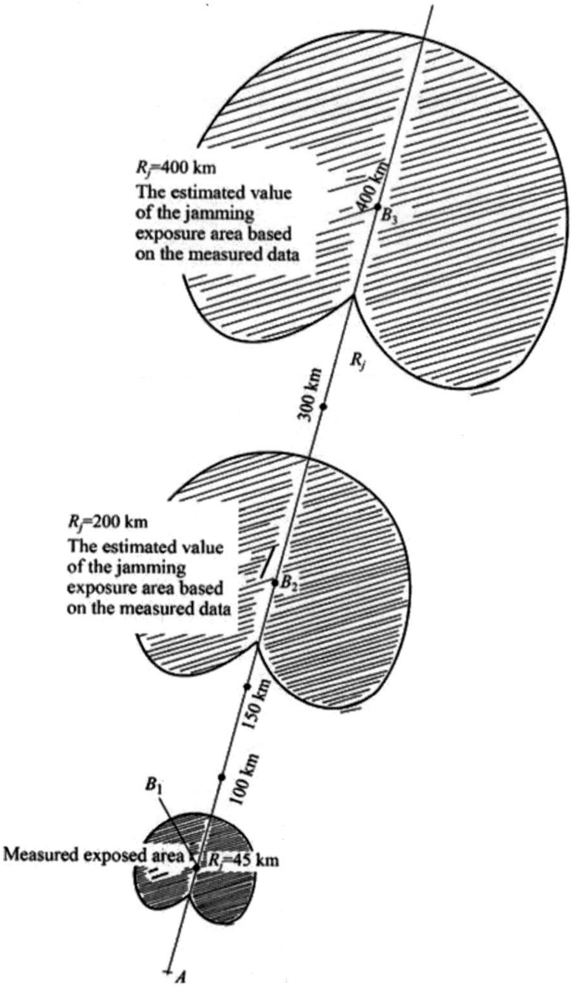


**Fig. 3.13** Measured jamming exposure area and estimated jamming exposure area for pulse compression radar

$R_j = 200\,\text{km}$; $B_3$ is the jamming exposure area estimated based on the experiment when $R_j = 400\,\text{km}$.

### *3.4.3 Measurement and Calculation of Jamming Suppression Coefficient of Pulse Compression Radar*

The equation of jamming suppression coefficients of pulse compression radar is Formula (3.18).

$$K_{jt} = \frac{4\pi P_j \cdot G_j \cdot R_{\min}^4}{P_t \cdot G_t \cdot \sigma \cdot R_j^2} \cdot \frac{L_j \cdot \gamma_j \cdot K_f}{L_t \cdot L_\gamma} \cdot \frac{G_t(\theta)}{G_t}$$

1. When entering in direction 15°,$P_t G_t = 8.4 \times 10^9$ W; $P_j G_j = 5 \times 10^6$ W; $\sigma = 3.5\,\text{m}^2$; $R_j = 45$ km; $R_{t\min} = 22.1$ km; $G_t/G_t(\theta) = -44$ dB; $\frac{L_j \cdot \gamma_j \cdot K_f}{L_t \cdot L_\gamma} \approx 0.8$
   Substitute into Formula (3.18) and get

$$K_{jt(1)} = \frac{12.6 \times 5 \times 10^6 \times 2.21^4 \times 10^{16} \times 0.4 \times 10^{-4} \times 0.8}{8.4 \times 10^9 \times 4.5^2 \times 10^8 \times 3.5} = 9\,\text{dB}$$

2. When entering in direction 45°, $P_t G_t = 6.4 \times 10^9$ W; $P_j G_j = 5 \times 10^6$ W; $\sigma = 3.5\,\text{m}^2$; $R_j = 45$ km; $R_{t\min} = 34.6$ km; $G_t/G_t(\theta) = -53$ dB; $\frac{L_j \cdot \gamma_j \cdot K_f}{L_t \cdot L_\gamma} \approx 0.3$
   Substitute parameters into Formula (3.18)

$$K_{jt(2)} = \frac{12.6 \times 5 \times 10^6 \times 3.4^4 \times 10^{16} \times 0.5 \times 10^{-4} \times 0.8}{6.4 \times 10^9 \times 4.5^2 \times 10^8 \times 3.5} = 9\,\text{dB}$$

3. This radar is a pulse compression radar modified from a conventional pulse radar. The peak power of the original radar is 1 MW; the pulse width is 2 μs; the average power is 0.9 kW. After changing to pulse compression radar, the peak power is 0.8 MW; the pulse width is 43 μs; the pulse width after compression is 0.3 μs, and other parameters remain unchanged.
   Substitute parameters into Formula (3.23)

$$K_{d(1)} = \frac{D}{D_0 \cdot K_0} = \frac{\frac{43}{0.3}}{\frac{2}{0.3} \times 1.5} = 11\,\text{dB}$$

   Substitute parameters into Formula (3.24)
   Get $K_{d(2)} = \frac{\overline{P_{t2}}}{\overline{P_{t1}}} = \frac{10}{0.9} \cong 10\,\text{dB}$.
4. Compare the jamming suppression coefficient required by the pulse compression radar in several cases:
   (a) When entering from direction 15°, the measured jamming suppression coefficient is 10 dB;
   (b) When entering from direction 45°, the measured jamming suppression coefficient is 10 dB;
   (c) From Formula (3.23), the radar gain factor is calculated to be 11 dB (the pulse compression coefficient is 22 dB);
   (d) From Formula (3.24), after the radar was changed to pulse compression radar, the average power increased by nearly 10 dB. As mentioned earlier, the multiple of the average power increment of the pulse compression radar is the jamming suppression factor required to jamming with it.

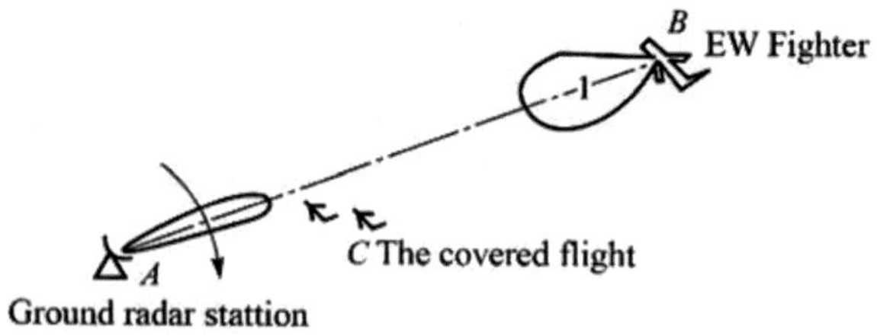


**Fig. 3.14** Schematic diagram of air-to-ground cover jamming

The jamming suppression coefficients obtained by these four methods are about 10 dB, and the pulse compression coefficient of the radar is 22 dB, which also satisfies the conclusion of $k_{jd} < D$.

## 3.5 Jamming Suppression Zone of the Airborne Jammer Against Ground Pulse Compression Radar

This type of jammer belongs to the air-to-ground covering jamming, and its jamming equation can still use Eq. (3.19)

Which is

$$P_j \cdot G_j = \frac{K_{jt} \cdot P_t \cdot G_t \cdot \sigma \cdot R_j^2}{4\pi \cdot R_{\min}^4} \cdot \frac{L_t}{L_j \cdot \gamma_j \cdot K_f} \cdot \frac{G_t}{G_t(\theta)}$$

The relationship between radar, jammer, and the covered target is shown in Fig. 3.14.

***Example 3.1*** Known radar parameters: Operating frequency in S-band; $P_tG_t = 96\,\text{dBW}$; $\sigma = 3.5\,\text{m}^2$; The transmitted pulse width is 43 μs; the compressed pulse width is 0.3 μs; $R_j = 200\,\text{km}$; $K_{jt}$ equals 10 dB; the antenna main/sidelobe ratio is shown in Table 3.5; the equivalent power of the jammer is $P_jG_j = 10^5$ W. Calculate the jamming exposure area for the radar.

Apply Formula (3.25)

Take $\frac{L_t}{L_j \cdot \gamma_j \cdot K_f} = 2$

$$R_{\min} = \left( \frac{K_{jt} \cdot P_t \cdot G_t \cdot \sigma \cdot R_j^2}{4\pi P_j \cdot G_j} \cdot \frac{L_t}{L_j \cdot \gamma_j \cdot K_f} \cdot \frac{G_t}{G_t(\theta)} \right)^{1/4}$$

Substitute known parameters into equation

$$R_{\min} = \left( \frac{10 \times 4 \times 10^9 \times 3.5 \times \left(2 \times 10^5\right)^2}{4\pi \times 10^5} \cdot 2 \cdot \frac{G_t}{G_t(\theta)} \right)^{1/4}$$

**Table 3.5** jamming exposure area with different θ and Rj, when PjGj=50dBW

| $\theta(0)$ | $G_t/G_t(\theta)$ (dB) | $R_j = 200\,\text{km}$ (km) | $R_j = 100\,\text{km}$ (km) |
|---|---|---|---|
| 1 | 0 | 17 | 11.9 |
| 5 | 10 | 30.6 | 21.4 |
| 8.5 | 20 | 54.8 | 38.1 |
| 13 | 40 | 170 | 119 |
| 18 | 56 | 420 | 294 |
| 27 | 46 | 238 | 167 |
| 45 | 55 | 408 | 286 |
| 90 | 55 | 408 | 286 |
| 180 | 52 | 187 | 131 |

$$= 1.7 \times 10^4 \times \left(\frac{G_t}{G_t(\theta)}\right)^{1/4}$$

The resulting data is shown in Table 3.5.

Draw jamming exposure area according to Table 3.5.

When the new radar is not jammed with, its detection range for $\sigma = 3.5\,\text{m}^2$ target is 400 km. When an airborne jammer jamming with an equivalent jamming power of 100 kW, its detection area is less than 100 km within the range of $\pm 10°$ . In the range of $+(15°–45°)$ and $-(15°–45°)$, the detection distance is less than 200 km; in the range of $180° \pm 20°$, the detection distance is less than 250 km. The total detection area has been reduced by about 15%, and it can be seen that the effect of this jammer is relatively poor. Only using multiple jammers to array jamming can get a better jamming-effect.

When five jammers are used at the same time and an airborne jammer is 10° apart, the jamming exposure area formed by it is shown in Fig. 3.16.

The horizontal line area is the jamming exposure area; the 400 km dotted area is the radar effective detection area.

***Example 3.2*** If the radar's parameters remain unchanged, increase the jammer's equivalent power to 60 dBW, substitute the parameters into Eq. (3.25), and the get data in Table 3.6.

When $R_i = 100\,\text{km}$, the jamming exposures area is small; when $R_j = 200\,\text{km}$, the jamming exposure area is large.

In Fig. 3.15, $P_jG_j = 50\,\text{dBW}$; in Fig. 3.17, $P_jG_j = 60$ dBW. Jamming exposure areas formed for the same type of radar.

The jamming exposure area in Fig. 3.15 is 3.2 times larger than that in Fig. 3.17.

**Table 3.6** jamming exposure area with different θ and Rj, when PjGj=60dBW

| $\theta(0)$ | $G_t/G_t(\theta)$ (dB) | $R_j = 200\,\text{km}$ (km) | $R_j = 100\,\text{km}$ (km) |
|---|---|---|---|
| 1 | 0 | 9.4 | 6.6 |
| 5 | 10 | 17 | 11.9 |
| 8.5 | 20 | 30.4 | 21.2 |
| 13 | 40 | 94.4 | 66.1 |
| 18 | 56 | 233.3 | 163.3 |
| 27 | 46 | 132.2 | 92.8 |
| 45 | 55 | 226.7 | 158.9 |
| 90 | 55 | 226.7 | 158.9 |
| 180 | 52 | 103.9 | 72.8 |

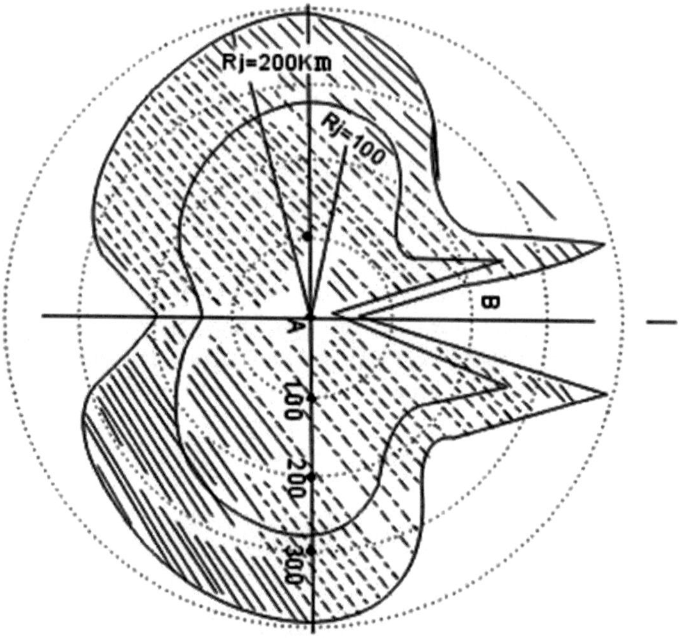


**Fig. 3.15** Schematic diagram of the jamming exposure area of the ground-based new pulse compression radar. Point A is the position of new ground pulse compression radar; point B is similar to the position of EA-6B jammer

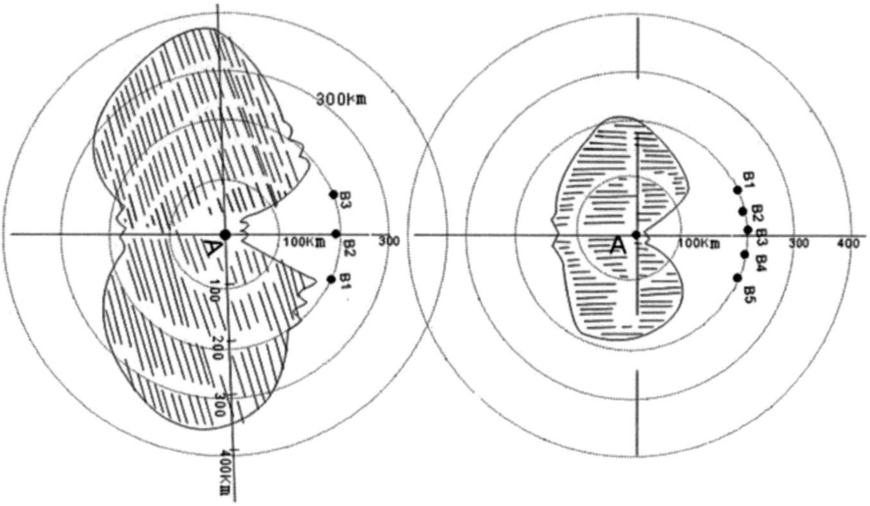


**Fig. 3.16** Schematic diagram of the jamming exposure area formed when five airborne jammers ($P_jG_j = 50$ dBW) jamming with the new pulse compression radar. A is the location of the radar; $B_{1,2,3,4,5}$ is the position of five airborne jammers

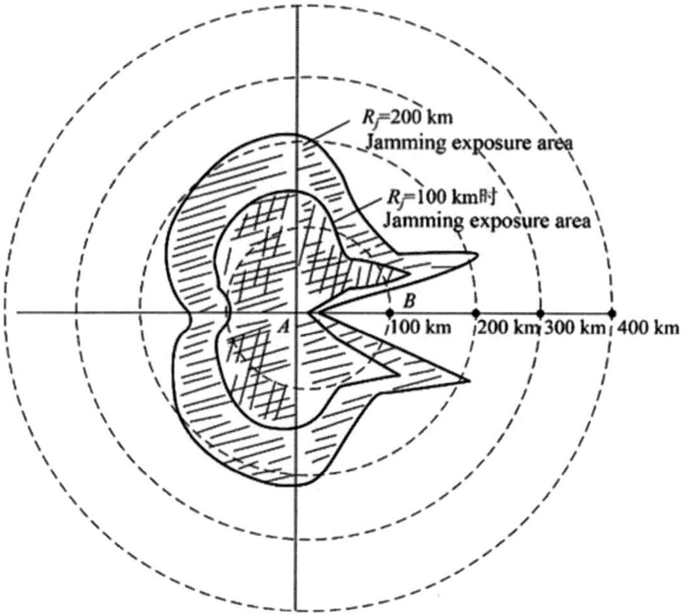


**Fig. 3.17** Shows the radar jamming exposure area. Point A is the position of the radar station; Point B is the position of the $E_{A\text{-}6\,B}$ jammer

# Chapter 4
# Jamming Techniques for Pulse-Doppler Radar

Pulse-Doppler radar is a new type of radar-based on the requirements of radar development. Airborne radar needs to be able to detect low-altitude moving targets. But ground objects or sea clutter affect the radar's low-altitude performance. The clutter of ground objects and the Doppler spectrum of low-altitude moving targets are separated in the frequency domain. According to the distribution characteristics of the clutter spectrum of the ground object and the Doppler spectrum of the moving target, the radar designers used a group of narrow-band filters to filter out the clutter of the ground object and extract the Doppler frequency information of the moving target. Therefore, this kind of radar enhances the ability to detect low-altitude moving targets. It has become a new type of radar and is widely used.

## 4.1 Jamming Principle for Pulse-Doppler Radar

Early warning aircraft radar tactically requires that its detection range be far, and it can also detect low-altitude moving targets. In order to increase the detection range of the radar, the radar uses a pulse compression radar (already analyzed in the above two sections). In order to find low-altitude moving targets, the radar should use the Pulse-Doppler system.

### *4.1.1 Operating Principle of Pulse-Doppler Radar*

The pulse Doppler system can detect the target's echo of a specific Doppler frequency shift and filter out other clutter and industrial jamming. The superior performance of the Pulse-Doppler system is achieved since it is based on high technical requirements. The current technical keys are:



X. Zhang et al., *Theory to Countermeasures Against New Radars*,
https://doi.org/10.1007/978-981-16-6715-2_4

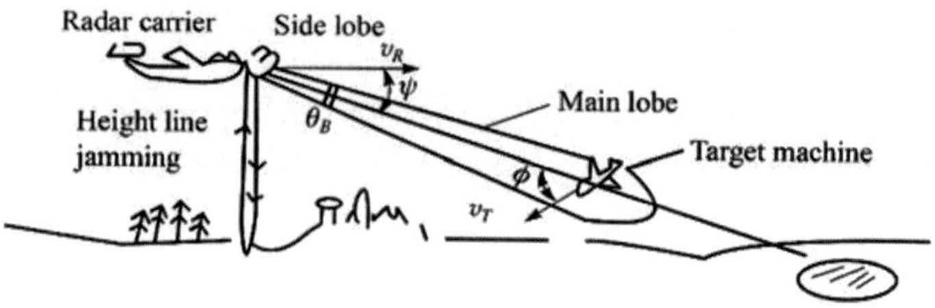


**Fig. 4.1** Source of the aircraft Doppler radar clutter

1. Very stable signal generation and amplification technology;
2. Very low sidelobe antenna technology
3. Advanced digital signal processing technology.

Taking airborne Pulse-Doppler radar as an example, we analyze the filtering of the signal and clutter spectrum of the Pulse-Doppler radar system.

1. Sources of airborne radar clutter jamming

As shown in Fig. 4.1, the flying speed of the radar carrier is $V_R$; The beamwidth of the radar antenna is $\theta_B$; The angle between the beam direction and the flight direction is $\psi$; the flying speed of the target is $V_T$; The angle between the flight direction of the target and the sight between the radar and the target is $\phi$.

(1) Clutter of the main lobe

The antenna beam of airborne radar may illuminate the background of the ground feature, but the ground feature has a relative motion concerning the radar. Therefore, the clutter of the main lobe formed by this part of the background reflection has a Doppler shift, and its magnitude is determined by Eq. (4.1), which is

$$f_{mL} = \frac{2V_R}{\lambda} \cdot \cos \psi \tag{4.1}$$

(2) Clutter of side lobe

Because the side lobes of the radar will also illuminate the ground, the echoes irradiated by the side lobes to the ground cause sidelobe jamming. Because the side lobe gain is small and points to different directions, the spectral density of the side lobes is also small, but its Doppler frequency shift range is scattered between $\pm f_{SL\max}$:

$$f_{SL\max} = \frac{2U_R}{\lambda} \tag{4.2}$$

(3) Height line clutter

Radiation of the radar vertically downward through the side lobes will cause large clutter. This is because the reflection is the strongest when it is perpendicular to the ground, and it is closest to the ground along with the flight height.

Because the flight flying horizontally has no radial velocity vertically down to the ground, there is no Doppler shift in the height line clutter.

The ground clutter of airborne Doppler radar is shown in Fig. 4.2.

Transmitted signal spectrum and received signal spectrum of airborne Doppler radar is shown in Fig. 4.3.

Doppler frequency of the target's motion projected in the direction of the main lobe of the beam is $f_T = \frac{2V_T}{\lambda} \cdot \cos\phi$.

The mathematical model of the clutter is that Pulse-Doppler radar performs spatial and frequency filtering on the target. Only when a clutter enters a distance gate at the same time with the target and is in the same narrow-band filter with the target at the frequency level, it can constitute jamming to the target.

The spatial geometric relationship is that assume that the period of radar pulse repetition is $t_\gamma$, the gate position distance from the transmitted pulse is $t_G$, and the gate width is $T_G$, shown in Fig. 4.4.

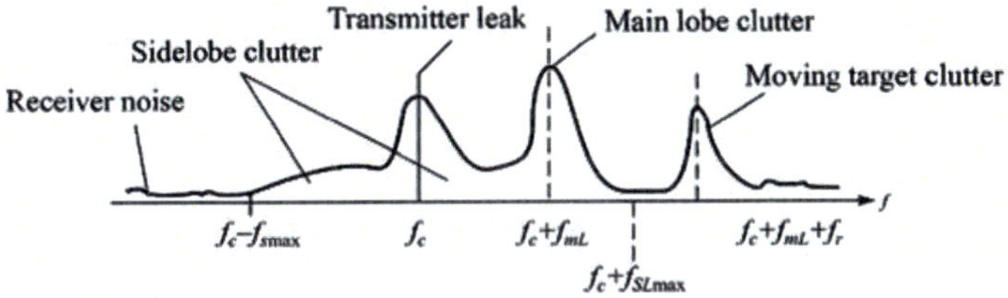


**Fig. 4.2** Ground clutter spectrum of airborne pulse Doppler radar

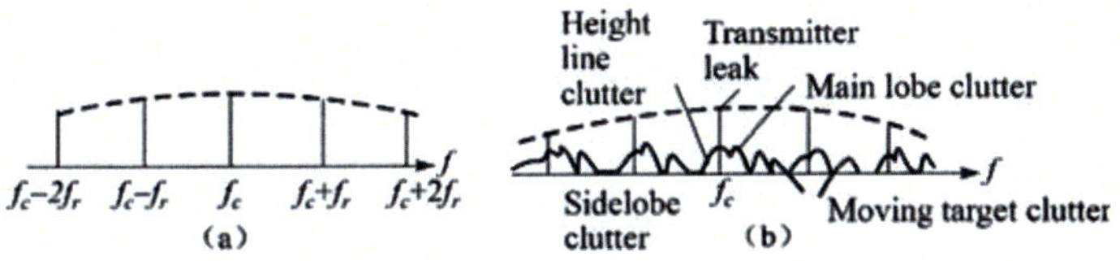


(a) Emission signal spectrum (b) Received signal spectrum

**Fig. 4.3** Transmitted signal spectrum and received signal spectrum of airborne Doppler radar

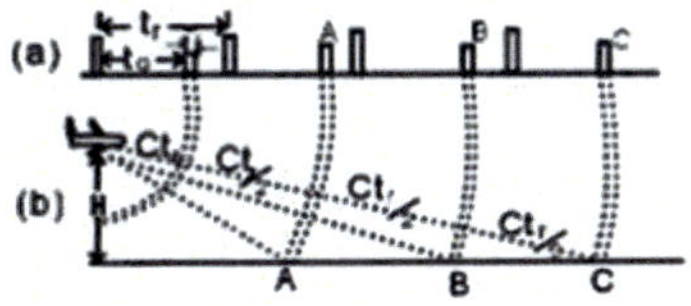


(a) Transmission pulse and range gate waveform diagram (b) The spatial relationship corresponding to the distance gate

**Fig. 4.4** The geometric relationship of the clutter area that can pass through the distance gate

Assume that the flight flies horizontally, and the height is $H\left(\frac{C \cdot t_G}{2} < H < \frac{(t_G + T_G) \cdot C}{2}\right)$. When t = 0, the transmitted pulse reflected at point A and return to the radar at $t_\gamma + t_G$ through wave gate A. After reflecting at B, the echo returns to the radar at time $2t_\gamma + t_G$ and passes through wave gate B. After reflecting at C, the echo returns to the radar at time $3t_\gamma + t_G$ and enters the wave gate C.

According to the radar's equation, the power of received clutter reflected by the small clutter area dA is:

$$dP_\gamma = \frac{P \cdot \lambda^2 \cdot [G(\psi, \theta)]^2 \cdot \sigma^0 \cdot dA}{(4\pi)^3 \cdot R^4 \cdot L} \tag{4.3}$$

where,

| | |
|---|---|
| P | is the transmit power of continuous wave radar; |
| $\lambda$ | is the radar's wavelength; |
| $G(\psi, \theta)$ | is the antenna gain in the direction of the tiny clutter area; |
| $\sigma^0$ | is the clutter's cross-sectional area per unit area on the ground; |
| R | is the sight from radar to tiny clutter area dA; |
| L | is the loss factor of propagation and system. |

According to the above analysis of clutter sources:

(a) Pulse-Doppler radar has the main clutter area, height line clutter area, sidelobe clutter area, and clutter-free area that only has noise and no ground clutter. The main lobe clutter area is the strongest, and the height line jamming is also strong. The target echo can be in the cluttered area or the clutter-free area, depending on the radial velocity of the target.

(b) When the modulus of moving target to radar carrier speed |V| > $V_R$, the doppler shift of echo is $|f_d| > 2V_R/\lambda$, and the moving target is in the clutter-free zone, in which the range is the largest at this time.

(c) When the modulus of moving target to radar carrier speed |V| > $V_R$, the doppler shift of echo is $|f_d| > 2V_R/\lambda$, and the moving target is in the sidelobe clutter region. The ultra-low sidelobe antenna is used to reduce the clutter level, increase the flying height, and increase the echo's power. It can detect useful echo signals in the sidelobe area.

2. Block diagram of a typical Pulse-Doppler radar

Because the Pulse-Doppler system uses spectrum separation technology to filter out the single spectral line of the desired moving target, the filtering of the clutter is mainly done in the frequency domain using various filters. Also, it cooperates with some time gates to select useful signal echo.

As shown in Fig. 4.5, it is the composition of typical Pulse-Doppler radar.

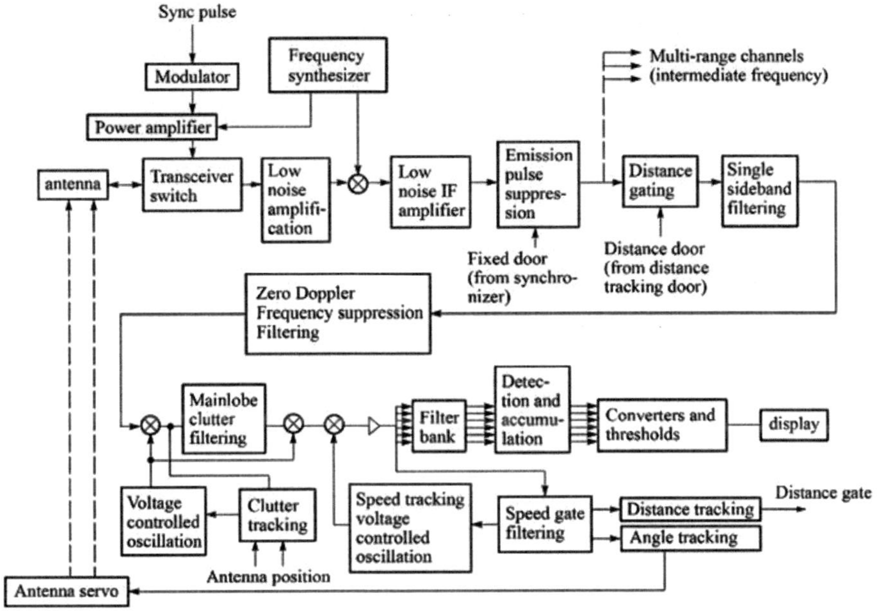


**Fig. 4.5** The principle block diagram of typical airborne pulse Doppler radar

(1) Single sideband Filtering

Its bandwidth is close to the pulse repetition frequency. Only one spectral line and its corresponding Doppler frequency shift component are filtered out from the echo so that the subsequent signal and clutter filtering is performed on a single spectral line. Generally, the single-sideband filter selects the centerline of the echo spectrum with high power. When multiple pulse repetitions are used, and the repetition frequency changes, only the position of this line remains unchanged.

Due to the narrow passband, which is only equivalent to a repetition frequency interval, the input intermediate frequency pulse signal passes through it with only one spectral line and becomes a continuous wave. The distance information cannot be extracted. So, the distance gate must be added before the single-sideband filter.

The output of a single spectral line greatly reduces the signal power (reduced by $d^2$ times, d is the duty factor of transmit pulse). However, the output of clutter and noise is also reduced. Therefore, the single-sideband filter does not reduce the signal-to-noise ratio, and the out-of-band rejection of the single-sideband filter requires more than 60 dB.

(2) Clutter filter of the main lobe

Doppler frequency of main lobe clutter is $f_{mL} = \frac{2V_R}{\lambda} \cdot \cos\psi$. When angle $\psi$ changes between $\psi_{\max}$ and $\psi_{\min}$, The corresponding frequency of main lobe

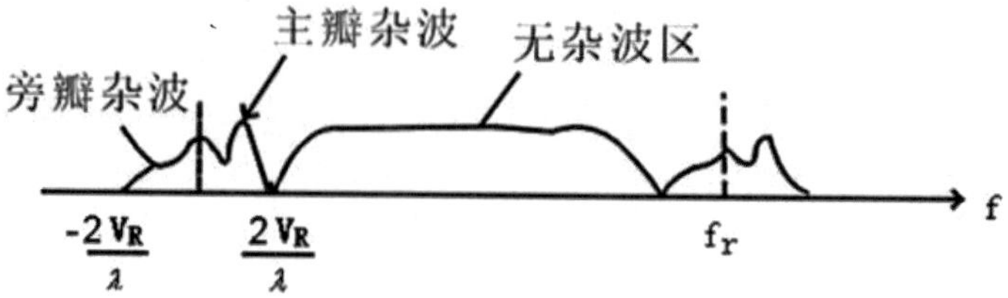


**Fig. 4.6** Frequency response characteristics of the filter without clutter

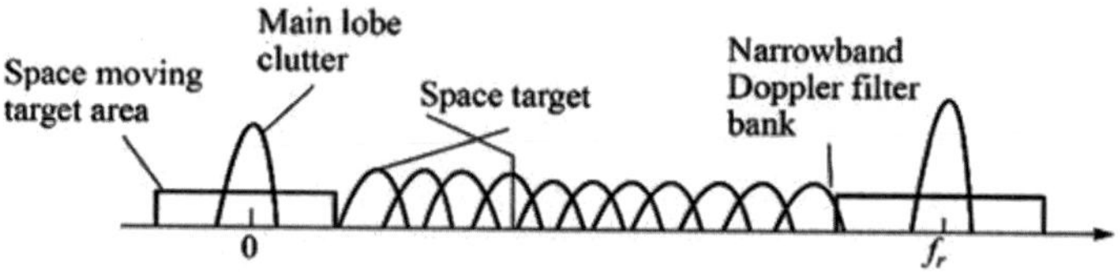


**Fig. 4.7** Schematic diagram of Doppler radar filter bank

clutter will change in range $\psi_{mL\max} - \psi_{mL\min}$. The clutter tracker can be used to filter the main lobe clutter, shown in Fig. 4.6.

When Pulse-Doppler radar uses pulse compression, single-sideband filtering cannot be used before pulse compression.

(3) Filter the height line clutter

Since the Doppler frequency of a carrier aircraft flying horizontally is zero, a single fixed frequency suppression filter is usually used, and it is filtered with a zero Doppler filter.

(4) Doppler filters bank

The Doppler filter bank is a set of adjacent narrow-band filter banks covering the target's expected Doppler frequency range in a pulse repetition frequency interval. It is shown in Fig. 4.7.

The speed of the target relative to the radar is different, the Doppler frequency shift is distinct, and the narrow-band filter that falls is also different, which achieves the measurement and resolution of the speed. At the same time, it also plays a role in suppressing clutter and improving sensitivity. It is an indispensable component to make Pulse-Doppler radar to have high performance. The Doppler filter bank can be set in the video or intermediate frequency, and the filter bank. Currently, it uses digital filtering. It has the advantages of small size, lightweight, high reliability, and flexible change of pulse interval, weighting, and bandwidth processing.

(5) Selection of Pulse repetition frequency

Generally, the selection of repetition frequency of high pulse makes the speed measurement unambiguous, which means that the repetition frequency is so high that the spectral lines of moving target generated by adjacent spectral lines do not overlap.

Maximum speed $V_{T\max}$ relative to the ground.

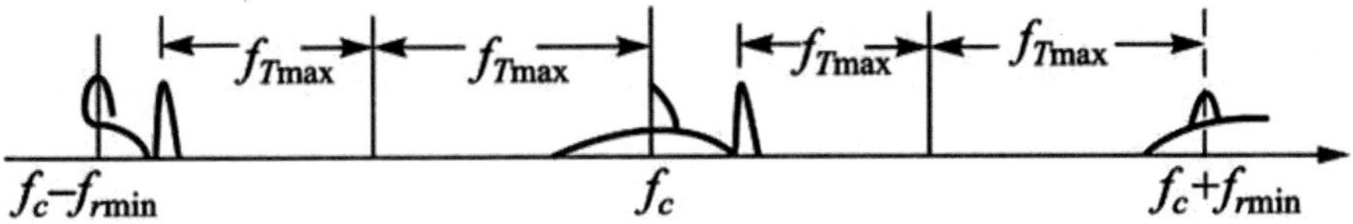


**Fig. 4.8** Pulse Doppler radar to overcome velocity ambiguity

The maximum Doppler frequency of the main beam moving target's echo relative to the center frequency of the main lobe clutter

$$\pm f_{T\max} = \pm 2V_{T\max}\lambda$$

Then the minimum value of the pulse's repetition frequency $f_{\gamma\min}$ is

$$f_{\gamma\min} = 2f_{T\max} = 4V_{T\max}/\lambda \tag{4.4}$$

Center frequency of main lobe clutter is $f_{mL} = \frac{2V_R}{\lambda} \cdot \cos\psi$.

The maximum speed of the target relative to the ground is $V_{T\max}$, then the maximum Doppler frequency of the target relative to the ground is $\pm f_{T\max} = \pm 2V_{T\max}/\lambda$. Doppler shift of the center of main beam clutter changes between $f_{mL\min}$ and $f_{mL\max}$. Therefore, the Doppler frequency range that the target may appear is between $(f_{mL\min} - f_{T\min})$ and $(f_{mL\max} + f_{T\max})$ (Fig. 4.8).

Minimum pulse's repetition frequency $f_{\gamma\min}$ is

$$f_{\gamma\min} = 2f_{\gamma\max} + f_{mL\max} - f_{mL\min} = \frac{4V_{T\max}}{\lambda} + \frac{2V_R[\cos\psi_{\min} - \cos\psi_{\max}]}{\lambda} \tag{4.5}$$

In the formula:

- $\psi_{\min}$ is the minimum scan angle of the velocity vector of the beam deviation;
- $\psi_{\max}$ is the maximum scanning angle of the velocity vector of the beam deviation;
- $V_{T\max}$ is the maximum speed of the target relative to the ground;
- $V_R$ is the flight speed of the radar carrier;
- $\lambda$ is the working wavelength for radar.

(6) Method for eliminating the ambiguity of distance measurement

When the pulse repetition is selected according to the above method, the ambiguity of speed measurement is avoided, but the ambiguity of distance measurement comes out. That is, when the pulse repetition frequency is high, the echo generated by a pulse may not be received until several cycles. It is shown in Fig. 4.9.

Use multiple discrete pulse repetition frequencies.

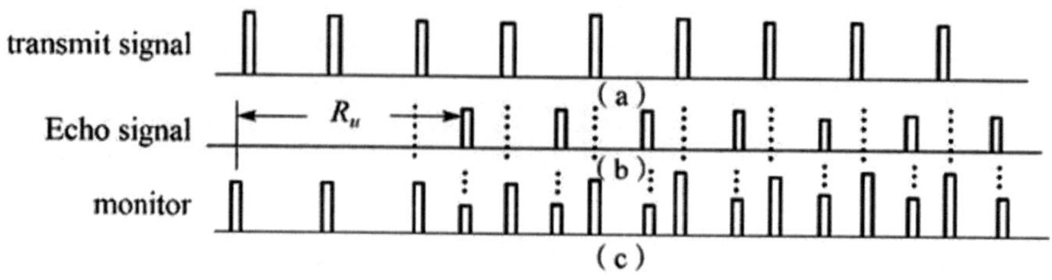


Fig. 4.9 Schematic diagram of distance measurement blurring

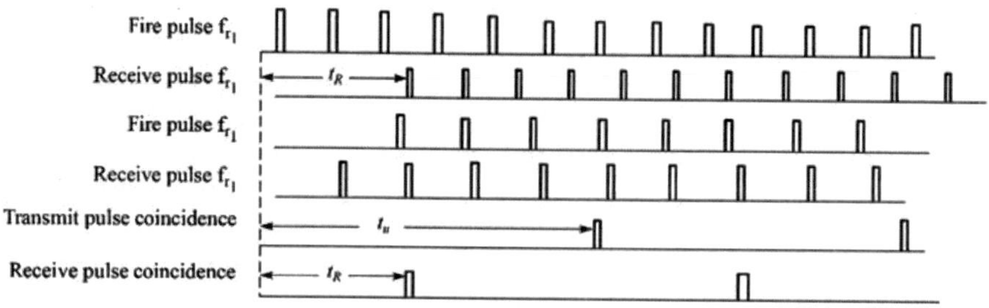


Fig. 4.10 Waveform diagram of double pulse repetition frequency to eliminate distance ambiguity

We commonly used two or three repetitions. Use two repetitions first.

If the radar transmitter works at repetition frequency $f_{\gamma 1}$ first and then works at $f_{\gamma 2}$, it works alternately like this. Through the memory comparison device, the two transmitted pulses coincide with each other, that the received pulse coincides with the received pulse. The delay $t_u$ between the transmitted pulse and the received pulse represents the true distance of the target, as shown in Fig. 4.10.

When using dual-frequency, the maximum distance $R_{u\max}$ without ambiguity can be achieved, determined by the maximum convention frequency $1/t_u$ of $f_{\gamma 1}$ and $f_{\gamma 2}$.

For example, when $f_{\gamma 1} = 10\,\text{kHz}$, $f_{\gamma 2} = 11\,\text{kHz}$, the maximum convention frequency $1/t_u = 1\,\text{kHz}$ and the maximum distance without ambiguity $R_{u\max} = 150\,\text{km}$.

For actual radar, due to multi-target and clutter jamming, the received echo cannot be directly overlapped. Still, the distance tracking gate is used to replace the echo for overlapping. With three or more repetition frequencies, the distance without ambiguity can be further extended.

In the case of multiple targets, as long as the speed or angle of these targets can be distinguished, the method of overlapping is useful. If the speed and angle are indistinguishable, overlapping may confuse.

In tracking radar while scanning, the range ambiguity must be resolved within the time the antenna beam sweeps over the target.

For tracking-while-scanning radar, the distance ambiguity must be resolved within the time that the antenna beam sweeps through the target. At this time, it can only be solved by adding equipment and fast algorithms, that is, the distance wave gate is used to fill the entire repetition period, and a parallel channel is used after each wave gate to reduce the measurement time.

***Example 4.1*** For a triple frequency system, set the distance gate width to T, repetition frequency is $t_1$ fulfilled with $m_1$ wave gates. Similarly, $t_2$ and $t_3$ are fulfilled with $m_2$ and $m_3$ wave gates. Let $m_1$, $m_2$ and $m_3$ be close prime numbers, and $m_1 > m_2 > m_3$, which $f_{\gamma 1} = \frac{1}{m_1}T < f_{\gamma 2} = \frac{1}{m_2}T < f_{\gamma 3} = \frac{1}{m_3}T$. The maximum ambiguity distance at this time is

$$R_{u\max} = \frac{C \cdot T}{2}(m_1 \cdot m_2 \cdot m_3) \tag{4.6}$$

If $T = 1\ \mu s$ and $m_1$:$m_2$:$m_3$ = 9:8:7, the maximum ambiguity distance is

$$R_{u\max} = 0.15 \times 9 \times 8 \times 7 = 75.6\,\text{km}$$

Doppler frequency without ambiguity $f_{du}$

$$f_{du} = N \cdot f_\gamma + f_d \tag{4.7}$$

The corresponding speed without ambiguity $V_u$ of the target relative to the radar

$$V_u = \frac{\lambda \cdot f_{du}}{2} \tag{4.8}$$

3. Range Measurement Performance of Pulse-Doppler radar

(1) Equation of radar distance:

$$R_0^4 = \frac{P_t \cdot G_t^2 \cdot \lambda^2 \cdot \sigma}{(4\pi)^3 \cdot R \cdot t^0 \cdot B_n \cdot F_n \cdot L} \tag{4.9}$$

In the formula,

$P_t$ is the radar pulse power;
$\sigma$ is the radar's equivalent reflection area of the target;
$R_0$ is the range with signal-to-noise ratio equal to 1;
R is the Bozeman constant;
$t^0$ is the absolute temperature;
$B_n$ is the receiver's bandwidth;
$F_n$ is the noise figure;
L is the feeder loss.

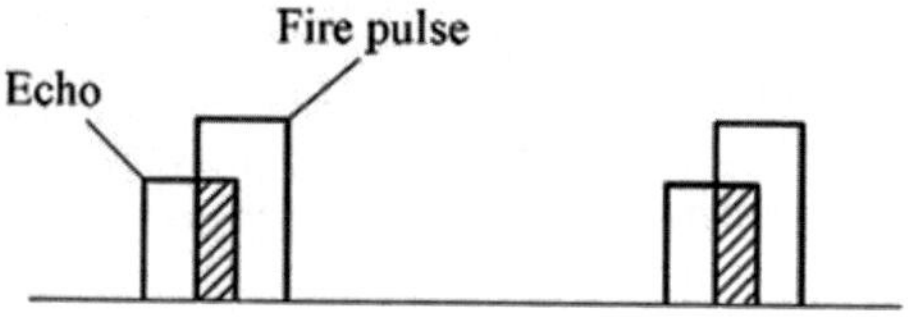


**Fig. 4.11** Schematic diagram of the shielding effect of the transmitted pulse

In Pulse-Doppler radar, because the received signal needs to pass through the selection of range gate, single-sideband filter, and Doppler narrowband filter bank, the power before the detection of the signal, clutter and noise have influenced. Therefore, the general radar range equation (4.9) must be corrected before it can be used in Pulse-Doppler radar.

In Pulse-Doppler radar, a higher repetition frequency is generally used, and the range is highly ambiguous. When the echo moves from a repetitive period to an adjacent period, the pulse "occlusion" of the echo will occur, as shown in Fig. 4.11, which is called the occlusion effect of the transmitted pulse. This effect of occlusion reduces the effective width of the echo and the effective power of the echo, thereby reducing the detection range of the radar.

(a) Bridging Effect of the range gate

When the target echo passes through the range gate, the echo may fall between adjacent range gates, that is, a bridge occurs, as shown in Fig. 4.12.

The effect of crossing also reduces the power of the echo and the radar range.

(b) Effects of single sideband filtering

The bandwidth of the single-sideband filtering is equal to the pulse repetition frequency. So, after the signal passes, only one spectral line is filtered out, and the echo power provided by the detection is greatly reduced. The bandwidth of each filter of the narrow-band filter (Doppler filter) group matches the width of the echo's spectral line and does not affect the power of an echo's spectral line.

After passing through the single-sideband filter, the power of the extracted signal center line is $d^2$ times lower than the pulse power.

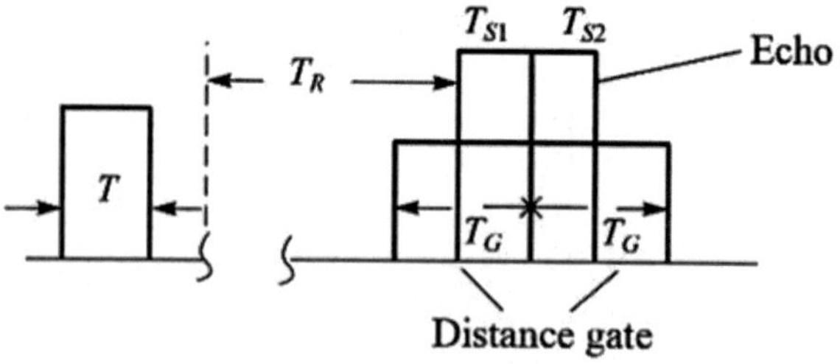


**Fig. 4.12** Crossover effect of distance gate. $T_G$ is the width of the range gate; $T_{S1}$ is the part where the echo coincides with the first crossed gate; $T_{S2}$ is the part where the echo corresponds with the second crossed gate

$$d = T/t_\gamma \tag{4.10}$$

Due to the occlusion effect, the effective width of the echo is reduced to $T_S$.

Set ds = $T_S$. Considering the single-sideband filtering, the influence factors of pulse occlusion distance are connected. After the connection and occlusion, the duty coefficient of the echo signal, then the effective echo signal detection power is reduced by a factor of $ds^2$

Besides, the noise power is also affected by the signal processing method. In ordinary radar, the noise power that equivalent to the radar input is $R \cdot t^0 \cdot B_n \cdot F_n$. In Pulse-Doppler, after gate selection, the noise power is reduced by a factor of $d_G = T_G/t_\gamma$.

$d_G$ is the duty coefficient of the range gate. After passing through the Doppler filter, the noise bandwidth $B_n$ is equal to the passband of a narrowband filter.

(2) Range equation of Pulse-Doppler radar

When the Doppler frequency of the moving target falls in the clutter-free zone, the radar's range is only limited by the system noise. The effective power of the signal before detection is converted to the radar input as

$$P_\gamma = \frac{P_t \cdot G_t^2 \cdot \lambda \cdot \sigma}{(4\pi)^3 R^4 \cdot L} ds^2 \tag{4.11}$$

The noise power before detection is referred to as the radar input

$$P_n = R \cdot t^0 \cdot B_n \cdot F_n \cdot d_G$$

Range of a Pulse-Doppler radar with a signal-to-noise ratio of 1

$$R_0^4 = \frac{P_t \cdot G_t^2 \cdot \lambda^2 \cdot \sigma}{(4\pi)^3 R \cdot t^0 \cdot B_n \cdot F_n \cdot L} \cdot \frac{ds^2}{d_G} \tag{4.12}$$

Let

$$\text{Let } D_p = (M + 1) \cdot \frac{ds^2}{d_G} \tag{4.13}$$

Is the number of adjacent equal-distance wave gates

Formula (4.13) can be written as

$$R_0^4 = \frac{P_t \cdot G_t^2 \cdot \lambda^2 \cdot \sigma \cdot D_p}{(4\pi)^3 R \cdot t^0 \cdot B_n \cdot F_n \cdot L \cdot (M + 1)} \tag{4.14}$$

Consider the relationship between pulse power and average power

$$R_0^4 = \frac{P_{av} \cdot G^2 \cdot \lambda^2 \cdot \sigma \cdot D_{av}}{(4\pi)^3 R \cdot t^0 \cdot B_n \cdot F_n \cdot L} \tag{4.15}$$

$$P_t = P_{av} \cdot t_\gamma / T_{\overline{p}} = P_{av}/d \tag{4.16}$$

$$D_{av} = ds^2/d \cdot d_G \tag{4.17}$$

(4.12), (4.14), and (4.15) are three expressions of the range when the Pulse-Doppler radar signal falls in the clutter-free zone, and the signal-to-noise ratio is 1.

Where $D_p$ and $D_{av}$ represent the coefficients of range loss of Pulse-Doppler radar when the pulsed transmit power is constant, or the average transmits power is constant.

If the Doppler frequency of the target echo of Pulse-Doppler radar falls in the cluttered area of side lobe and the signal-to-noise ratio is 1, the range is calculated as Eq. (4.18)

$$R_0^4 = \frac{P_{av} \cdot G^2 \cdot \lambda^2 \cdot \sigma \cdot D_{av}}{(4\pi)^3 \left[C + R \cdot t^0 \cdot B_n \cdot F_n\right] \cdot L} \tag{4.18}$$

In the formula, the clutter power of the narrowband filter is converted to the value of power of the radar input.

## 4.2 Analysis of Jamming Suppression Coefficient for Pulse Doppler Radar

### *4.2.1 Determination of Jamming Suppression Coefficient for Pulse-Doppler Radar*

Echo signal $P_\gamma$ received by the Pulse-Doppler radar

$$P_\gamma = \frac{P_t \cdot G_t^2 \cdot \lambda^2 \cdot \sigma}{(4\pi)^3 R^4} \tag{4.19}$$

Due to the effects of pulse occlusion, distance crossing, and Doppler filter, these effects are converted to the input of the receiver.

$$S_\gamma = P_\gamma \cdot \frac{d_G}{ds^2} \cdot B_n = \frac{P_t \cdot G_t^2 \cdot \lambda^2 \cdot \sigma}{(4\pi)^3 R^4} \cdot \frac{d_G}{ds^2} \cdot B_n \tag{4.20}$$

Radar's jamming power $P_N$ from human-made noise

$$P_N = \frac{P_j \cdot G_j}{4\pi R_j^2} \cdot \frac{\lambda^2}{4\pi} \cdot G_t(\theta) \cdot B_m \tag{4.21}$$

According to the definition of jamming suppression coefficient $K_{jN}$.

The jamming suppression coefficient is equal to the ratio of the jamming power at the input of the radar receiver to the power of the echo signal (at the minimum jamming exposure distance)

$$\begin{aligned} K_{jN} &= \frac{P_{jN}}{S_\gamma} = \frac{\frac{P_j \cdot G_j}{4\pi R_j^2} \cdot \frac{\lambda^2}{4\pi} \cdot G_t(\theta) \cdot B_m}{\frac{P_t \cdot G_t^2 \cdot \lambda \cdot \sigma}{(4\pi)R^4} \cdot \frac{d_G}{ds^2} \cdot B_n} \\ &= \frac{4\pi P_j \cdot G_j \cdot R^4 \gamma_j \cdot L_j \cdot G_t(\theta)}{P_t \cdot G_t \cdot \sigma \cdot R_j^2 \cdot L_t \cdot L_\gamma \cdot G_t} \cdot \frac{ds^2}{d_G} \cdot \frac{B_m}{B_n} \end{aligned} \tag{4.22}$$

where,

| | |
|---|---|
| $P_tG_t$ | is the equivalent power of radar radiation; |
| $\sigma$ | is the effective reflection area of target; |
| $P_jG_j$ | is the equivalent power of jamming radiation; |
| $G_t/G_t(\theta)$ | is the gain ratio of radar antenna main/side lobes; |
| $R_j$ | is the distance between jammer and radar; |
| R | is the distance between the covered target and the radar; (These parameters are the same as conventional radar). |
| $L_t, L_\gamma$ | are the feeder losses for radar antenna transmission and reception; |
| $\gamma_j$ | is the polarization loss; |
| $L_j$ | is the feeder loss of jammer; |
| $ds = T_S/t_\gamma$ | is the loss of Doppler radar pulse bridge echo occlusion; |
| $d_G = T_G/t_\gamma$ | is that after passing through the distance wave gate, the noise is reduced by a factor of $d_G$; |
| $B_n$ | is the bandwidth of the narrowband filter of Pulse-Doppler radar; |
| $B_m$ | is the spectrum width of the jammer. |

The broadband $\Delta f$ of the Pulse-Doppler radar filter bank according to Fig. 3.19

$$\Delta f = f_\gamma - 2f_{mL} = f_\gamma - 4\frac{V_R}{\lambda} \cdot \cos\psi \tag{4.23}$$

where

| | |
|---|---|
| $f_\gamma$ | is the pulse repetition frequency of the Pulse-Doppler radar; |
| $V_R$ | is the speed of the Pulse-Doppler radar carrier; |
| $\lambda$ | is the working wavelength of the radar; |
| $\psi$ | is the angle between the radar beam and the direction of flight of the carrier aircraft; |
| $f_{mL}$ | is the clutter frequency of the main lobe. |

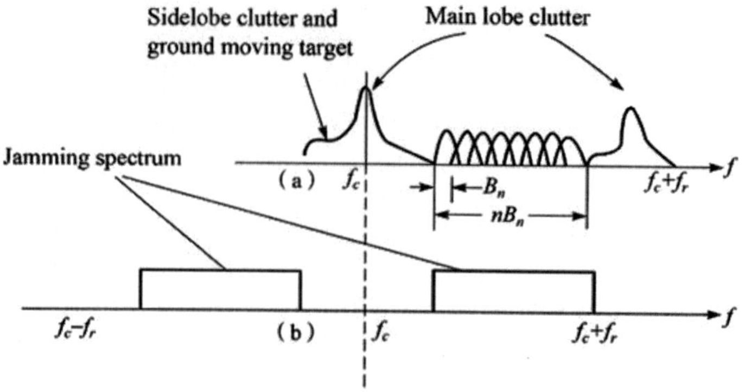


**Fig. 4.13** Correspondence between jamming spectrum and filter

The relative relationship between the jamming spectrum and the radar filter bank is shown in Fig. 4.13.

The filter bank of the radar is single-sideband, and the jamming spectrum is double-sideband. Because the jamming detection does not know whether the single-sideband of the radar is higher or lower, it jamming with the full jamming of the two sidebands.

Select $\Delta f \approx n \cdot B_n$

Select

$$B_m = 2n \cdot B_n = 2\Delta f = 2\left(f_Y - \frac{4V_R \cdot \cos\psi}{\lambda}\right) \tag{4.24}$$

For example: Speed of carrier $V_R = 300\,\text{m/s}$; $\psi = 45°$; $\lambda = 3\,\text{m}$; $f_r = 80\,\text{kHz}$.

Then $\Delta f = f_Y - 4\frac{V_R}{\lambda} \cdot \cos\psi = 80 - 27 = 53\,\text{kHz}$

$B_m = 106–159\,\text{kHz}$

If $B_n = 1\,\text{kHz}$ (Narrowband width of each filter), $T_S = 0.5\,\mu\text{s}$, $T_G = 1\,\mu\text{s}$, $ds = T_S/T_Y, dG = T_G/t_Y$,

So $\frac{ds^2}{dG} = \frac{T_S^2 \cdot f_Y^2}{T_G \cdot f_Y} = \frac{T_S^2 \cdot f_Y}{T_G} = \frac{0.25\times10^{-6}}{1} \times 80 \times 10^3 = 2 \times 10^{-2}$

$T_S$ is the pulse width after the transmission is blocked, and $T_G$ is the width of the radar's range gate.

In Formula (4.22), $K_j = \frac{4\pi P_j \cdot G_j \cdot R^4}{P_t \cdot G_t \cdot \sigma \cdot R_j^2} \cdot \frac{\gamma_j \cdot L_j}{L_t \cdot L_Y} \cdot \frac{G_t(\theta)}{G_t}$ is the jamming suppression coefficient of conventional pulse radar, generally, 3 dB is selected.

Compare Formula (4.22) with conventional pulse radar

$$K_{jN} = K_j \cdot \frac{ds^2}{dG} \cdot \frac{B_m}{B_n} = K_j \cdot \frac{T_S^2}{T_G} \cdot f_Y \cdot \frac{B_m}{B_n} \tag{4.25}$$

Substitute the parameters from "Example."

$$K_{jN} = 2 \times 2 \times 10^{-2} \times 106 = 4.2$$

Substituting Eq. (4.24) into Eq. (4.25), the jamming suppression coefficient for Pulse-Doppler radar is obtained

$$K_{jN} = K_j \cdot \frac{T_S^2}{T_G} \cdot f_\gamma \cdot 2\left(\frac{f_\gamma - \left(4\frac{V_R}{\lambda}\cos\psi\right)}{B_n}\right) \tag{4.26}$$

In the formula:

- $K_j$ is the jamming suppression coefficient required for conventional pulse radar;
- $T_S$ is the pulse width of the Pulse-Doppler radar after being blocked by the pulse;
- $T_G$ is the pulse width of Pulse-Doppler radar's range gate;
- $f_\gamma$ is the pulse repetition frequency of Pulse-Doppler radar;
- $V_R$ is the speed of the Pulse-Doppler radar carrier;
- $\psi$ is the angle between the Pulse-Doppler radar carrier and the moving direction of the target;
- $B_n$ is the bandwidth of narrowband filter of the Pulse-Doppler radar filter bank;
- $\lambda$ is the working wavelength for radar.

### 4.2.2 *Calculation of Equivalent Jamming Power for Pulse-Doppler Radar*

According to the jamming equation, $P_jG_j$ can be obtained (Jamming equivalent power)

$$P_jG_j = \frac{K_{jN} \cdot P_t \cdot G_t \cdot \sigma \cdot R_j^2}{4\pi R^4} \cdot \frac{L_t \cdot L_\gamma}{L_j \cdot \gamma_j} \cdot \frac{G_t}{G_t(\theta)} \tag{4.27}$$

where: the $\frac{P_t \cdot G_t \cdot \sigma \cdot R_j^2}{4\pi R_{\min}^4} \cdot \frac{L_t \cdot L_\gamma}{L_j \cdot \gamma_j} \cdot \frac{G_t}{G_t(\theta)}$ part is the same as the jamming equivalent power required to jamming with the conventional pulse radar. If this pulse radar is changed to Pulse-Doppler radar, the equivalent jamming power it needs is increased $K_{jN}$ times than that of the conventional pulse radar. This $K_{jN}$ is a function of $T_S, T_G, f_\gamma, V_R, \psi, B_n, \lambda$. If these parameters are known, it can be calculated. Generally speaking, as long as the jammer has a certain accurate frequency spot, the spot error is in the order of "kHz," and the jamming spectral width is close to twice the bandwidth of the Pulse-Doppler radar filter bank, which $K_{jN}$ is generally about 4–8 dB.

The modulation method of jamming is shown in Fig. 4.14.

$f_c'$ is the continuous wave signal output from the frequency-targeted receiver, $f_j$ is the video noise spectrum that spectrum width is $2f_{mL} \sim f - 2f_{mL}$.

The jamming spectrum is shown in Fig. 4.15.

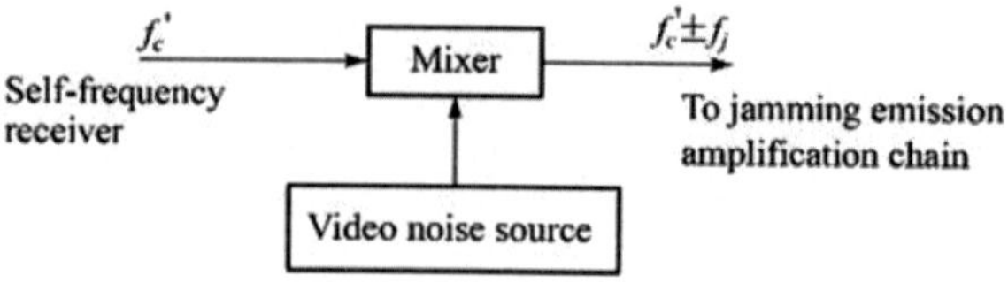


**Fig. 4.14** Schematic diagram of man-made noise jamming modulation

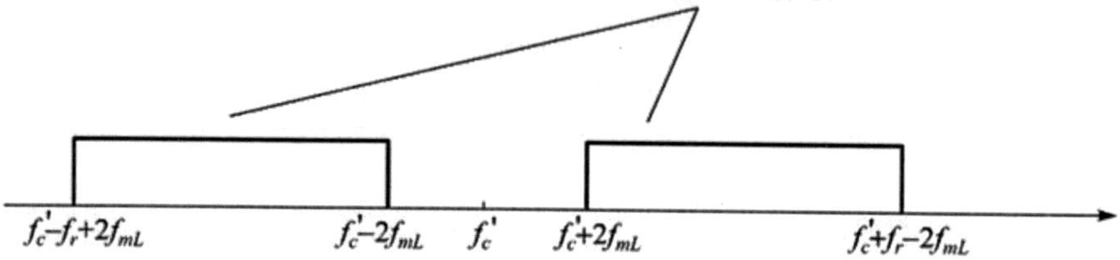


**Fig. 4.15** Schematic diagram of man-made jamming noise spectrum for Pulse-Doppler radar

## 4.2.3 Example of Jamming Suppression Coefficient of Pulse-Doppler Radar

Pulse-Doppler radar's parameters and jammer's parameters: $P_t = 100$ kW, $G_t = 33$ dB; $T_S = 0.5\,\mu s$; $T_G = 1\,\mu s$; $f_\gamma = 80\,kHz$; $B_n = 1\,kHz$; $R_{\min} = 1$ km; $P_j G_j = 1$ kW; $B_m = 50\,kHz$; $\sigma = 10\%^2$; $\gamma_j L_j / L_\gamma L_t \approx 1$.

Substitute parameters into Eq. (4.22)

$$K_{jN} = \frac{4\pi P_j \cdot G_j \cdot R_{\min}^4}{P_t \cdot G_t \cdot \sigma \cdot R_j^2} \cdot \frac{L_j \cdot \gamma_j}{L_t \cdot L_\gamma} \cdot \frac{G_t(\theta)}{G_t} \cdot \frac{ds^2}{dG} \cdot \frac{B_m}{B_n}$$

Jamming test of radar and jammer for main-lobe to main-lobe, $\frac{G_t(\theta)}{G_t} = 1$, $R_{\min} = R_j$

then

$$\begin{aligned} K_{jN} &= \frac{4\pi P_j \cdot G_j \cdot R_{\min}^2}{P_t \cdot G_t \cdot \sigma} \cdot \frac{T_S^2 \cdot f_r}{T_G} \cdot \frac{B_m}{B_n} \\ &= \frac{12.6 \times 10^3 \times 10^6}{10^5 \times 2 \times 10^3 \times 10} \times 0.25 \times 10^{-6} \times 8 \times 10^4 \times 50 \\ &= \frac{12.6}{2} \times 2 \times 10^{-2} \times 50 = 63\ (8\,\text{dB}) \end{aligned}$$

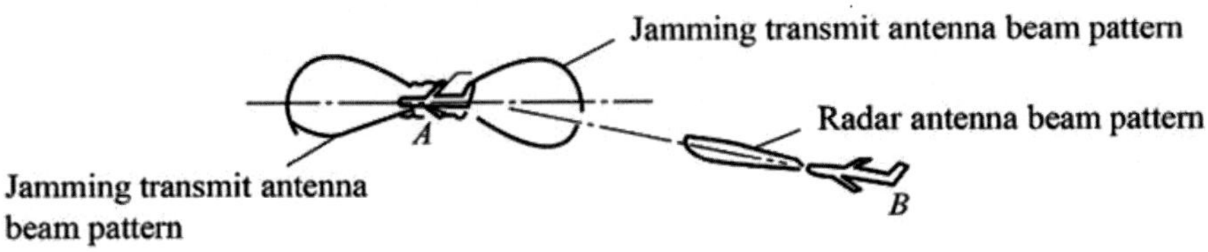


**Fig. 4.16** Schematic diagram of air-to-air jamming self-defense jamming exposure area. A is the target being attacked, and a self-defense jammer is installed on the aircraft; B is an attack aircraft, equipped with Pulse-Doppler radar

## 4.3 Jamming Exposure Area for Pulse-Doppler Radar

If the attack target is equipped with a self-defense jammer, and the attack aircraft is equipped with a Pulse-Doppler radar, it is shown in Fig. 4.16.

Assuming that the peak power of Pulse-Doppler radar is 60 kW; pulse width is 1 μs; antenna gain is 30 dB; relative motion speed of the two machines is 100 m/s;operating frequency is in X band; pulse occlusion width is 0.5 μs; The width of the range gate is 1 μs; the bandwidth of the narrow-band filter is 1 kHz; the angle between the Pulse-Doppler radar and the direction of movement of the carrier is $\phi = 30°$; the equivalent power of the jammer is 500 W; the bandwidth of the filter bank is 50 kHz, $\sigma = 2\,\text{m}^2$.

Substitute the parameters into the Formula (4.26) to calculate the jamming suppression coefficient

$$\begin{aligned} K_{jN} &= K_j \cdot \frac{T_S^2}{T_G} \cdot f_Y \cdot 2\left(\frac{f_Y - (4\frac{V_R}{\lambda}\cos\psi)}{B_n}\right) \\ &= 2 \times 0.25 \times 10^{-6} \times 60 \times 10^3 \times 2 \times \left(\frac{60 \times 10^3 - \frac{4\times 100}{3\times 10^{-2}\times 0.86}}{10^3}\right) = 2.9 \end{aligned}$$

Simplify Formula (4.27) into the self-defense jamming equation

$$P_j G_j = \frac{K_{jN} \cdot P_t \cdot G_t \cdot \sigma}{4\pi R_{\min}^2} \tag{4.28}$$

$$\begin{aligned} R &\approx \left[\frac{K_{jN} \cdot P_t \cdot G_t \cdot \sigma}{4\pi P_j G_j}\right]^{1/2} \\ &= \left[\frac{2.9 \times 60 \times 10^3 \times 10^3 \times 2}{12.6 \times 500}\right] \\ &= 240\,(\text{m}) \end{aligned}$$

The jamming exposure area of carrier A is shown in Fig. 4.17.

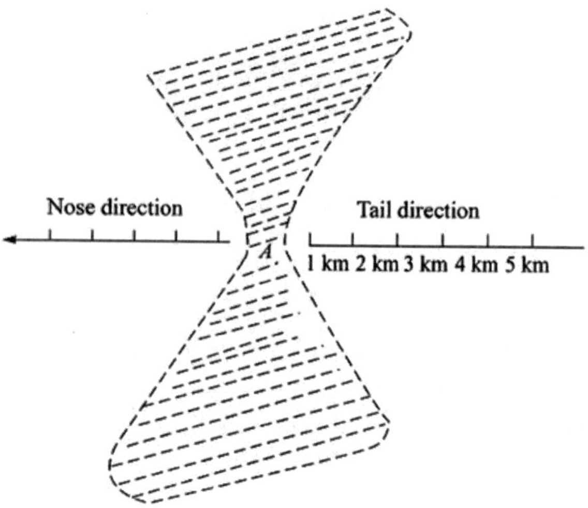


**Fig. 4.17** Schematic diagram of the jamming exposure area of the airborne self-defense jammer

A is the aircraft carrying the jammer; the area of the horizontal line is the jamming exposure area; the outside of the area of the horizontal line is the effective jamming area.

From the above discussion we could know that:

1. Pulse-Doppler radar is mainly designed to find low-altitude moving targets, according to the needs of tactics, not intended to resist human jamming.
2. Pulse-Doppler radar filters out main-lobe clutter, sidelobe clutter, and clutter of ground target through single-sideband filtering, filter bank, and narrow-band filtering, etc., which greatly improves the radar's ability to detect low-altitude moving targets.
3. Single-sideband filtering selects only one spectral line within the pulse, which greatly reduces the energy of the signal. Also, the shielding and bridging of the transmitted pulse cause a great loss of echo signal's energy. However, due to the narrowband filtering and range gates in the receiver, the noise of the receiver is greatly reduced. The overall signal-to-noise ratio of the receiver is improved, and the original detection distance of the radar is maintained.
4. For human-made spot jamming of noise, as long as the aiming frequency accuracy is maintained at the kHz level and the width of the jamming spectrum is on the order of tens of kHz to 100 kHz, the noise jamming suppression coefficient required for it is generally 4–8 dB.

In summary, the jamming technology to Pulse-Doppler radar can be realized, and the required jamming equivalent power is maintained at the same level as the jamming conventional radar.

# Chapter 5
# Unified Jamming Equation and "One-to-Many Systems" Jammer

In the previous chapters, synthetic aperture radar, pulse compression radar, and Pulse-Doppler radar were discussed separately, and the jamming equations and suppression coefficients required for these new radar systems were derived. In these radars, although their jamming equations are expressed in different ways, they have a close internal relationship. If we can find out its internal connection, we can derive the unified radar jamming equation by normalization, which can simplify the design process of the jammer, and then realize the One-to-many systems, so that the radar jammer can be multi-functional. Because radar has realized the integration of vigilance, multi-target tracking, rapid frequency anti-jamming, moving target tracking, and imaging detection. If a jammer can only effectively jamming with a certain system of radar, this will not only cause waste of the jammer's resources but also likely to delay the fighter and cause irreparable losses. Therefore, we need to actively analyze the characteristics and commonalities of jamming to various systems of radar. Find the unified equation of jamming, and make the theoretical preparation for the development of "one to many systems" multifunction jammer.

## 5.1 Unified Jamming Equation of Radar Jammer

In Chap. 2, we derived the jamming equation for synthetic aperture radar (2.20)

$$K_{j\Sigma\max} = \frac{4\pi P_j \cdot G_j \cdot R_{\min}^4}{P_t \cdot G_t \cdot \sigma_i \cdot R_j^2} \cdot \frac{G_t(\theta)}{G_t} \cdot \frac{L_j \cdot \gamma_j \cdot K_f}{L_d} \tag{5.1}$$

When expressing the jamming equation in terms of jamming equivalent power, Eq. (5.1) can also be written as

$$P_j \cdot G_j = \frac{K_{j\Sigma\max} \cdot P_t \cdot G_t \cdot \sigma_i \cdot R_j^2}{4\pi \cdot R_{\min}^4} \cdot \frac{G_t}{G_t(\theta)} \cdot \frac{L_d}{L_j \cdot \gamma_j \cdot K_f} \tag{5.2}$$



X. Zhang et al., *Theory to Countermeasures Against New Radars*,
https://doi.org/10.1007/978-981-16-6715-2_5

where $\sigma_i$ and the effective reflection area of the target refers to the effective reflection area of the target in the resolution area of the synthetic aperture radar in the radar antenna beam.

$P_t$ is the peak transmitter power of the radar. Other parameters have the same meaning as the conventional pulse radar

$$K_{j\Sigma\max} = \eta_d \cdot \eta_b$$

Equations of Conventional Pulse Radar:

$$P_\gamma = \frac{P_a \cdot G_a^2 \cdot \lambda^2 \cdot \sigma_a \cdot K_1}{(4\pi)^3 R_a^4} \tag{5.3}$$

$$R_a^4 = \frac{P_a \cdot G_a^2 \cdot \lambda^2 \cdot \sigma_a \cdot K_1}{(4\pi)^3 \cdot P_\gamma} \tag{5.4}$$

Equation of Synthetic Aperture Radar:

$$P_S = \frac{P_t \cdot G_t^2 \cdot \lambda^2 \cdot \sigma_i \cdot K_1}{(4\pi)^3 R_t^4} \tag{5.5}$$

$$R_t^4 = \frac{P_t \cdot G_t^2 \cdot \lambda^2 \cdot \sigma_i \cdot K_1}{(4\pi)^3 P_s} \tag{5.6}$$

Assuming that the maximum detection distance of conventional pulse radar and synthetic aperture radar is equal, that is, Formulas (5.6) and (5.4) are equal.

$$R_{a\max}^4 = \frac{P_a \cdot G_a^2 \cdot \lambda^2 \cdot \sigma_a \cdot K_1}{(4\pi)^3 \cdot P_{\gamma\min}}$$

$$R_{t\max}^4 = \frac{P_t \cdot G_t^2 \cdot \lambda^2 \cdot \sigma_i \cdot K_1}{(4\pi)^3 \cdot P_{S\min}}$$

set $R_{a\max} = R_{t\max}$:

$$\frac{P_a \cdot G_a^2 \cdot \lambda^2 \cdot \sigma_a \cdot K_1}{(4\pi)^3 \cdot P_{\gamma\min}} = \frac{P_{ta} \cdot G_t^2 \cdot \lambda^2 \cdot \sigma_i \cdot K_1}{(4\pi)^3 \cdot P_{S\min}} \tag{5.7}$$

If $G_a = G_t$; $P_{\gamma\min} = K \cdot T \cdot \Delta F_a \cdot N$, $P_{S\min} = K \cdot T \cdot \Delta F_t \cdot N/\eta_d \cdot \eta_b$.
In the formula:

| | |
|---|---|
| K | The Bozeman constant; |
| T | The absolute temperature; |
| $\Delta F_a$ | Bandwidth of the conventional pulse radar receiver; |
| N | Noise figure; |
| $\Delta F_t$ | Bandwidth of the synthetic aperture radar receiver; |

$G_a$ and $G_t$ Radar antenna gains;
$\sigma_a$ The equivalent reflection area of the target in the main beam of the radar antenna;
$\sigma_i$ The equivalent reflection area of the target in the resolution area of the synthetic aperture radar;
$K_1$ Radar feeder loss

Substitute the parameters into the (5.7), and simplify:

$$\frac{P_a \cdot \sigma_a}{\Delta F_a} = \frac{P_t \cdot \sigma_i \cdot \eta_d \cdot \eta_b}{\Delta F_t} \tag{5.8}$$

$$P_t = \frac{P_a \cdot \sigma_a \cdot \Delta F_t}{\Delta F_a \cdot \sigma_i \cdot \eta_d \cdot \eta_b} \tag{5.9}$$

$$\Delta F_a = \frac{1}{\tau};\ \Delta F_t = \frac{1}{\Delta\tau};\ \frac{\Delta F_t}{\Delta F_a} = \frac{\tau}{\Delta\tau}$$

$\tau/\Delta\tau$ is the radar range pulse compression factor, so that $\frac{\tau}{\Delta\tau} = \eta_d$
Plug into (5.9):

$$P_t = \frac{P_a \cdot \sigma_a \cdot \eta_d}{\sigma_i \cdot \eta_d \cdot \eta_b} = \frac{P_a \cdot \sigma_a}{\sigma_i \cdot \eta_b} \tag{5.10}$$

Substitute $P_t$ into (5.2):

$$\begin{aligned} P_j \cdot G_j &= \frac{K_{j\sum\max} \cdot P_t \cdot G_t \cdot \sigma_i \cdot R_j^2}{(4\pi)^3 R_{\min}^4} \cdot \frac{G_t}{G_t(\theta)} \cdot \frac{L_d}{L_j \cdot \gamma_j \cdot K_f} \\ &= \frac{\eta_d \cdot \eta_b \cdot \left(\frac{P_a \cdot \sigma_a}{\sigma_i \cdot \eta_b}\right) \cdot G_t \cdot \sigma_i \cdot R_j^2}{(4\pi)^3 R_{\min}^4} \cdot \frac{G_t}{G_t(\theta)} \cdot \frac{L_d}{L_j \cdot \gamma_j \cdot K_f} \end{aligned}$$

So:

$$P_j \cdot G_j = \frac{n_d \cdot P_a \cdot G_t \cdot \sigma_a \cdot R_j^2}{(4\pi)^3 R_{\min}^4} \cdot \frac{G_t}{G_t(\theta)} \cdot \frac{L_d}{L_j \cdot \gamma_j \cdot K_f} \tag{5.11}$$

In the formula:

$\eta_d$ The range pulse compression factor of the radar;
$\sigma_a$ The effective reflection area of the target in the irradiation area of the main radar lobe;
$P_a$ The peak power of the radar transmitter;
$R_{\min}$ The radar jamming exposure distance;
$G_t/G_t(\theta)$ The gain ratio of antenna main lobe/sidelobe;
$L_d$ Radar feeder loss (including rain and fog loss);

$L_j$ Feeder loss of the jammer (including rain and fog loss);
$K_f$ The ratio of the bandwidth of the radar receiver to the width of the jamming spectrum.

Formula (5.11) is the same as the pulse compression radar jamming equation.

After calculating and simplifying the jamming equation for SAR, it becomes the same formula as the pulse compression radar. Therefore, the jamming equation for the synthetic aperture radar can be unified into the jamming of the conventional pulse radar, as long as the required jamming suppression coefficient is calculated for radars of different systems. This unified jamming equation can be written as:

$$P_j \cdot G_j = \frac{K_{j\Sigma i} \cdot P_t \cdot G_t \cdot \sigma \cdot R_j^2}{4\pi R^4} \cdot \frac{G_t}{G_t(\theta)} \cdot \frac{L_d}{L_j \cdot \gamma_j \cdot K_f} \tag{5.12}$$

We call (5.12) the unified radar jamming equation. k is the jamming suppression coefficient in the unified jamming equation, and other parameters are the same as pulse radar. In the previous chapters, the radars of different systems have been analyzed, and different jamming suppression numbers are required. The existing radar system can be divided into four categories according to the required jamming suppression coefficient. For conventional pulse radar, rapid frequency radar, multi-carrier frequency radar, frequency diversity radar, etc., their jamming suppression coefficient can be taken as 3 dB.

### 5.1.1 *Calculation of Jamming Suppression Coefficient for Pulse Compression Radar*

There are four methods of calculation for pulse compression radar's jamming suppression coefficient:

1. If this pulse compression radar is based on the original conventional pulse radar, the pulse width is widened, and the peak power and repetition frequency are changed, the jamming suppression coefficient is $K_{d1}$. The derivation process is in (3.22).

$$K_{d1} = \frac{P_{t2}}{P_{t1}} \cdot \frac{\tau_1}{\tau_0 \cdot K_0} \cdot \left(\frac{F_2}{F_1}\right)^{1/3} \tag{5.13}$$

In the formula:

$P_{t2}$ the radar peak power after changing to pulse compression radar
$P_{t1}$ Peak power of the original radar
$\tau_1$ Transmitted pulse width after changing to pulse pressure
$K_0$ After switching to compressed radar, the transmitted pulse width and waveform loss factor are usually taken $K_0$ as 1.5.

If changed to pulse compression radar, the peak power and repetition frequency have not changed:

$$P_{t1} = P_{t2}; F_1 = F_2$$

So that:

$$K_{d1} = \frac{\tau_1}{\tau_0 \cdot K_0} = \frac{\tau_1/\tau_2}{\tau_0/\tau_2} \cdot \frac{1}{K_0} = \frac{D}{D_0 \cdot K_0}$$

In the formula:

- $\tau_1$ Transmitted pulse width
- $\tau_2$ Pulse width after compression
- D Pulse compression factor $D_0$—After the original radar is changed to a pulse compression radar, the loss factor caused by the receiver bandwidth change.

2. If only the change of maximum detection distance is known after the original radar changed to pulse compression radar (The original maximum detection distance of the same target is $R_{1\max}$. After changing to a pulse compression radar, its maximum detection distance is $R_{2\max}$), so that:

$$K_{d2} = \left(\frac{R_{2\max}}{R_{1\max}}\right)^4 \tag{5.14}$$

3. If only the transmitted pulse width of the pulse compression radar and the compressed pulse width is known, then:

$$K_{d3} = \left(\frac{\tau}{\Delta\tau}\right)^{0.6} \text{ or } K_{d3} = D^{0.6} \tag{5.15}$$

4. If we only know that the pulse compression radar has improved the sensitivity of the original radar n times, then $K_{d4}$ takes n value.

Take an example to verify the correctness of these four algorithms:

***Example 5.1*** The peak power of a conventional pulse radar is 1 MW; the emission pulse width is 2 μs; the detection range for a 2 $m^2$ target is 210 km. After changing to pulse compression radar, $R_{max}$ is 370 km; the average power is 900 W; peak power is 800 kW; transmit pulse width is 43 μs; pulse width after compression is 0.3 μs; sensitivity is improved by 12 dB; other parameters remain unchanged.

(a) Substitute parameters into (5.13)

$$K_{d1} = \frac{1}{0.8} \times \frac{43}{2 \times 1.5} = 12\,\text{dB}$$

(b) Substitute parameters into (5.14)

$$K_{d2} = \left(\frac{370}{210}\right)^4 = 10.4\,\text{dB}$$

(c) Substitute parameters into (5.15)

$$K_{d3} = \left(\frac{40}{0.3}\right)^{0.6} = 12.6\,\text{dB}$$

(d) The measured sensitivity is improved by 10–12 dB after changed to pulse compression radar. That is, $K_{d4}$ is 10–12 dB.
(e) After the actual flight test, the measured jamming suppression coefficient was 10.4 dB. It can be seen that the measured value of the jamming suppression coefficient of the pulse compression radar is consistent with the theoretically calculated value. That is, the value of $K_{d1}$, $K_{d2}$, $K_{d3}$, and $K_{d4}$ are the same, which are less than D.

The relationship with suppression coefficient under different repetition frequency and different speed.

### 5.1.2 *Calculation of Jamming Suppression Coefficient of Pulse-Doppler Radar*

In Chap. 4, the jamming suppression coefficient equation of Pulse-Doppler radar has been derived. It is a complex equation that relates to Pulse-Doppler radar occlusion pulse width, range gate, pulse width, pulse repetition frequency, speed of radar carrier, the angle between the carrier and moving target, Doppler filter bandwidth, radar working wavelength, etc.

In order for readers to have a basic concept of the value of the jamming suppression coefficient of pulse Doppler radar, for common pulse Doppler radar, some values are assumed and calculated. Calculate the relationship between the jamming suppression coefficient of pulse Doppler radar and the radar repetition frequency, as shown in Table 5.1.

**Table 5.1** The relationship between the jamming suppression coefficient and the different repetition frequency and speed

| $K_{jN}$ / $V_R$ (m/s) \ $f_r$(kHz) | 50 | 80 | 100 | 150 | 200 |
|---|---|---|---|---|---|
| 100 | 1.5 | 4.2 | 6.8 | 15.8 | 28 |
| 200 | 1.17 | 3.6 | 6.1 | 14.7 | 27.2 |
| 300 | 0.83 | 3.1 | 5.4 | 13.7 | 25.8 |

It can be seen from Table 5.1 that when the repetition frequency of pulse Doppler radar is in the range of 80–100 kHz, its jamming suppression coefficient is in the range of 5–10 dB. If the repetition frequency is in the range of 100–200 kHz, the jamming suppression coefficient is in the range of 12–14 dB. It can provide designers with a concept and range of the design value of pulse Doppler radar jamming suppression coefficient.

***Example 5.2*** Calculate the radar jamming suppression coefficient according to Formula (4.27):

$$K_{jN} = K_j \cdot \frac{T_S^2}{T_G} \cdot f_\gamma \cdot 2\left(\frac{f_\gamma - \frac{V_R}{\lambda}\cos\psi}{B_n}\right) \tag{4.27}$$

In the formula:

- $K_j$ Jamming suppression coefficient required by conventional pulse compression radar, $K_j = 1.5$ dB.
- $T_S$ The pulse width of pulse Doppler radar after occlusion, $T_S = 0.5\ \mu s$.
- $T_G$ Pulse Doppler radar range gate pulse width, $T_G = 1\ \mu s$.
- $f_\gamma$ Pulse Repetition Frequency of Pulse Doppler Radar, $f_\gamma = 50-200$ kHz.
- $V_R$ Relative velocity of pulse Doppler carrier, $V_R = 100$ m/s, 200 m/s, 300 m/s
- $B_n$ Narrowband filter bandwidth of the filter bank, $B_n = 1$ kHz
- $\psi$ 45°
- $\lambda$ Wavelength, $\lambda = 3 \times 10^{-2}$ m

The relationship between the jamming suppression coefficient and the repetition frequency of the pulse Doppler radar is shown in Fig. 5.1.

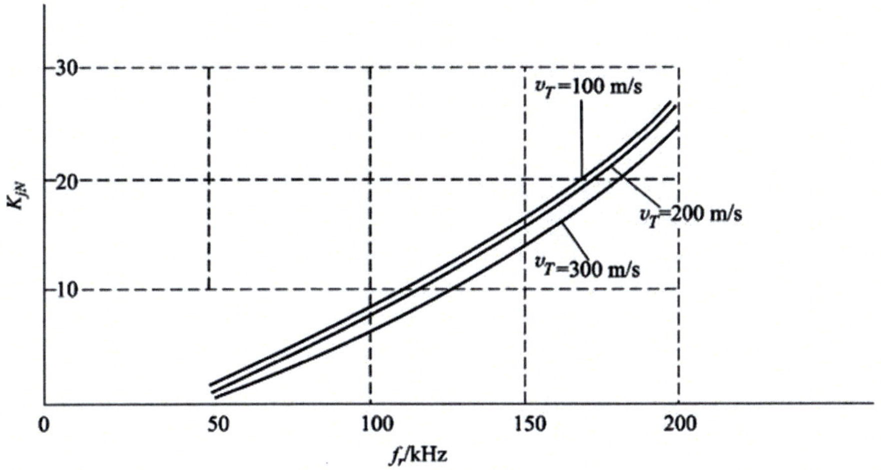


**Fig. 5.1** Value table of jamming suppression coefficient for Pulse-Doppler radar

When the Doppler frequency is below 50 kHz, take the value according to $K_{jN} \approx K_j$.

### 5.1.3 Calculation of Equivalent Power of Jamming Synthetic Aperture Radar

When calculating the equivalent power of synthetic aperture radar jamming, the first problem encountered is how to determine the jamming suppression coefficient. According to the derived synthetic aperture radar jamming Eq. (2.17):

$$K_{j\text{Lmax}} = K_j \cdot \eta_a \cdot \eta_b$$

In the formula:

$K_j$ The jamming suppression coefficient of conventional pulse radar. Generally, $K_j = 2\,\text{dB}$.
$\eta_a$ Gain coefficient of distance compression for synthetic aperture radar
$\eta_b$ Gain coefficient of azimuth compression for synthetic aperture radar.

In the unified radar jamming Eq. (5.11), the jamming suppression coefficient for synthetic aperture radar.

$$K_{j\Sigma i} = \eta_d$$

In the formula:

$\eta_d$ Gain coefficient of range compression for synthetic aperture radar, which is the same as $\eta_a$.

In (2.17) and (5.11), there is a parameter that must be clarified; otherwise, a large calculation error will occur. In Formula (2.17), $\sigma_i$ is the effective reflection area of the target. It refers to the effective reflection area of the target in a split area of the synthetic aperture radar. In Formula (5.11), $\sigma_a$ refers to the target equivalent reflection area of the internal illuminated area of half-power spot beam of the synthetic aperture. After clarifying these relationships, the jamming equivalent power calculated by (2.17) or (5.11) is the same.

***Example 5.3*** Known SAR parameters: $P_t = 300\,\text{W}$; $G_t = 17\,\text{dB}$; $\theta_{0.5} = 8°$; Transmitted pulse width $\tau = 33\,\mu\text{s}$; $\Delta\tau = 16\,\text{ns}$; $H = 7200\,\text{m}$; $R = 20\,\text{km}$; $V = 150\,\text{m/s}$; Angle reflectors represent point targets $\sigma_i = 20\,\text{m}^2$; Known Jammer parameters: $G_j = 24\,\text{dB}$; In the main lobe area of the radar, the power of the image cannot be acquired normally by the radar $P_j = 0.3$.

Using Eqs. (2.17) and (5.11), calculate the equivalent jamming power required to jamming with the main lobe of the SAR:

1. Calculation according to (2.17) formula:

$$P_j \cdot G_j = \frac{K_{j\Sigma\max} \cdot P_t \cdot G_t \cdot \sigma_i \cdot R_j^2}{4\pi R_{\min}^4} \cdot \frac{G_t}{G_t(\theta)} \cdot \frac{L_d}{L_j \cdot \gamma_j \cdot K_f}$$

where:

$$G_t/G_t(\theta) = 1;\ \frac{L_d}{L_j \cdot \gamma_j \cdot K_f} = 1;\ R_j = R_{\min} = 20\,\text{km};$$

Then, $P_j \cdot G_j = \frac{K_{j\Sigma\max} \cdot P_t \cdot G_t \cdot \sigma_i}{4\pi R_{\min}^2}$.

$$K_{j\Sigma\max} = \eta_a \cdot \eta_b;$$

Distance compression gain coefficient: $\eta_a = \frac{33\times10^{-6}}{16\times10^{-12}} = 33\,\text{dB}$
Azimuth compression gain factor: $\eta_b = f_\gamma \cdot T_\gamma = f_\gamma \cdot \theta_{0.5} \cdot \frac{R}{V} = 40.3\,\text{dB}$

$$\text{K}_\text{j} = 1;$$

Then: $K_{j\Sigma\max} = (1 + 33 + 403)\,\text{dB} = 743\,\text{dB}$
In practical work, synthetic aperture radar has nonlinear loss and bandwidth loss.

Experience has shown that the actual radar gain coefficient $K_{j\Sigma 0} = \left(K_{j\Sigma\max}\right)^{0.8}$

Calculate the gain coefficient of synthetic aperture radar:

$$K_{j\Sigma 1} = (743)^{0.8} = 59.4\,\text{dB}$$

Measured Gain coefficient of synthetic aperture radar:

$$K_{j\Sigma 1} = (743)^{0.8} = 59.4\,\text{dB}$$

Measured Gain coefficient of synthetic aperture radar $K_{j\Sigma 0}$:

$$K_{j\Sigma 1} = \frac{4\pi P_j \cdot G_j \cdot R_{\min}^2}{P_t \cdot G_t \cdot \sigma_i} = \frac{12.6 \times 0.3 \times 251 \times \left(20 \times 10^3\right)^2}{300 \times 50 \times 20} = 60.7\,\text{dB}$$

According to the calculation results and actual measurement results, the jamming suppression coefficient of synthetic aperture radar $K_{j\Sigma 0} \approx K_{j\Sigma 1}$. The equivalent jamming power is calculated according to (2.17)

2. Measured value: $P_{j1} G_{j1} = 251 \times 03 = 753\,(\text{W})$
3. According to the jamming equation of synthetic aperture radar:

$$P_{j2}G_{j2} = \frac{(K_{j\Sigma\max})^{0.8} \cdot P_t \cdot G_t \cdot \sigma_i}{4\pi R_{\min}^2} = (594 + 42 + 13)\,\text{dB} - 98\,\text{dB}$$
$$= 164\,\text{dBW}(437\,\text{W})$$

4. Calculation according to the unified equation:

$$P_{j2}G_{j2} = \frac{(K_{j\Sigma\max})^{0.8} \cdot P_t \cdot G_t \cdot \sigma_i}{4\pi R_{\min}^2} = (594 + 42 + 13)\,\text{dB} - 98\,\text{dB}$$
$$= 164\,\text{dBW}(437\,\text{W})$$

where: $n_d = n_a = 33$ dB; $(n_d)^{0.8} = 26.4$ dB

$\sigma$—The equivalent reflection area illuminated by the radar main lobe, the empirical value is 47 dB.

Then: $P_{j3}G_{j3} = (264 + 42 + 47)$ dB − 98 dB = 174 dBW(55 W).

The equivalent power error range obtained by the three methods is within ±1.4 dB.

Calculated $P_{j1}G_{j1}$ of the measured jamming's equivalent power is 75.3 W;

Calculated $P_{j2}G_{j2}$ of the synthetic aperture radar jamming's equivalent power is 43.7 W;

Calculate $P_{j3}G_{j3}$ of the unified equation jamming's equivalent power is 55 W.

It can be seen that it is credible to use the unified equation to calculate the jamming equivalent power required by several existing radar systems.

## 5.2 "One-to-Many Systems" Radar Jammer

The One-to-many systems mentioned here means that one jammer can effectively jam with the radars of various systems. Ordinary jammers are highly targeted. The radar systems are diverse, and the operating frequency is also very different. If a jammer can only effectively jamming with a certain radar, it will inevitably cause a wide variety of jammers, wasting resources, and may bungle the chance of winning a battle, which caused irreparable losses. We strive to develop a radar that can function against multiple systems. Incorporate integrated jammers for air-to-air reconnaissance, jamming, and positioning to meet the increasingly urgent need for electronic countermeasures. One machine to many systems, this possibility exists. First, the uniform equation of radar jamming has been derived in Sect. 5.1. According to this unified equation, no matter what kind of radar system, their maximum detection distance to the same target is the same, so the equivalent power to suppress jamming required by them is also the same.

To achieve effective jamming to radars of various systems in the unified radar jamming equation, it must satisfy six elements.

### 5.2.1 Essential Elements of 'One-to-Many Systems'

1. Frequency targeting

   For radars of various systems, the accuracy requirements of frequency targeting are different. For conventional pulse radar, pulse compression radar, and SAR, the aiming frequency error should be on the order of megahertz. In contrast, the Pulse-Doppler radar's aiming frequency error should be less than the kilohertz order. Therefore, the 'One-to-many systems' jammers need to adapt to changes with different radar systems.
2. Continuous-time of the jamming signal

   The jamming signal is required to be continuous noise suppression jamming. But in the jammer, it is also needed to be able to detect the radar signal frequency changes after the radar jamming at any time. That is, it is necessary to monitor the change of radar signal frequency. In this way, the jamming signal must have a certain gap to observe the shift of the radar signal. In order to solve the transmission and reception isolation of the jammer, there is also a separation of transmission and reception time. According to long-term verification, the time separation ratio between the jamming signal and the received signal should be greater than 9:1, and the same reception time should be maintained around one pulse repetition period of the jammed radar.
3. The direction of the jammer antenna

   In the airspace, the antenna of the jamming transmitter should be aligned with the direction of the jamming radar, and the jamming resources should be used effectively.
4. Jamming spectrum width

   The bandwidth of the radar receiver should be maintained to ensure that every spectrum of the radar signal can be suppressed by jamming. For synthetic aperture radar (SAR), $\Delta f/\Delta F = 0.8$; For other radars, $\Delta f/\Delta F = 0.3–0.5$.
5. Jamming suppression factor $K_{j\Sigma i}$

   There are different requirements for different systems. For the digital transmission system, the jamming suppression coefficient requires only a few tenths; for conventional pulse radars, it is required to be greater than 1; for pulse compression radars, Pulse-Doppler radars, and synthetic aperture radars, the values can be calculated according to special calculation equations.
6. Polarization

   Radar often uses different polarizations to obtain useful information, and also uses polarizations for anti-jamming. Therefore, there must be anti-jamming measures against radar polarization in jammers. Orthogonal polarization jamming is commonly used in jammers. It can randomly generate all kinds of polarization, which makes the radar's anti-jamming of polarization change fail.

   When a 'One-to-many systems' jammer meets these six conditions at the same time, it can effectively jam with any system of radar or other electronic equipment.

### 5.2.2 Composition of 'One-to-Many Systems' Jammer

Description of 'one to many system' jammer's components. Principle block diagram of one-to-many radar jammer is shown Fig. 5.2.

1. Receive tracking antenna
   Four beams are composed of an antenna reflection surface and four symmetrical feeds, two of which are azimuth beams and two elevation beams to form a tracking beam. The four-synthetic beam uses the received sight frequency and signals to measure the parameters.
2. Transmitting antenna
   Orthogonal polarization feed is used. The jamming signal is input by two channels, and the polarization input of the two channels is changed into a random polarization jamming signal, which can jam with the multi-polarization work of the radar and the anti-jamming radar with variable polarization.
3. Frequency spot and frequency guidance subsystem
   The sub-system includes an accurate frequency spot, and the accurate frequency spot is on the order of kHz. Generally, the frequency spot is on the order of MHz. Accurate frequency spot is mainly used to jamming with Pulse-Doppler radar. General aiming frequency is mainly used to conventional pulse radar, pulse compression radar, synthetic aperture radar, etc. These two frequency spot methods should use the signal characteristics of the jammed radar to distinguish adaptive, automatic conversion, difference frequency spot, deceptive jamming sighting, and its adaptive control circuit.
4. Accurate frequency spot
   The radiofrequency storage technology is used to target the frequency of the jammed signal accurately. Mainly used to jamming with Pulse-Doppler radar, and use high repetition rate of Pulse-Doppler radar to make the aiming frequency system convert.

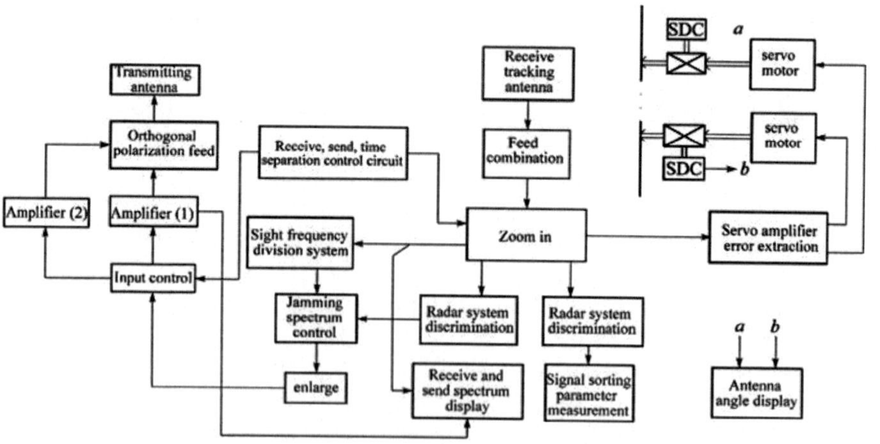


**Fig. 5.2** The principle block diagram of one-to-many radar jammer

5. Normal frequency spot and adaptive frequency spot conversion

In addition to Pulse-Doppler radar, most other radar systems use general aiming frequency guidance, such as conventional pulse radar, pulse compression radar, synthetic aperture radar, and agile frequency radar. They aim for frequency errors in the order of MHz and use channelized frequency guidance. When the frequency spectrum of the signal is relatively narrow, it uses one or two signals to receive and connect the narrow jamming spectrum. When the signal spectrum is wide, such as a synthetic aperture radar, it has multiple channels connected. It outputs a wide frequency spectrum, so that $\Delta f/\Delta F$ is kept in the vicinity of 0.5–0.8, which can not only effectively use jamming resources, but also not be easily resisted by radar.

(a) Difference frequency spot

The difference frequency spot mainly jamming with satellite communications. The frequency it receives is not the same as the frequency it transmits. Therefore, a mixer must be added when receiving the frequency of the satellite communication signal. When adjusting the local oscillator frequency to make the received signal noisy again, the local oscillator frequency $f_A$ is the frequency of the signal transmitted from the ground to the satellite. The signal sent after being mixed with $f_A$ or after being received on board is the frequency of the signal received by the ground satellite station. Modulating noise or other disturbing sounds on the local oscillator $f_A$ prevents the ground satellite receiving station from properly receiving useful information.

(b) Deception Jamming with Frequency Guidance

Deception jamming with frequency guidance can spoof radar, graphic spoofing, appropriate network attacks, or some electronic spoofing attacks in the future.

(c) Adaptive frequency spot and frequency guidance

Adaptive frequency spot and frequency guidance are inseparable from the adaptive control circuit. In adaptive control, adaptive frequency spot and frequency guidance cleverly use different characteristics of radar signals for adaptive control. For example, Pulse-Doppler radar requires an accurate frequency spot, and its signal characteristic is that the pulse repetition frequency is relatively high, generally ranging from tens of kilohertz to two or three hundred kilohertz. Using this feature, a band-pass filter from a dozen to two hundred and three hundred kilohertz is added to the input of its control circuit. When a Pulse-Doppler radar signal is an input, the adaptive control circuit automatically turns on the accurate frequency spot to perform precise aiming frequency on the signal. If the incoming signal is a conventional pulse radar, pulse compression radar, or synthetic aperture radar, their pulse repetition frequency is generally between two and three hundred hertz to two or three kilohertz. Connect a band-pass filter of tens of hertz to 5 kHz in front of their control circuit.

When this kind of signal enters, it will automatically turn on the channelization adaptive frequency spot and frequency guidance to make the jamming signal spectrum and signal spectrum match together.

6. Angle automatic passive tracking subsystem

   This part of the circuit has four channels, two of which are azimuth tracking systems, and the other two are elevation tracking branches. Each one has a microwave switch, amplifier, mixer, filter, splitter, detection amplifier, error detection, and servo amplifier. These four paths require symmetry. Besides, azimuth drive motor and azimuth servo drive mechanism, elevation servo motor, and elevation drive motor are also required. It can accurately and automatically passively track the radiation source.
7. Launch amplification chain

   To meet the needs of orthogonal polarization, the power amplifier is divided into two channels. There are a radiator and high-voltage power supply in the power amplifier branch. In order to reduce the static noise of the power amplifier, the gate or anode of the power amplifier is also controlled so that the gain is reduced as much as possible when no signal is transmitted, thereby reducing the noise. As required, a small power amplifier, microwave switch, and mixing filter are connected in front of the power amplifier to make the carrier frequency of the jamming noise source coincide with the input signal frequency.
8. Transceiver isolation control

   Transceiver isolation control is one of the key technologies for each jammer. This part has a time separation control circuit, transmitter microwave switch control circuit set, and receiver microwave switch control set. The transmitter does not work when receiving, and the receiver does not work when transmitting. In this way, it will not affect the normal operation of the receiver by its jamming signal.
9. Other circuits

   Other circuits include signal sorting sub-system, signal parameter display sub-system, signal and jamming spectrum display sub-system, servo angle display sub-system, medium and low voltage power supply sub-system, digital transmission information grid network interface, etc.

#### 5.2.2.1 Adaptive Frequency Spot and Frequency Guidance

In four different types of radar, the pulse width, repetition frequency, and signal spectrum are very different. You can use these differences to identify them and automatically select the corresponding parameters to jamming.

For Conventional Pulse radars: their pulse width is generally from a few tenths to two or three microseconds; the repetition frequency is a few hundred hertz to two or three kilohertz; the signal spectrum is from a few tenths to two or three megahertz.

For Pulse Compression radar: The pulse width is generally greater than 5 μs, and the signal spectrum width is 3–10 MHz.

For Pulse Doppler Radar: The general repetition frequency is in the range of 20–200 kHz.

For Synthetic aperture radar: Generally, the repetition frequency is within the range of 1000–2000 Hz, and the signal spectrum width is tens of megahertz to hundreds of megahertz.

In the frequency spot subsystem, there are five parts:

1. Channelized adaptive frequency guidance and frequency guidance subsystem see Fig. 5.3. It is mainly used to jamming with conventional pulse radar, pulse compression radar, synthetic aperture radar, rapid frequency radar, etc.
2. Accurate frequency spot subsystem: mainly used to jamming with Pulse-Doppler radar.

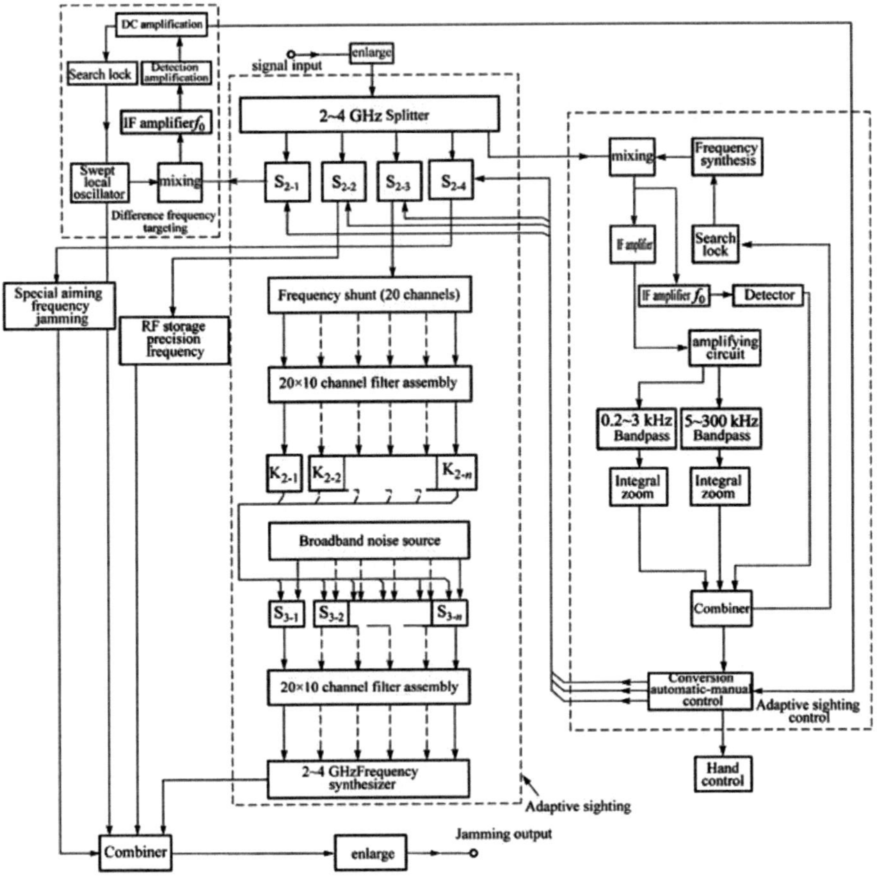


**Fig. 5.3** Block diagram of one-to-multi-system adaptive receiving and aiming frequency division system

3. Difference frequency spot subsystem: mainly used to jamming with satellite communication equipment.
4. Special frequency spot jamming: mainly used for deception jamming, video jamming and network deception suppression, etc.
5. Adaptive control and manual control of the adaptive frequency division system: it can automatically guide the corresponding jamming to the relevant radar frequency division system for different radar characteristics.

#### 5.2.2.2 Composition of Adaptive Frequency Spot and Frequency guidance

The composition of adaptive frequency spot and frequency guidance is shown in Fig. 5.3.

The adaptive frequency spot and frequency guidance subsystem are one of the key extensions of a 'one to many system' jammers. It consists of five major parts: (1) channelized adaptive frequency guidance subsystem; (2) radio frequency storage accurate frequency spot division System; (3) difference frequency spot division system; (4) special frequency spot jamming subsystem; (5) control subsystem.

#### 5.2.2.3 Channelized Adaptive Frequency Guidance

The components of this part include: amplifier, power splitter, control switch $K_{2\text{-}1.2.3.4}$, frequency splitter, S-band 20 × 10 filter bank (the same two groups), S-band wide noise source; control 2-1.1, .... 20 controllers and corresponding 20-channel $K_{2\text{-}1.2.3.4}$ microwave switch; 20-channel frequency combiner.

Working principle: When the microwave switch is turned on, the S-band signal is divided into 20 channels of 100 MHz through 20 frequency splitters, and then divided into 10 MHz or 5 MHz. The composition of the microwave shunt filter bank is shown in Fig. 5.4.

In Fig. 5.4, there are one S-band 20-channel frequency splitter and one frequency combiner; 2*200-channel Electronic filter: 2.000–2.010 GHz, 2.010–2.020 GHz, 2.020–2.030 GHz, etc. 2*20-channel Frequency splitter: 2.000–2.100 GHz, 2.100–2.200 GHz, etc. 200-channel Single pole double throw switch; 2–32.5 GHz, 2.5–3 GHz, 3–3.5 GHz, 3.5–4 GHz broadband noise sources; one S-band amplifier.

The process of adaptive frequency guidance:

1. If the input signal is the frequency of the conventional pulse radar signal at the channel i, the i-th controller has an output. When the channel $i$ switch $K_{2\text{-}i}$ is turned on, the noise spectrum $f_i$ is added to the frequency combiner through the channel i's filter and use an amplifier to output, which constitutes the frequency spot guided jamming.
2. If the input signal is multi-carrier frequency, such as $f_{i1}$, $f_{i2}$, $f_{i3}$, $f_{i4}$. The four signals pass through the corresponding four filters. It should be connected to

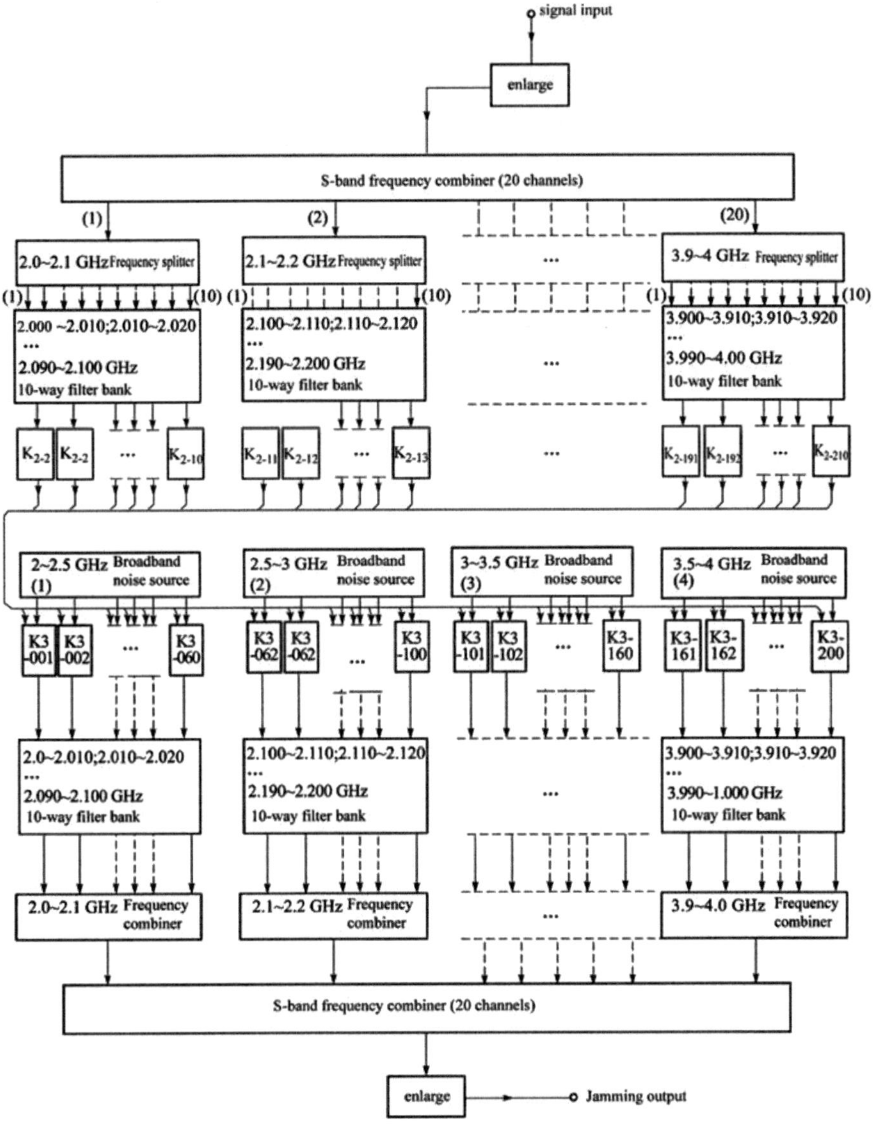


**Fig. 5.4** Block diagram of the microwave shunt filter bank

the corresponding four-channel controller and turn on the corresponding four-channel switch. $K_{3\text{-}i\ 1,2,3,4}$ is the corresponding noise source input and added to the corresponding four-channel filter. The four-channel jamming noise spectrum is added to the amplifier through a frequency combiner, then output.

3. Suppose the input signal is rapid frequency conversion. Set ten frequency modulation points, respectively $f_{i1,2,\ldots 10}$. There are ten corresponding filters, so the

corresponding controller has ten outputs and ten switches $K_{3\text{-}i1,2,...10}$ connected. The noise source is connected through the corresponding 10-channel filter, and the corresponding 10-channel noise signal is added to the amplifier through the frequency combiner, then output.

4. If the pulse compression radar signal is added, because its spectrum width is wider than that of the conventional pulse radar, 2–3 filters may be connected. The corresponding switch is turned on. The noise source is added to the corresponding filter through the corresponding switch, and then added to the frequency to the amplifier, then output the jamming signal.
5. If the added signal is a signal of a synthetic aperture radar, it may have a range of tens of megahertz to hundreds of megahertz because of the wide spectrum. The corresponding filter has several or dozens of receiving, so that the corresponding controller has an output, and controls the corresponding switch to turn on. The noise source is added to the corresponding filter through the corresponding switch, and then added to the frequency to the amplifier, then output the jamming signal.

Thus, it can be seen that the adaptive frequency guidance system can automatically perform frequency guidance/aiming at the jamming signal spectrum and properly cover the signal spectrum for conventional pulse radar, multi-carrier frequency radar, frequency conversion radar, and synthetic aperture radar. When necessary, manual intervention is also possible.

#### 5.2.2.4 Accurate Frequency Spot

As shown in Fig. 5.5, The Accurate frequency spot is mainly used to deal with Pulse-Doppler radar, which is characterized by a high signal repetition frequency and an adaptive control system. Using this feature, the switch $K_{2\text{-}2}$ is turned on. The radar signal is added to the radio frequency storage subsystem for accurate sighting, and the error is less than 1 kHz. The width of the jamming spectrum is adjusted to the range of 80–120 kHz, which can also be adjusted appropriately according to the field test.

#### 5.2.2.5 Difference Frequency Spot

The difference frequency spot jamming with satellite communication as shown in Fig. 5.6.

Let the ground satellite communication launch station be b1. The transmitted carrier frequency is f1. The ground user receiving station is b2. The received carrier frequency is f2. The ground jamming launch station is a1. The jamming frequency is fa; The ground receiving station is a2. The carrier frequency of the received signal is fb2. Receiver b2 is the same as a2.

Let the local oscillator frequency on the satellite be $f_{X0}$. The frequency received by the receiver under the satellite is $f_{B2}$. The transmitter frequency under the satellite

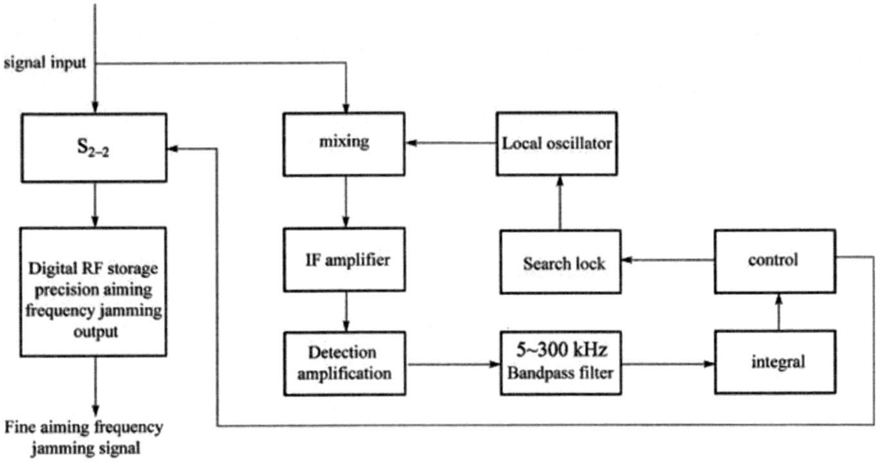


**Fig. 5.5** The principle block diagram of the precise aiming frequency adaptive control

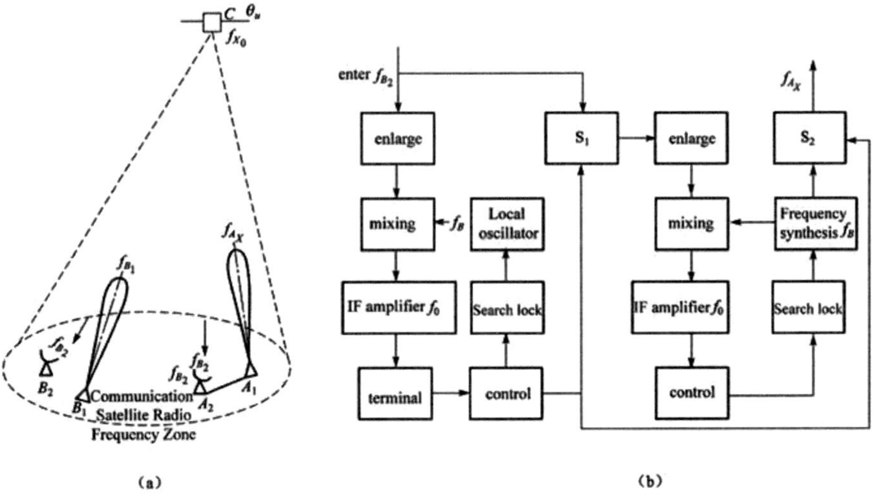


**Fig. 5.6** Schematic diagram of the composition of frequency division system for satellite communication

is $f_{B1}$. Then, $f_{B2} = f_{B1} \pm f_{X0}$. The local oscillator frequency of the under-satellite receiver is $f_B$. If Intermediate Frequency is $f_0$. Get:

$$\begin{cases} f_{B1} \pm f_{X0} - f_B = f_0 \\ f_{A1} \pm f_{X0} - f_B = 2f_0 \end{cases} \tag{5.16}$$

As long as the frequency synthesis's $f_B$ is adjusted properly, the terminal would have a display. At this time, the frequency of $f_{AX}$ is now aligned with $f_{B1}$. The transmission frequency after the frequency conversion of the equipment on the satellite is $f_{BX}$ so that the satellite receiving station on the ground is jammed. If the ground station $A_1$ adjusts some information on $f_{AX}$, its strength is stronger than the station $B_1$, then the $B_X$ receiver will receive this information.

#### 5.2.2.6 Special Frequency Spot Jamming

Special jamming includes network attacks, image deception jamming, audio deception jamming, and some other virtual deception jamming. These spoofing jammings are highly targeted and can only be achieved if the other party's electronic equipment is clearly understood. Therefore, its uncertainty is very large, you can only wait for the opportunity, and occasionally carry out once when the conditions are met.

#### 5.2.2.7 The Sensitivity of Receiver and Receiver's Instantaneous Bandwidth

The instantaneous bandwidth of the receiver should meet the maximum bandwidth of the signal. In the S-band, the spectral width of the signal can reach 200 MHz, and the receiver's ultimate sensitivity $P_\gamma$:

$$P_\gamma = K \cdot T \cdot \Delta f \cdot N \tag{5.17}$$

where:

K Bozeman coefficient
T Absolute temperature
$\Delta f$ Bandwidth of receiver
N Noise Figure.

$$P_\gamma = 1.38 \times 10^{-23} \times 300 \times 200 \times 10^6 \times 3 - 116\,\text{dBW}.$$

If the signal spectrum width is 600 MHz, $P'_\gamma$ should be −111 dBW.

If the receiver antenna gain is 30 dB, the receiver's actual working sensitivity $P_\gamma < P'_\gamma$.

$$P'_\gamma = \frac{P_t \cdot G_t(\theta)}{4\pi R_j^2} \cdot \frac{\lambda^2}{4\pi} \cdot G_j \tag{5.18}$$

where:

$P_t$ Peak power of the radar;
$G_t(\theta)$ Gain of radar side lobe level aligned with the jammer;
$\lambda$ Radar operating wavelength;
$G_j$ Gain of Receiving antenna of the jammer;
$R_j$ Distance from the jammer to the radar.

$P_{\gamma 1}$ for spaceborne SAR radar:

$$P_{\gamma 1} = \frac{5 \times 10^3 \times 1 \times 0.1^2 \times 10^3}{1.56 \times 10^2 \times 1.2 \times 10^{12}} = -96\,\text{dBW}$$

$P_{\gamma 2}$ for airborne early warning radar:

$$P_{\gamma 2} = \frac{10^6 \times 1 \times 0.1^2 \times 10^3}{1.56 \times 10^2 \times 16 \times 10^{10}} = -54\,\text{dBW}$$

$P_{\gamma 3}$ for airborne platforms:

$$P_{\gamma 3} = \frac{10^6 \times 1 \times 0.1^2 \times 10^3}{1.56 \times 10^2 \times 10^{12}} = -72\,\text{dBW}$$

$P_{\gamma 4}$ for spaceborne SAR in X band:

$$P_t = 5\,\text{km};\ G_t(\theta) = 1;\ R_j = 1200\,\text{km};\ \lambda = 3 \times 10^{-2}\,\text{m}$$

$$P_{\gamma 4} = \frac{5 \times 10^3 \times 1 \times 9 \times 10^{-4} \times 10^3}{1.56 \times 10^2 \times 1.2^4 \times 10^{12}} = -107\,\text{dBW}$$

If the X band receiver antenna gain is 37 dB, then $P_{\gamma 4} = -100\,\text{dBW}$

#### 5.2.2.8 Jamming Transmitter

The composition of the jamming transmitter is shown in Fig. 5.7.

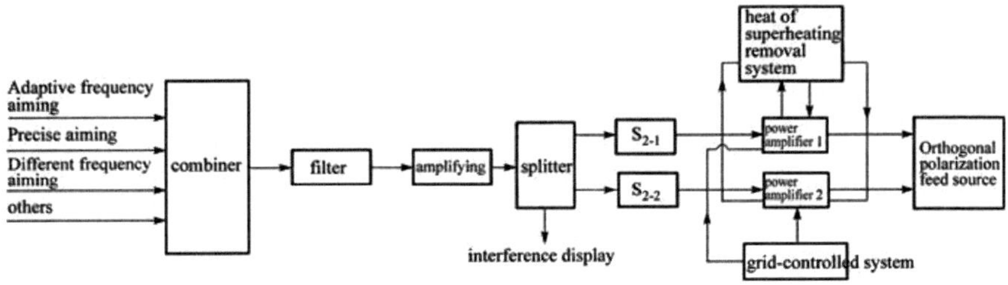


**Fig. 5.7** Composition principle diagram of interference transmitter

Jamming transmitter consists of combiner, filter, amplifier, splitter, on–off microwave switch, power amplifier, radiator, gate control subsystem.

Combiner: The adaptive aiming frequency signal, fine aiming frequency signal, differential aiming frequency guidance and other special aiming frequency deception jamming signals are added to the combiner respectively. According to different interference objects, the input signal is input to the input end of the filter, and the filter is a 2. 4 GHz band-pass filter.

Amplifier: Amplifies the signal of the interference source enough to drive the maximum output of the power amplifier, $K_{2-1,2}$ microwave switch, used for time division transceiver isolation. When the interference is emitted, $K_{2-1,2}$ is connected. When $K_{2-1,2}$ is disconnected at reception, select the transceiver time ratio, better than 9:1.

Power amplifier 1, 2: two gain should be balanced, the output power should be consistent, the asymmetry of the two phase difference should be less than 1 dB, two output should be added to the positive polarization feed, so that the interference radiation source can form a random polarization signal for anti-radar anti-jamming and multi-polarization.

The basic working principle of orthogonal polarization feed: the orthogonal polarization feed has two input ports, which are orthogonal to each other. If at two input ports, one input is sine wave ($\sin \phi$), the other input is cosine wave ($\cos \phi$), at the output port should be Li Shayu graphics. If the two paths are of the same magnitude, a circle should be formed. If the two amplitudes are different, an ellipse should be formed. If the phase shift of the two paths changes, the left-handed or right-handed changes can occur. If this vector is a polarized wave modulated by noise, it forms left-handed circular polarization or right-handed circular polarization or elliptical polarization or linear polarization. These different types of random polarization changes make the radar have the same probability of receiving the random polarization interference signal regardless of the polarization form. This kind of random polarization jamming can resist all variable polarization jamming and use multiple polarization radar receiving antenna. Macroscopic observation and test show that the polarization loss of the interference is 3 dB. In addition, when the interference signal becomes random polarization, it can turn a frequency modulation noise into amplitude modulation noise. It can compress the interference spectrum to the same bandwidth as the radar receiver, that is, the $K_f$ is increased to close to 1, and the equivalent power of the interference signal into the radar receiver is changed from $P_M K_f = 0.3 P_M$ to $1.4 P_M$, which is 4.7 times higher than that of the single tube. The actual test proves that the interference effect of orthogonal polarization is significantly improved than that of single tube interference. The use of orthogonal polarization feed can not only resist the anti-jamming and multi-polarization receiving of radar, but also improve the jamming effect. Therefore, it is necessary to change single-channel interference into double-channel orthogonal polarization interference.

#### 5.2.2.9 Precise Single Pulse Passive Tracking

As shown in Fig. 5.8 Angle automatic passive tracking subsystem includes: servo tracking antenna, azimuth and elevation four channels, amplifier, mixer (if the frequency is 2–4 GHz can not be used), filter, splitter, combiner, amplifier and detection video, error detection, servo effect, motor and transmission mechanism and angle data detection mechanism.

The servo passive automatic tracking subsystem adopts the amplitude-comparison single-pulse passive automatic tracking system, and the actual working sensitivity of the four-channel receiver is required to be $-100\,\text{dBW}$. Four-way amplifier to maintain balance, the imbalance is less than 1.5 dB. The tracking errors of azimuth and elevation are less than 0.25°. The gain of the receiving antenna is 30 dB in S band and 37 dB in x band. The angular tracking speed of azimuth and elevation is 0°–20°. azimuth search speed 0°–30°. Angle search speed 0°–20°.

The antenna and feeder should be shielded and isolated to make the sidelobe level $G_t(\theta)/G_t < -50\,\text{dB}$ near $90° \pm 20°$.

The azimuth rotation range is 0°–360°, and the elevation rotation range is 0°–90°. The receiving and microwave amplification, transmitter and heat dissipation

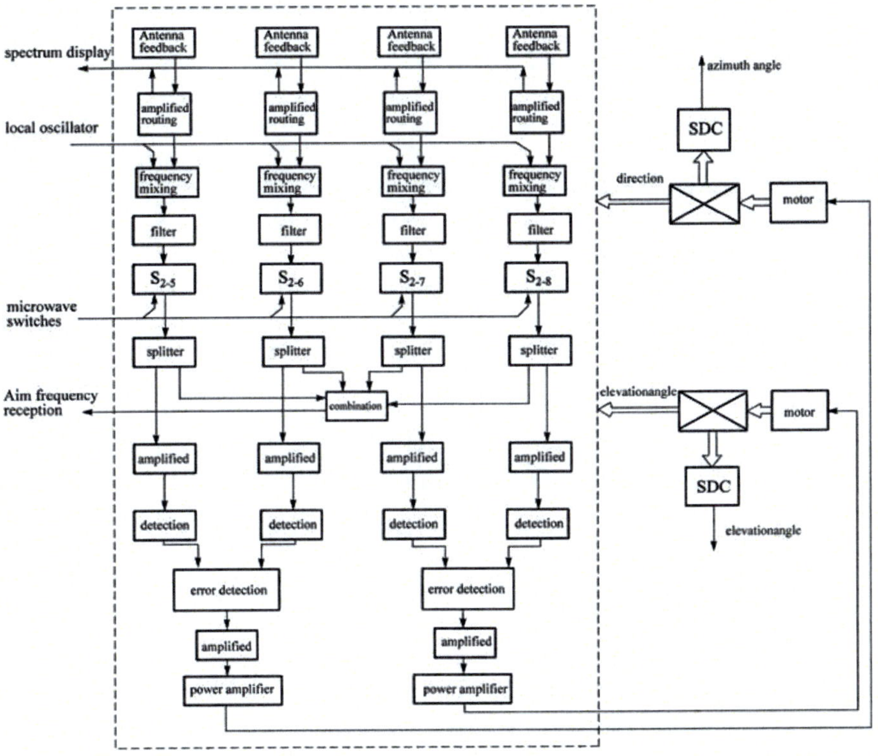


**Fig. 5.8** Composition principle diagram of angle tracking passive tracking subsystem

system rotate synchronously with the azimuth, and the RF high power output of the transmitter is added to the antenna pitch rotation feeder branch through the rotary joint.

The passband width of the band-pass filter is 2–4 GHz.

Four-way antenna feed (the central axis relative to the antenna surface) can move two points relative. A focus and formal focus to ensure the antenna symmetry and antenna gain, narrow beam working state. Another focus can broaden the antenna beam (antenna gain decreases) and deteriorate the symmetry. This state is mainly used as a search state, allowing the antenna beam to be doubled. The antenna gain is reduced by 6 dB.

Four-way microwave switch (K) is used as the time division between sending and receiving. When the interference signal is transmitted, the switch K is disconnected, and when the signal is received, the four-way switch K is connected, which is controlled by the time division circuit.

#### 5.2.2.10 Time-Separated Transceiver Isolation Control

The transceiver isolation control is one of the key technologies in the jammer. The receiving system of a single-machine to multiple-machine jammer requires high receiving sensitivity, wide instantaneous bandwidth and high isolation. Therefore, compared with other jammers, the transceiver isolation increases many difficulties. For a multi-system jammer transceiver isolation requirements to be unified design, multiple channels split to solve, and finally meet the system transceiver isolation requirements.

The spontaneous self-receiving of the jammer is mainly caused by the internal coupling of the transceiver subsystem, the reflection of the ground object at low elevation angle and the sidelobe coupling of the transceiver antenna. If other factors can be solved respectively, only the sidelobe coupling of the transceiver antenna is left.

Set the jammer receive antenna sidelobe receive transmit antenna sidelobe interference power is $P_{\gamma j}$.

$$P_{\gamma j} = \frac{P_t \cdot G_t(\theta)}{4\pi \gamma_j^2} \cdot \frac{\lambda^2}{4\pi} \cdot G_\gamma(\theta) \cdot K_i \tag{5.19}$$

In the formula:

| | |
|---|---|
| $P_j$ | is interference transmitter power or transmitter static noise power. |
| $G_\gamma(\theta)$ | is the antenna sidelobe gain near 90° of the transmitting antenna. |
| $\lambda$ | is the wavelength of interference signal. |
| $G_t(\theta)G_\gamma(\theta)$ | is the antenna sidelobe gain near 90° of the receiving antenna. |
| $\gamma_j$ | is the distance between the center points of the transceiver antenna. |
| $K_i$ | is the improvement coefficient of antenna isolation after adding absorbing material. |

If $P_j$ transmitter power is 1 kW, static noise is 1 mW

$$G_t(\theta) = G_t(90°) = G_\gamma(90°) = 0.1; \lambda = 0.1\,\text{m}; r_j = 4\,\text{m}$$

Put the parameters into (5.19).

$$P_{y(1)} = \frac{10^{-3} \times 0.1 \times 10^{-2} \times 0.1}{1.56 \times 10^2 \times 16} \cdot K_i \leq 10^{-11} \quad K_i \leq 2.5 \times 10^{-7}$$

This value is currently impossible to achieve. There is a time-division system; that is, it does not transmit when receiving and does not receive when transmitting. In this case, $P_j = 1\,\text{MW}$:

$$P_{\gamma j(2)} = \frac{10^{-3} \times 0.1 \times 10^{-2} \times 0.1}{1.56 \times 10^2 \times 16} \cdot K_i \leq 10^{-11} \quad K_i \leq \frac{10^{-11}}{2.5 \times 10^{-10}} = 0.04$$

Add some absorption and shielding materials between the transceiver antennas, so that the isolation coefficient between the transceiver system is less than 0.04.

Figure 5.9 is a block diagram of the principle of time division control. The time division circuit does not work in the receiving state. At this time, the switch $K_{1\text{-}1,2,3,4}$ are on, and switch $K_{2-5,6,\ldots 13}$ are closed.

When the jamming is transmitted, the output of the time-division circuit is as shown in Fig. 5.10.

The general receiving time $T_{\hat{E}\tilde{o}}$ is 1.8 to 2 ms; $T_{.¢}$ is the transmitting time 20 ms; $\tau_{1,2}$ are the non-transmitting and non-receiving time, generally take 50 μs.

When jamming with the agile radar, $T_{.¢}$ takes 0.8 T (radar pulse repetition period). The time division circuit is controlled by the external trigger of the radar signal, and its control time waveform is shown in Fig. 5.11.

In the transmitting state, 0.1 mW signal power is added to the input of the switch. If the switch is not open, it will saturate the circuit behind; if it is open, this signal will have little effect on the circuit behind. Plus $K_{2\text{-}5}$ and $K_{2\text{-}10}$ are connected in series;

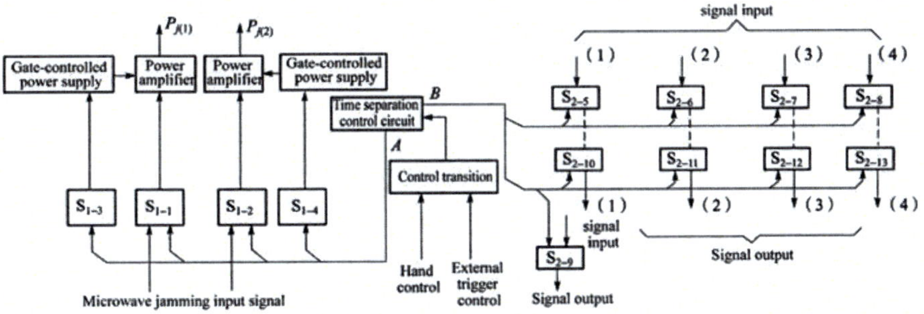


**Fig. 5.9** Functional block diagram of the composition of the time separation control of the transceiver antenna

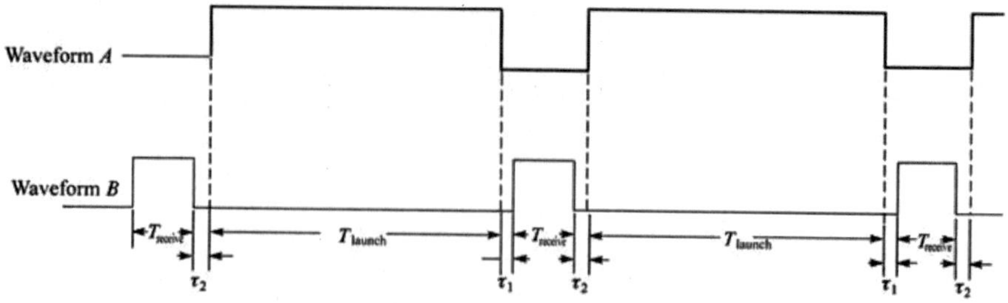


**Fig. 5.10** Time separation waveform

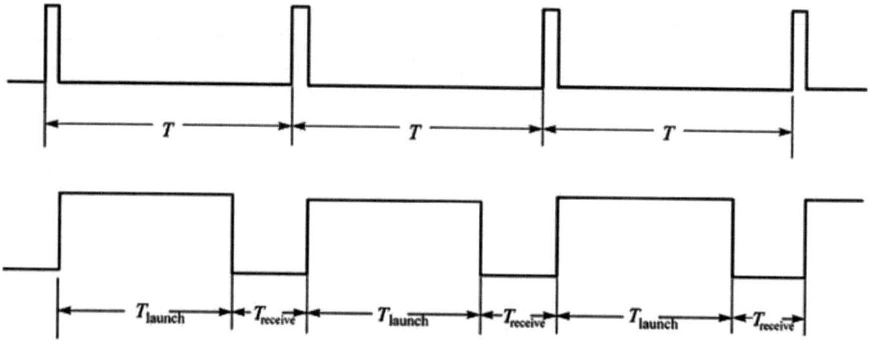


**Fig. 5.11** Waveform of jamming control against frequency-agile radar

$K_{2\text{-}6}$ and $K_{2\text{-}11}$ are connected in series; $K_{2\text{-}7}$ and $K_{2\text{-}12}$ are connected in series; $K_{2\text{-}8}$ and $K_{2\text{-}13}$ are connected in series. Their open-circuit series insertion loss can reach 100 dB. The insertion loss, when switched on, is 2 dB. Therefore, the two-stage switch in series does not affect signal amplification. However, it should be noted that the paired connection makes the four-channel signal amplification balanced to meet the requirement that the unbalance error is less than 1.5 dB. By gate control or positive control of the power amplifier tube, reduce the gate control voltage within 2 ms of the receiving time, that is, reduce the gain of the power amplifier tube to make the squelch output power less than 1 mW. The gate control voltage returns to normal during transmission, and then the power amplifier tube normally output power.

## 5.3 The Layout of Air-to-Space Integration of 'One-to-Many Systems' Jammer

The schematic diagram of the ground-to-space and ground-to-air jamming of 'one to many system' jammer is shown in Figs. 5.12 and 5.13.

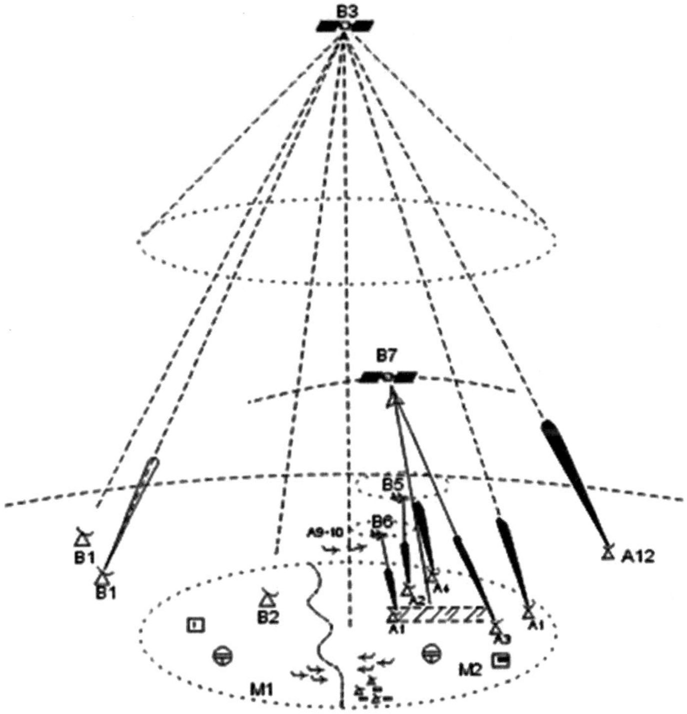


**Fig. 5.12** Schematic diagram of One-to-many systems ground-to-air-to-space integrated jamming system. A1,2,3,4—'One to many system' jammer, C1,2,3—Target protection jammer at our ground point, M1—Opponent's local combat control zone, M2—Our local combat control area, 1—Opponent's airport, 2—Our airport, 1—Opponent's Combat Command Center, 2—Our Combat Command Center, B1—Opponent's satellite communication transmitting station on the ground, B2—Opponent's satellite communication receiving station on the ground, B3—Opponent's communications satellite, B5—Opponent's early warning platform, B6—Opponent's early warning aircraft, B7—Opponent's low-orbit spaceborne SAR

In the near future, radar carried by airborne platforms may appear. Although early warning aircraft radars and space-borne radars have their advantages, they have also been well applied in the Gulf War and local wars. But as the war situation develops, they also show their shortcomings. For example, although the early warning aircraft can see a long distance and has the function of an air command post, according to current needs, its flying height is low, and the detection distance is only a few hundred kilometers. So for information warfare, its detection range is not enough. Although the detection distance of the spaceborne radar is far, the monitoring time for the target is too short, only a few tens of seconds to a few minutes. Therefore, a space platform with a height of 30–100 km is needed. The platform speed is within 3–8 Mach (or 0–3 Mach), and the target can be continuously monitored from hundreds to 1000 km.

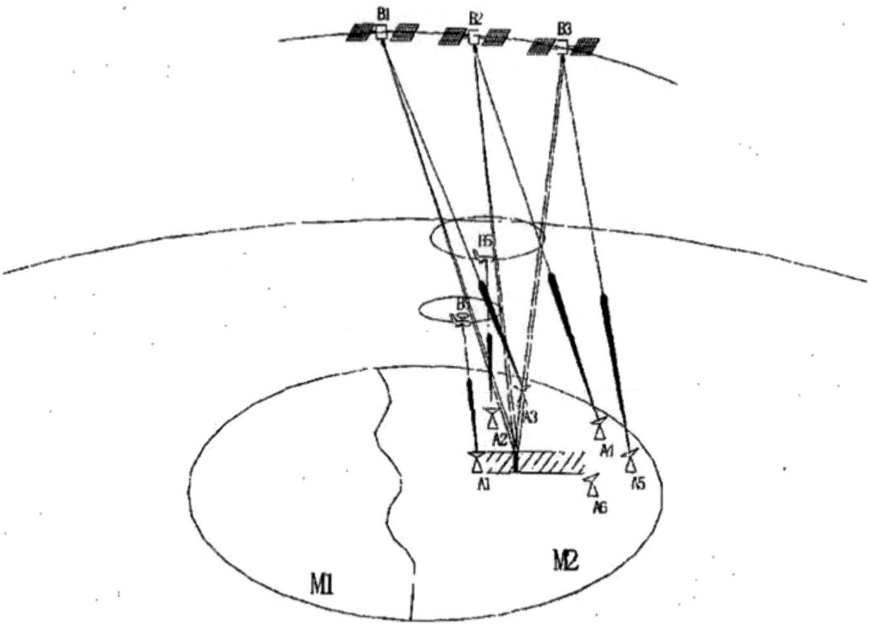


**Fig. 5.13** Schematic diagram of integrated one-machine-to-multi-system multi-star multi-machine. B1,2,3—Multi-satellite imaging system, B4—Airborne detection platform, B5—Airborne early warning machine system, M1—Opponent's control zone, M2—Our control zone, A1,2,3—One-to-many systems air-to-air integrated jammer

Now let's take three examples. The general idea is that the sending and receiving subsystems are installed on a platform; the receiving and transmitting extensions, power supplies, microwave devices, etc. are all installed on the azimuth turntable to rotate synchronously with the antenna subsystem; Two transmitting source passes through two rotating stage outputs high-power microwave jamming signal to orthogonal polarization feed to form orthogonal polarization random jamming.

The distance between A1/A2 and the center is about 200 km.

When the jammer is in the S-band, its equivalent jamming power is 70 dBW.

Figure 5.14 shows the necessity of a One-to-many systems-to-air-to-air integrated jamming system. In the past, a jammer was developed for early warning aircraft radar, a jammer for spaceborne radar, and a jammer for communication satellites. In this way, the wide variety of jammers is not conducive to the maintenance of the troops, and it is not convenient for mass production. According to the idea of a One-to-many systems air-to- space integration, only one of these types of jammers needs to be designed. It can not only jamming with early warning aircraft radars and space-borne radars, but also jamming with spaceborne communication satellites, data transmission satellites, remote control satellites, navigation satellites, and other satellites. As long as they are in the same frequency band, they can be effectively jammed.

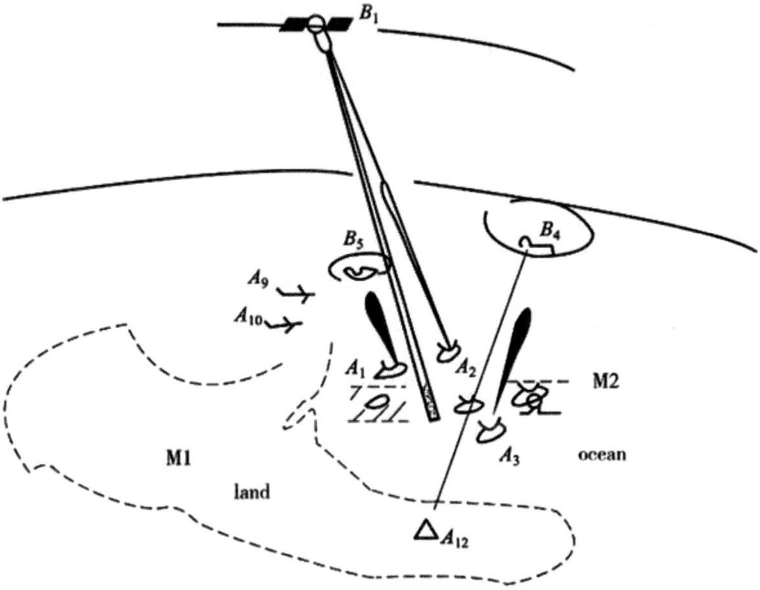


**Fig. 5.14** Ship-to-air system-to-multisystem (air-to-air) integrated jamming diagram. B1—Spaceborne SAR, B4—Radar warning for airborne platforms, B5—Airborne early warning radar, A1,2,3—Shipborne special jamming ships, which are used to cover the fleet formation

B4 is a kind of airport platform. Its height can rise from 30 to 100 km. The airborne platform can overcome the low flying height of the early warning aircraft so that the radar on the aircraft is limited by the curvature of the earth and cannot detect targets at longer distances. At the same time, it can also overcome the shortcomings of the low-orbit spaceborne radar to monitor the target for a short time. According to current predictions, the airport platform is likely to become the most suitable surveillance platform shortly. This platform can be used as a radar platform, regional navigation platform, data relay, and forwarding platform. It can also be used as a jamming platform to jamming with the enemy. This platform can monitor the surrounding radius of about 1000–1500 km. The area under surveillance is about 3–7 million square kilometers. As for local wars, as long as there are one to three platforms, the overall situation of regional wars can be monitored.

Schematic diagram of the jamming exposure area and the jamming effective area is shown in Fig. 5.15. As a special carrier-based jammer, the demand for fighting with the fleet is also increasing. For example, an aircraft carrier formation in the United States has X Aegis convoys, two E-2C early warning aircraft, X escort aircraft, and X escort helicopters. This forms an escort area of 360 km, as shown in Fig. 5.16. Also, the forecast satellite gives notification information, so that its aircraft carrier formation forms an electronic information network with both offensive and defensive capabilities. In this figure, if the surveillance of the airborne platform is added, the regional scale and continuity of surveillance will be stronger. The use of Fig. 5.16

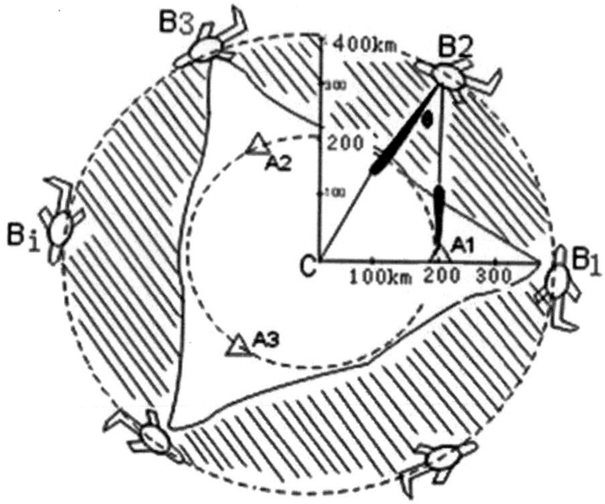


**Fig. 5.15** Schematic diagram of the jamming exposure area and the jamming effective area when the carrier-based dedicated ship shielding fleet is formed. C—Fleet Distribution Center; B1–B6—Early warning aircraft are flying around C to search for targets; A1–A3—Accompanying jamming ship centered on point C; Horizontal line—Jamming exposed areas; Inside horizontal line—Jamming effective area. $\theta_i$—The side lobe angle of the radar to the jammer when B is in a different position. E1, E2—Jamming screened fleet

ship-specific jamming system is the need for development. From the perspective of information warfare, the aircraft carrier fleet needs escort ships, escort aircraft, and early warning aircraft, as well as special carrier-based jammers, to ensure the safety of the fleet during offense and defense.

The aircraft carrier formation is an air-to-space integrated jamming system composed of Shipborne Special jamming ships. The jamming exposure area and jamming effective area formed by the jamming/cover and the double-layer ring configuration with Shipborne Special jamming ships are shown in Fig. 5.16.

In Fig. 5.16, A1–A6—Air-to-space integrated jammer with ring configuration; B1–B3—Surveillance radar or communication data transmission of the airborne platform; C—Distribution center of the aircraft carrier formation; D—Ground projection point of spaceborne SAR; E—Spaceborne SAR monitors aircraft carrier formations; H—Altitude of spaceborne SAR; Horizontal line—Jamming exposed areas; Bi—The airborne platform-based radar that conducts search and surveillance around the carrier formation centered on C; Dotted circle—Aircraft carrier formation distribution area, and also an effective jamming area for air-to-space jamming systems; Ellipse area—Effective area of single jammer jamming to spaceborne SAR; Middle cross—Spaceborne SAR imaging zone; N1–N3—Carrier formation screened by the jammer.

The jamming system consists of one jamming command ship and six special jamming ships. The first layer is a three-station circular configuration with a radius of about 200 km, and each jamming ship is separated by 120°. The second layer is a

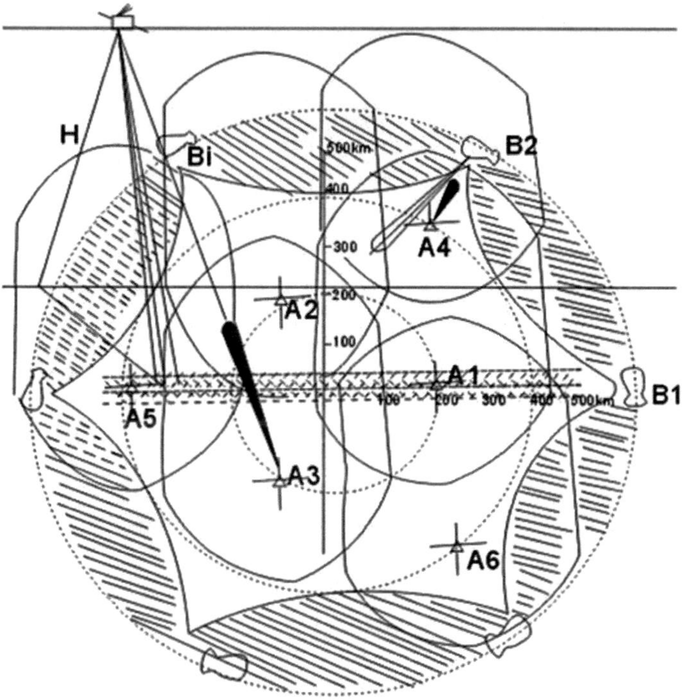


**Fig. 5.16** Schematic diagram of the jamming effective area and jamming exposure area formed by the special jamming ship of the aircraft carrier formation

three-station circular configuration, with a radius of about 400 km, and each jamming ship is separated by 120°. The system of six jamming ships is 60° apart in angle. Such a layout allows the aircraft carrier formation to detect at sea with or without spaceborne radar, and the jamming system can effectively jam. Each jamming ship can carry jammers in the P, L, S, C, X, and Ku bands, which enable the jamming system to handle multiple bands, multiple carrier frequencies, multiple directions, multiple systems, and multiple platforms. Under the condition of ensuring information, the aircraft carrier formation can have both offensive and defensive capabilities to ensure its safety. Special jamming ships are one of the development directions for ensuring the fleet's offensive and defensive capabilities under informationized combat conditions.

Each special jamming ship should also be equipped with warning radar, communication data transmission, various self-defense weapons, and etc.

# Chapter 6
# Analysis on Configuration of Radar Jammer and Protected Target

In informationized operations, radar plays a clairvoyant role. Radar is an important tool in long-range precision strikes. If we use multiple means to prevent the radar from finding the target, long-range precision strikes will not be possible.

Radar jammers protect different targets according to different platforms. Some are aimed at combat aircraft radars, protecting valuable and important military targets on the ground; some are aimed at early warning aircraft radars, covering our air force penetration; some are aimed at airborne imaging radars, protecting our important military deployments from being reconnaissance by imaging radars; some are aimed at enemy's satellite-based imaging radars and jamming with enemy imaging and reconnaissance systems to protect our important military installations. No matter which kind of target is covered, there is an jamming protection area and jamming exposure area. In this chapter, we mainly analyze the protection area and jamming exposure area of the jammer. It is closely related to the relevant parameters in the jamming equation.

## 6.1 Ground Jamming Angainstaircraft Radar to Protect Important Military Targets

Figures 6.1a and b show schematic diagrams of ground-to-air jammers, jammed targets, protected areas, and relative positions.

In the diagrams:

A—Center of Protected Target; B—Position of Jammer; C—Position of Jamming radar; $r$—radius of Protected target; d—The distance between the jamming station and the center of the protected target; $\sigma-$ Effective reflection surface of protected target; $R_C$—The distance between the radar and protected target; $R_j$—The distance between radar and jamming; H—the height of airborne platform; D—Projection point of airborne radar on the ground; $R$—The projection distance of $R_c$.

The power of the target echo signal received by the radar:



X. Zhang et al., *Theory to Countermeasures Against New Radars*,
https://doi.org/10.1007/978-981-16-6715-2_6

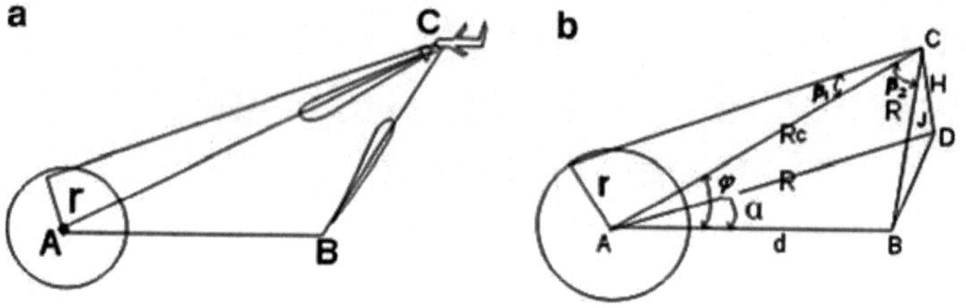


**Fig. 6.1** Relative position diagram of radar, jammer and protected target

$$P_2 = \frac{P_a G_a^2 \sigma L_d \lambda^2}{(4\pi) R_C^4} \quad (6.1)$$

The power of the jammer received by the radar:

$$P_1 = \frac{P_j G_j \lambda^2 K_1 K_f G_a(\theta)}{(4\pi)^2 R_j^2} \quad (6.2)$$

According to the definition of jamming suppression coefficient:

$$K_j = \frac{P_1}{P_2} = \frac{4\pi P_j G_j R_C^4 K_1 K_f L_j}{P_a G_a \sigma R_j^2 L_d} \cdot \frac{G_a(\theta)}{G_a} \quad (6.3)$$

$\beta_1$—The jamming fan due to the protected target has a certain radius.

$\beta_2$—The jamming fan due to the jamming station and protected target is not configured together.

If the jamming station can effectively cover the protected target, the minimum effective jamming fan $Q_{\min}$ is equal to:

$$Q_{\min} = 2(\beta_1 + \beta_2) \quad (6.4)$$

$$\beta_1 = \arctan \frac{\gamma}{R_C} \quad (6.5)$$

$$\beta_2 = \arcsin\left[\frac{d \sin\varphi}{\sqrt{R_C^2 + d^2 - 2dR_C \cos\varphi}}\right] \quad (6.6)$$

$G_a(\theta)$ is the sidelobe gain of the main radar line, according to the general empirical formula, within the range of $\beta_2 \geq 3\theta_{0.5}$ and $\beta_2 \leq 60$o:

$$G_a(\theta)/G_a \approx K_a \left(\frac{\theta_{0.5}}{2\theta}\right)^2 \quad (6.7)$$

Substitute (6.4), (6.5), (6.6), (6.7) into (6.3).
We get:

$$K_j = \frac{4\pi P_j G_j \theta_{0.5}^2 (R^2 + H^2)^2 \cdot K(\theta) X K_1 K_f L_j K_a}{\sigma P_a G_a \left[R^2 + H^2 + d^2 - 2d(R^2 + H^2)^{1/2} \cos \partial\right] L_d} \times \frac{1}{\left\{ \arctan\left[\frac{r}{(R^2+H^2)^{1/2}}\right] + \arcsin \frac{d \sin\left[\arccos\left(\frac{R \cos \partial}{(R^2+H^2)^{1/2} \cos \partial}\right)\right]}{\left[R^2+H^2+d^2-2d(R^2+H^2)^{1/2} \cos \partial\right]^{1/2}} \right\}^2} \quad (6.8)$$

In the formula:

| | |
|---|---|
| $\theta_{0.5}$ | Beamwidth at half power point of radar. |
| $K(\theta)$ | In the near area, the area where the radar mainlobe illuminates the target is reduced, which is equivalent to a multiple of the radar main lobe gain. Generally, $K(\theta)$ takes 5–10 times. |
| $K_1$ | Feeder loss factor of jammer. |
| $\beta = 60°$ | Ratio of radar receiver bandwidth to jamming spectrum. |
| $L_j$ | Loss factor of jamming polarization. |
| $K_a$ | Antenna side lobe coefficient. |
| $L_d$ | Feeder loss factor of radar transmitter. |
| $\alpha$ | The projection angle of the $\varphi$ in the azimuth plane. |

***Example 6.1:*** Take $K_1 = 0.5$; $K_a = 0.1$; $\theta_{0.5} = 1.5°$; $P_j G_j = 50\,\text{dBw}$;$H = 6\,\text{km}$; $L_d = 0.4$; $P_a G_a = 80\,\text{dBw}$; $d = 3\,\text{km}$; $r = 1\,\text{km}$;$K(\theta) = 9$; Substitute the parameters into the formula (6.8), the calculated jamming exposure area is shown in Fig. 6.2.

The horizontal line in the figure is the jamming exposure area (When the radar enters the horizontal zone, the protected target can be found). A—Center of

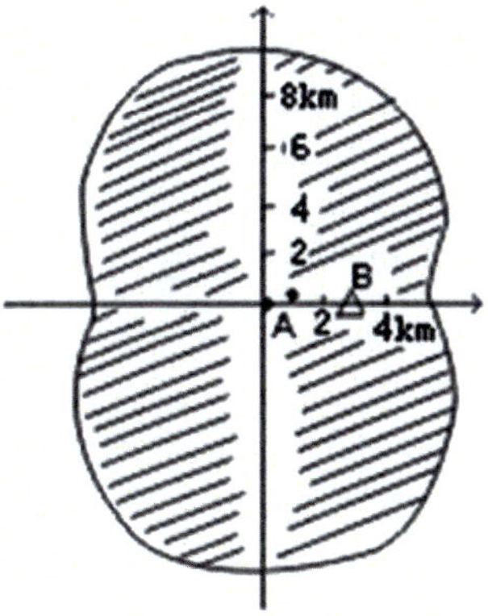


**Fig. 6.2** Jamming exposure area when the ground-to-air jamming station protects the target

**Fig. 6.3** Schematic diagram of the jamming zone formed by the three-station ring configuration

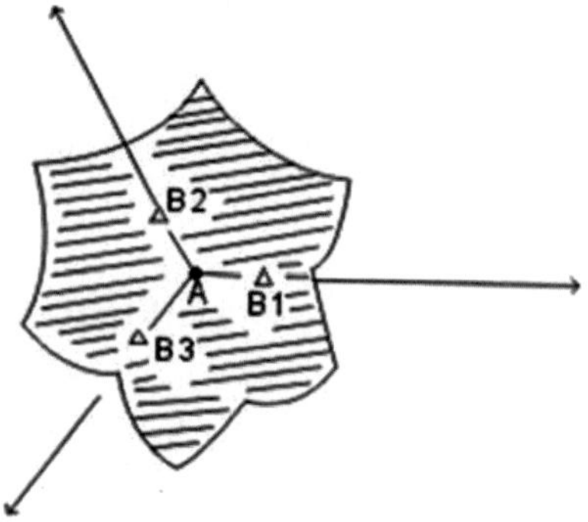


Protected Target; B—Position of Jammer. If the jamming station is a three-station ring configuration, the jamming exposure area formed is shown in Fig. 6.3.

A—Center of Protected Target; B1,2,3—The symmetrical ring configuration of the three jamming stations is the jamming exposure area.

If the jamming station is far away from the protected target, and then, (6.8) can be simplified into (6.9).

$$\text{That is: } K_j = \frac{4\pi P_j G_j R^4}{P_a G_a \sigma R_j^2} \cdot \frac{K_1 L_j K_f}{L_d} \cdot \frac{K_a \theta_{0.5}^2}{\left[\arctan\left(\frac{r}{R}\right) + \alpha_c\right]^2} \tag{6.9}$$

(6.9) can also be written as (6.10):

$$R = \left[\frac{K_j P_a G_a \sigma R_j^2}{4\pi P_j G_j} \cdot \frac{L_d}{K_1 L_j K_f} \frac{\left(\alpha_c + \arctan\left(\frac{r}{R}\right)\right)^2}{K_a \cdot \theta_{0.5}^2}\right]^{1/4} \tag{6.10}$$

$$R_j^2 = R^2 + d^2 - 2dRcos\alpha \tag{6.11}$$

$$\alpha_r = \arctan\left(\frac{r}{R}\right) \tag{6.12}$$

$$\alpha_c = \alpha_x - \alpha \tag{6.13}$$

The graphical solution of simultaneous equations is shown in Fig. 6.4

***Example 6.2:*** Given:

$P_j G_j = 50\,\text{dBw}$; $P_a G_a = 80\,\text{dBw}$; $d = 20\,\text{km}$; $r = 15\,\text{km}$; $K(\theta) = 1$; $L_d = 0.4$; $K_1 = 0.5$; $K_a = 0.1$; $\theta_{0.5} = 1.5°$.

Solve the simultaneous equations of (6.10), (6.11), (6.12), and (6.13), find the relationship curve between R and $\alpha_x$.

Take $K_j = 1$; $\sigma = 10^4\text{m}^2$. When solving $\alpha_x = 0°, \alpha_c = 0, \alpha = 0, R_j^2 = (R-d)^2$, $\alpha_\gamma = \tan^{-1}\left(\frac{15}{R}\right)$.

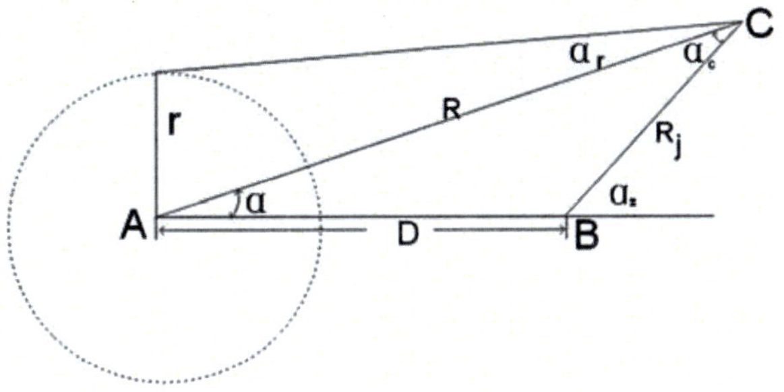


**Fig. 6.4** Schematic diagram of the jamming exposure area when the protected target has a large radius and is far away

Substitute (6.10) formula, solved $R = 48$ km.
When $\alpha_x = 90°$, Substitute (6.10) formula, solved $R = 80$ km.
When $\alpha_x = 180°$, Substitute (6.10) formula, solved $R = 55$ km.

Figure 6.5a, b is the schematic diagram of jamming exposure when $P_jG_j = 50\,\text{dBw}$, $P_aG_a = 80\,\text{dBw}$, $d = 20\,\text{km}$, $r = 15\,\text{km}$, $\theta_{0.5} = 1.5°$, $\sigma = 10^4\text{m}^2$

***Example 6.3:*** If the equivalent power of the jammer $P_jG_j = 60\,\text{dBw}$, other parameters remain unchanged, and the parameters are substituted into (6.10) to calculate the jamming exposure area, as shown in Fig. 6.6a, b.

$P_aG_a = 80\,\text{dBw}$; $d = 20\,\text{km}$; $r = 15\,\text{km}$; $\theta_{0.5} = 1.5°$; $\sigma = 10^4\text{m}^2$;

Or: When $P_jG_j = 50\,\text{dBw}$; $P_aG_a = 80\,\text{dBw}$; $d = 20\,\text{km}$; $r = 15\,\text{km}$; $\theta_{0.5} = 1.5°$; $\sigma = 10^3\text{m}^2$, the jamming exposure area is the same as Fig. 6.6a, b.

***Example 6.4:*** If $P_jG_j = 60\,\text{dBw}$; $P_aG_a = 80\,\text{dBw}$; $d = 20\,\text{km}$; $r = 15\,\text{km}$; $\theta_{0.5} = 1.5°$; $\sigma = 10^3\text{m}^2$, the jamming exposure area is the same as Fig. 6.7a, b.

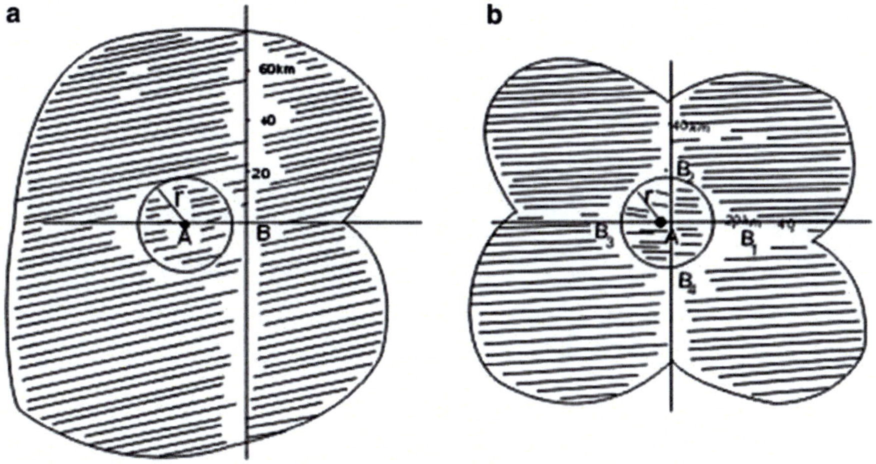


**Fig. 6.5** **a** Single station jamming exposure area **b** four-station ring configuration jamming exposure area (No. 1)

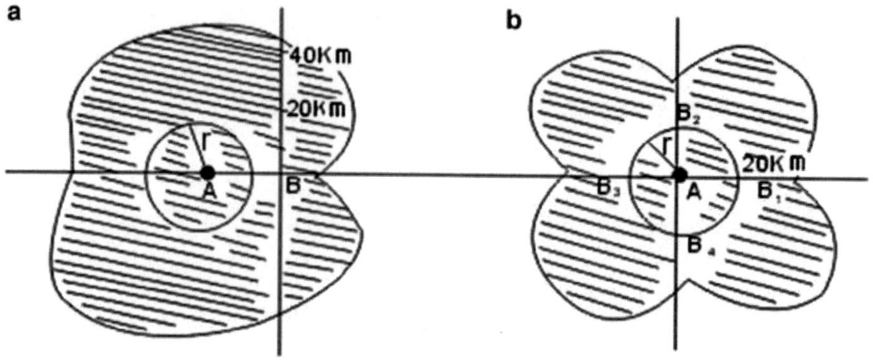


**Fig. 6.6** **a** Single station jamming exposure area **b** four-station ring configuration jamming exposure area (No. 2)

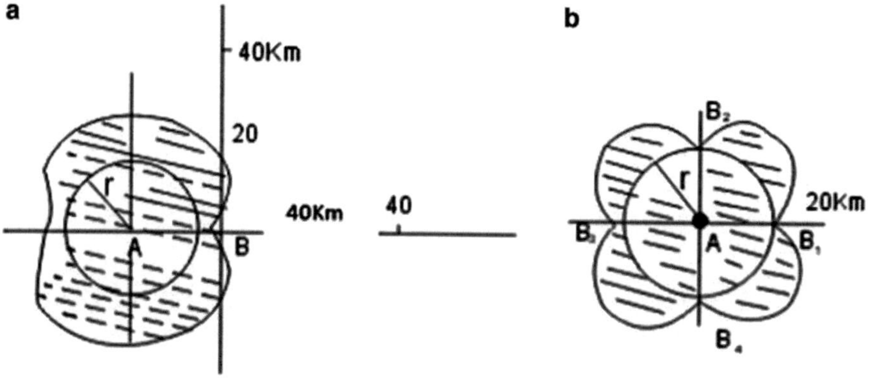


**Fig. 6.7** **a** Jamming exposure area of a single jammer **b** Jamming exposure area formed by the ring configuration of four jammers

In Example 6.4, When $P_j G_j = 50\,\text{dBw}$; $P_a G_a = 80\,\text{dBw}$; $d = 20\,\text{km}$; $r = 1\,\text{km}$; $\theta_{0.5} = 1.5°$; $\sigma = 10^4\text{m}^2$, other parameters remain unchanged, according to formula (6.12), $\alpha_\gamma \approx 0$.

When $\alpha_x = 0°$, $G_a/G_R(d, \beta) \approx 3 \times 10^3$.

Then, the R in formula (6.10) can be written as:

$$R_{(1)} = \left[ \frac{K_j P_a G_a \sigma R_j^2}{4\pi P_j G_j} \cdot \frac{L_d}{K_1 L_j K_f} \cdot \frac{G_a}{G_a(\alpha, \beta)} \right]^{1/4} \tag{6.14}$$

$$R_{(1)} = \frac{1 \times 10_8 \times 10^4 \times 8^2 \times 10^6 \times 5.3 \times 10^3 \times 3}{12.6 \times 10^5} = 30 \text{ at } 0°$$

$$R_{(1)} = 45\text{ km};\ G_a/G_R(d, \beta) = 3 \times 10^2 \text{ at } \alpha_x = 90°$$

$$R_{(1)} = 28\text{ km};\ G_a/G_R(d, \beta) = 10^2 \text{ at } \alpha_x = 180^\circ$$

The jamming exposure area is drawn from the data calculated in Example 4 is shown in Fig. 6.8a, b.

A) Given parameters $r = 1$ km; $P_jG_j = 50$ dBw; $d = 20$ km, the Jamming exposure area formed by a single jammer.
B) Given parameters $r = 1$ km; $P_jG_j = 50$ dBw; $d = 20$ km, the Jamming exposure area formed by four jammers.

***Example 6.5:*** If the equivalent power of the jammer $P_jG_j = 60$ dBw, other parameters remain unchanged, the jamming exposure area is shown as Fig. 6.9a, b. If $P_jG_j = 50$ dBw; $\sigma = 10^3\text{m}^2$, the jamming exposure area is also the same as Fig. 6.9a, b.

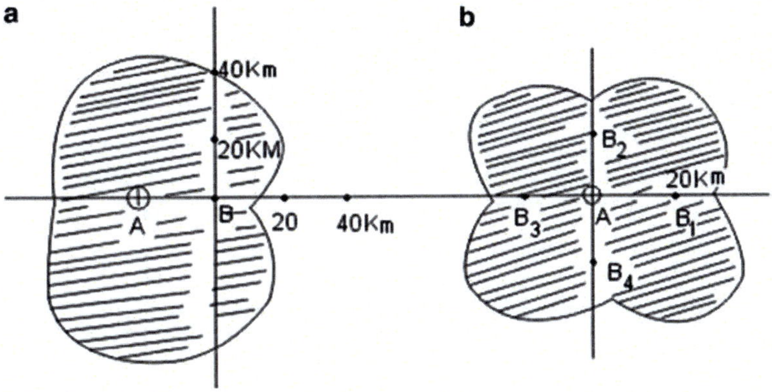


**Fig. 6.8** **a** Jamming exposure area of a single jammer. **b** Jamming exposure area formed by four jammers (No.1)

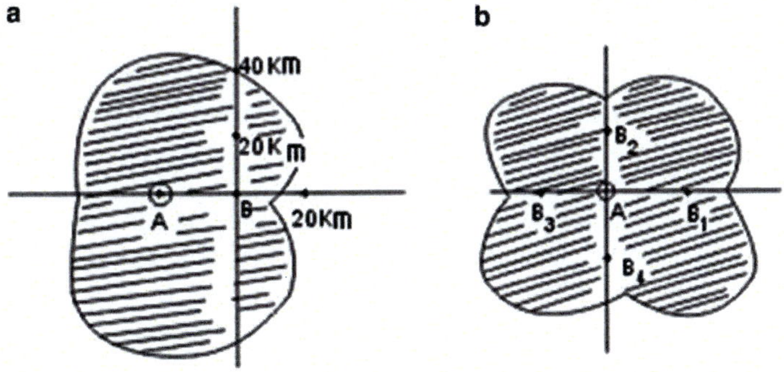


**Fig. 6.9** **a** Jamming exposure area of a single jammer. **b** Jamming exposure area formed by four jammers (No.2)

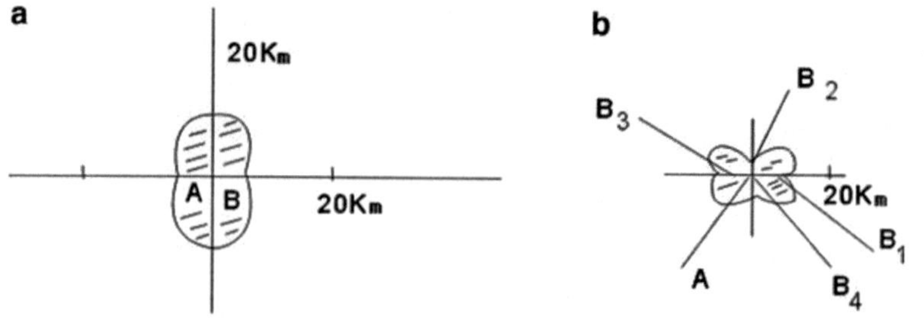


**Fig. 6.10** **a** Jamming exposure area of a single jammer. **b** Jamming exposure area formed by four jammers (No.3)

A) Given parameters $r = 1\,\text{km}$; $P_jG_j = 60\,\text{dBw}$; $\sigma = 10^4\text{m}^2$(or $r = 1\,\text{km}$; $P_jG_j = 50dBw$; $\sigma = 10^3\text{m}^2$), Jamming exposure area formed by a single jammer.
B) Given parameters $r = 1\,\text{km}$; $P_jG_j = 60\,\text{dBw}$; $\sigma = 10^4\text{m}^2$(or $r = 1\,\text{km}$; $P_jG_j = 50dBw$; $\sigma = 10^3\text{m}^2$), Jamming exposure area formed by four jammers.

If Figs. 6.2 and 6.3 are scaled down to uniform coordinates, the jamming exposure area is Fig. 6.10a, b

A) Given parameters $r = 1\,\text{km}$; $P_jG_j = 60\,\text{dBw}$; $d = 3\,\text{km}$; $\sigma = 10^4\text{m}^2$, Jamming exposure area formed by a single jammer.
B) Given parameters $r = 1\,\text{km}$; $P_jG_j = 60\,\text{dBw}$; $d = 3\,\text{km}$; $\sigma = 10^4\text{m}^2$, Jamming exposure area formed by four jammers.

If the radar parameters are fixed, the main factors affecting the size of the jamming exposure area are: the jamming equivalent power of the jammer $P_jG_j$; the main/sub ratio of the radar antenna $G_a/G_R(\theta)$; the radius of the protected target $r$; the effective reflection area of the protected target $\sigma$; the configuration distance d between the jammer and the protected target d, etc. From Figs. 6.5–6.10, the jamming exposure areas differ greatly. For example, in Fig. 6.5a, the radar equivalent radiation power $P_aG_a$ is 80dBw; the jammer equivalent radiation is 50dBw; the protected target radius r is 15 km; the center distance d between the jammer and the protected target is 20 km. In Fig. 6.10a, the radar parameters and jamming parameters have not changed. As long as the protected target radius r becomes 1 km and d become 3 km, the jamming exposure area does change greatly (See Figs. 6.5a–6.10a). Therefore, there are strict regulations on the configuration of ground-to-air jammers. For the typical configuration of the jammer with $P_jG_j = 50\,\text{dBw}$, the average radius of the jamming exposure area smaller than 10 km at $r \leq 1$ km, d = 3—5 km, $\sigma \leq 10^4\text{m}^2$. For the typical configuration of the jammer with $P_jG_j = 60dBw$, the average radius of the jamming exposure area smaller than 15 km at $r \leq 1$ km, d = 5—15 km, $\sigma \leq 10^4\text{m}^2$.

If the equivalent reflection area of the protected target has a large change, the radius of the protected target has a large change, or the configuration distance has a large change, it is the best to use the radar jamming equation to make a new estimate of the jamming exposure area. Select the configuration scheme of the average radius of the best jamming exposure area. Make a limited number of jammers give full play to the effective protection of the jammers on the target. If no selection is made, or the position of the jammer and the protected target is not configured according to the jammer's regulations, the jamming exposure area must be increased to affect the role of the jammer.

## 6.2 Jamming Exposure Area of Ground Jammer Against Early Warning Aircraft Radar

The distance between the ground jammer and the early warning radar is far away. Relative to the height of the early warning aircraft, $R_j$(or $R$) $\gg$ H, H is negligible. Jamming exposure area $R_{\min}(\theta)$:

$$R_{\min}(\alpha,\beta)=\left[\frac{K_{j\Sigma}P_aG_a\sigma R_j^2}{4\pi P_jG_j}\cdot\frac{L_d}{K_1L_jK_f}\cdot\frac{G_a}{G_a(\alpha,\beta)}\right]^{1/4} \tag{6.15}$$

### 6.2.1 *Theoretical Calculation of Jamming Exposure Area for Early Warning Aircraft Radar*

The radar installed on the early warning aircraft is generally pulse compression and Pulse-Doppler radar. The calculation of the jamming suppression coefficient of the pulse pressure radar has been specifically analyzed in Chapter 3. The jamming suppression coefficient of Pulse-Doppler radar has been specifically analyzed in Chapter 4. It will not be repeated here. The following is an analysis of the jamming exposure area.

***Example 6.6:*** Basic parameters of early warning radar: Peak power = 1 MW; Antenna gain = 40Db; Equivalent reflective area of the detected target $\sigma = 5\text{m}^2$; $L_d/\gamma_jL_jK_f = 2$; Transmitted pulse width = $40\mu s$; Pulse width after compression $= 0.3\mu s$; Pulse compression factor after conversion $D = \frac{40}{0.3} = 21.5dB$; Jamming suppression factor $K_{j\Sigma} = 15dB$; $G(\theta)/G \cong K\left(\frac{\theta_{0.5}}{2\theta}\right)^2$; The equivalent jamming power of the jammer $P_jG_j = 70dBw$; Substitute the parameters into the formula (6.16) to calculate Table 6.1. Where $\theta$ includes the value of $\alpha,\beta$ $\frac{G_t(\theta)}{G_t} = \frac{G_t(\alpha,\beta)}{G_t}$. Because the early warning aircraft is far away from the jammer $R$, $R_j \gg H$, so $\beta$ is generally in $0° \sim 3°$, and $G_t(\alpha,\beta) \approx G_t(\alpha)$.

The dotted line is the detection area when the early warning aircraft radar does not jamming.

**Table 6.1** Jamming Exposure Area for different PRI and Rj

| $\theta$ | $G_t(\theta)/G_t$ (dB) | $R_{\min(1)}(R_j = 100\,\text{km})$ | $R_{\min(2)}(R_j = 200\,\text{km})$ | $R_{\min(3)}(R_j = 300\,\text{km})$ |
|---|---|---|---|---|
| 0° | 1 | 3.3 | 4.6 | 5.6 |
| 10° | −44 | 42 | 60 | 72.9 |
| 30° | −52 | 59 | 82 | 100 |
| 45° | −54 | 72 | 100 | 122 |
| 90° | −50 | 56 | 78 | 95 |
| 180° | −50 | 56 | 78 | 95 |

The horizontal line area is the detection area after the radar is jammed (also called jamming exposure area).

A is the location of the jamming station.

B1,2 is the position of early warning aircraft.

C is the target being protected (Fig. 6.11).

After being disturbed, the effective detection area of the early warning aircraft radar has been reduced by more than 90%.

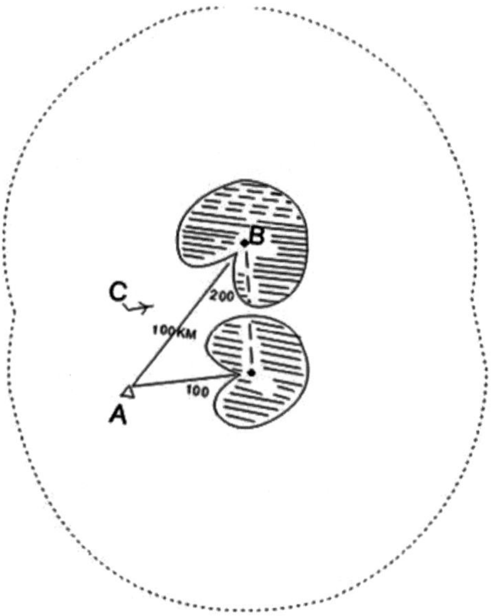


**Fig. 6.11** The early warning aircraft radar jamming exposure area drawn according to Table 6.1

### 6.2.2 *The Measured Jamming Exposure Area for the Early Warning Aircraft Radar*

The radar parameters and jammer parameters are the same as the data in Example 6–2.

The schematic diagram of the jamming test in flight is shown in Fig. 6.12.

Where:

B- Participating radar; A—Participating jammer; The parameters of the two devices are shown in Example 6–2. Route (1) enters from 0°; Route (2) Enter from 15°; Route (3) Enter from 45°; The distance $R_j$ between radar and jamming is 45 km. The target being protected is a drone (the equivalent reflection area is 3.5 m$^2$).

The test method is arranged according to Fig. 6.12, and the parameters of radar/jammer and the actual parameters of the radar reaching the side lobe antenna are tested on-site. Specific steps are as follows:

(1) The drone first enters the starting point at 60 km from the route (1), and the radar searches for the target. When the drone entered the radar headspace from the starting point, the radar did not find the target. The minimum detection distance of the radar is 12 km, so the minimum jamming exposure distance is considered to be less than 12 km.

(2) The drone enters from the 15° direction of route (2). The radar did not find the target when the target at the starting point. The target was found until 22.1 km. The radar finds the target only after finding the target in two consecutive turns. Make three consecutive entries on the route (2), and find the root mean square value.

(3) The drone enters from the 45° direction of the route (3). The radar did not find the target when the target at the starting point. The target was found until 34.4 km. Make three consecutive entries on the route (3), and find the root mean square value.

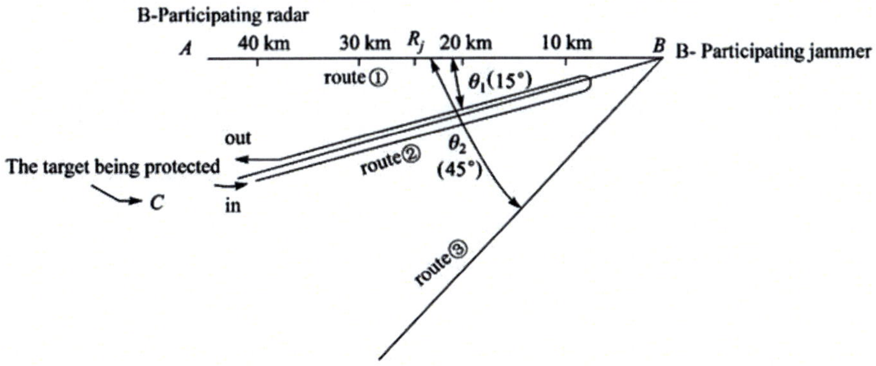


**Fig. 6.12** Schematic diagram of actual jamming flight test of jammer and radar

**Table 6.2** The actual test and estimated jamming exposure distance with different route

| Jamming effect | Flight path | Measured burn-through distance (km) | Measured elevation point of burn-through point | Actual burning point $G_t(\theta)/G_t$ (dB) | $R_j = 300km$ Estimated value (km) | Root mean square (km) |
|---|---|---|---|---|---|---|
| Tertiary jamming | 0° direction | Less than 12 | | | 31.2 | Less than 12 |
| Tertiary jamming | 15° | 18.1 | 3.05° | −43 | 57.5 | 22.1 |
| | | 22.1 | 2.36° | −44 | | |
| | | 23 | 2.52° | −44 | | |
| Tertiary jamming | 45° | 34.4 | 1.48° | −53 | 90 | 34.6 |
| | | 32.5 | 1.78° | −53 | | |
| | | 37 | 1.44° | −53 | | |

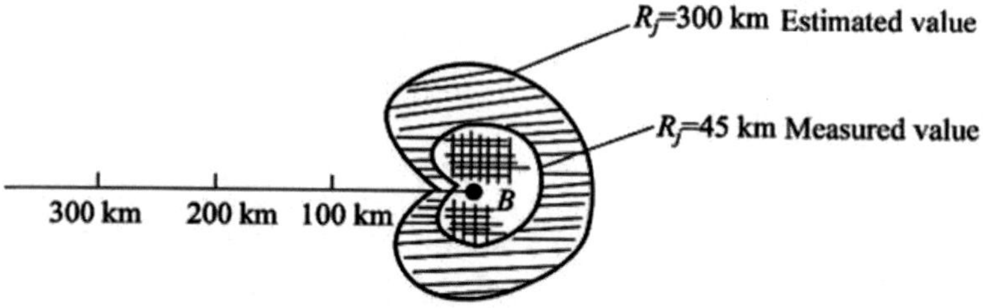


**Fig. 6.13** Measured value and estimated value of the jamming exposure area of the early warning aircraft radar

(4) The actual test and estimated jamming exposure distance are shown in Table 6.2

According to Table 6.2, the measured jamming exposure area and the calculated jamming exposure area can be drawn in Fig. 6.13

When $R_j = 300$ km, Fig. 6.12 is the theoretically calculated value, and Fig. 6.13 is the actual measured value. The jamming exposure areas of the theoretical value and measured value are basically the same.

### *6.2.3 The Jamming Exposure Area Formed by Shipborne Jammer covering the fleet*

***Example 6.7:*** Assume that the radius of the fleet formation is 50 km, the distance between the jammer and the early warning aircraft radar is 200 km, and the effective reflection area is $10{,}000^2$ m. The configuration is shown in Fig. 6.14.

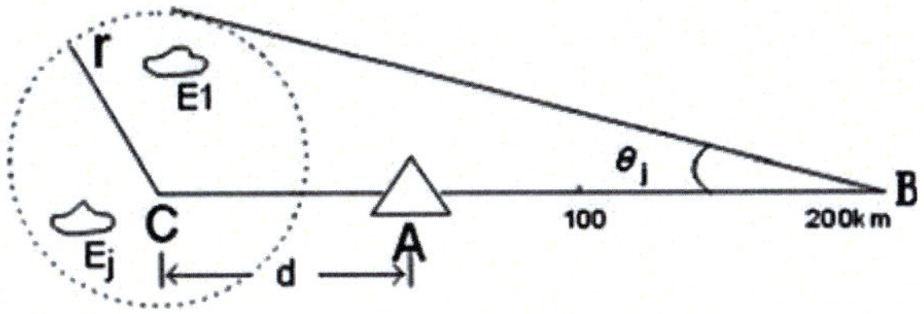


**Fig. 6.14** Shipborne jammer covering the fleet

A—position of the jammer; B—position of the early warning aircraft radar; C—Center of the protected target; d—the distance between point A and point C; r—the radius of the protected target, all protected ships are in the park with the radius of r; E—Protected target.

Analysis of jamming exposure area:

Known parameters of early warning aircraft radar: P-band frequency; peak power is 1 MW; antenna gain is 21.5 dB; transmit pulse width 1 is 3 μs; pulse width after compression is 0.25 μs; $\sigma = 10^4 m^2$; $R_j = 200\,\text{km}$; $r = 50\,\text{km}$; $P_j G_j = 50\,\text{dBw}$; $R_j \gg H$.

Due to the larger radius of the protected target, $G_a/G_a(\alpha, \beta) \approx G_a/G_a(\alpha, \theta_i)$.

Then (6.16) can be written as

$$R_{\min}(\alpha, \theta_i) = \left[ \frac{K_{j\Sigma} P_a G_a \sigma R_j^2}{4\pi P_j G_j} \cdot \frac{L_d}{K_1 L_j K_f} \cdot \frac{G_a}{G_a(\alpha, \theta_i)} \right]^{1/4} \tag{6.16}$$

The graphical relationship of (6.17) is shown in Fig. 6.15

According to Fig. 6.15, when $\alpha_x = 0°$, $\alpha = 0°$, $\theta_i = \tan^{-1} \frac{r}{R_j + d}$.

(1) when $\alpha_x = 0°$, $\theta_{i0} = \tan^{-1} \frac{50}{200+62.5} = 11°$
(2) when $\alpha_x = 30°$, $\overline{AD_i} = 200 \times 0.86 = 172\,\text{km}$

$$\overline{BD_i} = 200 \times 0.5 = 100\,\text{km}$$

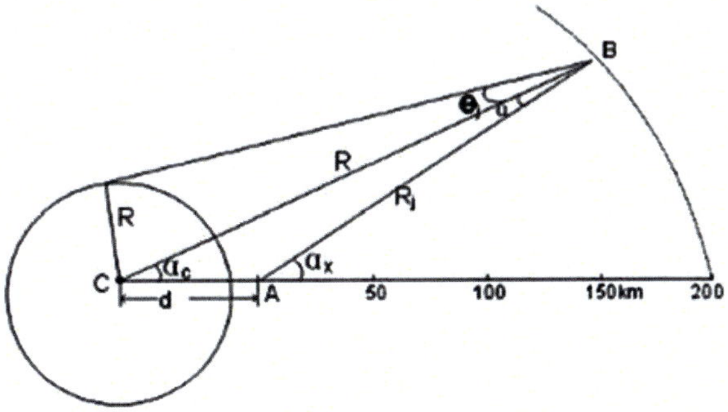


**Fig. 6.15** Shipborne jammer covering the fleet

$$\alpha_c = \tan^{-1} \frac{\overline{BD_i}}{d + \overline{AD_j}} = 23°;$$

$$\alpha = \alpha_x - \alpha_c = 30° - 23° = 7°$$

$$R = \left[\overline{BD_i} + (d + AD_i)^2\right]^{1/2} = 254\,\text{km}$$

$$\theta_{i(1)} = \sin^{-1} \frac{r}{R} = 11.8°$$

(3) when $\alpha_x = 45°, \overline{AD_{i(1)}} = R_j \cos 45° = 141.8km$

$$\overline{BD_{i(1)}} = 141.8\,\text{km}$$

$$\alpha_c = \tan^{-1} \frac{\overline{BD_{i(1)}}}{d + \overline{AD_{j(1)}}} = 35°;$$

$$\alpha = \alpha_x - \alpha_c = 45° - 35° = 10°$$

$$\theta_{i(2)} = \sin^{-1} \frac{r}{R} = \sin^{-1} \frac{50}{247} = 12°$$

(4) when $\alpha_x = 60°, \overline{AD_{i(2)}} = R_j \cos 60° = 100km$

$$BD_{i(2)} = R_j \sin 60° = 172\,\text{km}$$

$$\alpha_c = \tan^{-1} \frac{172}{162} = 48°;$$

$$\alpha = \alpha_x - \alpha_c = 60° - 48° = 12°$$

$$\theta_{i(2)} = \sin^{-1} \frac{r}{R} = \sin^{-1} \frac{50}{247} = 12°$$

(5) when $\alpha_x = 60°, \overline{BD_{i(4)}} = 200\,\text{km}$

$$\overline{AD_{i(2)}} = 0$$

$$R = \left(R_j^2 + d^2\right)^{1/2} = 209\,\text{km}$$

$$\alpha_c = \tan^{-1}\frac{R_j}{d} = 76°; \alpha = 90° - 76° = 14°$$

$$\theta_{i(4)} = 12°$$

(6) when $\alpha_x = 90°$ 时, $\overline{BD_{i(4)}} = 0$

$$\overline{AD_{i(2)}} = 200\,\text{km}$$

$$R = 200 - d = 150\,\text{km}$$

$$\alpha_c = 180°; \alpha = 0°$$

$$\theta_{i(5)} = \tan^{-1}\frac{r}{R_j - d} = 20°$$

Organize the changes of $\alpha_c$, $\theta_i$ and $G_a/G_a(\alpha, \theta_i)$ caused by the change of e into tabular data, see Table 6.3.

$G(\theta)/G \cong K\left(\frac{\theta_{0.5}}{2\theta}\right)^2$, take $\theta_{0.5} = 4°$, $\theta = \alpha + \theta_i$, K = 0.2, $K_{j\Sigma} = 10dB$, $L_d/K_1 r_j K_f \cong 4$.

According to the calculation data, the jamming ship can cover the fleet.

The jamming exposure zones drawn according to Table 6.3 are shown in Figs. 6.16a and b.

**Table 6.3** jamming exposure distance with different angles

| $\alpha_x$ | 0° | 30° | 45° | 60° | 90° | 180° | Remarks |
|---|---|---|---|---|---|---|---|
| $\alpha$ | 0° | 7° | 10° | 12° | 14° | 0° | |
| $\theta_i$ | 11° | 11.8° | 12° | 12° | 12° | 20° | |
| $G_a/G_a(\alpha, \theta_i)$ | 14 dB | 22 dB | 24 dB | 25 dB | 25.5 dB | 23 dB | |
| $R_{\min(1)}(\alpha, \theta_i)$ | 89 km | 138 km | 159 km | 167 km | 175 km | 151 km | $P_j G_j = 50\,\text{dBw}$ |
| $R_{\min(2)}(\alpha, \theta_i)$ | 49.4 km | 76.7 km | 88.3 km | 92.8 km | 97.2 km | 83.9 km | $P_j G_j = 60\,\text{dBw}$ |

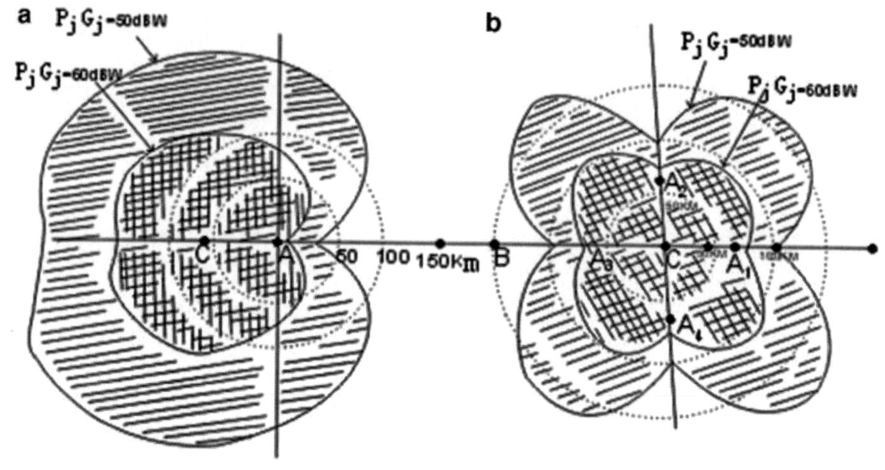


**Fig. 6.16** A Single jamming ship shielding fleet formation jamming exposure area $R_j$ = 200km; $d$ = 62km;$r$ = 50km. **B** Four Jamming Ships Ring Configuration Covering Fleet Formation Jamming Exposure. $R_j$ = 200km; $d$ = 62km; $r$ = 50km

## 6.3 Analysis on the Jamming Exposure Area and Jamming Effective Area of Ground-to-Air Jammers Against Airborne SAR

### *6.3.1 Analysis and Calculation on Jamming Effective Area for Airborne SAR*

The schematic diagram of airborne SAR imaging on the ground is shown in Fig. 6.17.

B—SAR Carrier.

$\overline{BB'}$—Flight Line of SAR Carrier.

D—Projection point of SAR carrier.

$\overline{DD'}$—Ground projection line of SAR carrier.

C—SAR imaging center.

$\overline{CC'}$—The centerline of the SAR imaging zone.

Horizontal line—SAR imaging zone.

**Fig. 6.17** Schematic diagram of airborne SAR imaging

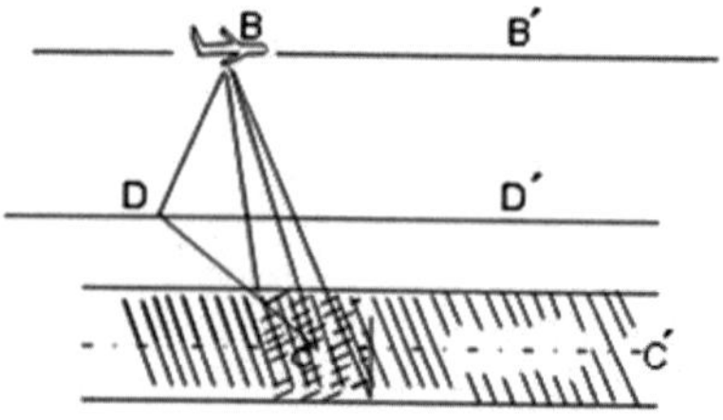

The schematic diagram of airborne SAR when it is jammed is shown in Fig. 6.18a, b.

A—Position of the jammer; B—Position of the radar carrier; C—The projection point of the carrier; E—The vertical point between the centerline of the irradiation and the parallel line of the jammer; H—Carrier height; R—Distance between carrier radar and irradiation center; $R_j$—Distance between jamming and carrier.

Set: $\overline{BD'} = H$, $\overline{DC} = d$, $\overline{CE} = da$, $EA = r$. According to the Pythagorean Theorem: $R^2 = H^2 + d^2$; $R_{j0}^2 = R_j^2 - H^2$; $R_{j0}^2 = (d + da)^2 + r^2 = R_j^2 - H^2$:

$$\begin{cases} R_j^2 = (d + da)^2 + r^2 + H^2 \\ R^2 = H^2 + d^2 \\ R_j^2 = \frac{4\pi P_j G_j R^4}{K_{j\Sigma i} P_t G_t \sigma} \cdot \frac{G_t(\theta)}{G_t} \cdot \frac{L_j r_j K_f}{L_d} \end{cases}$$

Solve the above equation:

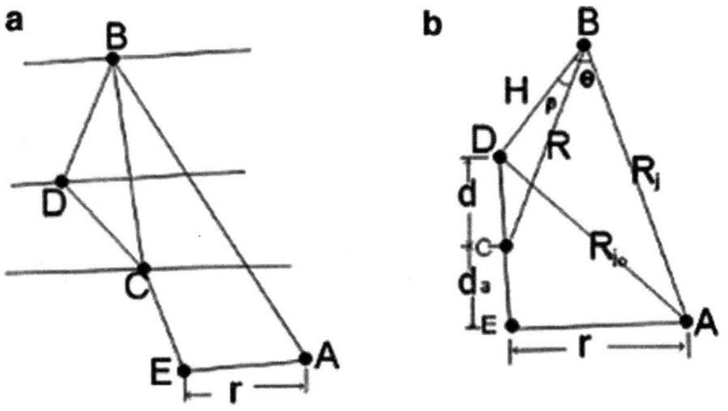


**Fig. 6.18** **a** Schematic diagram of onboard jamming **b** geometry of airborne SAR when jammed

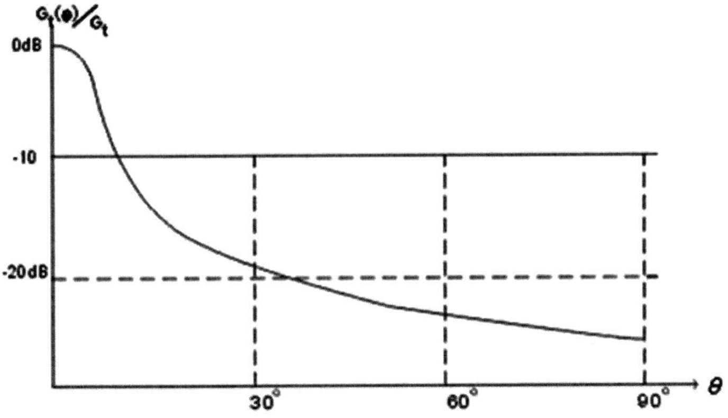


**Fig. 6.19** Relation curve between $G_t(\theta)/G_t$ and $\theta$

$$r^2 = \left[\frac{4\pi P_j G_j (H^2 + d^2)^2}{K_{j\Sigma i} P_t G_t \sigma} \frac{G_t(\theta)}{G_t} \cdot \frac{L_j r_j K_f}{L_d}\right] - (d + da)^2 + H^2) \quad (6.17)$$

***Example 6.8:*** Known Parameters: $P_j G_j = 45\,\text{dBw}$; $P_t G_t = 42\,\text{dBw}$; $K_{j\Sigma i} = 20\,\text{dB}$; $\sigma = 40dB$; $H = 7.2\,\text{km}$; $d = 18.7\text{km}$; $\frac{L_j r_j K_f}{L_d} = 1$; $\theta_{0.5} = 8°$; $G_t(\theta)/G_t = K\left(\frac{\theta_{0.5}}{\theta}\right) \cong 0.4\left(\frac{8}{2\theta}\right)^2$,$\sigma = 10^4$.

Data relationship between Gt (θ)/Gt and θ is shown in Table 6.4. Relation curve between Gt (θ)/Gt and θ is shown in Fig.6.19.

If $d_a = 0$, substitute $G_t(45°)/G_t = -23\text{Db}$ into (6.18), and calculated $r_{(1)} = 39$ km.

If $d_a = 18$ km, substitute $G_t(\theta)/G_t = \frac{G_t(90°)}{G_t} = -25\,\text{dB}$ into (6.18), and calculated $r_{(2)} = 3.2\text{km}$.

If $d_a$ move 18 km back, substitute $G_t(\theta)/G_t = -13\text{Db}$ into (6.18), and calculated $r_{(3)} = 28\text{km}$.

In Fig. 6.18b, $\beta$ is the elevation angle of the airborne SAR towards the imaging center, generally 20–57° (Fig. 6.19). When the airborne SAR is imaging for illuminating far-range targets, take 57°. Now $da$ is zero, which is +18 km and −18 km. The three data are plotted on the graph to show the jamming effective area and jamming exposure area of the airborne SAR, as shown in Fig. 6.20. Taking the jammer "A" as the center, solve the intersection point of the ground projection line of the SAR carrier track (1) and $R_{j01}$ to obtain $r_{(1)}$; solve the intersection point of the ground

**Table 6.4** Data relationship between Gt (θ)/Gt and θ

| $\theta$ | 0° | 8° | 16° | 32° | 45° | 90° |
|---|---|---|---|---|---|---|
| $G_t(\theta)/G_t$ | 1 | −10 dB | −16 dB | −20 dB | −23 dB | −25 dB |

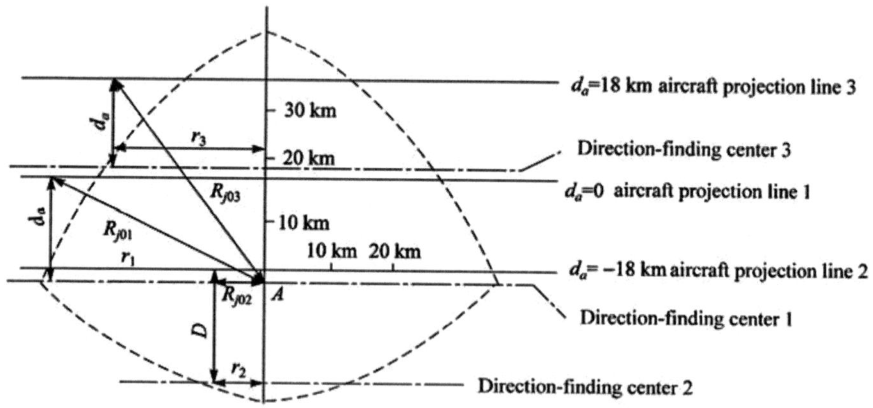


**Fig. 6.20** Airborne SAR jamming effective area and jamming exposure area schematic diagram (ground projection area)

projection line of the SAR carrier track (2) and $R_{j02}$ to obtain $r_{(2)}$; solve the intersection point of the ground projection line of the SAR carrier track (3) and $R_{j03}$ to obtain $r_{(3)}$.

In the figure, point A is the position of the jammer. Inside the dotted line is the jamming effective area. Outside the dotted line is the jamming invalid area.

***Example 6.9:*** If the imaging angle of the airborne SAR in Example 1 is 23°, d = 3 km, and other parameters remain unchanged.

Substitute the data into (6.18):

When $d_a = 0$, $G_t(\theta)/G_t = 4 \times 10^{-2}$

$$r_{(1)} = \left[\frac{12.6 \times 3 \times 10^4 (7.2^2 + 3^2)^2 \times 10^{12} \times 0.5 \times 8 \times 10^{-2}}{1.6 \times 10^7 \times 10^4} - (7.2^2 + 3^2) \times 10^6\right]^{1/2}$$
$$= 17km$$

When $d_a = 20\text{km}$, $G_t(\theta)/G_t = 8 \times 10^{-2}$

$$r_{(2)} = \left[\frac{12.6 \times 3 \times 10^4 (7.2^2 + 3^2)^2 \times 10^{12} \times 8 \times 10^{-2}}{1.6 \times 10^7 \times 10^4} - (7.2^2 + 23^2) \times 10^6\right]^{1/2}$$
$$= 12km$$

According to the data of $r_{(1)}$ and $r_{(2)}$, the jamming effective area described in Example 6–9 can be drawn roughly.

In the figure, point A is the position of the jammer. Inside the dotted line is the jamming effective area.

It can be seen from Fig. 6.20 and Fig. 6.21 that two factors have a greater influence on the effective area of airborne SAR jamming: The first is when the airborne SAR is imaging irradiation, when its working elevation $\beta$ is small, the effective area of jamming is small; When $\beta$ is large, the jamming effective area is large (under the same conditions of other parameters); the second is that the sidelobe characteristics

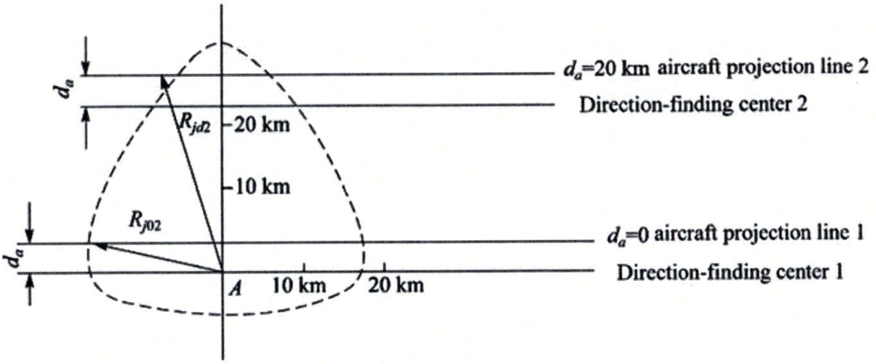


**Fig. 6.21** Schematic diagram of the effective area of airborne SAR jamming at angle hours

of the airborne SAR antenna have a greater impact on the jamming effective area. This is one of the key elements of radar jamming and anti-jamming.

***Example 6.10:*** Known parameters of airborne SAR: $P_t = 2\,\text{kw}$; $G_t = 30\,\text{dB}$; $K_{j\Sigma i} = 20\,\text{dB}$; $\sigma = 40\,\text{dB}$; $P_t G_t = 50\,\text{dBw}$ or 60dBw; working in the X-band, the azimuth of the beamwidth of the radar antenna is 4°; elevation angle is 6°; the characteristics of $G_t(\theta)/G_t$ are shown in Table 6.5.

Calculate $d = H \tan\beta = 8 \times 1.75 = 14\,\text{km}$ and $R = 16.1\,\text{km}$ based on known data.

When $d_a = 0$, substitute the parameter into (6.18):

$$r_{(1)} = \left[\frac{12.6 \times 10^5 \times (16.1 \times 10^3)^4}{100 \times 2 \times 10^5 \times 3 \times 10^4} \times 10^{-2} - (16.1 \times 10^3)^2\right]^{1/2} = 34\,\text{km}$$

When $d_a = -14\,\text{km}$:

$$r_{(2)} = \left[\frac{12.6 \times 10^5 \times (16.1 \times 10^3)^4}{6 \times 10^{11}} \times 10^{-3} - 6.4 \times 10^7\right] = 8.9\,\text{km}$$

When $d_a = 14\,\text{km}$:

$$r_{(3)} = \left[\frac{12.6 \times 10^5 \times (16.1 \times 10^3)^4}{6 \times 10^{11}} - ((d + d_a)^2 + H^2)\right]^{1/2} = 24\,\text{km}$$

The effective area of jamming to airborne SAR drawn based on the calculated data is shown in Fig. 6.22.

A—Position of Jamming; d—The vertical distance between the projection line of the carrier's flight path and the centerline of the measuring tape; $P_j G_j = 50\,\text{dBw}$; Inside the dotted line—the jamming effective area.

***Example 6.11:*** Change $P_j G_j = 60$ dBw, other parameters remain unchanged, radar parameters also remain unchanged, the calculated effective jamming area. Substituting the parameters into the (6.18) formula yields:

When $d_a = 0$:

$$r_{(1)} = \left[\frac{12.6 \times 10^6 \times (16.1 \times 10^3)^4 \times 0.3 \times 10^{-2}}{6 \times 10^{11}} - (16.1 \times 10^3)^2\right]^{1/2} = 63\,\text{km}$$

When $d_a = -14\,\text{km}$:

**Table 6.5** Table of the side lobe characteristics of $\beta = 60°$ with 8 km in height

| $\theta$ | 0° | 4° | 8° | 16° | 32° | 45° | 90° |
|---|---|---|---|---|---|---|---|
| $G_t(\theta)/G_t$ | 0 dB | −10 dB | −16 dB | −22 dB | −29 dB | −31 dB | −32 dB |

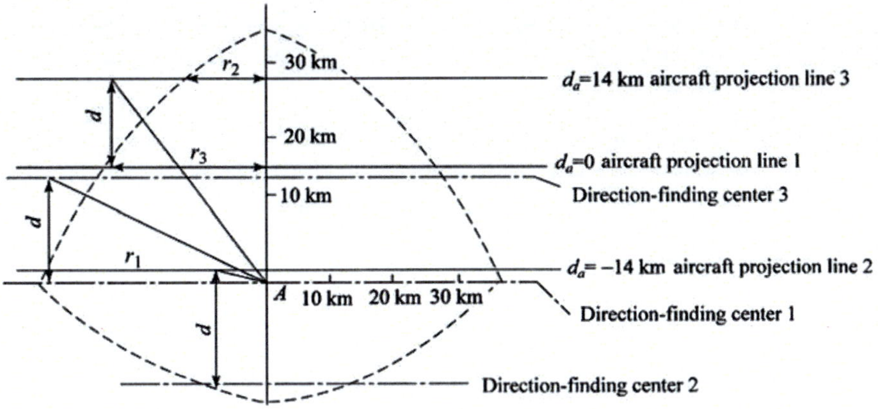


**Fig. 6.22** Airborne SAR, X-band jamming effective area schematic

$$r_{(2)} = \left[\frac{12.6 \times 10^6 \times (16.1 \times 10^3)^4 \times 0.3 \times 10^{-3}}{6 \times 10^{11}} - 6.4 \times 10^7\right]^{1/2} = 19.1\,\text{km}$$

When $d_a = 14\,\text{km}$:

$$r_{(3)} = \left[\frac{12.6 \times 10^6 \times (16.1 \times 10^3)^4 \times 0.3 \times 10^{-2}}{6 \times 10^{11}} - 8.48 \times 10^8\right]^{1/2} = 59.4\,\text{km}$$

The jamming effective area plotted by the calculation data of Example 4 is shown in Fig. 6.23.

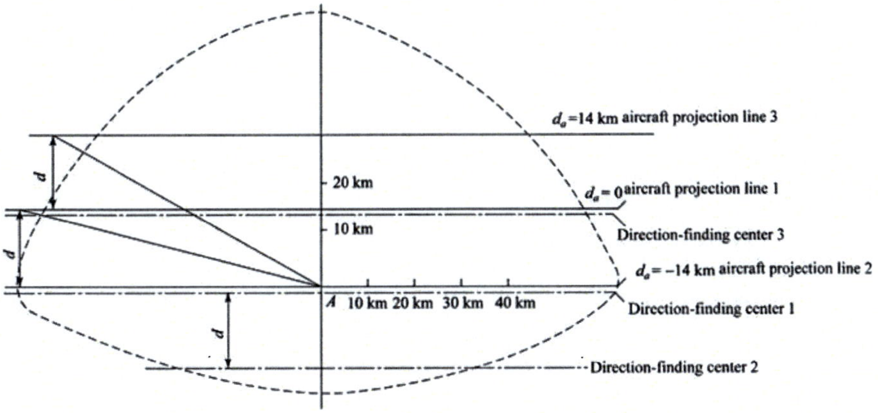


**Fig. 6.23** Schematic diagram of X-band jamming effective area of airborne SAR

As can be seen from Fig. 6.23, after the increase of $P_j G_j = 60\,\text{dBw}$ (the effective area of jamming in the dotted line area), the effective area of jamming to airborne SAR increases significantly.

***Example 6.12:*** If the airborne SAR works at a small angle of $\beta$ (23°), the jamming effective power significantly reduces its jamming effective area. When $\beta$ is 23°, d = 3 km is calculated, and other parameters remain unchanged.

Substitute it in (6.18), yield:
When $d_a = 0$:

$$r_{(1)} = \left[\frac{12.6 \times 10^5 (8^2 + 3^2)^2 \times 10^{12} \times 10^{-2} \times 4}{6 \times 10^{11}} - (8^2 + 3^2) \times 10^6\right]^{1/2} = 19.3\,\text{km}$$

When $da = 20\,\text{km}$:

$$r_{(2)} = \left[\frac{12.6 \times 10^5 (8^2 + 3^2)^2 \times 10^{-2} \times 8}{6 \times 10^{11}} - (8^2 + 23^2) \times 10^6\right]^{1/2} = 16.3\,\text{km}$$

If $P_j G_j = 60\,\text{dBw}$, Substitute it in (6.28), yield:
When $d_a = 0$:

$$r'_{(1)} = \left[\frac{12.6 \times 10^6 (8^2 + 3^2)^2 \times 10^{12} \times 10^{-2}}{6 \times 10^{11}} - (8^2 + 3^2) \times 10^6\right]^{1/2} = 32\,\text{km}$$

When $da = 20km$:

$$r'_{(2)} = \left[\frac{12.6 \times 10^6 (8^2 + 3^2)^2 \times 10^{12} \times 10^{-2}}{6 \times 10^{11}} - (8^2 + 23^2) \times 10^6\right]^{1/2} = 22.4\,\text{km}$$

### 6.3.2 *Steps and Methods for Measuring the Jamming Effective Area to Airborne SAR*

Many examples are given in "6.2.4", and the jamming effective areas of the airborne SAR in Figs. 6.20, 6.21, 6.22, 6.23 and 6.24 are calculated. The size and shape of these jamming effective areas are related to the radar's equivalent radiated power, the radar's transmitted pulse width, working system, beamwidth, antenna sidelobe characteristics, etc. It is also related to the equivalent reflection area of the target in the irradiation area of the radar antenna beam. The equivalent radiated power of the jammer, carrier height, antenna beam elevation angle, polarization loss, feeder error, jamming antenna alignment error, and many other factors. Therefore, it is difficult to accurately determine the values of these parameters in the calculation, which will

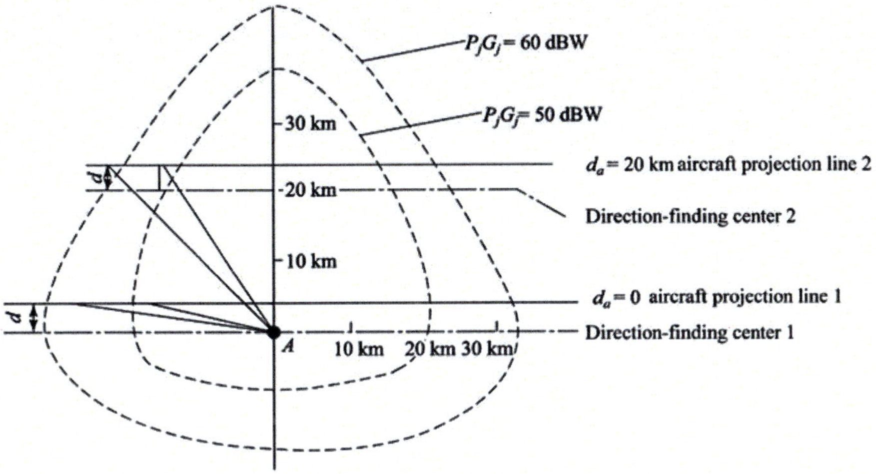


**Fig. 6.24** Airborne SAR, X-band jamming schematic diagram at small $\beta$ angles

bring a large calculation error. If you want to get a more accurate jamming effective area for the airborne SAR, you must do an actual flight test to measure the actual jamming effective area. Thereby correcting the value error of some parameters to provide help for the theoretical design and layout of jammers.

Test steps:

1. Measurement of the characteristics of the equivalent radiated power, pulse width, and antenna side lobe of the participating airborne SAR.
2. Test the equivalent reflective area of the protected target.
3. Test the jammer's equivalent jamming power, aiming frequency error, angle tracking error.
4. Determine the flying height of the aircraft and the value of the elevation angle $\beta$.
5. Determine the flight path so that the centerline of the reference imaging zone passes through the position of the jammer.
6. Identify the various tables that need to be used to format the test data.
7. Quipped with necessary radar to measure the trajectory and instantaneous coordinate position of the tested SAR carrier.
8. Transmit airborne SAR images to the ground command center in time.

Test method:

1. Flight according to the pre-set route, as shown in Fig. 6.25.

In Fig. 6.25: $A_1$—Position of Jammer; $A_2$—Position of radar; B—SAR Carrier; H—Carrier height; $\beta$—Elevation angle of the radar beam; Di—Projection point of SAR; d—When the elevation angle is equal to $\beta$, the vertical distance between the projection line and the centerline of the reference measuring zone.

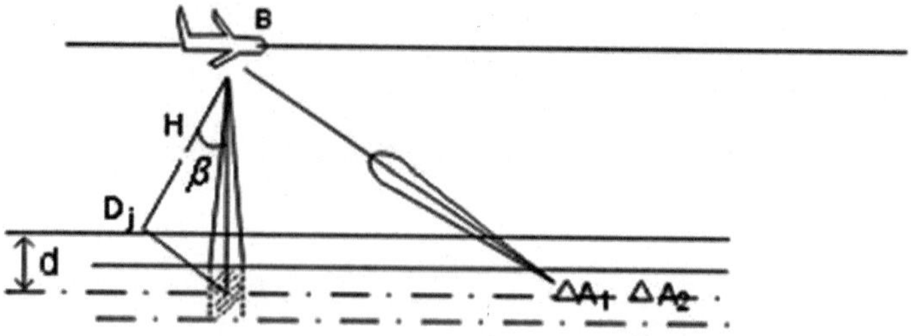


Fig. 6.25 Measured airborne SAR jamming effective area schematic

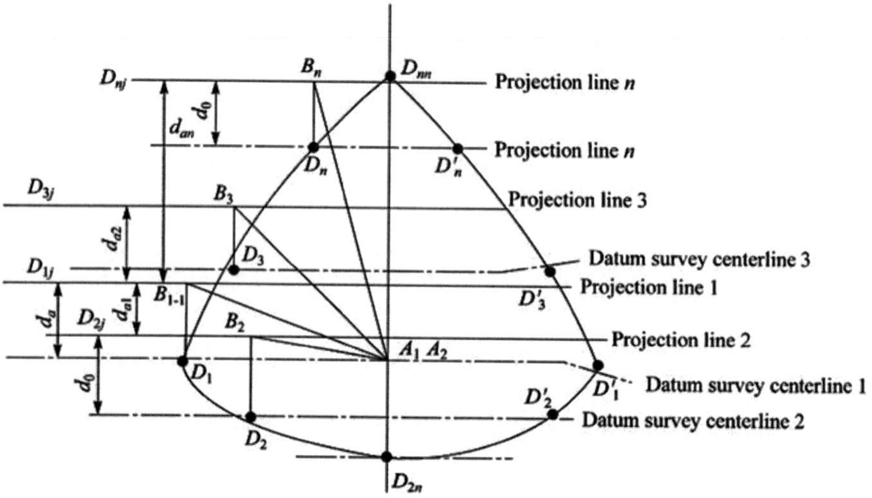


Fig. 6.26 Flight test airborne SAR effective jamming diagram

2. The projection line diagram of the proposed carrier is shown in Fig. 6.26.
3. Airborne SAR is flying according to Di route. The jammer began a reconnaissance search. Turn on the transmitter after finding the test SAR signal. Then radar (A2) tracks carrier SAR.
4. Airborne SAR transmits images to the command post at the right time. After the radar is disturbed, the image is gradually blurred. When the third level of jamming is reached, let radar A2 give the position of the SAR carrier and record it as B1-1. Repeat (3) and (4) the flight, and record the data of B1-2 and B1-3, respectively. The average B1-1 to B1-3 data yields B1-1 points. Draw a vertical line at point B1-1 on the centerline of the reference measurement to get the point D1. The D1A1 distance is taken as the effective jamming distance of the reference imaging centerline.
5. The airborne SAR changes the flight path, flying according to D2i, repeating the steps (4) to get the point B2. Connect point B2 with A1. From point B2,

make a vertical line on the centerline (2) of the reference measurement to get the point D2. Move the route D2i, and gradually measure D2n.

6. Change the airborne SAR route to D3i. Repeat step (4) to get the point B3. Connect point B3 to A1. From point B3, make a vertical line on the reference measurement centerline (3) to get the point D3.
7. Change the airborne SAR route to Dni. Repeat step (4) to get the point B3. Connect point Bn to A1. From point Bn, make a vertical line on the reference measurement centerline (n) to get the point Dn. Move the route Dni, and gradually measure Dnn.
8. Connect the measured D2n, D1, D1, D3, Dn, and Dnn in Fig. 6.26 with a dotted line in sequence, and draw the dotted line from D2n to Dnn.
9. According to the principle of symmetry, draw the points of D2‘, D1‘, D3‘ and Dn‘ and connect them with dotted lines. The dashed circles connected to the entire D2n, D2, D1, DB, Dn, Dnn, Dn‘, D3‘, D1‘, and D2‘ are measured airborne SAR jamming effective areas. Outside the dotted area is the invalid jamming area.
10. Compare the measured effective area of airborne SAR jamming with the theoretically calculated jamming effective area to find the difference. Look for the error of some parameter values in the calculation, so that the system can obtain the optimized effect.

Change jammer power or radar parameters

According to the steps and methods of "6.3.1" and "6.3.2", the jamming area to the airborne SAR with different jamming equivalent power is measured. The effective jamming areas of airborne SAR measured by different radar parameters and different jamming equivalent powers are added to the instruction manuals of some jammers to use these jamming devices better and effectively.

## 6.4 Analysis and Calculation of Jamming Effective Area F Spaceborne SAR

The jamming to spaceborne SAR is similar to that for airborne SAR. It's just that the spaceborne SAR has a higher flying height, a wider imaging range, and the equivalent jamming power required for jamming is greater.

The schematic diagram of jamming to spaceborne SAR is shown in Fig. 6.27.

A—Position of Jammer; B—Position of spaceborne SAR; D—Projection point of spaceborne SAR; C—The imaging center of the main lobe region of the spaceborne SAR.

Draw Fig. 6.27 into a geometric figure as shown in Fig. 6.28

H—Satellite height; Rj—Distance between jammer and radar.

According to Fig. 6.27, the following equations can be listed:

$$d = \overline{DC} = H \tan \beta \tag{6.18}$$

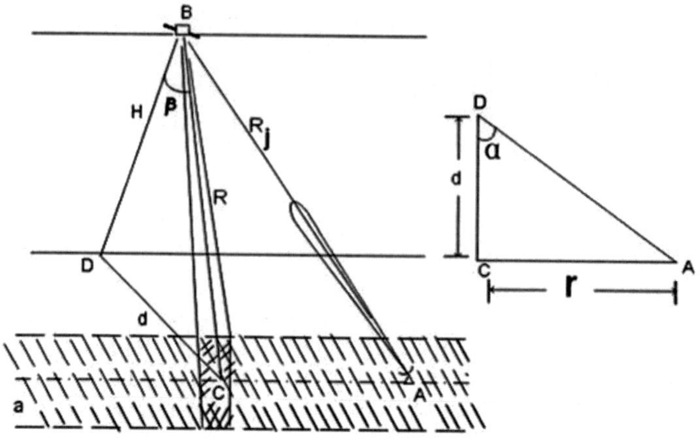


Fig. 6.27 Schematic diagram of jamming to spaceborne SAR

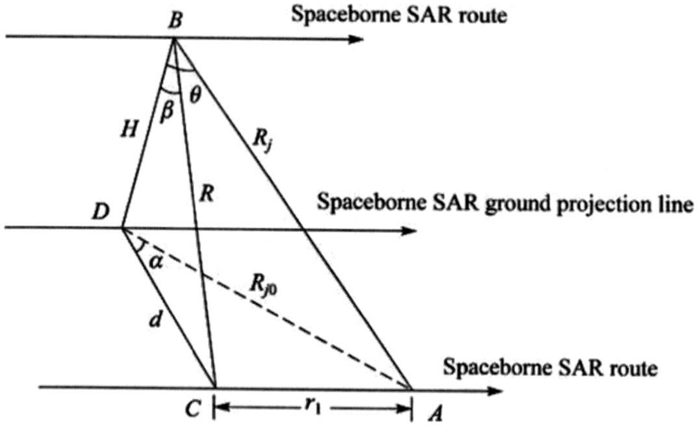


Fig. 6.28 Spaceborne SAR jamming geometry line

$$R = (H^2 + H^2 \tan^2 \beta)^{1/2} = H(1 + \tan^2 \beta)^{1/2} \tag{6.19}$$

$$r_1 = d \tan \alpha \tag{6.20}$$

$$R_{j0} = (r^2 + d^2)^{1/2} = d(1 + \tan^2 \alpha)^{1/2} \tag{6.21}$$

$$\text{or} = H \tan(1 + \tan^2 \alpha)^{1/2} \tag{6.22}$$

$$R_j = (H^2 + R_{j0})^{1/2} = \left[H^2 + H^2 \tan^2 \beta (1 + \tan^2 \alpha)\right]^{1/2}$$

$$= H\left[1 + \tan^2 \beta(1 + \tan^2 \alpha)\right]^{1/2} \tag{6.23}$$

According to the jamming equation of the radar

$$P_j G_j = \frac{K_{j\Sigma i} P_t G_t \sigma R_j^2}{4\pi R^4} \cdot \frac{G_t}{G_t(\theta)} \cdot \frac{L_d}{L_j r_j K_f} \tag{6.24}$$

Substituting (6.19) and (6.23), we get:

$$P_j G_j = \frac{K_{j\Sigma i} P_t G_t \sigma H^2\left[1 + \tan^2 \beta(1 + \tan^2 \alpha)\right]}{4\pi H^4(1 + \tan^2 \beta)^2} \cdot \frac{G_t}{G_t(\theta)} \cdot \frac{L_d}{L_j r_j K_f}$$

$$\text{or} = \frac{K_{j\Sigma i} P_t G_t \sigma\left[1 + \tan^2 \beta(1 + \tan^2 \alpha)\right]}{4\pi H^2(1 + \tan^2 \beta)^2} \cdot \frac{G_t}{G_t(\theta)} \cdot \frac{L_d}{L_j r_j K_f} \tag{6.25}$$

In the formula, k is the jamming suppression coefficient obtained by jamming synthetic aperture radar. It has two calculation methods. If $\sigma$ is the equivalent reflection area of the radar resolution area, $K_{j\Sigma i}$ is the gain coefficient of two-dimensional compression. If $\sigma$ is the radiation equivalent reflection area of the main lobe area of the radar antenna, $K_{j\Sigma i}$ is the distance compression gain coefficient. No matter which of these two algorithms is used, the result is $K_{j\Sigma i} = (D)^{0.8}\sigma$ or $K_{j\Sigma i} = (Dn)^{0.8}\sigma_i \cong 10^6 \sim 10^7$. In the formula,$\sigma$ is the equivalent reflection area of the radiation area of the main lobe of the radar antenna; $\sigma_i$ is equivalent reflection area in radar resolution zone; D is radar range compression area; $\eta$ is the gain coefficient of radar azimuth compression. The 0.8th power is due to the processing amplification of the radar and the loss caused by the nonlinearity.

***Example 6.13:*** $P_j G_j$ is the equivalent radiated power of SAR, and the value is 80dBw in the L band.

$\beta$ is the angle between the elevation angle of the spaceborne SAR antenna, and the value is 25°.

$\alpha$ is the azimuth angle of the SAR projection on the ground, and the value is 40°.

H is the height of spaceborne SAR, and the value is 660 km.

$G_t/G_t(\theta)$ is 42 dB.

$L_d$ is the radar feeder plus atmospheric loss, and the value is 0.12. $L_j$ is the jamming polarization loss, which is 0.5. $r_j$ is jamming plus atmospheric loss, which is 0.4. $k_f$ is ratio of radar bandwidth to jamming spectrum, which is 0.6. $L_d/L_j r_j k_f = 1$.

Substituting parameters into (6.25) formula:

$$P_j G_j = \frac{10^7 \times 10^8 \times \left[1 + 0.46^2(1 + 0.88^2)\right]}{4\pi R^4} \cdot 1.5 \times 10^4 \times 1$$

$$= 2.8 \times 10^6 = 64\,\text{dBw}$$

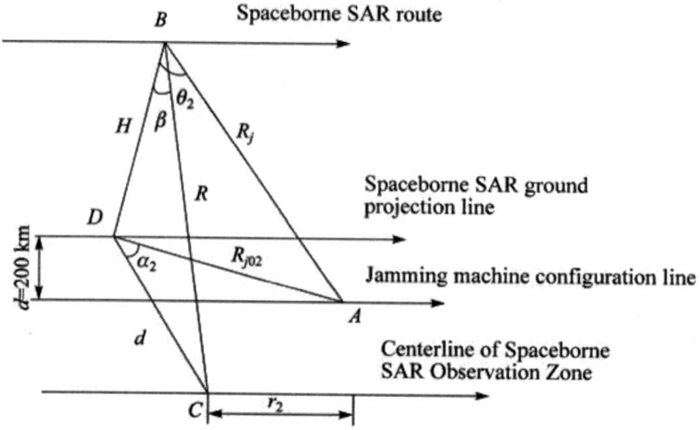


**Fig. 6.29** Picture of spaceborne SAR moving 200 km towards A

According to the known parameters, the jamming effective area of the spaceborne SAR is configured, as shown in Fig. 6.28. The jammer is equipped in the center. Line of the SAR imaging zone and the parameters are substituted into the formula (6.21) to calculate $r_1 = d \tan \alpha = 303 \times \tan 40° = 254.5$ km.

Substitute the parameters into the (6.19) formula to calculate $d = H \tan \beta = 660 \times 0.46 = 306.6$ km.

If the elevation angle $\beta$ of the spaceborne SAR is unchanged, its route is translated 200 km towards the jammer. The radar equivalent radiated power and the jammer equivalent radiated power are unchanged, and the value of $r_2$ is obtained, as shown in (6.29).

It can be seen from Fig. 6.29 that $R_j$ decreases and $G_t/G_t(\theta)$ increases. (6.25) can be written as:

$$P_j G_j = \frac{K_{j\Sigma i} P_t G_t \sigma \left[1 + \tan^2 \beta (1 + \tan^2 \alpha_2)\right]}{4\pi H^2 (1 + \tan^2 \beta)^2} \cdot \frac{G_t}{G_t(\theta_2)} \cdot \frac{L_d}{L_j r_j K_f} \tag{6.26}$$

Divide (6.25) by (6.26) and simplify to (6.27).

$$\frac{\left[1 + \tan^2 \beta (1 + \tan^2 \alpha_2)\right]}{\left[1 + \tan^2 \beta (1 + \tan^2 \alpha)\right]} \cdot \frac{G_t}{G_t(\theta_2)} = 1 \tag{6.27}$$

In Fig. 6.28, $\alpha = 40°$ and $G_t(\theta) = 1$. In Fig. 6.39, $\alpha_3$ is less than $\alpha$. Simplify (6.30) into (6.31), take $G_t(\theta_3) = 0.93$.

$$\tan \alpha_3 = \left\{ \frac{\left[1 + \tan^2 \beta (1 + \tan^2 \alpha)\right] G_t(\theta_3) - (1 + \tan^2 \beta)}{\tan^2 \beta} \right\}^{1/2} \tag{6.28}$$

Substitute the parameters into (6.31), and calculate $\tan \alpha_3 = 0.4$.

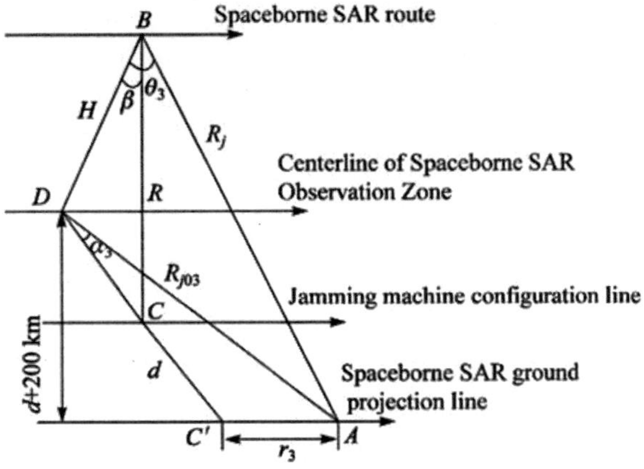


**Fig. 6.30** Spaceborne SAR moving 200 km away in the far A direction

$$r_3 = (d + 200)\tan\alpha_3 = 202\,\text{km}$$

In $G_t(\theta)$, it contains the changes of the elevation side lobe and azimuth side lobe of the SAR antenna, and has a greater impact on the jamming area. In the calculations of (6.26) and (6.31), the error of the value of $G_t(\theta)$ may be large, because there is no accurate measurement data of the side lobe characteristics of the antenna at this moment. In order to calculate more accurately in the future, the azimuth angle and elevation angle of the SAR antenna sidelobe (that is, $G_t(\alpha, \beta)$) must be accurately measured. Otherwise, it is difficult to calculate more accurately. It is best to actually measure the effective jamming area of the spaceborne SAR and correct some parameter values.

According to the analysis and calculation of Figs. (6.28), (6.29), and (6.30), the data of $r_{1,2,3}$ is obtained. The jamming effective area and jamming exposure area of spaceborne SAR can be drawn with the jamming station as the center. This spaceborne SAR, which works in the L-band, $P_tG_t = 80$ dBw, uses a pulse compression system, is about 660 km above the ground, and the calculated equivalent power of the jammer is 64dBw.

A—position of jammer (The effective jamming area of this jamming station is greater than $100{,}000^2$ km). Inside the dotted line is the jamming effective area. Outside the dotted line is the jamming exposure area (SAR effective imaging area).

To achieve a stable and reliable jamming effect on SAR, it is necessary to ensure six jamming elements: (1) the frequency aiming is accurate; (2) the direction of the jamming antenna is accurate; (3) the jamming signal is continuous in time; (4) there are countermeasures against variable polarization anti-jamming; (5) Ensure that the jamming signal spectrum covers the radar signal spectrum; (6) The jamming suppression coefficient meets various radar system requirements. As long as these six jamming elements are guaranteed to meet the design requirements, as far as the current test results are concerned, it is difficult for radars of various systems to resist such suppressed complex jamming. Although another deceptive jamming can also have a certain jamming effect, it is more targeted and requires more jamming object parameters to be understood. Radar is more resistant to such deceptive jamming. Therefore, when designing jammers, clutter suppression type jamming (jamming that meets the six elements of jamming) should be the main, and deceptive jamming is the supplement. For example, spoofing jamming or image spoofing jamming to SAR, it can only have to spoof jamming under limited conditions. It can be known from the jamming Eq. (6.7).

$$P_j G_j = \frac{K_{j\Sigma i} P_t G_t \sigma R_j^2}{4\pi R^4} \cdot \frac{G_t}{G_t(\theta)} \cdot \frac{L_d}{L_j r_j K_f}$$

For spaceborne SAR, when it works in a certain imaging zone, R is unchanged, and $R_j$ is changed. The time relationship between the target echo received by the radar and the deceptive jamming signal is shown in Fig. 6.32.

It can be seen from Fig. 6.32 that the AND gate circuit is connected in series in the SAR receiver circuit, and the range gate signal controls AND gate. The range gate works synchronously with the target echo signal, and it is difficult for the deceptive signal to have the opportunity to synchronize. Therefore, most of the output of the AND gate is the target echo signal output. The deceptive jamming signal is considered to be removed so that the anti-jamming is very effective.

Spaceborne SAR is currently one of the most widely used and most effective means of reconnaissance, and has played a very good role in both the Gulf War and the Iraq War. However, it also has its weaknesses. Therefore, appropriate jamming measures can be taken to jamming with it effectively. An "Earth-to- space jammer " can make it impossible to conduct imaging reconnaissance in an area of 100,000$^2$ kms. This is one of the effective means for developing countries to deal with powerful people.

***Example 6.14:*** On the basis of Example 6.8, the $\beta$ of the spaceborne SAR changed from 25° to 55°, and other parameters of the jammer and radar did not change. The effective area of jamming to this spaceborne SAR can be calculated.

According to formula (6.19):

$$d = H \tan\beta = 660 \times \tan 55° = 924\,\text{km}$$

According to formula (6.25):

$$P_j G_j = \frac{K_{j\Sigma i} P_t G_t \sigma \left[1 + \tan^2\beta(1 + \tan^2\alpha)\right]}{4\pi H^2 (1 + \tan^2\beta)^2} \cdot \frac{G_t}{G_t(\theta)} \cdot \frac{L_d}{L_j r_j K_f}$$

Known parameters: $P_j G_j = 64\,\text{dBw}$; $P_t G_t = 80\,\text{dBw}$; $K_{j\Sigma i} = 70\,\text{dB}$; $\beta = 55°$; $G_t/G_t(\theta) = 42\,\text{dB}$; $\frac{L_d}{L_j r_j K_f} = 1$.

Formula (6.25) can be written as formula (6.32).

$$\tan\alpha = \frac{\left[\left(\frac{4\pi H^2(1+\tan^2\beta) P_j G_j}{K_{j\Sigma i} P_t G_t \sigma} \cdot \frac{G_t(\theta)}{G_t} \cdot \frac{L_j r_j k_j}{L_d}\right) - (1+\tan^2\beta)\right]^{1/2}}{\tan\beta}$$; Solution: $\alpha = 17°$.

Substitute the parameters into the formula (6.32) and calculate $\tan\alpha = \left[\frac{12.6\times 43.6\times 10^{10}\times 3\times 2.8\times 10^{6} 0.7\times 10^{-4}}{10^{7}\times 10^{8}} - 3\right]^{1/2}/1.42 = 0.3$

According to (6.32):

$$r_{(1)} = d\tan\alpha = 924 \times 0.3 = 277km$$

If the spaceborne SAR route moves 200 km towards the jammer,
Substitute parameters (6.28)

$$\tan\alpha = \frac{\left\{\left[1+\tan^2\beta(1+\tan^2\alpha)\right]G_t(\theta_2) - (1+\tan^2\beta)\right\}^{1/2}}{\tan\beta}$$

Known parameters:$\beta = 55°$; $\alpha = 17°$; $G_t(\theta_2) \approx 1$;
So: $\tan\alpha_2 = \frac{[1+2(1+0.1)-(1+2)]^{1/2}}{1.42} = 0.3$
Solution: $\alpha_2 = 17°$.
According to (6.20):

$$r_{(2)} = (d - 200)\tan\alpha_2 = 217km$$

If the spaceborne SAR route moves 200 km away from the jammer.
According to formula (6.31)

$$\tan\alpha_3 = \frac{\left\{\left[1+\tan^2\beta(1+\tan^2\alpha)\right]G_t(\theta_3) - (1+\tan^2\beta)\right\}^{1/2}}{\tan\beta}$$

Take $G_t(\theta_3) = 0.95$; $\beta = 55°$; $\alpha = 17°$, Substitute (6.31)

$$\tan\alpha_3 = \frac{\left\{\left[1+1.42^2(1+0.3^2)\right]\times 0.95 - (1+1.42^2)\right\}^{1/2}}{1.42} = 0.2$$

Solution: $\alpha_3 \cong 11°$

$$r_{(3)} = (d+200)\tan\alpha_3 = (924+200)\times 0.2 = 224km$$

According to $r_{(1),(2),(3)}$, the jamming effective area plotted at $\beta = 55°$ is shown in Fig. 6.33.

As can be seen from Fig. 6.21 and Fig. 6.23, as the elevation angle $\beta$ of the spaceborne SAR increases, the effective jamming area of the radar jammer increases. It can be seen that when designing the jammer, the jammer designed according to the small elevation value can meet all the jamming effective areas of the spaceborne SAR within the range of elevation angle. The range of elevation angle of the spaceborne SAR is generally from 25° to 57°.

***Example 6.15:*** Known parameters of spaceborne SAR: $H = 660\,\text{km}$; $\beta = 25°$; $P_t = 5\,\text{km}$; $G_t = 55\,\text{dB}$; $\tau/\Delta\tau = 40\,\text{dB}$; $K_{j\Sigma i} = 32\,\text{dB}$; $\sigma = 40\,\text{dB}$; $\theta_{0.5} = 0.3°$; $\sigma = 40\,\text{dB}$; $L_d/L_j r_j L_j = 0.5$; Working in X band, $\alpha = 30°$.

Substitute parameters into (6.25):

$$\begin{aligned} P_j G_j &= \frac{K_{j\Sigma i} P_t G_t \sigma\left[1+\tan^2\beta(1+\tan^2\alpha)\right]}{4\pi H^2(1+\tan^2\beta)^2} \cdot \frac{G_t}{G_t(\theta)} \cdot \frac{L_d}{L_j r_j K_f} \\ &= \frac{1.4\times 10^3 \times 5\times 10^3 \times 10^5 \times 10^4\left[1+0.46(1+0.58^2)\right]}{12.6\times 6.6^2 10^{10} \times (1+0.46^2)^2} \times 4\times 10^4 \times 0.5 \\ &= 78\,\text{dBw} \end{aligned}$$

Configure according to Fig. 6.28; the jammer is placed on the centerline of the measuring tape.

Substitute the parameters into (6.21):

$$r_{(1)} = d\tan\alpha = 303 \times 0.58 = 175.7\,\text{km}$$

Substitute the parameter into (6.19) to calculate $d = H\tan\beta = 660 \times 0.46 = 306.6\,\text{km}$.

If the elevation angle of the spaceborne SAR does not change, and its route shifts 150 km in the direction of the jammer, the radar equivalent radiated power and jamming equivalent radiated power remain unchanged (similar to the configuration in Fig. 6.29).

According to Fig. 6.29, it can be seen that when $R_j$ reduced and $G_t/G_t(\theta)$ increased, the formula (6.25) can be written as:

$$P_j G_j = \frac{K_{j\Sigma i} P_t G_t \sigma\left[1+\tan^2\beta(1+\tan^2\alpha)\right]}{4\pi H^2(1+\tan^2\beta)^2} \cdot \frac{G_t}{G_t(\theta)} \cdot \frac{L_d}{L_j r_j K_f} \tag{6.29}$$

Divide (6.25) by (6.33), simplify to (6.34):

$$\frac{\left[1+\tan^2\beta(1+\tan^2\alpha_2)\right]}{\left[1+\tan^2\beta(1+\tan^2\alpha)\right]} \cdot \frac{G_t(\theta)}{G_t(\theta_2)} = 1 \tag{6.30}$$

In this configuration, $\alpha = 30°$; $G_t(\theta) = 1$; $G_t(\theta_2)$ should be $G_t(\alpha_i, \beta_i)$; $\alpha_i$ decrease with the increase of $\beta_i$.

At this angle, the gain of the side lobe of the radar antenna changes slowly, temporarily taking $G_t(\theta_2)$ to be 1.1. And simplify (6.34) into:

$$\tan\alpha_3 = \frac{\left\{\left[1+\tan^2\beta(1+\tan^2\alpha)\right]G_t(\theta_3) - (1+\tan^2\beta)\right\}^{1/2}}{\tan\beta} \tag{6.31}$$

Substitute the parameter $\beta = 25°$, $\alpha = 0.3°$, and $G_t(\theta_2) = 1.1$ into the formula (6.35), we get:

$$\tan \alpha_2 = 0.7$$

$$r_{(2)} = (d - 150) \times \tan \alpha_2 = 109\,\text{km}$$

If the $\beta$ value of the spaceborne SAR elevation angle is unchanged, the route will move 150 km away from the jammer, and the equivalent radiated power of the radar and jammer and other parameters will remain unchanged. Find the value of $r_{(3)}$.

From Fig. 6.30, we can see that r is reduced, g is increased, and (6.25) can be written as

$$P_j G_j = \frac{K_{j\Sigma i} P_t G_t \sigma \left[1 + \tan^2 \beta (1 + \tan^2 \alpha_3)\right]}{4\pi H^2 (1 + \tan^2 \beta)^2} \cdot \frac{G_t}{G_t(\theta_3)} \cdot \frac{L_d}{L_j r_j K_f} \tag{6.32}$$

Divide (6.25) by (6.32), simplify to (6.33):

$$\frac{\left[1 + \tan^2 \beta (1 + \tan^2 \alpha)\right]}{\left[1 + \tan^2 \beta (1 + \tan^2 \alpha_3)\right]} \cdot \frac{G_t(\theta_3)}{G_t(\theta)} = 1 \tag{6.33}$$

Substitute the parameter $\alpha = 30°$, $G_t(\theta) = 1$, and $G_t(\theta_3) = 1$ into the simplified formula (6.37), we get:

$$\tan \alpha_3 = \frac{\left\{\left[1 + \tan^2 \beta (1 + \tan^2 \alpha)\right] G_t(\theta_3) - (1 + \tan^2 \beta)\right\}^{1/2}}{\tan \beta} = 0.36$$

Substitute the parameters into (6.31) formula and calculate

$$r_{(3)} = (d + 150) \tan \alpha_3 = 162.4\,\text{km}$$

$G_t(\theta)$ contains the changes of the elevation side lobe and azimuth side lobe of the SAR antenna, which has a great influence on the effective area of jamming. In the calculations of (6.35) and (6.37), the error in the value $G_t(\theta)$ is large. Therefore, the calculated error of the effective area of the jamming to the spaceborne SAR is large, and it needs to be corrected continuously in the actual test.

According to the calculated data of $r_{(1),(2),(3)}$ in X-band, the effective jamming area of X-band ground-based jammers to X-band spaceborne SAR can be drawn, as shown in Fig. 6.34.

A—position of the jammer. Inside the dotted line is the jamming effective area. Outside the dotted line is the jamming exposure area.

***Example 6.16:*** In the X band, the parameters of the jammer and radar are unchanged, but b is 55°, and d can be calculated according to the above calculation.

$$d = H \tan\beta = 660 \times 1.42 = 924\,\text{km}$$

According to (6.25):

$$P_j G_j = \frac{K_{j\Sigma i} P_t G_t \sigma\left[1 + \tan^2\beta(1 + \tan^2\alpha_3)\right]}{4\pi H^2(1 + \tan^2\beta)^2} \cdot \frac{G_t}{G_t(\theta_3)} \cdot \frac{L_d}{L_j r_j K_f}$$

Plugin with $P_j G_j = 78\,\text{dBw}$; $P_t G_t = 92\,\text{dBw}$; $K_{j\Sigma i} = 72\,\text{dB}$; $\beta = 55°$; $G_t/G_t(\theta) = 3 \times 10^4$; $\frac{L_d}{L_j r_j K_f} = 0.5$, we get:

$$\tan\alpha = \frac{\left[\left(\frac{4\pi H^2(1+\tan^2\beta)P_j G_j}{K_{j\Sigma i} P_t G_t \sigma} \cdot \frac{G_t(\theta)}{G_t} \cdot \frac{L_j r_j k_j}{L_d}\right) - \left(1 + \tan^2\beta\right)\right]^{1/2}}{\tan\beta}$$

$$= \frac{\left[\left(\frac{12.6\times6.6^2\times10^{10}(1+2)\times7\times10^7}{1.4\times10^7\times1.4\times10^9} \cdot 0.30 \times 10^{-4} \times 2\right) - (1 + 2)\right]^{1/2}}{1.4}$$

$$= 0.25$$

$$r_{(1)} = d \tan\alpha = 924 \times 0.25 = 231\,\text{km}$$

If the spaceborne SAR route shifts 200 km towards the jammer, substitute the parameter (6.28):

$$\tan\alpha_2 = \frac{\left\{\left[1 + \tan^2\beta(1 + \tan^2\alpha)\right]G_t(\theta_2) - (1 + \tan^2\beta)\right\}^{1/2}}{\tan\beta}$$

$$= \frac{\{[1 + 2(1 + 0.04)] - (1 + 2)\}^{1/2}}{1.42} = 0.2$$

$$r_{(2)} = (d - 200) \times 0.2 = 144\,\text{km}$$

If the spaceborne SAR route moves 200 km away from the jammer, substitute the parameter (6.31):

$$\tan\alpha_3 = \frac{\left\{\left[1 + \tan^2\beta(1 + \tan^2\alpha)\right]G_t(\theta_3) - (1 + \tan^2\beta)\right\}^{1/2}}{\tan\beta}$$

$$= \frac{\{[1 + 2(1 + 0.04)] - (1 + 2)\}^{1/2}}{1.4} = 0.2$$

$$r_{(3)} = (d + 200) \times 0.2 = 224\,\text{km}$$

According to the $r_{(1),(2),(3)}$ data and plotted in the X-band, the jamming effective area when $\beta = 55°$ is shown in Fig. 6.35.

From Fig. 6.35, it can be seen that the jamming effective range increases as the elevation angle $\beta$ increases.

We plotted Figs. 6.31 and 6.32 on one graph, and plot Figs. 6.33 and 6.34 on one graph. It was found that the parameters of the radar and jammer are the same, and the jamming exposure area is quite different. (Fig. 6.36)

Parameters in L band: $P_t = 5\,\text{kw}$, $G_t = 42\,\text{dB}$, $P_j G_j = 64\,\text{dBw}$.

Parameters in X band: $P_t = 5\,\text{kw}$, $G_t = 55\,\text{dB}$, $P_j G_j = 74\,\text{dBw}$.

Due to the inaccurate selection of radar antenna sidelobe characteristic parameters and the selection errors of other parameters, it is difficult to calculate the effective area of ground-to-space SAR jamming effectively. When conditions permit, it is best to use the measured data for correction.

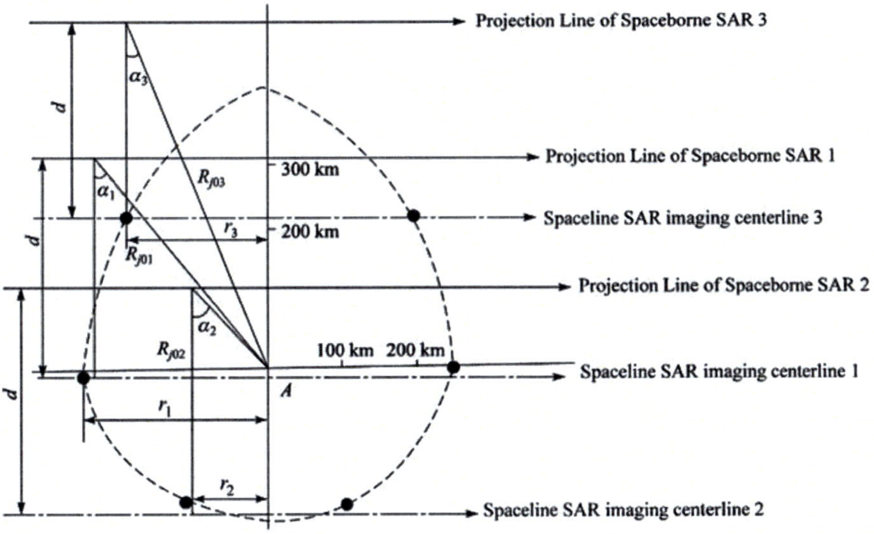


**Fig. 6.31** Schematic diagram of effective and exposed areas of jamming to spaceborne SAR

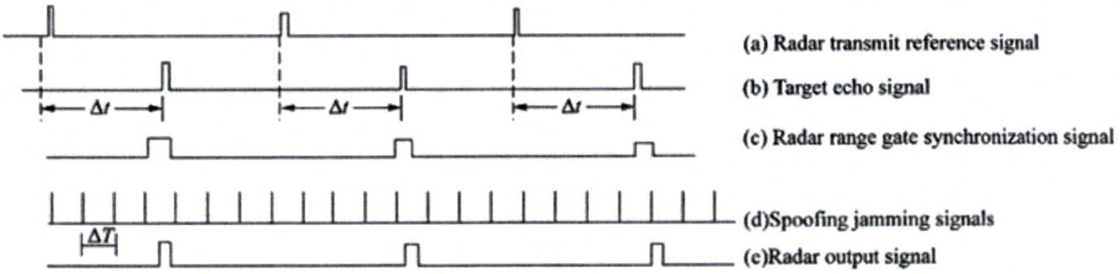


**Fig. 6.32** Schematic diagram of anti-jamming circuit of radar synchronization range gate

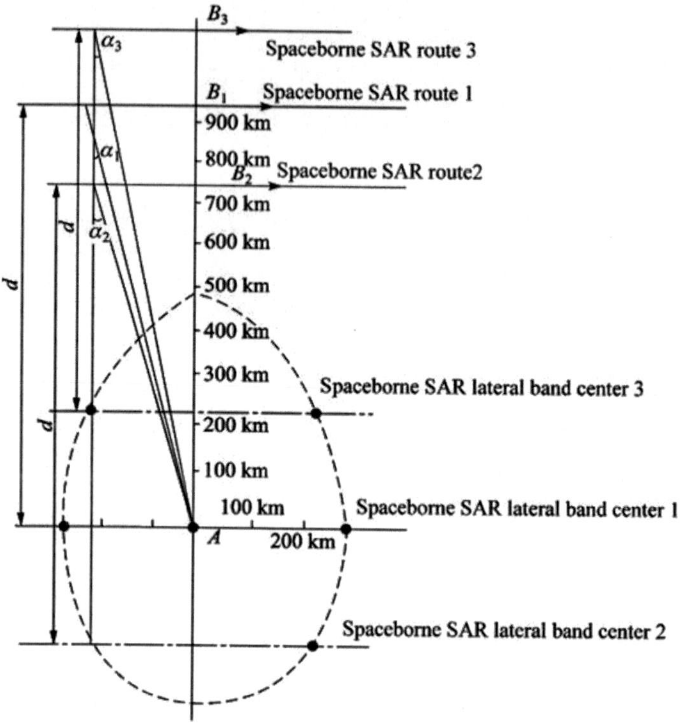


**Fig. 6.33** Jamming effective area at $\beta = 55°$ (other parameters unchanged)

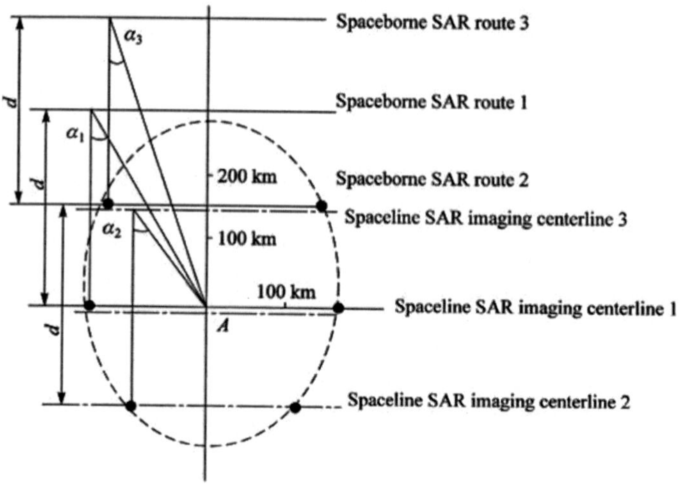


**Fig. 6.34** Spaceborne SAR X band, jamming effective area at $\beta = 25°$

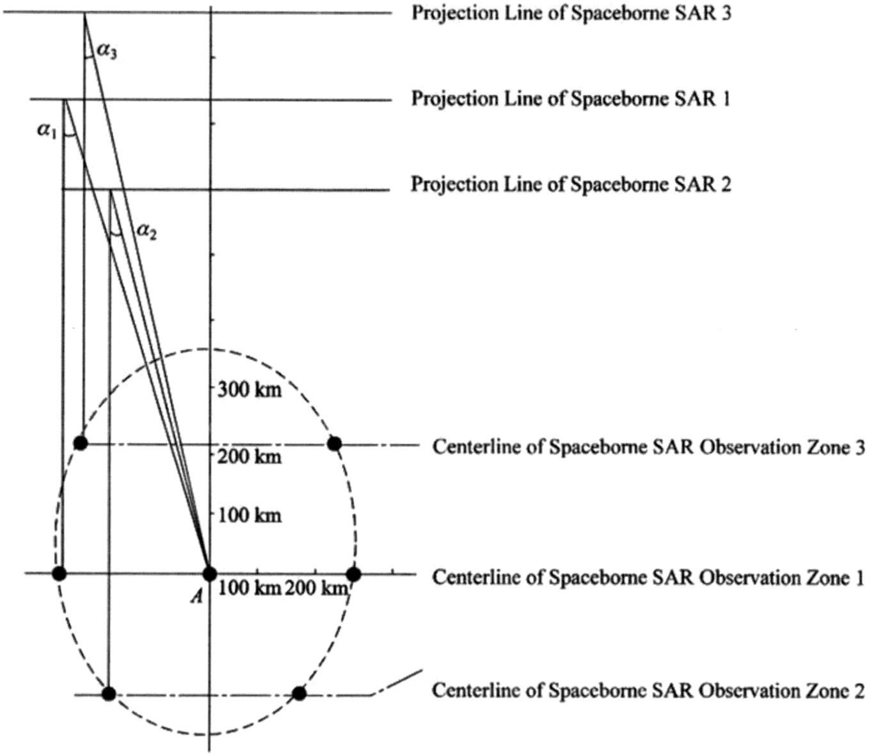


**Fig. 6.35** X-band $\beta = 55°$ jamming schematic diagram

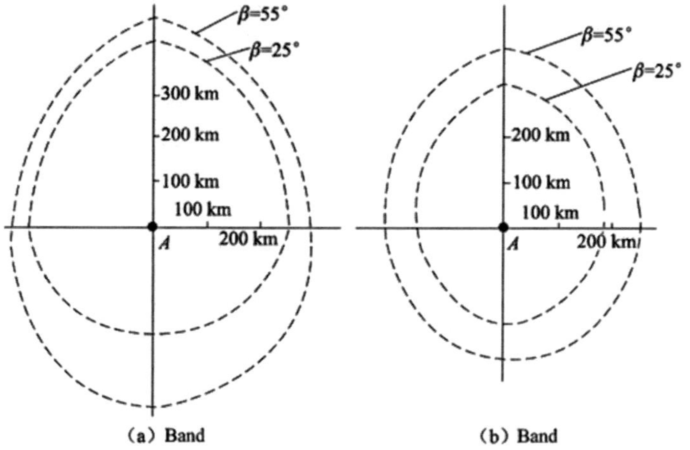


**Fig. 6.36** Comparison of effective jamming area when L and X band elevation angle are different

# Chapter 7
# Jamming Techniques for Bistatc Radar

In order to effectively anti-stealth, anti-jamming, anti-radiation weapon attacks, and the discovery of low-altitude targets, modern radars are developing in the direction of bistatic and multistatic radars. In this chapter, according to the basic working principle of bistatic radar, the method of jamming to bistatic radar is proposed, and the jamming equation of jamming bistatic radar is derived. According to the basic performance of existing radars and jammers, some parameters were set to calculate the jamming exposure area of jamming bistatic radar, and the characteristics of jamming bistatic radar were analyzed. The results show that the jamming technology and jamming equipment of the bistatic jamming radar and the jamming monostatic radar are the same. Still, there is a difference in the beam width or omnidirectional antenna of the jamming transmitting antenna. The increase in power brings certain difficulties to the development of jammers.

## 7.1 Basic Working Principle of Bistatic Radar

The bistatic radar system consists of a transmitting station, a receiving station, and a time–frequency synchronization system, and its working block diagram is shown in Fig. 7.1.

The atomic clock, which is strictly synchronized in frequency and phase, provides reference signals for the transmitting station and the receiving station, respectively, as the radar operating frequency reference and phase reference of the two stations. The transmitting station works in the normal sending and receiving mode, and at the same time, transmits its position, working frequency code, transmitting trigger, working video, and target data information to the receiving station in real-time through the time–frequency synchronization system. And the time synchronization between the receiving station and the sending station is achieved by comparing the synchronization pulses.



X. Zhang et al., *Theory to Countermeasures Against New Radars*,
https://doi.org/10.1007/978-981-16-6715-2_7

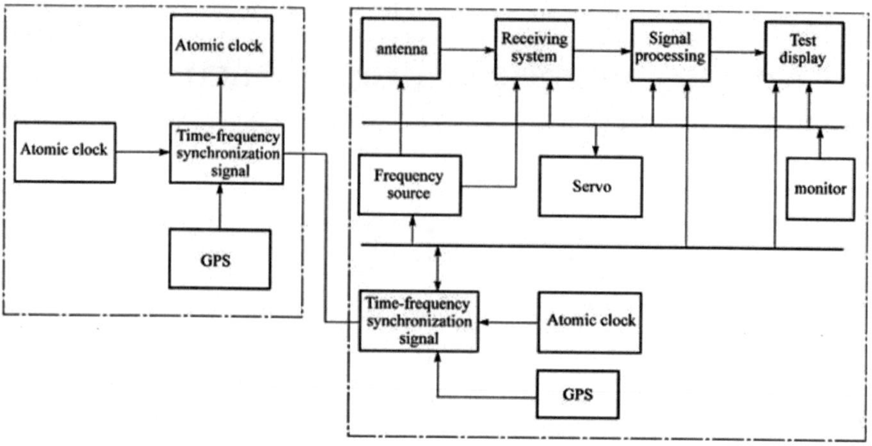


**Fig. 7.1** Block diagram of the working principle of the bistatic radar system

When the transmitted pulse is irradiated to the target, it generates lateral scattering of electromagnetic energy; the receiving station receives the scattered echo of the target through the planar array antenna. After the multi-channel receiver, A/D sampling, and digital pulse compression, the signal processing system Digital beamforming technology is used to form multiple beams in real-time. According to the angle of view of the transmitting station, the information of the receiving and sending distance and the baseline distance performs a real-time pulse search for the transmitted pulse within a certain angle of the receiving azimuth and achieves spatial synchronization to the target. After the terminal processes the received signal, the terminal collection, display correction processing, and data fusion processing are completed. The target point signals sent by the receiving station, the transmitting station, and the target recognition system are processed for relevant fusion and then sent for display.

The detection range of bistatic radar can be expressed by Eq. (7.1).

$$(R_t R_r)^2 = 239.4^4 \times \frac{P_t \tau G_t G_r \sigma_b F_t^2(\theta) F_r^2(\theta)}{f^2 T_s D_o C_B L_{tr} L_s} \quad (7.1)$$

In the formula:

$R_t$—The distance from the transmitting station to the target.
$R_r$—The distance from the receiving station to the target.
$G_t$—The gain of the transmitting antenna.
$G_r$—The gain of the receiving antenna.
$P_t$—The power of the transmitter.
$\tau$—The width of the transmitting pulse; $F_t^2(\theta) \cong F_r^2(\theta) \approx 1$.
$\sigma_b$—The cross-sectional area of the target bistatic radar; when the bistatic angle $\theta < 135\text{o}, \sigma_b = \sigma_m$; when bistatic angle $\theta > 135\text{o}, \sigma_b > \sigma_m$.
$f$—The operating frequency of the bistatic radar.

Fig. 7.2 Change chart of bistatic radar detection range

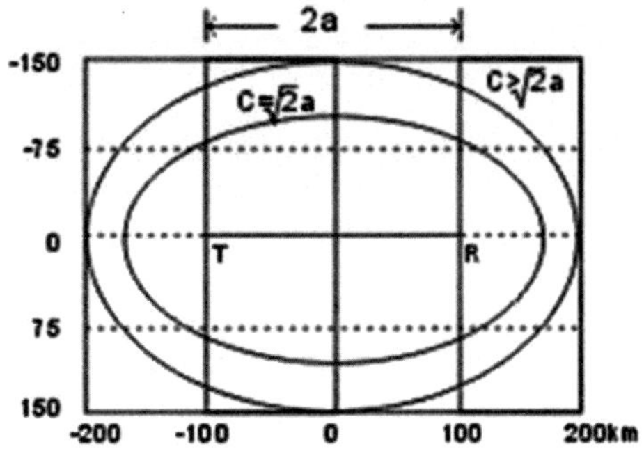


$Ts(ok)$—The equivalent noise temperature equivalent to the input of the receiver;

$D_o$—The detection capability factor.

$C_B$—The bandwidth correction factor.

$L_{tr}$—The propagation loss of the signal from the transmitter to the target and then to the receiver.

$L_S$—The system loss of the bistatic radar.

When the parameters of the system are determined, formula (7.1) can be written as.

$\sqrt{R_t \cdot R_r} = C$, C is a constant. Make sure that both $R_t$ and $R_r$ are within sight.

The trajectory of the point that meets the above conditions that are, the outer edge of the plane detection range, the equation of the trajectory in rectangular coordinates is

$$(x^2 + y^2) - 2a^2(x^2 - y^2) = c^4 - a^4 \quad (7.2)$$

If $C = \sqrt{2}a$, the area detected by bistatic radar is shown in Fig. 7.2.

If $C > \sqrt{2}a$, the detection range curve of the bistatic radar is an elliptical waist-like convex oval line. In real bistatic radar, generally, when $C > \sqrt{2}a$, it constitutes an elliptical detection range. Point T is the position of the transmitting station; point R is the position of the receiving station.

## 7.2 Jamming Equation and Jamming Exposure Area for Bistatic Radar

As shown in Fig. 7.3, T—the position of the radar transmitter; R—the position of the radar receiver; B—the covered target; M—the jammer carrier; H—the jammer carrier's height; $R_j$—the distance between the jammer and the radar receiver; $R_N$—the distance between the jammer and the radar transmitter; $\beta$—the width of wave speed of the transmitting antenna of the jammer; $\theta$—the angle of the radar receiving antenna sidelobe aligned with the jamming direction.

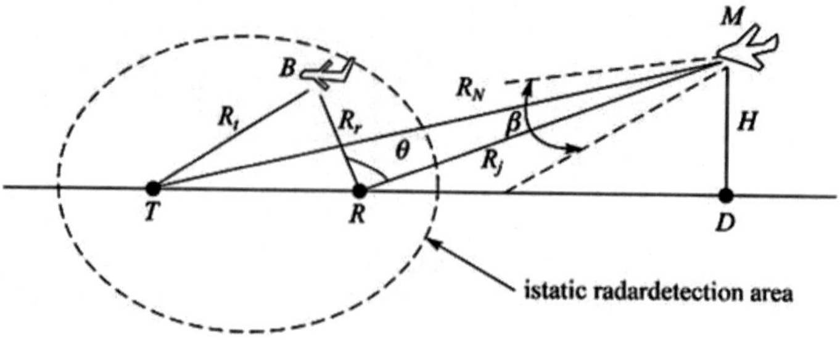


**Fig. 7.3** Schematic diagram of jamming to bistatic radar

The receiver of the bistatic radar station, the power of the received radar echo signal is $P_s$

$$P_s = \frac{P_t G_t \lambda^2 \sigma_b G_r}{(4\pi)^3 R_t^2 R_r^2} \tag{7.3}$$

The power of the received jamming signal $P_N$

$$P_N = \frac{P_j G_j \lambda^2 G_r(\theta)}{4\pi R_j^2 4\pi} \tag{7.4}$$

Jamming suppression coefficient $K_j = \frac{P_N}{P_S}$, get

$$K_j = \frac{4\pi P_j G_j R_t^2 R_r^2}{P_t G_t \sigma_b R_j^2} \frac{G_r(\theta)}{G_r} \tag{7.5}$$

### 7.2.1 Self-Defense Jamming to Bistatic Radar

$$P_j G_j = \frac{K_j P_t G_t \sigma_b R_j^2}{4\pi R_t^2 R_r^2} \cdot \frac{G_t}{G_t(\theta)} \cdot \frac{L_d}{r_j L_j} \tag{7.6}$$

In the formula:
$L_d$—Feeder loss and atmospheric loss of radar transmitter;
$K_f$—Jamming spectrum/signal bandwidth;
$r_j$—Jamming polarization loss;
$L_j$—The feeder loss and atmospheric loss of jamming transmitter.
When $R_r = R_j$; $\frac{G_r(\theta)}{G_r} = 1$;
Then

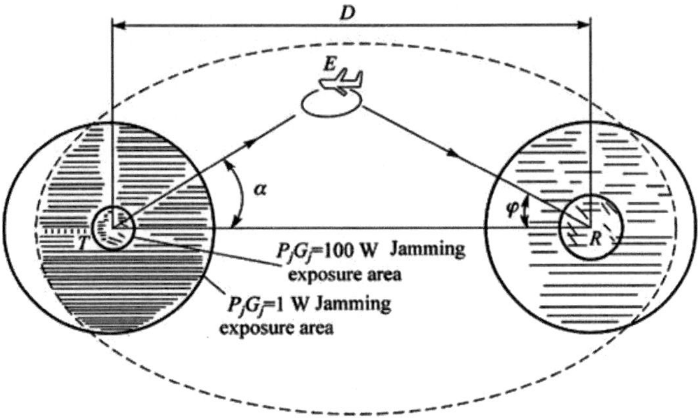


**Fig. 7.4** Jamming exposure area to bistatic radar during self-defense jamming

$$K_j = \frac{4\pi P_j G_j R_t^2}{P_t G_t \sigma_b} \tag{7.7}$$

$$P_j G_j = \frac{K_j P_t G_t \sigma_b}{4\pi R_r^2} \cdot \frac{L_d K_f}{r_j L_f} \tag{7.8}$$

For self-defense jammers, the jamming exposure area of bistatic radar is the same as that of monostatic radar:

$$R_{T\,\min} = \left[\frac{K_j P_t G_t \sigma_b}{4\pi P_j G_j}\right]^{1/2} \left[\frac{L_d K_f}{r_j L_f}\right]^{1/2} \tag{7.9}$$

$R_{T\,\min}$—the jamming exposure distance.

If $K_j = 10$; $P_t G_t = 10^9$ watt; $\sigma_b = 10\text{m}^2$; its jamming exposure area is shown in Fig. 7.4.

T—The position of the transmitter; R—The position of the receiving station; E—The target position; inside the dotted circle—the effective detection area when the bistatic radar is not disturbed; the horizontal line area—the jamming exposure area after the radar jamming, and it is also the effective detection area after the radar has jammed.

### 7.2.2 Exposure Area in Stand-Off Jamming to Bistatic Radar

As shown in Fig. 7.5.

$$R_r^2 = D^2 + R_t^2 - 2DR_r \cos\alpha \tag{7.10}$$

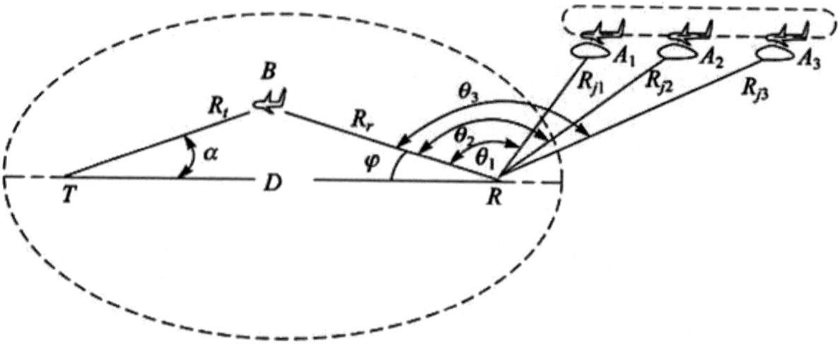


**Fig. 7.5** Jamming exposure area of bistatic radar during shielded jamming

Substituting Eq. (7.8) into Eq. (7.5), get

$$K_j = \frac{4\pi P_j G_j}{P_t G_t \sigma_b R_j^2} \cdot \frac{G_r(\theta)}{G_r} \cdot R_t^2 \left[ R_t^2 + D^2 - 2DR_t \text{Cos}\alpha \right] \tag{7.11}$$

A1,2,3 is the position of the jamming aircraft, $R_{j1}$,$R_{j2}$, $R_{j3}$ are the distance of the jamming aircraft, and B is the target being covered.

The formula (7.11) can be changed to

$$\frac{K_j P_t G_t \sigma_b R_j^2}{4\pi P_j G_j} \cdot \frac{G_r}{G_r(\theta)} = R_t^4 - 2DR_t^3 \text{Cos}\alpha + D^2 R_t^2 \tag{7.12}$$

Let

$$\frac{K_j P_t G_t \sigma_b R_j^2}{4\pi P_j G_j} \cdot \frac{G_r}{G_r(\theta)} = M_i \tag{7.13}$$

Set $K_j = 10$; $P_t G_t = 10^9$ w; $\sigma_b = 10$ m$^2$; $P_j G_j = 10^5$ w; the antenna side lobe of the radar receiving station is above $\theta > 60$o, $\frac{G_r}{G_r(\theta)} = 3 \times 10^4$; 1 at $\theta = 0$o, D is selected to be 400 km, Calculate three kinds of jamming exposure areas based on 100, 200, and 300 km. Substitute Eq. (7.13) into Eq. (7.12)

$$R_t^4 - 2DR_t^3 \text{Cos}\alpha + D^2 R_t^2 - M_i = 0 \tag{7.14}$$

Equation (7.14) is mathematically analyzed. There should be four solutions. From a physical sense, there should be a pair of imaginary numbers, a negative number, a positive number (this is the only useful solution).

When α = 90°, Eq. (7.14) can be written as

$$R_t^4 + D^2 R_t^2 - M_i = 0 \tag{7.15}$$

According to the quadratic solution, we get

$$R_t = \left[\frac{-D^2 \pm \sqrt{D^4 + 4M_i}}{2}\right]^{1/2} \tag{7.16}$$

When $R_j$ is equal to $R_{j1}$ which is 150 km, substitute the parameter into (7.16) to get $M_{i1} = 0.535 \times 10^{20}$;

Similarly, when $R_{j2}$ is 200 km, $M_{i2}$ is $0.95 \times 10^{20}$.

when $R_{j3}$ is 200 km, $M_{i3}$ is $2.14 \times 10^{20}$.

Substitute $M_{i1}$ into Eq. (7.16), get $R_{t1} = 18.1km$.

Substitute $M_{i2}$ into Eq. (7.16), get $R_{t2} = 24.3km$.

Substitute $M_{i3}$ into Eq. (7.16), get $R_{t3} = 36.3km$.

When $\alpha = 0°$, substitute into Eq. (7.16)

Get $R_t^4 - 2DR_t^3 + D^2R_t^2 - M_i = 0$,

Substitute parameter $M_{i1}$, the solution for $R_{t1}(0\text{o}) = 20.8km$.

Substitute parameter $M_{i2}$, the solution for $R_{t2}(0\text{o}) = 26.5km$.

Substitute parameter $M_{i3}$, the solution for $R_{t3}(0\text{o}) = 41km$.

When $\alpha = 180°$, substitute into Eq. (7.16)

Substitute parameter $M_{i1}$, the solution for $R_{t1}(180\text{o}) = 17km$

$$R_t^4 + 2DR_t^3 + D^2R_t^2 - M_i = 0$$

Substitute parameter $M_{i2}$, the solution for $R_{t2}(180\text{o}) = 23km$.

Substitute parameter $M_{i3}$, the solution for $R_{t3}(180\text{o}) = 34km$.

Let the radar transmitting station as the center. The jamming exposure area when $R_{j1} = 150km$ is shown in Fig. 7.6 curve (1); the jamming exposure area when $R_{j2} = 200km$ is shown in Fig. 7.6 curve (2); the jamming exposure area when $R_{j3} = 300km$ is shown in Fig. 7.6 curve (3).

If the receiving station is the center, there is also another group of the jamming exposure area.

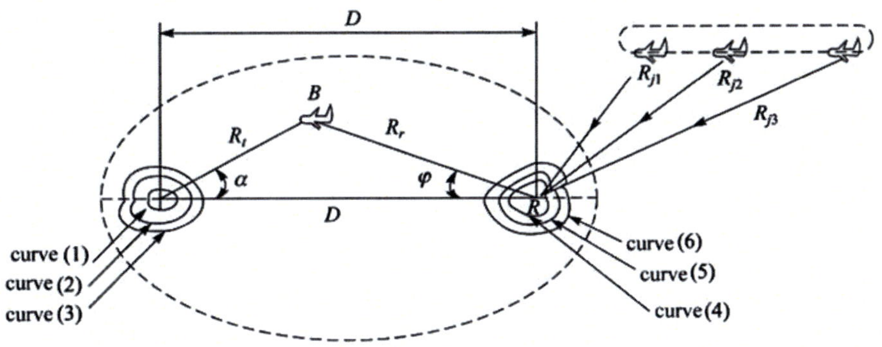


**Fig. 7.6** Jamming exposure area to bistatic radar

$$R_t^2 = D^2 + R_r^2 - 2DR_r\text{Cos}\varphi \tag{7.17}$$

Substituting Eq. (7.17) into Eq. (7.5), we get

$$R_r^4 - 2DR_r^3\text{Cos}\varphi + D^2R_r^2 - M_i = 0 \tag{7.18}$$

If other parameters remain unchanged, when $\varphi = 90°$, solve the Eq. (7.18) and get $R_{r1} = 18.1\text{km}$, $R_{r2} = 24.3\text{km}$, $R_{r3} = 36.3\text{km}$.

When $\phi = 0°$, $R_{r1}(0°)$ is 20.8 km, $R_{r2}(0°)$ is 26.5 km, $R_{r3}(0°)$ is 41 km.

When $\phi = 180°$, $R_{r1}(180°)$ is 17 km; $R_{r2}(180°)$ is 23 km; $R_{r3}(180°)$ is 34 km.

When taking the receiving station as the center, it may make $\theta = 0°$ in a certain direction. Substitute $\frac{G_r}{G_r(\theta)} = 1$ into Eq. (7.13) at this time to get

$M_{i1}(0°) = 0.1783 \times 10^{16}$; $M_{i2}(0°) = 0.3166 \times 10^{16}$; $M_{i3}(0°) = 0.7133 \times 10^{16}$.

Substitute $M_i(0°)$ into Eq. (7.18)

$$R_r(0°) \cong \frac{\sqrt{M_i(0°)}}{D} \tag{7.19}$$

We get $R_{r1}(0°) = 105\text{m}$, $R_{r2}(0°) = 562\text{m}$, $R_{r3}(0°) = 844\text{m}$.

Jamming exposure area centered on the receiving station: The jamming exposure area when $R_{j1} = 150\text{km}$ is shown in Fig. 7.6 curve (4); the jamming exposure area when $R_{j2} = 200\text{km}$ is shown in Fig. 7.6 curve (5); the jamming exposure area when $R_{j3} = 300\text{km}$ is shown in Fig. 7.6 curve (6).

The area Inside curve (1),(4) is the jamming exposure area when $R_{r1}(180°)$.

The area Inside curve (2),(5) is the jamming exposure area when $R_{r1}(180°)$.

The area Inside curve (3),(6) is the jamming exposure area when $R_{r1}(180°)$.

Plot the data table of the jamming exposure area for bistatic radar, see Table 7.1

**Table 7.1** Jamming exposure area for bistatic rada

| | | | | |
|---|---|---|---|---|
| Curve (1) | $R_{t1}$:18.1 km | $R_{t1}(0°)$:20.8 km | $R_{t1}(180°)$:17 km | |
| Curve (2) | $R_{t2}$:24.3 km | $R_{t2}(0°)$:26.5 km | $R_{t2}(180°)$:23 km | |
| Curve (3) | $R_{t3}$:36.3 km | $R_{t3}(0°)$:41 km | $R_{t3}(180°)$:34 km | |
| Curve (4) | $R_{r1}$:18.1 km | $R_{r1}(0°)$:20.8 km | $R_{r1}(180°)$:17 km | $R_{r1}(\theta)$:105 m |
| Curve (5) | $R_{r2}$:24.3 km | $R_{r2}(0°)$:26.5 km | $R_{r2}(180°)$:23 km | $R_{r2}(\theta)$:562 m |
| Curve (6) | $R_{r3}$:36.3 km | $R_{r3}(0°)$:41 km | $R_{r3}(180°)$:34 km | $R_{r3}(\theta)$:844 m |

## 7.3 Analysis on Jamming Exposure Area for Bistatic Radar

1. During self-defense jamming, the area exposed to bi-base and single-base jamming is the same, and the radius of the area exposed to jamming is determined by the formula (7.9). The jamming exposure area is a circle with the transmitting station as the center or a circle with the receiving station as the center.
2. In the case of shielded jamming, single-base radar is a set of jamming exposure areas centered on the radar station. In contrast, bistatic radar is two sets of jamming exposure areas centered on the transmission station and the reception station, as shown in Fig. 7.6.
3. The jamming exposure area is $\overline{TR}$ axis-symmetrical.
4. For jamming exposure distance, if α and φ are variables, when α and φ are 90°, use $R_{T,R(90°)}$ as the reference; When α and φ are 0°, use $R_{T,R(0°)}$ as the reference. $R_{T,R(0°)} > R_{T,R(90°)}$. when α and φ are 180°, use $R_{T,R}(180°)$as the reference. $R_{T,R(180°)} < R_{T,R(90°)}$
5. The jamming exposure area increases when the $R_j$ distance increases, and decreases when the $R_j$ decreases.
6. In the direction of the jammer, there is a deep concave point in the jamming exposed area centered on the receiving station (or centered on the transmitting station), and the specific value can be calculated by Eq. (7.18).
7. In terms of jamming technology, there is not much difference between jamming bistatic radar and jamming monostatic radar. Only for the beamwidth of the jamming transmitting antenna, wide beam, or omnidirectional antenna beam is required when jamming with the bistatic radar. If the required equivalent jamming power is the same, it is necessary to increase the power of the jamming transmitter, which makes it more difficult to develop high-power jammer.
8. For radars with the same equivalent emission energy, the effective detection area of bistatic radar is larger than that of single base radar. For the jammer, under the condition of the same equivalent jamming power, its jamming effective area/jamming exposed area ratio will also increase, which is also beneficial to jamming.

## 7.4 Selection of Beamwidth of the Jamming Antenna When Jamming Against Bistatic Radar

### *7.4.1 Selection of Beamwidth of Antenna Azimuth*

The selection of the beam is shown in Fig. 7.7.

T—The location of the bistatic radar transmitter;

R—The location of the bistatic radar receiver;

A—The position of the jammer;

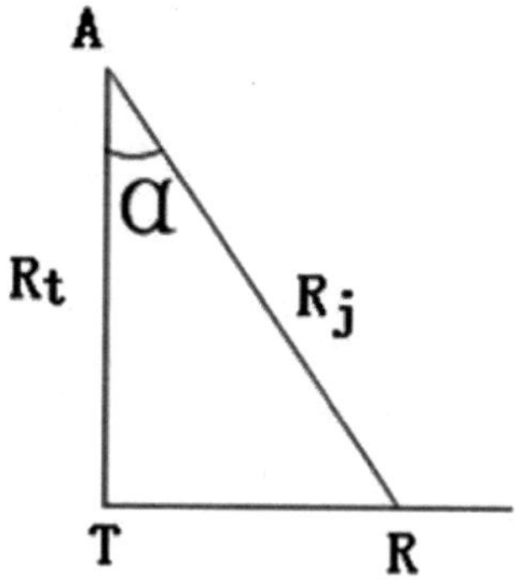


**Fig. 7.7** Schematic diagram of jamming bistatic radar

**Table 7.2** relationship between αmax and TR ($R_t = 200$km)

| $\overline{TR}$(km) | 30 | 40 | 50 | 60 |
|---|---|---|---|---|
| $\alpha_{\max}$(°) | 17° | 23° | 29° | 32° |

**Table 7.3** relationship between αmax and TR ($R_t = 150$km)

| $\overline{TR}$(km) | 30 | 40 | 50 | 60 |
|---|---|---|---|---|
| $\alpha_{\max}$(°) | 23° | 28° | 36° | 44° |

$R_t$—The distance between jammer and transmitter;
$R_j$—The distance between the jammer and the receiver.
If $R_t$ is perpendicular to $\overline{TR}$.
The maximum beamwidth of the transmitter antenna of jammer should be $\alpha_{\max}$

$$\alpha_{\max} = 2\tan^{-1}\left(\frac{\overline{TR}}{R_t}\right) \tag{7.20}$$

Let $R_t$ be 200 km and 150 km, and the relationship between $\alpha_{\max}$ and $\overline{TR}$ is shown in Table 7.2 and Table 7.3.

If the distance between the transmitter and receiver of the bistatic radar is within 60 km, when the distance between the jammer and the radar transmitter is greater than 120 km, the beamwidth of the transmitting antenna of jammer can be 60°.

### 7.4.2 *Selection of Beamwidth of Jamming Transmitting Antenna Elevation*

As shown in Fig. 7.8

**Fig. 7.8** Schematic diagram of jamming bistatic radar (elevation beam width selection)

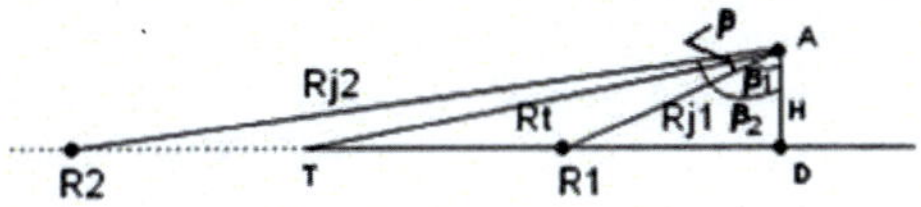


T—The location of the bistatic radar transmitter; $R_1$—The first location of the bistatic radar receiver; $R_2$—The possible receiver's second position for bistatic radar; A—The position of the jammer; H—The height of the jammer.

$$\beta_1 = \tan^{-1}\left(\frac{\overline{DR_1}}{H}\right) \tag{7.21}$$

$$\beta_2 = \tan^{-1}\left(\frac{\overline{DR_2}}{H}\right) \tag{7.22}$$

The beamwidth of jamming transmitting antenna

$$\beta = (\beta_2 - \beta_1) = \tan^{-1}\left(\frac{\overline{DR_2}}{H}\right) - \tan^{-1}\left(\frac{\overline{DR_1}}{H}\right) \tag{7.23}$$

Example: if H is 8 km, $\overline{DR_1}$ is 60 km, $\overline{DR_2}$ is 180 km.
Substitute parameters into formula (7.23) and get

$$\beta_{\max} = 87.4^\circ - 82.5^\circ = 4.9^\circ$$

Select the beamwidth of the jamming transmitting antenna: azimuth angle is 50°; elevation is 6°;
Then the maximum gain of the jamming transmitting antenna is:

$$G = \frac{25000}{\alpha\,\beta} = 19dB$$

## 7.5 Jamming Against Multi-Base Radar

### 7.5.1 Cover Jamming for Multi-Base Radar

The jammer is designed to jamming with bistatic radar. If the multi-base radar is designed based on the bi-base radar, the multi-reception station configuration is shown in Fig. 7.9.

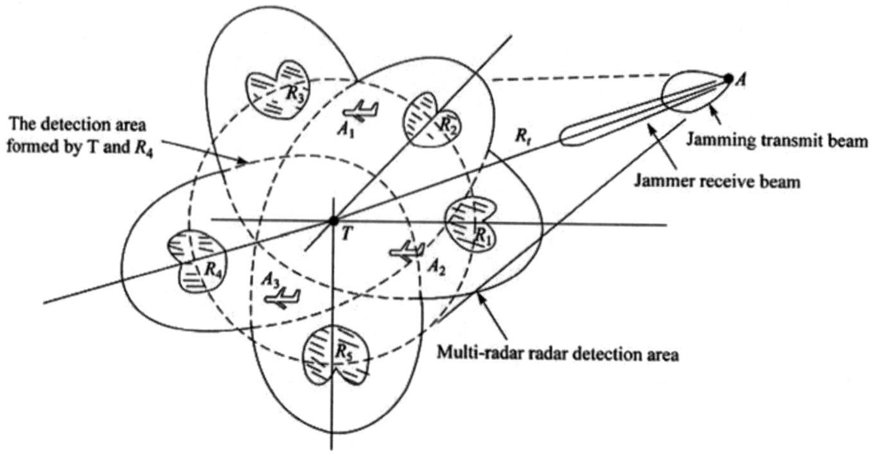


**Fig. 7.9** Schematic diagram of multi-base radar jamming exposure area

T is the multi-base radar transmitter, R1,2,3,4,5 are the positions of the multi-base radar receiver; A is the position of the jammer; A1,2,3 are the targets that the jammer cover; The horizontal line area centered on R1,2,3,4,5 is the jamming exposure area after jamming. That is after the multi-base radar has jammed, the target A1,2,3 enters the horizontal line area, the multi-base radar could find. If the target is beyond the horizontal line, the multi-base radar cannot find A1,2,3. Therefore, multi-base radar can also be jammed.

The size of the jamming exposure area and the required jamming equivalent power can be calculated using (7.6). The receiving antenna of the jammer always automatically tracks the radar transmission direction and simultaneously drives the jamming transmitting antenna to the radar receiving antenna. Because the beamwidth of the transmitting antenna can cover all R1,2,3,4,5 multi-base receiving antennas, this jamming exposure area can be formed.

## 7.5.2 *Self-Defense Jamming to Multi-Base Radar*

If the multi-base radar is still configured according to Fig. 7.9, the $B_1$ plane enters the detection area of the multi-base radar with a self-defense jammer, the jammer's receiving, transmitting, and antenna beams are all omnidirectional beams. Figure 7.10.

The multi-base radar effective detection area T is the position of the multi-base radar transmitter; $A_{1,2,3,4,5}$ is the position of the multi-base radar receiver; B1 is the target and carries the self-defense jammer. After the multi-base radar is jammed, the horizontal line area is the jamming exposure area (The specific calculation uses (7.8).

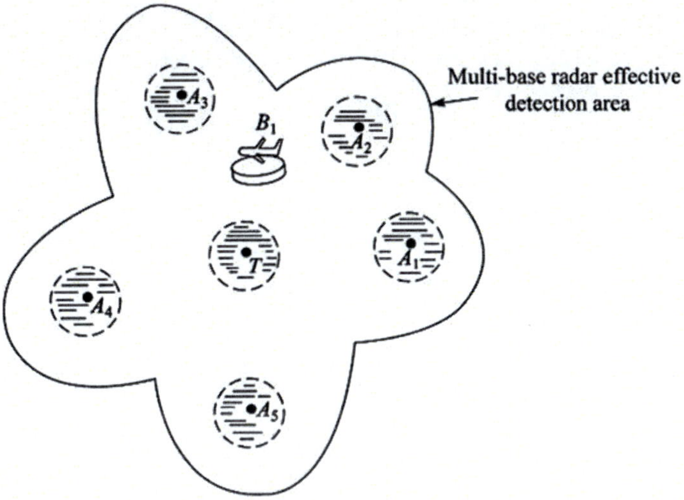


**Fig. 7.10** Schematic diagram of self-defense jammer jamming to multi-base radar

## 7.6 Block Diagram of the Bistatic Radar Jammers' Compostion

The structure of the bistatic radar jammer is the same as that of the single-base radar, except that the beamwidth of the transmitting antenna is different. For a single-base radar jammer launch antenna, a narrow beam or a wide beam can be used. While for a dual (multiple) base radar jammer, a wide beam must be used, and the beam width must cover all the locations where the transmitter and receiver are placed. The azimuth beamwidth is calculated using (7.20), and the elevation beam is calculated using (7.23). Figure 7.11.

The composition of the bistatic radar jammer includes:

Receiving antenna, transmitting antenna, mono-pulse passive tracking antenna, and passive tracking system, receiving aiming frequency sub-system, transmitter sub-system, and orthogonal polarization feed, etc. Features of bistatic radar jammer:

(1) It is required that the receiving antenna beam is narrow and the precision of mono-pulse passive tracking is relatively high;
(2) The transmitting antenna beam is relatively wide, depending on the layout of the bistatic radar.

Other parts are similar to single-base radar jammers.

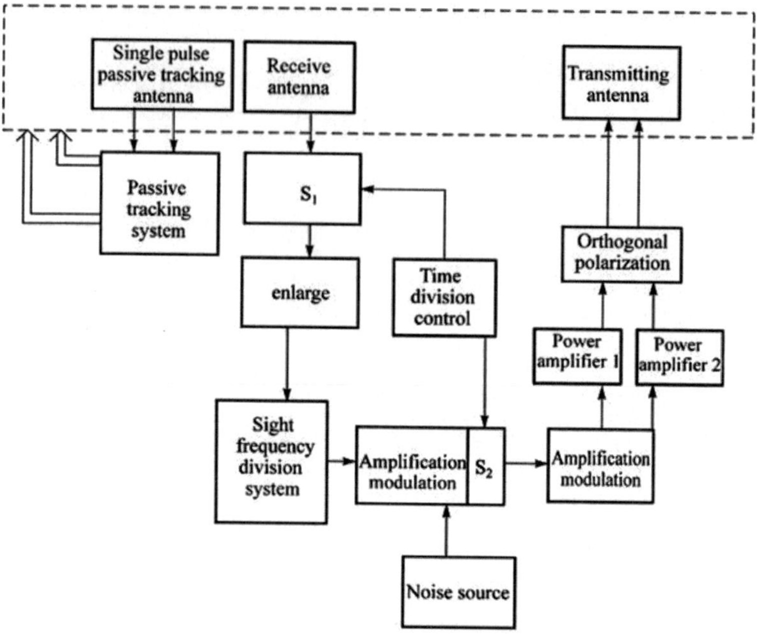


**Fig. 7.11** Principle block diagram of the composition of the bistatic radar jammer

## 7.7 Jamming Against Bistatic Airborne SAR

Bistatic SAR is an ideal anti-jamming measure. It makes the jamming unable to find the direction of the jammed receiving system, making it difficult to implement effective directional jamming. However, bistatic SAR also has difficulties in self-defense, such as time synchronization, space synchronization, frequency synchronization, and phase synchronization, which are all problems that are difficult to achieve by separately installing transceivers on two carriers. Although GPS timing devices can achieve time synchronization, GPS is easily disturbed, causing time synchronization to be damaged, which greatly affects space, frequency domain, and phase. Moreover, there is a certain intensity of suppression jamming, which will make bistatic SAR unable to effectively extract the image information of the target, which greatly affects the effect of bistatic SAR.

For anti-jamming, the airborne SAR installs the sending and receiving on two platforms, called bistatic SAR, as shown in Fig. 7.12.

Point A is the position of the jammer; $B_1$ is the airborne transmitting SAR; $B_2$ is the airborne receiving SAR; Point C is the protected target; $\sigma_t(x)$ is the common irradiation area of the main beam of the bistatic SAR antenna.

The intensity of the signal received by $B_1$ in the irradiation area is $P_a$

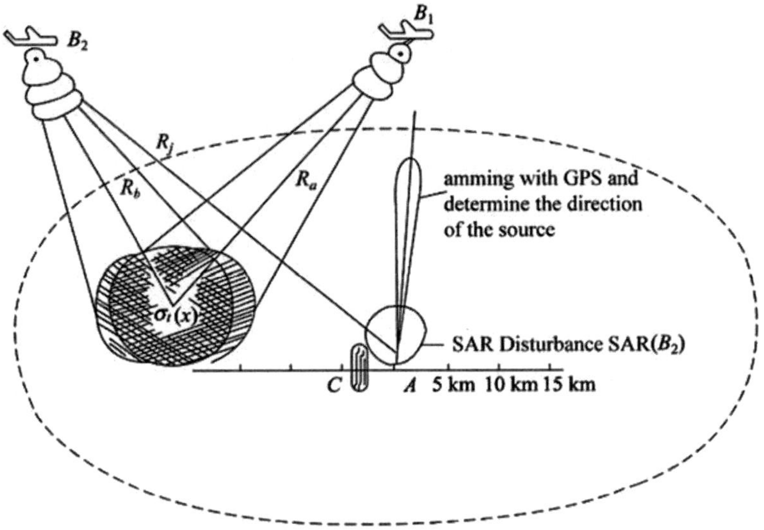


**Fig. 7.12** Schematic diagram of jamming dual-base SAR

$$P_a = \frac{P_t G_t}{4\pi R_a^2}\sigma_t(x) \tag{7.24}$$

$$P_b = \frac{P_a}{4\pi R_b^2} \cdot \frac{\lambda^2}{4\pi} G_b \tag{7.25}$$

Substitute Eq. (7.24) into Eq. (7.25):

$$P_b = \frac{P_t G_t G_b \lambda^2 \sigma_t(x)}{(4\pi)^3 R_a^2 R_b^2} \tag{7.26}$$

Signal strength $P_{bj}$ that $B_2$ received jamming

$$P_{bj} = \frac{P_j G_j}{4\pi R_j^2} \cdot \frac{\lambda^2}{4\pi} G_b(\theta) \tag{7.27}$$

Jamming/signal ratio $K_{j\Sigma}$ received at the output of the $B_2$ receiving antenna

$$K_{j\Sigma} = \frac{P_{bj}}{P_b} = \frac{\frac{P_j G_j}{(4\pi)^2 R_j^2} \cdot \frac{\lambda^2}{4\pi} G_0(\theta)}{\frac{P_t G_t G_b \lambda^2 \sigma_t(x)}{(4\pi)^3 R_a^2 R_b^2}} \cdot \frac{L_j r_j}{L_d K_f}$$

After simplified,

$$K_{j\Sigma} = \frac{4\pi P_j G_j R_a^2 R_b^2}{P_t G_t \sigma_t(x) R_j^2} \cdot \frac{G_b(\theta)}{G_b} \cdot \frac{L_j r_j}{L_d K_f}. \tag{7.28}$$

In the formula:

$K_{j\Sigma}$—The jamming suppression coefficient required by jamming SAR;
$P_j G_j$—The equivalent jamming power of the jammer;
$P_t G_t$—The equivalent power of SAR transmission;
$R_a$—The distance from the center of the irradiation area to $B_1$;
$R_b$—The distance from the center of the irradiation area to $B_2$;
$R_j$—The distance from the jammer to $B_2$;
$\sigma_t(x)$—The equivalent reverse area of the common irradiation area of the main lobes of antennas of the transmitting SAR and receiving SAR;
$\frac{G_b(\theta)}{G_b}$—The main/sidelobe gain ratio of the receiving SAR antenna;
$L_j$—The jamming feeder loss;
$r_j$—The polarization loss;
$K_f$—The ratio of transmitter spectrum width to receiver bandwidth;
$L_d$—The feeder loss of the transmitter;

The jammer uses dual-frequency jamming, with a frequency of L-band, actively jamming with the GPS receiver on the $B_1$ transmitting SAR, which prevents the GPS on the $B_1$ from working properly and makes the reference time garbled.

(1) The equivalent power $P_{jb}G_{jb}$ required to jamming with GPS

$$P_{jb}G_{jb} = K_j P_{tb} G_{tb} \left(\frac{R_j}{R_t}\right)^2 \tag{7.29}$$

In the formula:

$K_j$—The jamming suppression coefficient of GPS jamming which is 42 dB;
$P_{tb}G_{tb}$—The radiated equivalent power on the navigation satellite, which is taken as 10kw.
$R_j$—The distance from the jammer to $B_1$, and the value is 100 km;
$R_t$—The distance from the navigation satellite to $B_1$, and the value is 20000 km;
Substitute into (7.6) to get

$$P_{jb}G_{jb} = 1.5 \times 10^4 \times 10^4 \left(\frac{100}{20000}\right)^2 = 35dBw$$

Take the antenna gain of 15dB and the transmitter power of 100 watts;

(2) Jamming with B2 receiving SAR

$$P_j G_j = \frac{K_{j\Sigma} P_t G_t \sigma_t(x) R_j^2}{4\pi R_a^2 R_b^2} \cdot \frac{G_b}{G_b(\theta)} \frac{L_d \cdot K_f}{L_j r_j} \tag{7.30}$$

In the formula:

$K_{j\Sigma}$—Jamming suppression coefficient, $K_{j\Sigma} = 20\text{dB}$;

Known parameters: $P_t G_t = 15\text{Kw}$; $\sigma_t(x) = 40\text{dB}$; $\frac{G_b}{G_b(\theta)} = 20\text{dB}$; $R_j = 50\text{km}$; $R_a = 30\text{km}$;$R_b = 40\text{km}$; $\frac{L_j r_j}{L_d K_f} = 1.5$. Substituting the parameters into formula (7.30), the calculation is:

$$P_j G_j = \frac{10^3 \times 1.5 \times 10^4 \times 10^4 \times 5^2 \times 10^8 \times 10^2 \times 3}{12.6 \times 9 \times 10^8 \times 1.6 \times 10^9} = 35\text{dBw}$$

If the antenna gain is 5 dB, the jamming transmitter is 1kw.

According to the analysis and calculation, take the value of the jammer.

Jamming antenna: azimuth angle is 120°; elevation is 60°; transmitting antenna gain is 5 dB; transmitter power is 1000 watts; the azimuth angle is 3 × 120°; elevation is 60°. Figure 7.13

With a GPS jammer, after jamming with the other party's GPS, its time synchronization is disrupted, so that the air-domain synchronization, frequency-domain synchronization, and phase synchronization of the receiver and transmitter are destroyed, $\sigma_t(x)$ decreases, and $K_{j\Sigma}$ decreases. Also, with the suppression of a certain amount of noise jamming, its receiving SAR cannot effectively extract the target image, and the bistatic base loses its required effectiveness.

Also, the bistatic airborne SAR can also jamming through other means, such as jamming with the digital transmission system of the bistatic SAR. It can also use radar data to find and receive SAR. Guided by the direction of the radar, aim the

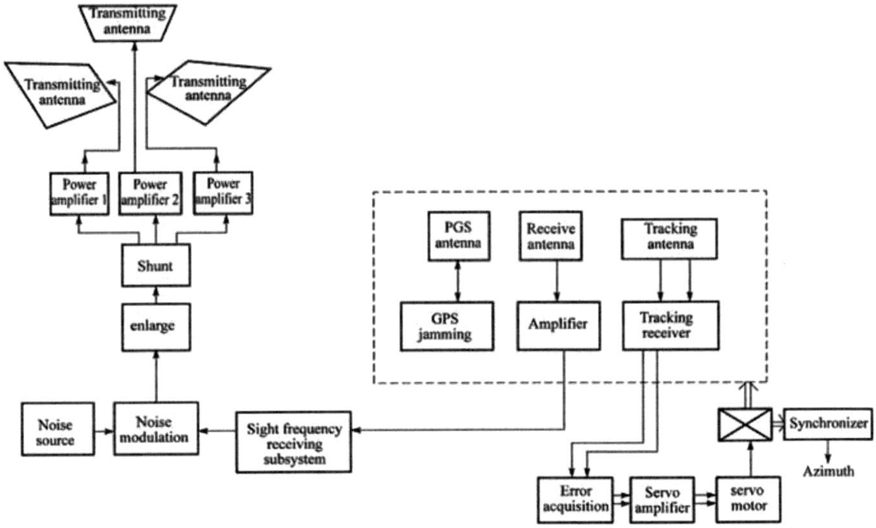


**Fig. 7.13** Paired SAR jammer composition diagram

jammer's antenna at the receiving SAR, so that a smaller jamming power source can be used, which can effectively jam with the bistatic SAR.

Bistatic SAR extracts useful imaging information, but it is subject to many restrictions. For example, it requires the flight balance of the receiver and transmitter; the accuracy of the center point of the imaging area is high, and the time synchronization of the receiver and transmitter is better; the reference frequency must be stable.

When there is no jamming, it is difficult to process useful images. However, there are some human jammings. It is more difficult to process useful images. Therefore, human jamming poses a great threat to bistatic SAR.

In short, bistatic radar and bistatic SAR do not cause too much difficulty in radar jamming. Instead, the radar itself causes a lot of trouble and difficulty after it becomes bistatic. Originally, the invention of radar was dual-base. Due to its positioning, coordinate conversion, and erecting calibration…, there were a lot of inconveniences in the application. Through improvement, the dual base becomes a single base, which is more convenient and applicable. For some anti-jamming or other purposes, it has been changed from a single base to a dual base, but it also faces the threat of jamming. Bistatic radar faces its difficulties, complex electromagnetic environment, and human-made jamming…

The survival and development of bistatic radar and bistatic SAR are faced with many technical problems, so how to develop it in the future which needs further in-depth analysis and research. However, you cannot relax from the perspective of jamming and take good measures to deal with it. As long as you have such equipment, you should deal with it effectively.

# Chapter 8
# Broadband Antenna Technology

With the emergence and use of new systems radar technologies such as Pulse-Doppler radar, pulse compression radar, fast-beam agile phase array radar, SAR radar, ISAR radar, and MIMO radar, countermeasures against traditional jamming methods have been applied with more effective ECCM methods in addition to their performance improvements. Besides, special designs, such as ultra-low sidelobe antennas, sidelobe concealment, sidelobe cancellation, polarization agility, fast beam scanning, etc., have been adopted for anti-reconnaissance and anti-jamming measures against electronic warfare, which poses a greater challenge to electronic countermeasures.

As we all know, an antenna is a transceiver device for electromagnetic waves. Electromagnetic waves are emitted from here and converge on here again. Since the advent of the antenna, modern civilization has begun. There are many types of antennas in various military electronic equipment (radar, communication, navigation, friend or foe identification, electronic warfare) and various TV receiving antennas, radio antennas, mobile phone antennas, walkie-talkie antennas, GPS antennas, microwave relay station antenna, etc. With the advancement and rapid development of computer technology and material technology, phased array antennas have been put into engineering use. The research and application of adaptive antenna and digital beamforming technology have accelerated the pace. And many new technologies appeared, like printed antenna technology, smart antenna technology, multi-function antenna technology, shared aperture technology, and integrated technology. The development of the antenna theory is reflected in the in-depth study of the numerical calculation method, the electromagnetic theory, the application of microelectronic technology, the improvement of various manufacturing techniques in technology, materials, and means, and the update of debugging, testing technology and instruments. The antenna technology research object has changed from the original single antenna technology to the array antenna and expanded from the array antenna to the antenna system formed by multiple antennas and multiple array combinations. Considering the overall relationships among the antenna, antenna-feeder, antenna-carrier platform, and the environment, we have developed a product that matches the latest requirement of modern electronic equipment with high performance, i.e., low power consumption,



X. Zhang et al., *Theory to Countermeasures Against New Radars*,
https://doi.org/10.1007/978-981-16-6715-2_8

small size, lightweight, high reliability, and good appearance. With different platforms, functions, working environments, and performance requirements, the specific implementation methods, shapes, and sizes of antennas have also changed. Many types of antennas have appeared. Electronic warfare includes a warning, electronic intelligence, electronic information support, and electronic jamming, and it is based on the characteristics of reconnaissance electromagnetic signals. According to the threat status and level of the signal, it takes corresponding countermeasures. Its broadband antenna has a great influence on its direction-finding performance and the efficiency of the transmitted signal. The electromagnetic waves radiated by it are one-way transmission. One of the important features is the wide frequency band and large angle coverage, which is compatible with electronic equipment development. Therefore, it is determined that the antenna used should also have characteristics of wideband.

New technologies compatible with electronic warfare jamming technology have been developed to effectively jamming and destroy various radars and electronic equipment's normal work. The functions of electronic equipment are becoming more and more powerful. Due to the developments in modern radar, communications, electronic warfare, navigation, as well as using more flexible jamming waveforms, it is required to increase the equivalent radiated power of the jamming system, receiver sensitivity, transceiver isolation, and variable polarization jamming, and also the ability to intercept information from low-intercept radars and jamming with broadband high-power radars. As the system's operating bandwidth continues to increase, this requires the receiving and transmitting antennas and receiving and transmitting antenna arrays to be oriented and effectively receive or radiate the bandwidth to be wider to meet the requirements of high resolution and imaging of modern radar. It makes the antenna and antenna array play an important role in determining the characteristics of electronic equipment. At the same time, it has promoted the further development of broadband antenna technology and developed broadband multi-beam antennas, broadband phased array antennas, composite antennas with arrays and large-aperture reflecting surfaces, broadband DBF antenna technology, and broadband antenna systems to meet modern electronic equipment requirements.

## 8.1 Major Indicator of the Broadband Antenna

In fact, the main indicators of broadband antennas are consistent with the general antennas. It also has specific requirements and conventions. The antenna's bandwidth refers to its main indicator, such as gain, main lobe width, sidelobe level, input impedance, polarization characteristics, phase distribution, etc. When they all meet the frequency range of the design requirements, the broadband antenna may have a large instantaneous bandwidth coverage or a small instantaneous bandwidth during specific operations. Generally, the antenna indicators change with frequency, so the antenna bandwidth is determined by the indicators' frequency characteristics.

If several indicators are required simultaneously, the most stringent requirements should be taken as the basis for determining the antenna bandwidth.

(1) notation of antenna bandwidth

Relative bandwidth notation: the ratio of the difference between the highest operating frequency and the lowest operating frequency of the antenna to the center frequency within the antenna operating frequency range, which is:

$$R = \frac{f_h - f_l}{f_0} \quad R = 2\frac{\Delta f}{f_0} \tag{8.1}$$

where, $f_h$ and $f_l$ are the upper limit frequency and the lower limit frequency within the operating frequency band,

$f_0 = \frac{f_h + f_l}{2}$ is usually expressed as a percentage.

Antenna bandwidth ratio: the ratio of the upper limit frequency to the lower limit frequency of the antenna operating frequency band

$$BW = \frac{f_h}{f_l} \tag{8.2}$$

For example, the operating frequency range of the antenna is 2–18 GHz, and its bandwidth ratio is 9:1

Octave bandwidth notation: the ratio of the upper limit frequency to the lower limit frequency within the antenna's operating frequency band is usually a power of two relationships. Ley p be the power

$$BW = 2^p \tag{8.3}$$

If $p = 1$, it is called 1 octave, and if $p = 2$, it is called 2 octaves. In the frequency range of 2–8 GHz, BW = 4, that is, the bandwidth is 2 octaves, and so on. In general, narrow-band antennas mostly use relative bandwidth notation, while broadband antennas usually use bandwidth ratio notation, and ultra-wideband antennas use octave notation. A relative bandwidth of about or more than 10% is defined as a broadband radar in a radar system, and a relative bandwidth of 25% and above is considered an ultra-wideband. In the electronic warfare system, the bandwidth ratio is greater than or equal to 1.5 as a broadband antenna. When the bandwidth is greater than 3 octaves, it is considered to be an ultra-wideband antenna. The common frequency range of electronic warfare is 50 MHz–40 GHz. Of course, it can be extended at both ends in the past. In engineering development, the maximum antenna bandwidth is 220:1.

(2) Antenna impedance and standing wave

The concept of impedance is particularly useful for mid- and low-frequency antennas. It is easy to determine a pair of input points where the impedance is a single value. At higher frequency bands or microwave frequencies, when there may be multiple impedance values, and it is almost impossible to directly measure the antenna's impedance, measuring the standing wave coefficient or reflection loss is used to calculate the input impedance of the antenna. This is because

$$Z_i = Z_c \frac{1 + \Gamma}{1 - \Gamma} \tag{8.4}$$

where $Z_c$ is the characteristic impedance of the transmission line, and $\Gamma$ is the reflection coefficient. The voltage standing wave coefficient VSWR is often used to indicate the matching of the antenna with the feeder and the system, which is

$$VSWR = \frac{1 + |\Gamma|}{1 - |\Gamma|} \tag{8.5}$$

The antenna radiates energy into free space and forms an electromagnetic wave that can propagate forward, and the space wave impedance is 377 Ω. The antenna also needs to transmit RF information with the following transmission line. The commonly used transmission line is a 50 Ω system. When the antenna forms a system, the various RF components and beamforming networks must be designed to be close to 50 Ω input/output ports to ensure that the antenna system can be connected to the 50 Ω receiver or transmitter as closely as possible. Therefore, for an antenna system, its input impedance is very important to the system engineer, and it directly affects the efficiency of the input or output of the antenna energy. For the free space wave impedance of 377 Ω and the transmission feeder of 50 Ω, no matter what kind of design measures of impedance transformation that the antenna adopts, each antenna radiating unit's input impedance requires 50 Ω as the reference. Since it is difficult to accurately calculate the self-impedance and mutual impedance characteristics of each radiating antenna element. Theoretically, theoretical analysis is mainly used to diagnose and guide engineering design. The antenna's input impedance (including the real resistance component and the imaginary reactance component) is a complex number. Whether it is a single antenna or an array antenna, its impedance characteristic is for the antenna and the array input or output port, which is a frequency function. When the antenna and the surrounding environment have strong electromagnetic coupling, its impedance characteristics are determined by the antenna's structure, electrical size, feeding network form, and the environment, especially in the broadband range. When the input impedance $Z_A$ of the antenna does not match the characteristic impedance of the transmission line $Z_0 = 50\ \Omega$, a standing wave is formed on the transmission line. The standing wave ratio indicates the degree of mismatch between the antenna impedance and the transmission line impedance. In electronic warfare equipment, the input voltage standing wave ratio of a single antenna or antenna array $VSWR \le 2.5$ is most commonly used. For a transmission system without protection such as a circulator, it is often required that the standing wave of the transmitting antenna $VSWR \le 2$. For antennas applied in narrow frequency bands, the standing wave characteristic $VSWR \le 1.2$.

**Table 8.1** Relationship between VSWR, reflection coefficient and reflected power

| VSWR | Reflection coefficient | Transmission power (%) | Reflected power (%) |
|---|---|---|---|
| 1 | 0 | 100 | 0 |
| 1.05 | 0.025 | 99.94 | 0.06 |
| 1.1 | 0.045 | 99.97 | 0.23 |
| 1.15 | 0.07 | 99.51 | 0.49 |
| 1.2 | 0.09 | 99.18 | 0.82 |
| 1.5 | 0.2 | 96 | 4 |
| 2.0 | 0.34 | 89 | 11 |
| 2.5 | 0.43 | 81.5 | 18.5 |
| 3.0 | 0.5 | 75 | 25 |
| 4.0 | 0.6 | 64 | 36 |
| 5.0 | 0.67 | 56 | 44 |

When VSWR = 1, the system is completely matched and unlikely to be realized in a real project;

When VSWR $\leq$ 1.5, the system matching is good; when VSWR $\leq$ 2, the system matching is good;

When VSWR $\leq$ 3, the matching degree can still be used;

When VSWR $\geq$ 3, it is considered that the matching is relatively poor. For a specific ultra-wideband reconnaissance system, the antenna's standing wave coefficient can be allowed to be less than or equal to 4. The relationship between VSWR, reflection coefficient, and reflected power is shown in Table 8.1.

Besides, for the waveguide or ridge waveguide transmission line, its characteristic impedance is no longer 50 Ω. When designing an antenna using this type of transmission line, it should be specifically considered or transitioned to 50 Ω impedance. Simultaneously, the characteristic impedance of the transmission line of the same size for the higher-order mode is very different from that of the basic model, and it should be specially considered for special use.

(3) Antenna pattern

The antenna pattern describes the distribution of electromagnetic field strength in space, which is a three-dimensional figure. In engineering, patterns in two mutually perpendicular planes in the direction of the antenna's maximum radiation are usually used to represent the antenna's directionality. Regarding the antenna's installation position, the patterns of the two principal planes of the azimuth plane and the elevation plane are respectively expressed. The specific parameters describing the patterns are: Lobe width (generally the main beam half-power width. There are special regulations under special requirements, such as 10 dB beamwidth, zero value width), sidelobe level, head-tilt (the angle at which the maximum beam radiation direction deviates from a predetermined value), front to back ratio (the ratio of the maximum radiation direction level to its reverse radiation level), crossing point drift,

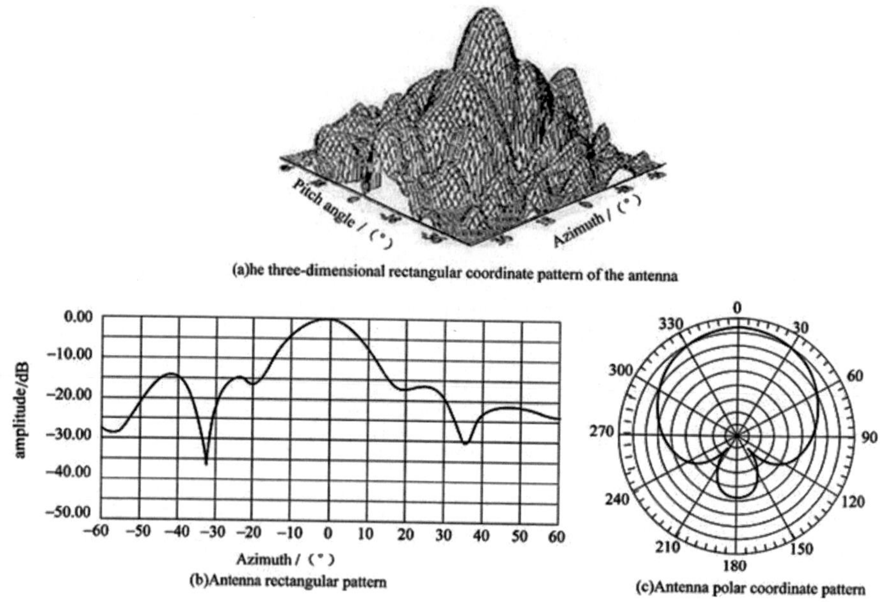


**Fig. 8.1** Representation of three antenna patterns

zero-depth of difference beam, zero drift, etc. The pattern most intuitively reflects the spatial distribution of the electromagnetic field's size, forming a free space distribution pattern. The three-dimensional figure describing it is called a radiation pattern, abbreviated as a pattern. Generally, the description of the spherical coordinate system (φ, θ) is: A = F (φ, θ) or the description of the rectangular coordinate system and the polar coordinate system of the main plane is shown in Figure 8.1b.

With different electronic equipment, the requirements for the antenna's pattern are also different. The most commonly used antenna patterns are:

(a) Omnidirectional pattern: the azimuth plane is covered by 360°, the figure is a circle, and the pitch plane is inverted V-shaped;
(b) Acicular beam pattern: the beamwidth of the azimuth plane is equal to the beamwidth of the pitch plane;
(c) Fan beam pattern: the beamwidth of the azimuth plane is narrow, and the beamwidth of the pitch plane is wide;
(d) Shaped beam pattern: the beamwidth of the azimuth plane is narrow, the beam of the pitch plane satisfies a specific pattern such as the cosecant square relationship, or only the beam of the azimuth plane is shaped, or both main planes are shaped;
(e) Multi-beam pattern: there are multiple beams distributed according to a certain rule in the azimuth plane, and the pitch plane can be narrow beam or multi-beam, vice versa. Such as orthogonal beam, and sum and difference beam, mono-pulse tracking/direction-finding beam, simultaneous multi-beam, etc.

For an oblique 45° polarized or circularly polarized antenna, in order to fully understand the characteristics of the antenna pattern, the response patterns of its E and H polarization planes are described in the same principal plane (e.g., azimuth/pitch plane), and they are often drawn on the pattern of the same main plane (such as azimuth plane/pitch plane). In addition, circularly polarized antennas are often represented by the main plane's polarization pattern to obtain the wide-angle polarization distribution of the antenna, as shown in Fig. 8.2. For some applications, the antenna sometimes requires knowledge of the orthogonal polarization distribution in addition to the matching polarization. The antenna pattern is not only a function of space but also a function of frequency. For the antenna pattern in the wideband range, the relevant parameters will have corresponding changes. According to the system's different usage, the antenna's requirements for the sidelobe level are different. In general, a single antenna's sidelobe level is required to be greater than or equal to 15 dB. It is sufficient when the array antenna's sidelobe level is greater than or equal to 13 dB, and the omnidirectional antenna fluctuation is less than ± 3dB. Moreover, the pattern of different types of antennas or feeding methods is quite different. For specific applications, the optimal selection and special design are selected according to the requirements.

Because the direction of the sidelobe is usually the direction that does not need to radiate (or receive) energy. So, the lower the antenna sidelobe level is, the weaker the antenna radiated energy in the unwanted directions is, or the stronger the ability to suppress unwanted signals is in these directions. In the application of broadband antennas, in addition to the amplitude and polarization patterns, the phase pattern is sometimes an important indicator required by the system. It is closely related to the type of antenna, operating frequency band, and spatial distribution status. It represents the change of the antenna's phase plane with the spatial angle range under broadband conditions. Generally, it is explained by the phase fluctuation range allowed by the system within the required angle range of the antenna. If the allowable range of phase fluctuation is larger, the effective angle range is larger, and it has a certain relationship with its amplitude pattern. In practice, the allowable phase fluctuation is about 10°. Figure 8.3 shows a typical phase pattern of a wide beam antenna. Its phase

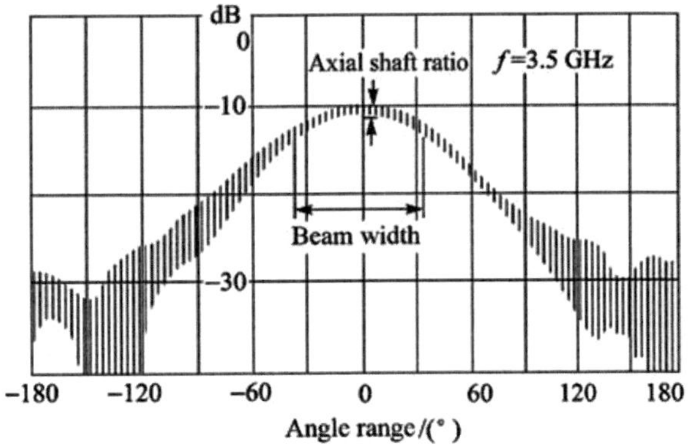


**Fig. 8.2** Antenna polarization pattern

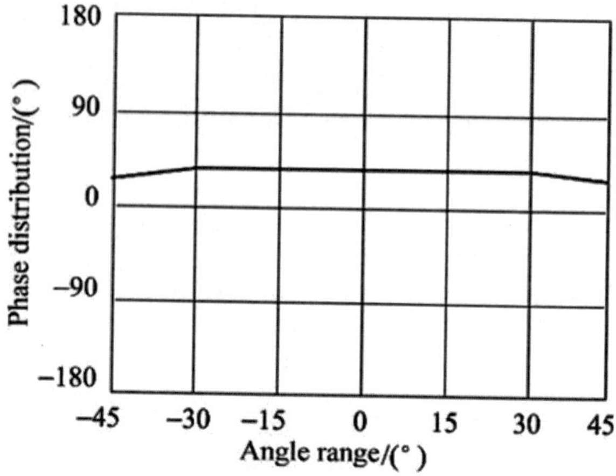


**Fig. 8.3** Antenna phase pattern

distribution is relatively uniform in the range of 90°. When the amplitude pattern is narrow, the phase distribution range becomes smaller; when the amplitude pattern is zero, the phase pattern changes in a reversed-phase.

On the other hand, the front-to-rear ratio refers to the ratio of the field value of the maximum radiated (or received) force (set 0° as the direction) to the maximum field value in the direction of 180 ± 60°. Usually expressed by the symbol F/B, the decibel value of the front-to-rear ratio is:

$$F/B = 20\ \lg \frac{\text{Maximum value of the front zone}}{\text{Maximum value of the back zone}} (\text{dB}) \tag{8.6}$$

(4) Antenna gain

The antenna gain G and directivity D are closely related to physical quantities, and the relationship is:

$$\text{G} = \eta.\text{D} \tag{8.7}$$

where η is the antenna's efficiency, which represents the efficiency of the antenna in energy conversion. When the antenna has no metal loss, dielectric loss, mismatch loss, and leakage loss, η is equal to 1. Usually, η is less than 1. The efficiency of loading antennas or electrically small antennas in engineering is relatively low. The antenna's directivity means the ratio of the antenna's intensity value in the direction of maximum radiation intensity to the average radiation intensity. In the spherical coordinate system, it can be expressed as:

$$D = \frac{4\pi}{\int_0^{2\pi} \int_0^{\pi} F^2(\theta, \phi) \cdot \sin\theta \cdot d\theta d\phi} \tag{8.8}$$

F(θ,φ)is the normalized radiation pattern of the peak value in the formula. This value can be calculated theoretically based on the calculated pattern, or it can be obtained by the graphical digital integration method based on the measured pattern. Therefore, directivity depends on the antenna pattern. Antenna gain G(φ,θ) is a quality factor that describes the antenna. It represents the maximum radiation intensity ratio in a given direction to the maximum radiation intensity in the same direction of an ideal parameter antenna with the same input power. The antenna gain is a function related to the spatial position (φ,θ). In a general antenna pattern, the level value in the direction of maximum radiation is recorded as the antenna gain $G_{A\,=}$ $G(\varphi_0,\theta_0)$(dBi)). The letter i indicates that the gain is relative to the isotropic radiator. And the half-wavelength oscillator is also used as a reference in the project. That is, the gain unit is (dBd), but $G_{A\,=}$ $G(\varphi_0,\theta_0)$(dBi) − 2.15. The directivity coefficient of the aperture antenna is usually estimated by the beamwidth of the two main planes of the antenna:

$$D = \frac{DB}{H_p(^\circ) \times E_p(^\circ)} \tag{8.9}$$

where $H_p$ and $E_P$ are the beamwidths of the main plane of the antenna

For uniform rectangular aperture: DB = 32383

When the E plane is uniform, and the H plane is cosine distributed: DB = 35230

For uniform circular caliber: DB = 33709

When the amplitude has a gradient of 12 dB: DB = 38933

For Gaussian and cos q(θ) rectangular beams: DB = 41250

For example, for a half-wave oscillator that Hp = 78°, Ep = 360°, D = 1.47, it is very close to the theoretical value of 1.64

For cone-beam: DB = 52525

The most direct expression for gain is

$$G = \frac{4\pi}{\lambda^2} A \cdot \eta_1 \tag{8.10}$$

where λ is the operating wavelength, $\eta_1$ is the aperture efficiency, and A is the aperture area.

A1 = A . η1 is the equivalent aperture. For broadband high-performance feed parabolic antenna η1, its value is greater than or equal to 50%. For the pyramid horn antenna, aperture efficiency is 70%. For general aperture antennas, DB = 26000 is sufficient.

(5) Polarization of antenna

The antenna's polarization refers to the orientation of the electric field vector of the antenna in the direction of maximum radiation. When the orientation of the electric field vector is unchanged, the antenna is linearly polarized. Taking the horizontal plane as a reference, linearly polarized antennas are divided into horizontally polarized antennas, vertically polarized antennas, and oblique polarized antennas (such as antennas commonly used for 45° polarized antennas).

When the electric field vector's orientation changes with time and the endpoint orbit is a circle, it is also called a circular pole antenna. When viewed in the direction of radio wave propagation, the clockwise direction is called right circular polarization, and the counterclockwise direction is called left circular polarization. If the orbit of the endpoint of the rotation vector is not round enough, it is called an elliptically polarized antenna. In order to express this characteristic of the antenna, the ratio of the long axis to the short axis of the polarized ellipse is generally called the axial ratio (Ar), and the expression is:

$$A_r = 20 \log \frac{a}{b} \tag{8.11}$$

The antenna radiates a large part of the input energy in the main polarization, but it may also radiate unwanted energy in other polarization directions. People call this unwanted radiation cross-polarized radiation. For linearly polarized antennas, the cross-polarization direction is orthogonal to the main polarization direction. For horizontal polarization, its cross-polarization is exactly vertical polarization. For circular polarization, cross-polarization is the opposite component of the main polarization. The cross-polarization component of right-hand circular polarization is left-hand circular polarization and vice versa. The circular polarization of the antenna is generated by two linearly polarized waves with the same amplitude, perpendicular to each other, and with a 90° phase difference. Any two circularly polarized waves with the same amplitude and opposite rotation can generate a linearly polarized wave.

The circularly polarized antenna can detect the circularly polarized wave with the same rotation direction, and, at the same time, detect the linearly polarized wave with any orientation with a 3dB polarization mismatch loss. Still, it cannot receive the counterrotated circularly polarized wave. Circularly polarized antennas, such as planar spiral antennas, are often used in airborne radar warning systems and homing guidance equipment. An oblique 45° linearly polarized antenna can detect any circularly polarized wave and linearly polarized wave except the linearly polarized wave orthogonal to it with a certain polarization mismatch loss, which is mostly used for shipborne and ground electronic equipment. There are two ways to generate circular polarization, such as self-radiating or receiving circularly polarized wave antennas and the use of feed conversion to obtain circularly polarized antennas. Commonly used polarization methods in broadband antenna technology are horizontal polarization, vertical polarization, left oblique 45° polarization, right oblique 45° polarization, left-hand circular polarization, and right-hand circular polarization. In order to consider the main polarization characteristics of multiple electromagnetic signals, it is preferable to lose 3 dB power level in the application. The polarization isolation of wideband antennas is generally around 20 dB. In good cases, it could be up to 30 dB, and only a dozen dB at small apertures. It is only a dozen dB at a small caliber. N electronic warfare equipment, the receiving antenna's axial ratio is generally less than or equal to 3 dB, and it may be larger for the transmitting antenna. In high-precision direction-finding arrays, the axial ratio is required to be close to 0 dB, and the axial ratio of circular polarization is required to be about 3 dB within the range of $-30°\sim+30°$. The meander antenna that realizes double circular polarization with the same aperture is a circularly polarized antenna with good performance. For electromagnetic information in the sidelobes area outside the main lobe, the importance of antenna polarization is less prominent.

Engineering development requires that the antenna's main electrical specifications, such as gain, lobe width, sidelobe level, voltage standing wave ratio, and polarization characteristics within the specified frequency range, must meet the technical specifications. Especially in the broadband range, it is very difficult to make each indicator very high, and sometimes it is impossible or unnecessary. For example, there is a certain relationship between gain and beamwidth. For the same antenna, the gain is high, and the beamwidth becomes narrow. Otherwise, other technologies must be used. In fact, the antenna's broadband characteristics mainly depend on the frequency characteristics of various technical indicators. They are the pattern bandwidth, gain bandwidth, impedance bandwidth, and polarization bandwidth. It may be that the deflection of the main lobe is too large, the main lobe is split or atrophy, the sidelobes are increased, the front-to-rear ratio is decreased, and the antenna phase fluctuations are increased within a specified angle range. When the orientation diagram deteriorates not to meet the design requirements, the pattern bandwidth is limited. The gain has a lower value in the low-frequency section. When it falls below the allowable value, the gain bandwidth is limited, limiting the lower operating frequency. The antenna's impedance bandwidth is the wide band of the frequency band whose standing wave ratio is lower than a certain value. It is the most basic index in antenna engineering. For small electric antennas, antenna matching is a key issue. For circularly polarized antennas, the polarization bandwidth is a very important technical indicator. It determines the polarization bandwidth when the axial ratio is less than a specified value within the main lobe's width. In short, the bandwidth of the antenna is a comprehensive technical indicator. In order to meet the needs of the system, sometimes a compromise is required. In engineering design, the narrowest bandwidth must be used as the benchmark. Within this frequency band, the antenna specifications can meet the requirements. Especially for the broadband antenna system, it involves the antenna layout, the impact of the carrier platform, the radome, passive network, active network, amplitude, and phase characteristics, electromagnetic compatibility, power supply, self-test circuit, component heat dissipation, reliability, and environmental adaptability and many other technologies. Various single broadband antennas are used as array elements connected to the corresponding array elements through special broadband amplitude and phase conversion networks to generate input and output relationships of specific functions. Some of these antenna arrays are used as direction finding and positioning arrays to complete predetermined direction finding and positioning function. For example, multiple helical antennas are used in the direction-finding system of an interferometer to form a direction-finding antenna array; multiple horn antenna arrays on electronic reconnaissance satellites improve direction-finding accuracy and angular resolution. Some antenna arrays are used to transmit power. With the array's help, a relatively high gain is generated, the beamwidth is compressed, and a specific area is covered. The in-phase spatial focusing is completed at the same time, resulting in a relatively high-power synthesis. In fact, the antenna array itself may have active microwave devices, and the array aperture itself can be received or transmitted. Due to the system's requirements, the array should be located at the transmitter or receiver. Both the functions must be clear at the beginning of the development. In fact, the focus of the two developments is different. If both the receiving and sending functions are considered, the array is first designed comprehensively, and then the switching function is set in the network. The scanning beam formed by the array wave composed of the same array element has an outline in the distribution pattern of the unit pattern, which is enhanced compared to the single-caliber electromagnetic compatibility capability. The array elements can be arranged in a linear array, a conformal array, and a planar array. The beamforming network is used to feed them, and the electrical scanning network is used to scan their sub-beams. According to different beamforming methods, multi-mode antennas, multi-beam antennas (one-dimensional multi-beam, two-dimensional multi-beam), phase array antennas, and other electrical scanning forms antennas are formed. There are many ways to feed the antenna array. The energy of each array element in the array antenna is radiated to a specific direction to form the beam's shape in the usual sense, the pointing of the maximum value of the beam, and the side level and zero points of the beam. The aperture of the array antenna has a great expansion relative to the unit antenna, so the beam they form is narrow and high in gain. In order to cover the predetermined airspace, such as 45° × 45° or 90° × 45°, it is necessary to enable the beam to achieve fast scanning, phase control by electronic drive or electronic beam switching between multiple-beam ports, or perform mechanical scanning through servo rotation, or use a mixture of the two scanning methods. A system composed of an array antenna can simultaneously search, monitor, and track multiple maneuvering targets. To achieve these functions and meet the needs of electronic systems in the broadband range, it is necessary to design various broadband arrays and feeding methods to compensate and solve the mutual coupling of antenna array elements, control the amplitude and phase balance of each channel. Many technical problems need to be solved to realize the project. The following is introduced from several aspects.

## 8.2 The Technology of Multi-mode Antenna

In broadband array antennas and feeder networks, through proper network weighting, the amplitude of each array element is the same, and the phase is a specific distribution method. The arrangement of the antenna array is evenly distributed in a circle. In this way, the phase relationship information corresponding to the azimuth plane angle is generated one by one. The corresponding phase is detected by subsequent devices such as the receiver to measure the target direction.

### 8.2.1 *Small Aperture Broadband Adcock Antenna Array*

Figure 8.4 shows a schematic diagram of a continuous array of omnidirectional elements. Where R is the array radius, θ is the normal direction of the wavefront PQ (referred to as the wavefront direction), α is the opening angle of the arc between an array element A and array element B, and the diameter of the array element B coincides with the directional radial of θ = 0° azimuth. If the array element S on the azimuth ray passing through the array diameter is used as the reference point, the exciting current of any array element A is:

$$\mathrm{I}_\alpha = e^{jK\alpha} \tag{8.12}$$

where, K = 0, ± 1, ± 2 …

According to the theory of geometric optics, the phase of the electric field radiated by the array element A to the wavefront plane PQ is:

$$\Phi_\alpha = \mathrm{K}\alpha - \mathrm{kR}[1 - \cos(\alpha - \theta)]$$

where k = 2π/λ, λ is the wavelength. From this, the far-field pattern of the circular array is obtained:

$$\begin{aligned} F_{(\theta,k)} &= \int_0^{2\pi} e^{j\Phi\alpha} d\alpha \\ &= \int_0^{2\pi} e^{-j(kR-K\theta)} \cdot e^{j[K(\alpha-\theta)+kR\cos(\alpha-\theta)]} d\alpha \end{aligned} \tag{8.13}$$

According to cylindrical function theory:

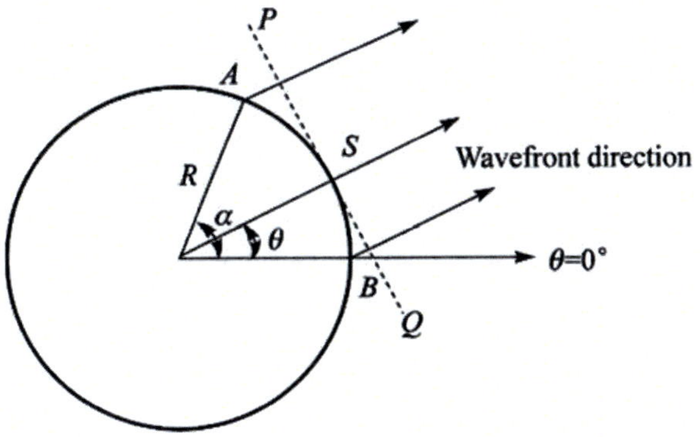


**Fig. 8.4** Schematic diagram of an omnidirectional circular array

$$2\pi j^K J_K(kR) = 2\pi \left(e^{j\frac{\pi}{2}}\right)^K J_K(kR)$$
$$= \int_0^{2\pi} e^{j(K\phi + kR\cos\phi)} d\phi$$

According to the formula (8.13):

$$F_{(\theta,K)} = 2\pi e^{j\left(\frac{K\pi}{2} + K\theta - kR\right)} J_K(kR)$$

Move the phase reference point from point S in Fig. 8.1 to the center of the circle:

$$F_{(\theta,K)} = 2\pi e^{j\left(\frac{K\pi}{2} + K\theta\right)} J_K(kR) \tag{8.14}$$

The formula (8.14) is the K-order phase mode of the omnidirectional continuous distribution circular array. When the omnidirectional elements of a circular array are uniformly distributed, the array element excitation can be regarded as a continuously distributed sampling excitation. According to the sampling theorem, as long as the circumferential sampling rate is much larger than the number of excitation phase' cycles (i.e., K value) of the circular array on the circumference, the discrete array of array elements can well reproduce the electrical performance of the continuous array of circular array elements. That is, its pattern can still be expressed by formula (8.14). In fact, when the excitation current of discrete array elements is:

$$I_i = e^{j\frac{2K\pi i}{N}} \tag{8.15}$$

where i is the array element number, i = 0, 1, 2, 3, …N − 1; N is the number of array elements), the electric field pattern is:

$$F_{d(\theta,K)} = N J_K(kR) e^{j\left(\frac{K\pi}{2} + K\theta\right)} + N j^{N-|K|} J_{N-|K|}(kR) e^{j\left(\frac{KN}{|K|} - K\right)\theta} \tag{8.16}$$

Considering that the problem we are studying, $|K| = 0, 1$, is far less than N, so the second term in formula (8.16) is much smaller than the first term, omitting the second term and we get:

$$F_{d(\theta,K)} \approx N J_K(kR) e^{j\left(\frac{K\pi}{2} + K\theta\right)} \tag{8.17}$$

Obviously, the expression of formula (8.17) and formula (8.14) are the same, so Eq. (8.17) is the K-order phase mode expression of the omnidirectional discrete distributed circular array.

The basic principle of the sine and cosine signals formed by the 0th and 1st modes is as follow:

When K = 0 and 1, according to formula (8.17):

$$F_{d(\theta,0)} = NJ_0(kR) \tag{8.18}$$

$$F_{d(\theta,1)} = NJ_1(kR)e^{j(\frac{\pi}{2}+\theta)} \tag{8.19}$$

It can be seen from Eqs. (8.18) and (8.19) that the amplitude of the 0th mode is much larger than that of the 1st mode (note kR ≪ 1), which is the main difference between the omnidirectional and the directional element (usually can be achieved only in the high-frequency band) DBD antenna array (The amplitudes of the available phase modes of the directional element DBD antenna array are nearly equal). The velocity of the 0th mode is appropriately attenuated so that its amplitude is equal to the amplitude of the 1st mode. Then the signal of the 1st mode is phase-shifted by φ, and then input into the same addition network to obtain a composite signal:

$$\begin{aligned} F_{\Sigma(\theta.\phi)} &= A\left[1 + e^{j(\frac{\pi}{2}+\theta+\phi)}\right] \\ &= 2A\cos\left(\frac{\pi}{4} + \frac{\theta+\phi}{2}\right)e^{j(\frac{\pi}{4}+\frac{\theta+\phi}{2})} \end{aligned} \tag{8.20}$$

where $A = NJ_1(kR)$.let φ = 0, π, and subtract the obtained information by square law detection:

$$\begin{aligned} S_{d(\theta)} &= \left|F_{\Sigma(\theta,\pi)}\right|^2 - \left|F_{\Sigma(\theta,0)}\right|^2 \\ &= 4A^2\left[\cos^2\left(\frac{\theta}{2} + \frac{3\pi}{4}\right) - \cos^2\left(\frac{\theta}{2} + \frac{\pi}{4}\right)\right] \\ &= 4A^2\sin\theta \end{aligned}$$

Similarly, let $\phi = \frac{\pi}{2}, \frac{3\pi}{2}$:

$$\begin{aligned} C_{d(\theta)} &= \left|F_\Sigma\left(\theta, \frac{3\pi}{2}\right)\right|^2 - \left|F_\Sigma\left(\theta, \frac{\pi}{2}\right)\right|^2 \\ &= 4A^2\cos\theta \end{aligned}$$

If we take N = 8 (8-element circular array), note that kR ≪ 1, using the series expression of the 1st order Bessel function:

$$J_1(kR) = \sum_{m=0}^{\infty} \frac{(-1)^m}{(1+m)!m!}\left(\frac{kR}{2}\right)^{1+2m}$$

Omit the small terms that more than three times of kR. The above two formulas can be simplified to

$$S_{d(\theta)} = 4^3(kR)^2\sin\theta \tag{8.21}$$

$$C_{d(\theta)} = 4^3 (kR)^2 \cos\theta \tag{8.22}$$

The phase detector has the function of phase-shifting 0, π/2, π, 3π/2 for one of the two input signals. Therefore, Eqs. (8.21) and (8.22) are equal levels of 0. The first mode is input to the same phase detector, then the four output signals are subtracted from each other, and the two types are compared to obtain:

$$tg\,\theta = S_{d(\theta)}/C_{d(\theta)} \tag{8.23}$$

As we all know, the dipole antenna has a "∞" pattern on the E plane, while the H plane has an omnidirectional pattern. Adcock once proposed to form a binary array by two reverse feed dipoles to obtain a "∞" shaped H-plane pattern. Using a circular array composed of 8 omnidirectional monopoles or dipoles, the antenna array contains two sets of back-feeding Omni-directional elements (each group has three pairs of back-feeding omni-directional element), each forming a "∞"-shaped H-plane pattern in the azimuth plane, and these two are orthogonal to each other in the H plane, which improves the ability of the wide-band angle discrimination. Since the angle measurement method of this antenna is still based on the "∞"-shaped H-plane pattern, it is still traditionally called the Adcock antenna. Circular array antenna 0th-order mode and orthogonal "∞" shaped H-plane pattern.

The 8-element Adcock antenna array comprises 8 broadband monopoles (or dipoles) and a feed network, as shown in Fig. 8.5. In the figure, $F_{0(\theta)}$ is obviously the 0th phase mode in the previous section, and $F_{s(\theta)}$ and $F_{c(\theta)}$ are mutually orthogonal "∞" shaped H-plane patterns. The corresponding amplitude excitation phases of each pattern corresponding to the array element are shown in Table 8.2.

According to Fig. 8.5 and Table 8.2:

$$\begin{aligned} F_{0(\theta)} &= \sum_{i=0}^{7} e^{ikR\cos\left(\frac{i\pi}{4}-\theta\right)} \\ &= 2\left\{\cos\left[kR\cos\left(\theta-\frac{\pi}{2}\right)\right]+\cos\left[kR\cos\left(\theta-\frac{\pi}{4}\right)\right]\right. \\ &\quad \left.+\cos[kR\cos\theta]+\cos\left[kR\cos\left(\theta+\frac{\pi}{4}\right)\right]\right\} \end{aligned} \tag{8.24}$$

$$\begin{aligned} F_{s(\theta)} &= \sum_{i=1}^{3} e^{jkR\cos\left(\frac{i\pi}{4}-\theta\right)} - \sum_{i=5}^{7} e^{jkR\cos\left(\frac{i\pi}{4}-\theta\right)} \\ &= 2\left\{\sin\left[kR\cos\left(\frac{\pi}{4}-\theta\right)\right]+\sin\left[kR\cos\left(\frac{\pi}{2}-\theta\right)\right]-\sin\left[kR\cos\left(\frac{\pi}{4}+\theta\right)\right]\right\}e^{j\frac{\pi}{2}} \end{aligned} \tag{8.25}$$

$$F_{c(\theta)} = \sum_{i=-1}^{1} e^{jkR\cos\left(\frac{i\pi}{4}-\theta\right)} - \sum_{i=3}^{5} e^{jkR\cos\left(\frac{i\pi}{4}-\theta\right)}$$

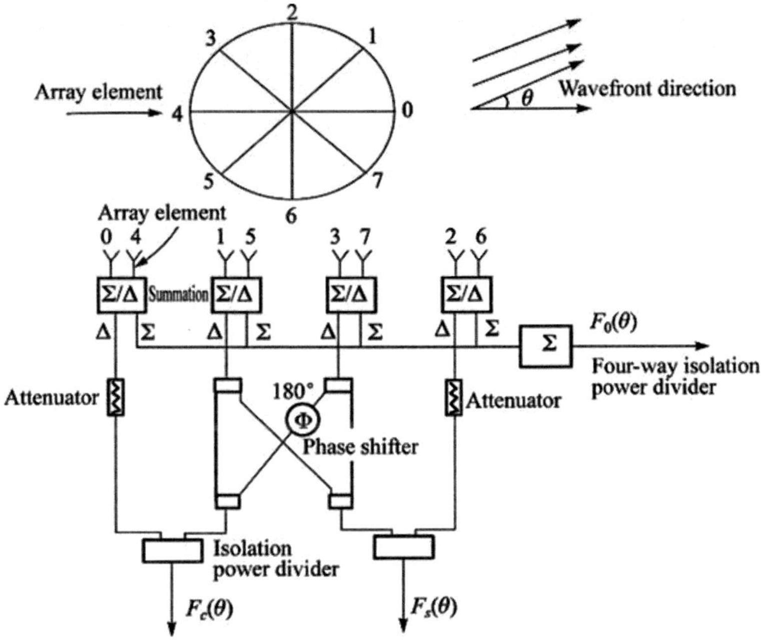


**Fig. 8.5** Schematic diagram of the 8-element Adcock antenna array

**Table 8.2** Equal amplitude excitation phases corresponding to array elements

| Direction number | Array element | | | | | | | |
|---|---|---|---|---|---|---|---|---|
| | 0 | 1 | 2 | 3 | 4 | 5 | 6 | 7 |
| $F_{0(\theta)}$ | 0° | 0° | 0° | 0° | 0° | 0° | 0° | 0° |
| $F_{s(\theta)}$ | / | 0° | 0° | 0° | / | 180° | 180° | 180° |
| $F_{c(\theta)}$ | 0° | 0° | / | 180° | 180° | 180° | / | 0° |

$$= 2\left\{\sin\left[kR\cos\left(\frac{\pi}{4}-\theta\right)\right] + \sin[kR\cos\theta] + \sin\left[kR\cos\left(\frac{\pi}{4}+\theta\right)\right]\right\}e^{j\frac{\pi}{2}} \tag{8.26}$$

Note that kR $\ll$ 1, expand the formulas (8.24), (8.25), (8.26) with Taylor series, and omit the higher-order terms that more than three times:

$$F_{0(\theta)} \approx 8 - 2(kR)^2 \approx NJ_0(kR)\big|_{N=8} = F_{d(\theta,0)}\big|_{N=8} \tag{8.27}$$

$$F_{s(\theta)} \approx (2+2\sqrt{2})kR\sin\theta \cdot e^{j\frac{\pi}{2}} \tag{8.28}$$

$$F_{c(\theta)} \approx (2+2\sqrt{2})kR\cos\theta \cdot e^{J\frac{\pi}{2}} \tag{8.29}$$

Obviously $F_{0(\theta)}$ is omnidirectional. $F_{0(\theta)}$ and $F_{s(\theta)}$ $F_{c(\theta)}$ are two orthogonal "$\infty$" patterns, and the amplitude of $F_{0(\theta)}$ is much greater than $F_{s(\theta)}$ and $F_{c(\theta)}$.

The principle of $F_{0(\theta)}$, $F_{s(\theta)}$ and $F_{c(\theta)}$ forming a sine and cosine signal.

Suppose that an appropriate attenuation η is introduced in the $F_{0(\theta)}$ channel, so that:

$$\left|\eta F_{0(\theta)}\right| = \left|F_{s(\theta)}\right|_{\max} = \left|F_{c(\theta)}\right|_{\max} = B$$

Introducing the phase shift of $\frac{\pi}{2}$ in one of the two channels (which can be achieved with a quadrature coupler), there is a composite signal:

$$F_{0+s}(\theta) = \left[\eta F_0(\theta)e^{j\frac{\pi}{2}} + F_s(\theta)\right] = B(1+\sin\theta)e^{j\frac{\pi}{2}}$$

$$F_{0-s}(\theta) = [\eta F_0(\theta) - F_s(\theta)] = B(1-\sin\theta)$$

$$F_{0+c}(\theta) = \left[\eta F_0(\theta)e^{j\frac{\pi}{2}} + F_c(\theta)\right] = B(1+\cos\theta)e^{j\frac{\pi}{2}}$$

$$F_{0-c}(\theta) = [\eta F_0(\theta) - F_c(\theta)] = B(1-\cos\theta)$$

The signals represented by the above four formulas are subtracted by square law detection:

$$\begin{aligned} S_a(\theta) &= |F_{0+s}(\theta)|^2 - |F_{0-s}(\theta)|^2 \\ &= 4^2(1+\sqrt{2})^2(kR)^2\sin\theta \end{aligned} \tag{8.30}$$

$$\begin{aligned} C_a(\theta) &= |F_{0+c}(\theta)|^2 - |F_{0-c}(\theta)|^2 \\ &= 4^2(1+\sqrt{2})^2(kR)^2\cos\theta \end{aligned} \tag{8.31}$$

From basic principles, the expressions of each voltage pattern are obtained, and the sine and cosine functions used for quantization coding are also given. The following discussion is based on these expressions.

From Eqs. (8.21), (8.22), (8.25) and (8.31), the shape of the sine and cosine signal has nothing to do with kR. That is, it has nothing to do with the operating frequency and array radius. In principle, a circular array with a very small array radius can be designed without affecting the direction-finding performance. However, as the array radius becomes smaller and the array element spacing becomes smaller, the mutual coupling between the array elements is bound to become stronger, thereby affecting the array's performance. Therefore, the minimum array radius mainly depends on the mutual coupling situation between the array elements.

Suppose the gain of the array element is 0 dBi (omnidirectional element). Since the amplitude diagram of the 0th order mode is non-directional, the gain of the 0th order mode can be determined as 0 dBi by avoiding the method of obtaining the gain by spatial integration. Therefore, the gain corresponding to the sine and cosine signal of the 8-element DBD antenna array is:

$$D_d = \frac{4^3(kR)^2}{\left[8 - 2(kR)^2\right]^2} \approx (kR)^2 \tag{8.32}$$

The gain corresponding to the sine and cosine signal of the 8-element Adcock antenna array is:

$$D_a = \frac{4^2(1+\sqrt{2})^2(kR)^2}{\left[8 - 2(kR)^2\right]^2} \approx (1+\sqrt{2})^2(kR/2)^2 \tag{8.33}$$

The gain of these two antenna arrays is very low. In order to ensure the sensitivity of the system, the receiver is required to have the highest possible sensitivity. For the number of elements in the circular array, it is obtained from Eqs. (8.17), (8.21), (8.22), (8.30) and (8.31). That is, the expressions of $F_d(\theta, 0) = F_{0(\theta)}$ and $F_d(\theta, 1)$ used in the solution are made processing of approximate. According to the cylinder function theory, the greater the number of array elements are, the better the approximation of (8.17) is, and vice versa. However, the larger N is, the more complicated the system is. In fact, it is better to choose an 8-element array after compromise. If the electrical index of the antenna array can be reduced, a smaller number of elements can also be used.

The feeding network of the DBD antenna array is more complicated than the feeding network of the Adcock antenna array. Besides, it is also known from (8.21), (8.22), (8.30) and (8.31) that the DBD antenna array based on the 0th-order and 1st-order mode has the same direction-finding performance as the Adcock antenna array, so it is more appropriate to adopt the Adcock antenna array. But for the Adcock antenna array, when the electrical performance of the feed network is given, its ultimate direction-finding accuracy is also determined.

To further improve the direction-finding accuracy, the above two systems have certain limitations. A large-distance interferometer or a large-radius circular array Doppler system is required. These two arrays exemplify its advantages in terms of direction-finding accuracy when it is electrically small because, in principle, these two systems are still based on the comparison. The relationship between phase and azimuth is kd $\sin\theta$ (d is the array element spacing) for interferometers. When d is less than half a wavelength, the azimuth changes by 360°, and the phase change are less than 360°. For the Doppler system, if the antenna locates at the center of the circle and is used as a reference, the relationship between phase and azimuth is:

$$kR\cos\left(\frac{2\pi i}{N} - \theta\right) \quad \text{((is the array element index, N is the array element number))} \tag{8.34}$$

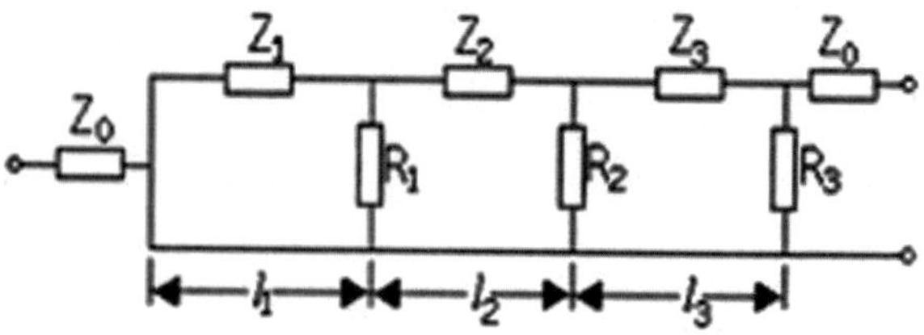


**Fig. 8.6** Power divider circuit structure diagram

When R is less than half a wavelength, and the fixed azimuth angle is θ, i is sequentially connected from the switch from 0 to N − 1, and its phase change is also less than 360°. For Adcock antennas, a 360° change in azimuth is equivalent to a 360° change in phase. Regarding the attenuation of the 0th mode, in the formation of sine and cosine signals by the phase detector at f and d, due to the phase detector's inherent phase error, reflection, and imperfect isolation of the phase detector quadrature coupler. Only when the amplitudes of the two signals that are to be phase-detected are equal, the highest phase-detection accuracy will be achieved. Therefore, the antenna arrays introduced here all require the receiver to have the function of automatically changing the 0th mode attenuation with frequency (amplitude tracking), so that the double signal input to the phase detector or quadrature coupler has equal amplitude. Of course, the better the phase detector and quadrature coupler's performance, the lower the accuracy of receiver amplitude tracking is. In fact, for the ideal phase detector and ideal quadrature coupler, the amplitude tracking of the receiver is not required because instead of attenuating the 0th mode, the formulas (8.18), (8.19), (8.20), (8.27), (8.28), (8.29) can also be used to get sine and cosine signals. For the convenience of implementation in the project, four reverse phase power splitters and an array element antenna are arranged at the center of the array to replace the sum difference and a four-channel isolated power synthesizer. Hence the components in the network are reduced to three types which are two-way isolated power splitter, 6 dB attenuator and reflected isolated power splitter and antenna array.

The entire antenna array uses a broadband monopole antenna as the array element. Considering the mutual coupling between the antennas, the smaller the radius of the circle, the stronger the mutual coupling, and it is irrelevant to the position of each unit on the circle. After theoretical analysis and corresponding experimental research, it was determined that eight array elements were evenly distributed on a circle with a radius of 20 cm. An antenna was located in the center of the array, and the monopole height was 220 mm. The antenna and microwave components were connected by phase consistency. The inverting isolated power divider and the isolated power divider are the same, but there is an inverting part inside. To meet the needs of the broadband operation, the microwave devices used must have broadband characteristics. For the design of the power splitter, a three-stage two-channel power divider is used. The geometric length in Fig. 8.6 is:

$$l_1 = l_2 = l_3 = \lambda_0/\left(4\sqrt{\varepsilon_e}\right) \tag{8.35}$$

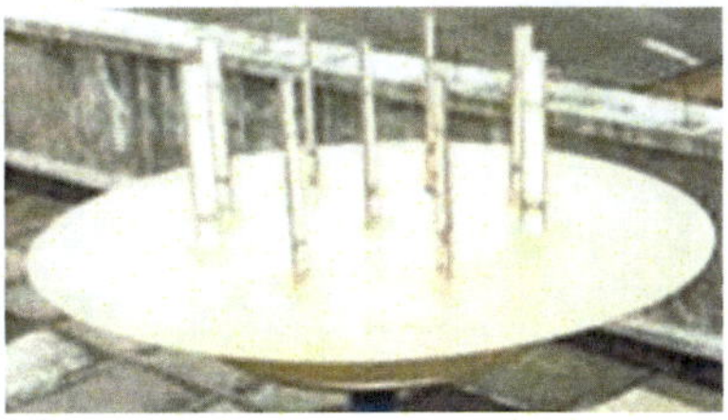

**Fig. 8.7** Adcock antenna array

where $\varepsilon_e$ is the effective dielectric constant of the microstrip line, which is made of double-sided copper-clad polytetrafluoroethylene, and its power divider part is the transitional form of the microstrip and the flat plate. If an equal-amplitude inverted output is required, the flat line can be ingeniously reversed and connected. The 3 dB directional coupler adopts a symmetrical three-stage quarter-wavelength coupler, and its even-mode impedance value $z_{e1} = 1.21$, $z_{e2} = 3.41$ is to reduce the mutual influence between the volume and adjacent lines. It is used in the power splitter and the coupler Slotted broken line method. The attenuator uses a π-type network of concentrated components. In order to make the phase of each channel in phase, a compensation line is added to maintain the good tracking characteristics of the channel phase. The amplitude is relatively easy, but debugging work is also necessary for addition to the entire design. The whole antenna array works in the 3:1 frequency range to get better direction-finding accuracy (Fig. 8.7).

Note: adcock is the inventor of the small caliber direction-finding antenna array. DBD (Digital Bearing Discriminator) digital phase detector.

### 8.2.2 Four-Arm Spiral Antenna

The multi-arm helical antenna adopts different excitation methods to produce different working modes and forms various radiation patterns suitable for direction finding, tracking, etc. For the N-arm helical antenna's excitation, due to the fact that it is an n-arm equiangular spiral antenna, the antenna arm is n rotationally symmetrical elements, as shown in Fig. 8.8. When the antenna is rotated by 2π/n around the axis, the structure remains unchanged. There is a circle with a diameter smaller than the wavelength as the feeding area in the antenna's center. In the feeding area, the arm's edge can be regarded as the n-terminal network terminal. Let $i_1, i_2 \ldots i_n$ be the current fed into the antenna arm with a given feeding method, the current flowing out of the feeding area is positive, and a current vector can represent the excitation:

$$I = (i_n,\ i_1,\ i_2 \ldots i_{n-1})$$

The current fed into the antenna arm should satisfy the conservation law of current:

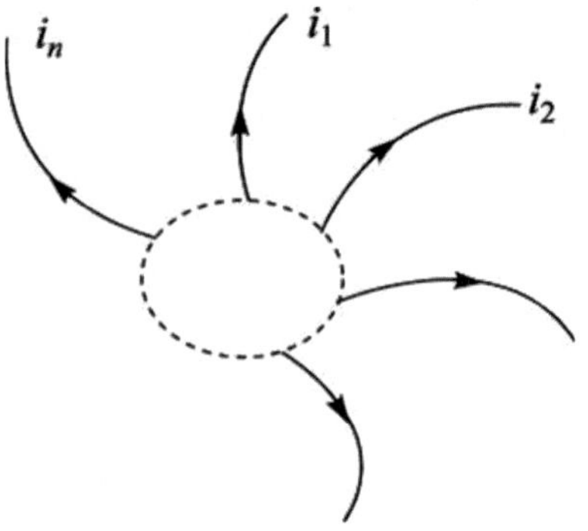


**Fig. 8.8** Schematic diagram of current feeding

$$\sum_{m=1}^{n} i_m = 0 \tag{8.36}$$

According to the antenna structure's rotational symmetry, when the excitation is rotated by 2π/n angle, it is equivalent to the antenna rotated by 2π/n angle. That is, the radiation field is rotated by 2π/n angle in space.

The current vector after rotation is:

$$I' = (i_1, i_2, \ldots i_n) \tag{8.37}$$

The rotation can be expressed linearly in the vector space:

$$I' = PI \tag{8.38}$$

The matrix P corresponding to this transformation should be a square matrix of order n × n. Find the eigenvalues of the matrix P, and then you can get the eigenvectors corresponding to each eigenvalue. Find the eigenvalue of P, that is, find the root of the Eigen equation:

$$\det(\lambda E - P) = 0 \tag{8.39}$$

$$\lambda_K = e^{j2\pi k/n} \quad (k = 1, 2 \ldots n) \tag{8.40}$$

The eigenvector can be obtained from the eigenvalues:

Let $i_1 = 1$, the eigenvector is:

$$^{n}I_K = \left(1, e^{j2\pi k/n}, e^{j2\pi k2/n}, \ldots, e^{j2\pi k(n-1)/n}\right) \tag{8.42}$$

Because when k = n, the formula (8.42) is not satisfied, so k can only take 1, 2, ..., (n − 1). Considering the situation of characteristic feeding and its mode, the infinite multi-arm planar logarithmic spiral antenna is excited by the Δ-phase excitation

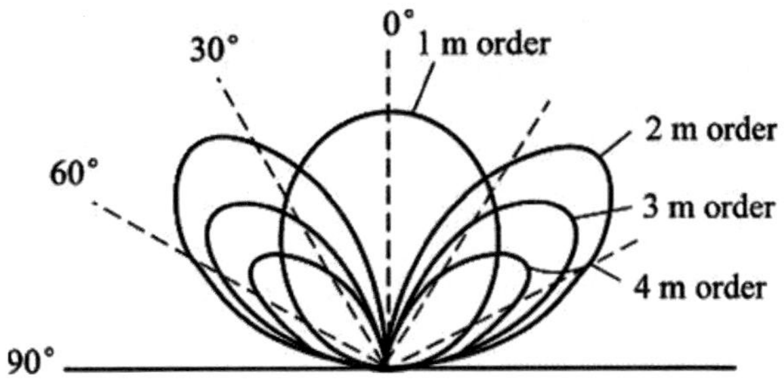


**Fig. 8.9** Radiation pattern of the second order

source with infinitely different phases of each arm, and the solution is:

$$A(\theta)e^{-j4(\theta)}\frac{1}{r}e^{j\left[m\left(\phi+\frac{2}{\pi}\right)-\beta\cdot r\right]} \tag{8.43}$$

It can be simply written as $F(\theta,\phi)=C(r,\theta)e^{jm\phi}$. m is the modulus. For different modulus m, the amplitude pattern is shown in Fig. 8.9. That is, the first-order mode is the largest radiation in the axial direction. The regional mode is a cone-beam with zero axial radiation. The relationship between the total field phase and the azimuth angle $\varphi$ is related to the modulus. For the first mode, the phase change in the azimuth plane has a 1:1 relationship with the angle $\varphi$, and the second mode is a 2:1 relationship, etc.

Examining the characteristic feed at n = 4, corresponding to (8.42), there are:

$$^{4}I_1=(1,j,-1,-j) \tag{8.44}$$

$$^{4}I_2=(1,-1,1,-1) \tag{8.45}$$

$$^{4}I_3=(1,-j,-1,j) \tag{8.46}$$

$^{4}I_1$ results from a field rotated by a π/2 angle equal to the phase change π/2, in the solution of $e^{jm\phi}$.Only the solution of m = 4 k + 1 (k = 0,1, 2… can satisfy the above relationship between rotation and phase change, so the field generated by the characteristic feed should be:

$$^{4}F_1=\sum_{K}{}^{4}C_1(r,\theta)e^{j(4k+1)\phi} \tag{8.47}$$

In other words, this feeding may stimulate the first-order mode, the fifth-order mode shown in Fig. 8.9. The field (m = 4 k + 2) generated by the characteristic feed can also have 2nd order mode, 6th order mode, etc. The field generated by characteristic feeding (m = 4 k + 3) can have 3rd order mode, 7th order mode, etc.

The mode with m equal to a negative value is equivalent to a circularly polarized field in the opposite direction. The spiral winding direction only has a circularly polarized field conducive to the same direction of rotation. Therefore, the mode field where m is equal to a negative value is very small. For m = 2, the characteristic feed is:

$$^{2}I_1 = (1, -1) \tag{8.48}$$

In other words, it is a general reverse-phase feed, and the exciting field should be:

$$^{2}F_1 = \sum_K {}^{2}C_1(r, \theta)e^{j(2k+1)\phi} \quad (k = 0, \pm 1, \pm 2 \ldots) \tag{8.49}$$

That is, it is possible to excite all odd-numbered modes, with m = 1, 3 as the advantage, because these two fields are added in phase or reverse phase alternately in the vertical mode plane. This causes a symmetrical change in beam rotation. Therefore, the relative size of these two fields can be estimated from the case of rotational symmetry. The densely wound spiral has better rotational symmetry than the weakly wound spiral, indicating that its m = 1 mode is dominant.

Since the eigenvectors of unequal feature roots are linearly independent, and

$$\sum_{l=1}^{n-1} e^{j2\pi(p-q)l/n} = \frac{1 - e^{j(p-q)2\pi n/n}}{1 - e^{j(p-q)2\pi/n}} = \begin{cases} 0(p \neq q) \\ n(p = q) \end{cases} \tag{8.50}$$

The feature vector has orthogonality:

$$^{n}I_P \cdot {}^{n}I_Q = \begin{cases} 0\ (p \neq q) \\ n\ (p = q) \end{cases} \tag{8.51}$$

Therefore, these feature vectors can form the basis of a vector space. Any excitation current vector can be decomposed into a linear combination of n − 1 basis vectors. While each characteristic feed corresponds to a mode field of $^{n}F_k$, the field generated by any current vector I can be described by a linear combination of mode fields of $^{n}F_k$. Any current vector I conforming to Eq. (8.36) can be decomposed into n-1 linear combinations of characteristic feed Eq. (8.42):

$$I = \sum_{p=1}^{n-1} a_p^n I_p \tag{8.52}$$

According to Eq. (8.51),

$$a_p = \frac{1}{n} I \cdot {}^{n}I_p^* \tag{8.53}$$

For a two-arm spiral fed by a general reverse phase, due to the excitation of a higher-order mode (2 k + 1), the pattern lobe's bandwidth's width in the antenna direction to maintain frequency-independent characteristics becomes significantly smaller. In order to obtain pure primary mode radiation, a multi-arm spiral fed as follows can be used. Let the number N of spiral arms be an even number (N = 4, 6, 8 …), let the excitation current $I_N$ of the spiral arm equal the sine of the terminal position $\phi_N$ of the spiral, and the terminal position angle is defined as the angle between the internal end and the inner end of the antenna arm, then

$$I_N = i_0\left(Sin0^\circ, Sin\frac{2\pi}{N}, Sin2\frac{2\pi}{N}, \ldots, Sin(N-1)\frac{2\pi}{N}\right) \tag{8.54}$$

The characteristic feed corresponding to formula (8.42) is

$$^{N}I_P = (1, e^{j\frac{2\pi}{N}P}, e^{j\frac{2\pi}{N}2P}, \ldots, e^{j\frac{2\pi}{N}(N-1)P})$$

Then $I_N$ can be decomposed into a linear combination of $^{N}I_p$

$$I_N = \sum_{P=1}^{N-1} a_p \cdot {}^{N}I_P$$

$$a_p = \frac{1}{N}I_N \cdot {}^{N}I_P^* = \frac{i_0}{N}\left(\text{Sin}0^\circ, \text{Sin}\frac{2\pi}{N}, \ldots, \text{Sin}(N-1)\frac{2\pi}{N}\right)\begin{pmatrix} e^{-j\frac{2\pi}{N}P} \\ e^{-j\frac{2\pi}{N}2P} \\ \vdots \\ e^{-j\frac{2\pi}{N}(N-1)P} \end{pmatrix}$$

$$= \begin{cases} \frac{i}{j} & (p = 1) \\ \frac{i}{2j} & (p = N-1) \\ 0 & (P \neq 1, N-1) \end{cases} \tag{8.55}$$

So,

$$I_N = \frac{i_0}{2j}({}^{N}I_1 - {}^{N}I_{N-1}) \tag{8.56}$$

It can be seen that with this kind of excitation, only the first-order mode and N − 1 order mode can be excited in the spiral region. Therefore, pure primary mode radiation can be achieved within a bandwidth of approximately (N − 2): 1 or greater, resulting in a stable pattern. Suppose the N − 1 order mode is suppressed by absorbing the load or terminating the resistive load at a specific antenna arm radius. In that case, pure first-order mode radiation can be obtained in a fairly wide bandwidth, greatly improving the helical antenna's frequency-independent characteristics. For example, in a six-arm helical antenna, the excitation current vector is:

$$I_N = \frac{\sqrt{3}}{2}i_0(0, 1, 1, 0, -1, -1) \tag{8.57}$$

That is, connect the 2nd, 3rd, and 5th, 6th arms of the spiral together, and then feed in reverse. The first and fourth arms are parasitic excitations. Terminating the resistive load outside the antenna arm to suppress the fifth-order mode, a stable first-order mode pattern can be obtained within a 4:1 or greater bandwidth. For an eight-arm helical antenna, the excitation current vector can be taken as:

$$I_N = (0, 0.707, 1, 0.707, 0, -0.707, -1, -0.707) \cdot i_0 \tag{8.58}$$

A 180° hybrid connector and two three-way power splitters are used to feed the excitation network so that a stable primary mode pattern can be obtained within a bandwidth of 7:1 or greater. The logarithmic spiral is a frequency-independent radiator, and the mathematical expression of the points on the line expressed in polar coordinates is:

$$r = r_0 \exp(a\phi)$$

Expressed in units of wavelength are:

$$r' = r_0 \exp[a(\phi - \phi_0)] \quad \phi_0 = (1/a)\ln(f_2/f_1) \tag{8.59}$$

The change in frequency corresponds to the change in $\phi_0$, which is equivalent to turning the original antenna by an angle $\Delta\phi$, that is, the same angle turns the radiation pattern $\Delta\phi$. The pattern is the distribution of the field on a large spherical surface with radius R, that is

$$E(\theta, \varphi, R) = \frac{e^{-jkR}}{R}\left[\hat{\theta} P_\theta(\theta, \varphi) + \hat{\varphi} P_\varphi(\theta, \varphi)\right] \tag{8.60}$$

$\hat{\theta}$ and $\hat{\varphi}$ are the unit vector of the coordinate direction. In other words, when the frequency changes, $P_\theta$, $P_\varphi$ both satisfy the following relationship:

$$P(\theta, \varphi, f/\tau) = P(\theta, \phi - \Delta\phi, f) \tag{8.61}$$

In theory, frequency-independent antennas of infinite size work in infinitely wide frequency bands. In order for the frequency-independent antenna to be practically applied, the antenna must meet the current cut-off principle, so that the limited-size antenna has the same electrical characteristics as the infinite structure in the required frequency band. The high end of the frequency is determined by the size of the central area of the structure, and the outer diameter of the structure determines its low end. Equiangular spiral antenna belongs to this type of antenna. In order to achieve the required low frequency coverage, the outer diameter $r_2$ of the spiral is required to be large enough, otherwise, the data truncation will produce unwanted reflections, which will degrade the performance of the antenna such as circular polarization and beam width. In order to appropriately reduce the antenna aperture $r_2$ without causing the censoring effect, the end of the logarithmic spiral can be used to connect

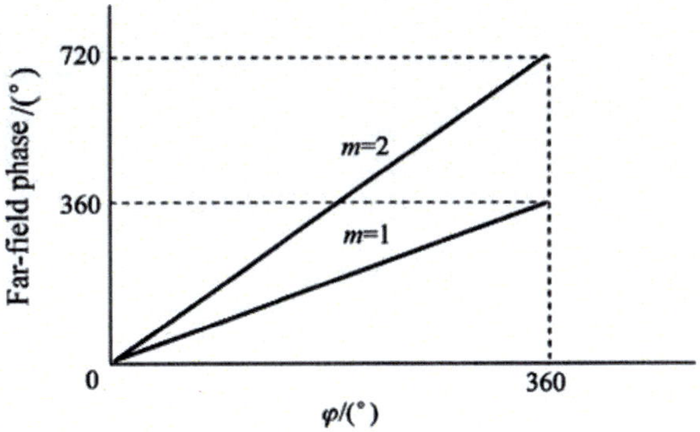


**Fig. 8.10** Phase relationship diagram

a logarithmic spiral that grows in inverse proportion, a sinusoidal modulation spiral or a resistive load to terminate the spiral. According to the radiation belt theory, if the helix is a 3-wire or multi-wire, $M_1$-mode and $M_2$-mode information can be obtained by proper feeding. For the six-arm and eight-arm spirals, their main advantage over the four-arm is that $M_1$ and $M_2$ are resistant to high-order distortion, which can improve the potential of bandwidth. Both flat and conical spiral radiation are circular polarized broadband radiators. Due to the relatively large volume of the cone type, the processing is complicated and difficult, especially its $M_1$-mode and $M_2$-mode phase centers are separated, and should be paid enough attention in practical applications (Fig. 8.10).

Helical antennas use the center feed in broadband operation because the outer feed will easily excite higher-order radiation modes. The center-fed planar four-arm spiral has four feeding points in the center. It can be known from the analysis that the phase centers of the M1 mode and M2 mode of the four-arm helical antenna almost coincide, and the "reference plane", that is, the plane of M1 and M2 in phase, rotates around the antenna axis as the frequency changes. In order to eliminate the rotation of the "reference plane", the corresponding compensation technology can be adopted at the rear end of the radiator, or a hyperbolic spiral plus an appropriate length of transmission line inserted at the M1 port. The mode (M1 mode) and differential mode (M2 mode) signals are the three characteristic feeds of the four-arm helical antenna. That is, ${}^4I_1$ is the adjacent arms differed by 90° (relative to the direction of rotation of the spiral), ${}^4I_2$ is the adjacent arms differed by 180°, and ${}^4I_3$ is the adjacent arms differed by 270°. It can be realized by the feed network in Fig. 8.11, which consists of a 90° phase shifter and a 3 dB directional coupler. The mode forming network is divided into two types: asymmetric mode and symmetric mode. The dual-mode four-arm helical antenna includes a four-arm helical radiator and a mode-forming network part, providing two types of circularly polarized wave sum-forming network and symmetric mode-forming network, as shown in Fig. 8.11. It is relatively simple to form networks with asymmetric modes. In an ideal mode forming network, the M port's excitation voltage provides the same feeding voltage amplitude to each antenna's arm through the mode forming network, and the phase is determined by m. For the m-th mode of the n-arm spiral, the feeding voltage of each arm of the

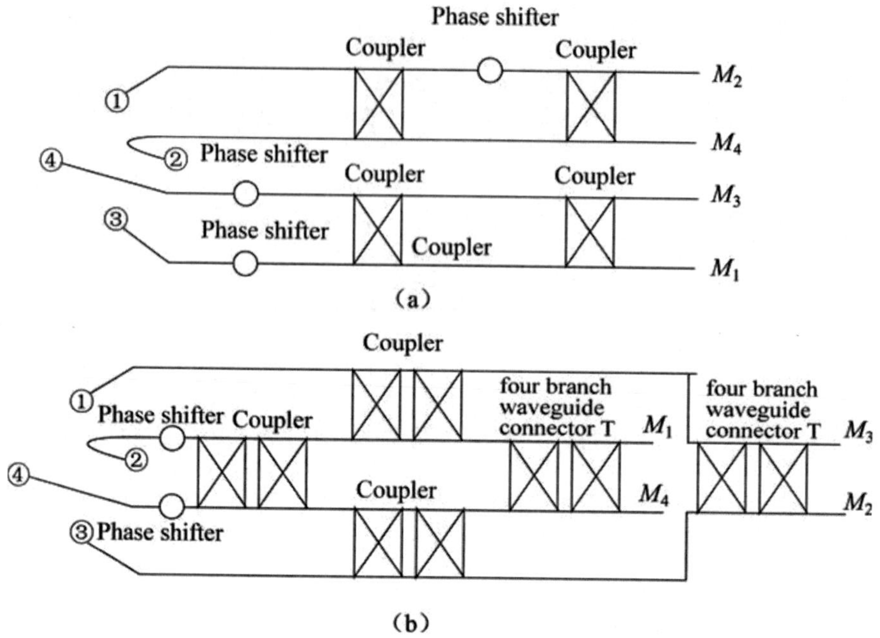


**Fig. 8.11** **a** Asymmetric mode forming network **b** Symmetric mode forming network

antenna is:

$$V_i = \exp(2\pi mi/n) \quad i = 1, 2 \ldots, n \tag{8.62}$$

Since each element in the mode-forming network is not ideal, an error occurs in the amplitude and phase of the voltage of each antenna's arm. It is expressed by a set of voltages Vi (i = 1,2,…n), which can be expressed as a linear combination of mode $M^m$ (m = 1,2, …n):

$$\begin{pmatrix} M_1 \\ M_2 \\ \vdots \\ M_n \end{pmatrix} = \frac{1}{n} \begin{bmatrix} e^0 \; e^{-1} \; \ldots \; e^{-(n-1)} \\ e^0 \; e^{-2} \; \ldots \; e^{-2(n-1)} \\ \vdots \\ e^0 \; e^{-n} \; \ldots \; e^{-n(n-1)} \end{bmatrix} \begin{pmatrix} V_1 \\ V_2 \\ \vdots \\ V_n \end{pmatrix} \tag{8.63}$$

The above formula gives the values of the required modes and their respective jamming modes, and the radiation pattern can be calculated with an appropriate combination of basic radiation. Due to the jamming mode's existence, the required mode pattern and circular polarization performance become worse.

As can be seen from the above formula, in the asymmetric mode forming network, the lowest mode in the jamming mode for M1 mode is M2, and the lowest mode in the jamming mode for M2 is M1. In an asymmetric mode forming network, the M2

**Fig. 8.12** Typical amplitude pattern

mode gives an even symmetrical excitation voltage. At this time, only even-order modes exist, and the amplitude of all odd-order modes is zero. So, the lowest order of the M2 mode is M4. The M1 mode gives an odd symmetrical excitation voltage. At this time, only odd-order modes exist, and all even-order amplitudes are zero. So, the lowest jamming mode for the M1 mode is M3. Coupled with the antenna aperture limitation and the attenuation of the high-order mode current on the spiral, the anti-jamming of the symmetric mode forming network is better than that of the asymmetric mode forming a network, which is clearly shown when the antenna is in broadband operation. The dual-mode four-arm helical antenna is designed according to the frequency range of 2–4 and 2–8 GHz. The four-arm logarithmic helical and symmetrical modes are used to form a network. The typical amplitude patterns of the M1 and M2 modes of the antenna are shown in Fig. 8.12. The root means square value of the zero deviation of the M2 mode within 4:1 bandwidth is not greater than 3°. Infinitely balanced feeding can be used to obtain the second mode of excitation easily. It only needs to connect the sheath of the coaxial cable that is wound around one of the antenna arms at the vertex of the antenna feed to the opposite arm's simulation cable. The core wire need to be connected to the simulation cable of the adjacent arm. For the tapered four-arm helical antenna, the spiral elevation angle is selected to be $\alpha = 45°$. By feeding in the manner as mentioned earlier, a circularly polarized omnidirectional coverage pattern can be obtained in a wide frequency band. The sum-mode and differential-mode signals can be compared to obtain the tracked target's azimuth, and the appropriate combined network can be used to generate the far-field amplitude and phase characteristics of different modes to form a variety of directional patterns suitable for direction-finding tracking.

### *8.2.3 Multi-mode Log-Periodic Antenna Array*

For the multi-arm helical antenna, the circular polarization information required for a two-dimensional direction finding can be generated in a relatively wide range. A radiation aperture is a plane. Still, it is a blind zone for another reverse-rotation circular polarization. At the same time, its phase reference plane rotates with frequency. Different frequencies must be tracked on the phase plane. Besides, the gain of the entire antenna array is relatively low than around zero decibels. The multi-mode log-periodic antenna array adopts the azimuth phase mode direction finding, pitch plane, and difference amplitude comparison methods have the multi-polarized two-dimensional direction-finding tracking function, and the gain is improved.

The array form of the antenna adopts a monopole log-periodic antenna as the array element. Multiple array elements are distributed on a circle. The maximum direction formed by the array element antennas all points to the circle's same central axis direction to form a tapered antenna array. Known from antenna theory, a linearly polarized antenna can receive any form of polarization other than the one perpendicular to itself. Suppose two or more linearly polarized antennas are cross-combined in one plane. In that case, the antenna's polarization characteristics can be greatly improved, and any form of polarized waves can be received at the same time. On the other hand, if a two-element antenna array is used for simultaneous sum and difference arrays, a sum and difference output can be obtained. Multiple sums and difference outputs with different spatial angles can be obtained using multiple pairs of two-element array linearly polarized antennas and their sum and difference devices. In this way, considering the antenna's sum, difference output, polarization characteristics, and broadband working requirements, the formed antenna array must be a tapered array with a dedicated mode feed network, as shown in Fig. 8.13. The figure uses six monopole log-periodic antennas, distributed on the cone's surface, and the circumference is 60°. They are E-plane arrays of log-periodic antennas. Each element is installed perpendicular to the cone surface, and the new angle of the cone can be obtained by a large number of theoretical analysis and experiments. With the different angles of the cones, the level of the cross point of the E-plane pattern of the same element antenna will change. In this way, the cone angle can be calculated according to different needs. Simultaneously, considering that the pattern of the elementary antenna is upturned, the antenna array is used for tracking and monitoring, and both the sum and difference beam levels are considered, the relative beam crossing level is required to be around 2 dB. On the other hand, it is also needed to determine the specific position of the array element antenna on the surface of the cone. When the beam is formed between the opposing elements, the interval between the opposing

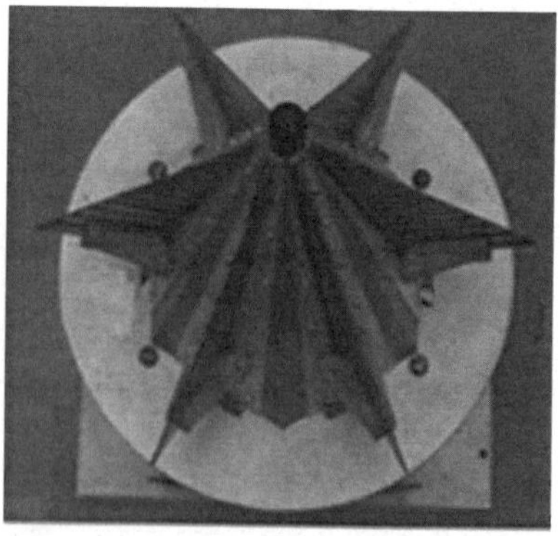

**Fig. 8.13** Multi-mode log-periodic antenna array

elements on the diameter is controlled at a maximum of 0.7λ (λ is the operating wavelength of the corresponding vibrator). Then the cone angle is larger, the array element antenna must be close to the cone vertex, and the cross-level of the E-plane pattern between the element antennas is higher. On the contrary, the unary antenna is farther away from the cone's apex, and the crossover level is lower. The relative pair of the array element antenna must be on the same plane. Also, the feeding problem of the unary antenna must also be considered. The domestic semi-rigid coaxial cable F-50-Φ3.6 model's diameter is relatively uniform, and the phase consistency is high. Considering that the six-monopole log-periodic antennas are all fed near the top of the cone, if a semi-rigid coaxial cable of φ3.6 is used, it cannot be arranged the top of the cone at all. For this purpose, a thin stripline is used, which passes under the groove on the cone's surface and is subtly turned into a microstrip line at the top of the antenna to realize feeding to each array element antenna. The network used in this antenna array is an asymmetric network, which includes seven 3 dB spatial couplers, two 4.77 dB spatial couplers, three 900 phase shifters, and an 1800 phase shifter. The block diagram of its feed network is shown in Fig. 8.14. As can be seen from the figure, this network is relatively complex. The cone-shaped hexagonal pyramid

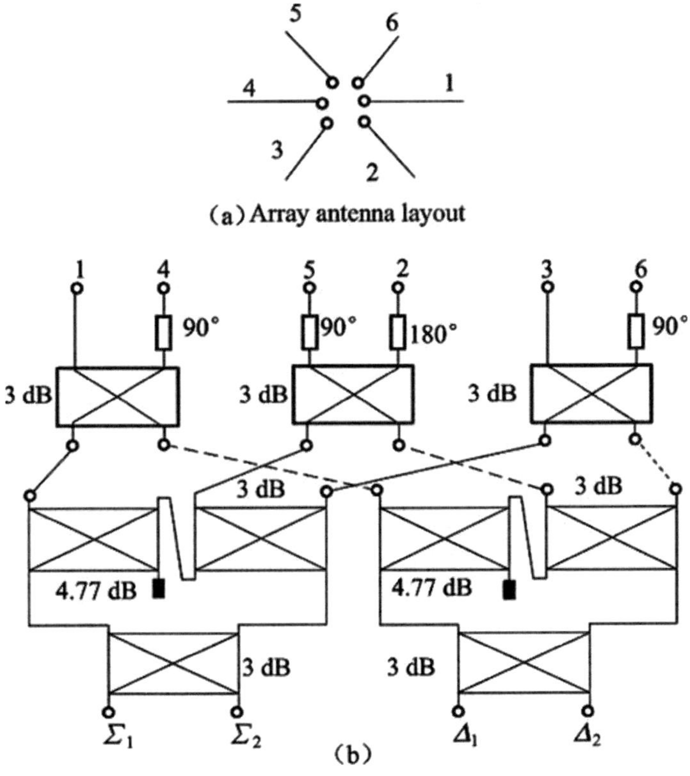


**Fig. 8.14** Block diagram of the feed network

**Table 8.3** Excitation phase relationship of the array elements

| Input phase (°) | Output port | | | | | |
|---|---|---|---|---|---|---|
| | 1 | 2 | 3 | 4 | 5 | 6 |
| $\sum_1$ | 0 | 60 | 120 | 180 | 240 | 300 |
| $\sum_2$ | 90 | 30 | 330 | 270 | 210 | 150 |
| Δ1 | 0 | 240 | 120 | 0 | 240 | 120 |
| Δ2 | 90 | 210 | 330 | 90 | 210 | 330 |

monopole log-periodic antenna array is a circular spatial array. See the circumferential mode analysis in the previous section. As the number of linearly polarized element antennas in a circular array increases, the theoretical analysis results can be approximated. But with the increase in the number of unary antennas, the network becomes very complicated. Still, the array itself is difficult to process, and feeding the array element antenna becomes difficult. It can be seen from the figure that the phase relationship between the feed port and the array element antenna, and the relative phase relationship between the antenna array element and the arrangement and distribution of the array element antenna in space. Antenna array elements No. 1–No. 4 differ by 1800, spaces No. 2–No. 5 differ by 1800, and elements No. 3–No. 6 differ by 1800. Table 8.3 corresponds to the excitation phase relationship of the array elements.

Let the electric fields generated by each element antenna on the cone be $E_1(\varphi,\theta)$, $E_2(\varphi,\theta)$, $E_3(\varphi,\theta)$, $E_4(\varphi,\theta)$, $E_5(\varphi,\theta)$, $E_6(\varphi,\theta)$. In an ideal situation, the field's amplitude generated by the six array antennas in space is the same. When feeding at the $\sum_1$ and $\sum_2$ ports, the current phase on the six array antennas is 1–4 array antennas, 2–5 array antennas, 3–6 array antennas differ by 180°, and the space of the array antenna difference. This makes the total phase difference of antenna elements 1–4 be 0° or 360°. That is, the antenna elements of array elements 1–4 are in phase. This is true for array antennas 2–5 and unary antennas 3–6, so that the total field at and near the antenna axis is:

$$\begin{aligned}
\vec{E}(\phi,\theta) &= \vec{E}_1(\phi,\theta) + \vec{E}_2(\phi,\theta) + \cdots + \vec{E}_6(\phi,\theta) \\
&= \left[\vec{E}_1(\phi,\theta) + \vec{E}_4(\phi,\theta)\right] + \left[\vec{E}_2(\phi,\theta) + \vec{E}_5(\phi,\theta)\right] \\
&\quad + \left[\vec{E}_3(\phi,\theta) + \vec{E}_6(\phi,\theta)\right] \\
&= \left[\vec{E}_1(\phi,\theta) + \vec{E}_4(\phi,\theta)\right] + \left[\vec{E}_1(\phi,\theta) + \vec{E}_4(\phi,\theta)\right]e^{j\alpha 1} \\
&\quad + \left[\vec{E}_1(\phi,\theta) + \vec{E}_4(\phi,\theta)\right]e^{j\alpha 2} \\
&= \vec{E}_1(\phi,\theta)\left(1 + e^{j\varphi 14}\right)\left[1 + e^{j\alpha 1} + e^{j\alpha 2}\right]
\end{aligned} \tag{8.64}$$

In the formula, α1 and α2 are the circumferential phase difference between the antenna element pairs No. 1–4, No. 2–5 and No. 3–6, and $\varphi 14$ is the phase difference introduced by the space path of the 1–4 antenna pair. It can be seen from Eq. (8.64) that this is an expression of the sum pattern, which shows that the sum pattern is obtained by feeding at $\sum_1$ and $\sum_2$. When feeding at the $\Delta_1\Delta_2$ port, as can be seen from Table 8.3 and Fig. 8.14, the current difference between the unit antenna pair No. 1–4; No. 2–5; No. 3–6 is 0°, that is, the relative unary antennas are in phase. Considering the spatial arrangement of the six element antennas, the total phase difference of the relative element antennas in space is 180°, so the total field at and near the antenna axis is:

$$\vec{E}(\phi,\theta) = \vec{E}_1(\phi,\ \theta)\left(1 - e^{j\varphi 14}\right)\left[1 + e^{j\alpha 1} + e^{j\alpha 2}\right] \tag{8.65}$$

$\vec{E}(\phi,\theta) = 0$ in the axial direction of the cone. Due to the amplitude of the actual network and phase errors, and the wave of the array antenna and various connectors, this causes the existence of zero depth of the differential beam and zero value drift. A general helical antenna can only receive a single-rotation circularly polarized wave. A linearly polarized antenna can receive any other form of polarized wave except perpendicular to itself. Thus synthesizing the linearly polarized antenna can realize the multi-polarization function. It can be seen from the phase relationship in Table 8.3 that when the port is fed, the current phases obtained on the six array antennas are: 0°, 60°, 120°, 180°, 240°, and 300°. Obviously, this is consistent with the spatial distribution of the six array antennas. It is a left-handed circular polarization. Similarly, when feeding at q, the current phases obtained on the six array antennas are 90°, 30°, 330°, 270°, 210°, and 150°. This can also be written as 0°, 300°, 240°, 180°, 120°, 60°. Obviously, this is exactly the opposite direction of the six array antennas' spatial distribution, indicating that they form right-handed circular polarization. For the differential beam, when feeding at $\Delta_1$, the phase distribution on the six-element antenna is 0°, 240°, 120°, 0°, 240°, 120°, which can be written as: 720°, 600°, 480°, 360°, 240°, 120°, which is the right-hand circular polarization of the difference beam. At the same time, it can be seen that when feeding at $\Delta_2$, a left-hand circularly polarized wave of a different beam is obtained. In this way, when the system is used for reception, two kinds of circularly polarized information with opposite rotation directions are obtained on the sum and difference ports. Thus the two kinds of circularly polarized waves can synthesize elliptically polarized waves and linearly polarized waves in any direction etc. The detection of the amplitude and phase values of each of the sum and difference ports of the four orthogonal circular polarizations of four ports $\sum_1$, $\Delta_1$ and $\sum_2$, $\Delta_2$, the multi-mode and multi-polarization characteristics of the system are realized.

The feeder of the monopole log-periodic antenna is still made of double-sided copper-clad reinforced polytetrafluoroethylene glass wire with $\delta = 1.5$ mm thickness. In order to avoid the influence of the transmission line on the radiation pattern, a strip line is used. The characteristic impedance is 50 ohms, the height b of the cavity is 0.6 mm, and the top of the antenna is converted into a microstrip. The angle $\psi =$ 370 of the cone surface of the cone relative to the antenna element. The design of the phase shifter is one of the main tasks of network design. al is the length of the uniform transmission line. Section $l$ is the coupling line shorted by the terminal. The input voltage is $V_0$ and the phase shift amount is $\Delta\varphi$. There are:

$$V_1 = V_0 e^{-j\phi_1}$$

$$V_2 = V_0 e^{-j\phi_2} \quad \theta = \frac{2\pi}{\lambda} l \quad \phi_1 = a\theta$$

$$\Delta\phi = \phi_1 - \phi_2 \tag{8.66}$$

$$\phi_2 = \cos^{-1}\left[\frac{(z_{0e}/z_{0o}) - \tan^2\theta}{(z_{0e}/z_{0o}) + \tan^2\theta}\right]$$

$z_{0e}$, $z_{0o}$ are the odd and even mode impedances of the coupled line:

The traditional method is to use no compensation section, taking $l = \lambda_0/4$ ($\lambda_0$ is the center wavelength), a = 2.5, $\rho = z_{0e}/z_{0o} = 1.8$. A C-shaped phase shifter is used to shift the phase by 45°, and the same combination reaches a 90° phase shift. Otherwise, to achieve the same phase shift value, it is very difficult to achieve ρ or the influence of processing errors is large, then:

$$\Delta\phi = \phi_1 - \phi_2 = 90^\circ \pm 4^\circ$$

For a 90° phase shifter, if a zigzag section is used, the gap is too narrow, and the processing error is too large. In practice, two 45° zigzag sections are used for composition. With the existing processing technology, it is best to eliminate or reduce the design's theoretical error. For this reason, we propose an improved C-type phase shifter, as shown in Fig. 8.15b. Take the compensation section ρ = 1.14 and C-type section ρ = 1.78:

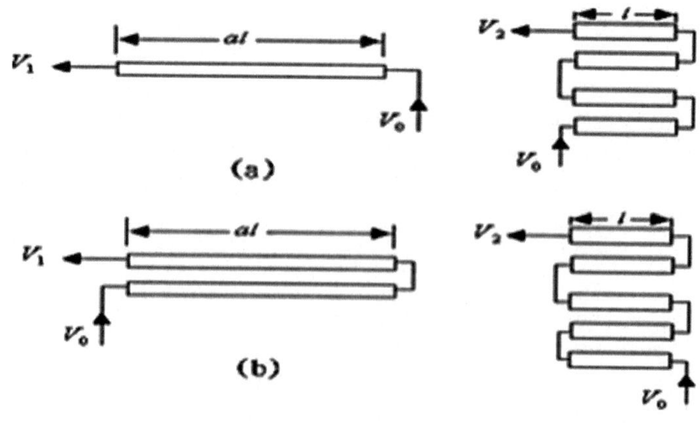


**Fig. 8.15** **a** Traditional phase shifter circuit diagram **b** Improved phase shifter circuit diagram

$$\Delta\phi = \phi_1 - \phi_2 = 90^\circ \pm 0.3^\circ$$

The 180° phase shifter in the network is realized by connecting two 90° phase shifters. In the antenna array's network design, the requirements on the amplitude and phase are relatively high, so the three-section symmetric form of the spatial coupler is used to realize the 3 and 4.77 dB coupler circuits. Since 4.77 dB directional coupler has no ready-made formulas and charts to be checked, multiple Chebyshev approximations are used to derive various formulas. Besides, the allowable coupling errors of the two couplers within the required bandwidth are obtained to find the sexual impedance of its model:

3 dB directional coupler:

Bandwidth ratio is 3, ripple R = 0.03 dB z0 = 50 Ω

$$\bar{z}_2 = 3.18399 \quad \bar{z}_1 = \bar{z}_3 = 1.15384$$

4.77 dB directional coupler:

Bandwidth ratio is 2.6, ripple R = 0.03 dB z0 = 50 Ω

$$\bar{z}_2 = 2.4 \quad \bar{z}_1 = \bar{z}_3 = 1.1191$$

The feed network is a relatively complicated feed system. The amplitude and phase of each channel are strictly controlled. The number of connectors is minimized and implemented in the form of a stripline. With the help of a new type of monopole log-periodic antenna, a tapered array can be skillfully formed, and the problem of the feeding of array elements is solved. For multi-mode multi-polarized array antennas larger than six elements, such as an eight-element antenna array, the arrangement's angular distance is smaller. At this time, it is 45°, and the network becomes an asymmetric network. But the relevant input and output relationships remain corresponding. Also, according to the antenna element in the working frequency band, other antennas can be used.

## 8.3 The Technology of Multi-beam Antenna

There are multiple methods for technically generating multi-beam radiation patterns, such as digital methods and optical frequency technologies, in addition to radiofrequency (RF) and intermediate frequency (IF). Here we focus on RF beamforming. A multi-beam antenna is an antenna that can form many beams from the same aperture surface. The difference between it and other antennas, such as an electronic scanning array, is that each beam has a port, and usually, the ports are isolated from each other. If a separate transmitting or receiving system is connected to each port, independent working conditions can be obtained simultaneously in many directions. On the other hand, a single transmitting or receiving system can also be connected to each beam port through a multi-way switch to obtain a sequential scanning beam.

A multi-beam antenna using an RF beamformer has the advantage that no frequency conversion device is required. Therefore, it is cheaper than IF, digital, and optical frequency methods and has simple implementation methods. In fact, many antennas' structures, such as reflectors and lenses, can form multiple beams. At this time, it only needs to replace an array with a separate feed, and each array unit can generate multiple beams.

The beamforming technology in the RF field includes two main aspects. One is a quasi-optical type, which is composed of a reflector or a lens and a feed, and the other is an orthogonal network comprising microwave components. The antenna's form can be divided into continuous aperture surface and array, although many beam antennas have the mixed characteristics of these two types. A further classification is based on the number of Fourier transforms (FT) in the device. Any beamformer must complete the FT transformation of the energy distribution from a single feed to the required aperture distribution. There are FT transforms of different orders from the feed to the wavefront in the objective path. This order determines whether the feed is a phase array or simply a combination of feed points. Generally, the more the FT order is, the stronger the function is.

A key factor in multi-beam antennas is wide-angle scanning, which is usually limited by one or more factors. The antenna's geometry determines the performance within this scanning range, and the beam width, beam spacing, and beam intersection level are all important parameters. The system loss components roughly are taper loss, overflow loss, resistance loss (including active and passive parts) of the aperture surface, in addition to processing errors and some additional losses related to beam spacing.

### 8.3.1 Bootlace Lens Multi-beamformer Bootlace

The weight factor of the solid dielectric lens is realized through the lens partition processing. The early working modes dealt with dielectric lenses in an analog manner. There are three basic lens formers, namely the Ruze lens, which contains the propagation of rays that parallel to the axis in the lens. The P–Q path in Fig. 8.16 can be formed in a variety of ways, such as a waveguide, coaxial line, three-plate ribbon, or microstrip line. The electrical length P–Q occurs as the intersection with the longitudinal axis Y of the lens changes. Utilizing the trajectories of ray that passing through P, Q, and the origin of the lens, the two simultaneous equations solve the lens thickness and refractive index and define the inner contour as an ellipse. In this case, assume that F1 and F2 are the only two focus. For general point C, this aberration phase or the path difference from the lens's plane wave can be expressed by a polynomial with y as a variable.

For long-focus lenses, the term above $y^3$ can be ignored. Moving C toward or away from the lens in the radial direction, like secondary focusing, can reduce the phase error. Increasing the focal length of the lens can also reduce aberrations. If the lens connection is no longer parallel to the axis, as shown in P–O′ in Fig. 8.16,

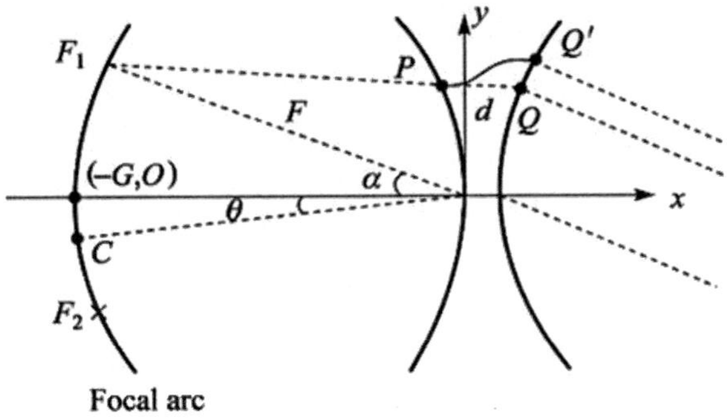


**Fig. 8.16** Constraint lens geometry

this constitutes a bootlace lens. It expands the Ruze design of the flat surface with three variable parameters, namely the lens profile, thickness, refractive index, and introduces the fourth parameter, the relative spacing of the front and rear units. The length of p–Q′ determines the refractive index. This fourth additional parameter makes the third accurate focus exist, that is, the lens axis (− G, O) point, which further reduces the lens aberration. The selection of G is to minimize the phase error in the secondary focusing of the Ruze lens. For a lens with special scanning requirements, if it is well defined, the opening angles of $F_1$ and $F_2$ and the normalized focal length G/F need to be defined. α is usually the required maximum scanning angle, because the phase error reaches a minimum between ± α, and increases rapidly outside this range. Amplitude control and multiple reflections mainly depend on the contour shape, adopting the same contour shape on both sides and allowing the centerline between the two sides of the lens to be separated. The array can be a linear array or a curved array. There is three focus on the confocal arc called the bootlace lens. If the lens's focal arc is a circular arc line, the two contour lines are different. At this time, the lens is called a Rotman lens. Discussed here is a multi-beam transmit antenna array with a bootlace lens former.

When a multi-beam antenna is used in a transmitting array, increasing the power of synthesis is one of the most important indicators. Therefore, the linear array is usually not used instead of the curved array, because the arc array will reduce the synthesis efficiency due to the inconsistent direction of the radiation element. Reducing the feeding phase error of the radiating element can increase the power of synthesis. In many planar lenses, the booteace lens has a relatively small aperture phase difference. The entire multi-beam transmit antenna array includes: (1) Beam port, used to introduce jamming signals into the lens; (2) Variant booteace lens, used to provide a specific amplitude and phase distribution to the linear array to generate a beam in a specified direction of space; (3) Array element port, providing an interface for the input port of the traveling wave tube amplifier; (4) Connecting cable, connecting the array element port and the radiating element; (5) Radiation element, radiating the jamming radio frequency signal in the specified direction.

The lens-fed multi-beam antenna array is made using the focusing principle of the lens. Figure 8.17 is the schematic diagram, where $\sum_1$ is the feed arc, $\sum_2$ is the focal arc, W is the extension cable, and PQ is the linear array. When a point

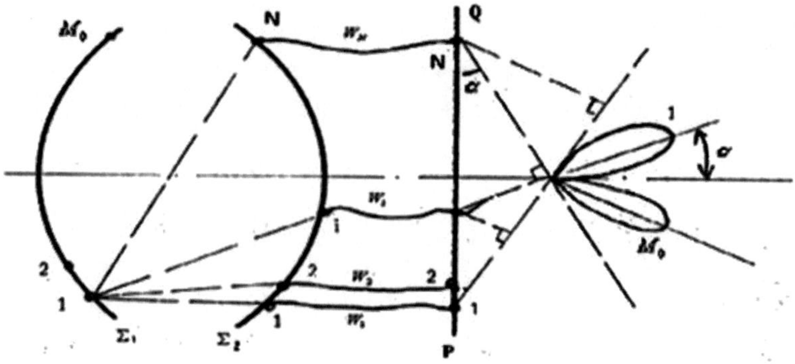


**Fig. 8.17** The working principle of the lens

source is fed from point 1 of the feed arc $\sum_1$, the wave is received by the receiving elements 1, 2, … N on the focusing arc, and passes through the compensation cables $W_1$, $W_2$, … $W_i$ … $W_N$ to the radiating elements of the linear array and radiate out as plane waves. The wavefront is at an angle of α with the aperture of the array, so a beam 1 with a direction α is formed. When the feed moves on the feed arc, the beam direction also moves accordingly. When the feed moves to point $M_0$, beam $M_0$ is formed. If $\sum_1$ is on the feed arc, $M_0$ feed points are set as required. Then, $M_0$ parallel beams can be formed in space. In order to suppress the appearance of grating lobes, the element spacing of a uniform linear array is usually taken as $1/2\,\lambda_{min}$ (where $\lambda_{min}$ is the minimum operating wavelength). Increasing the array element spacing within a certain range can increase the array gain. However, if the number of beams remains unchanged, the beam intersection level will be reduced. It will undoubtedly reduce the jamming power at the intersection. Increasing the number of beams can offset this trend, but it will make the switch matrix complicated. Generally taken in the project:

$$d = 1.2/(2\lambda_{min})$$

In order to increase the radiated power, the diameter D should be as large as possible, but the coverage range of the elevation angle should also be considered. Assuming that the 3 dB beamwidth hat f = 18 GHz is greater than 20°, the following can be obtained:

$$D = 68 \cdot \frac{\lambda}{2\theta_{0.5}} = 56\,(\text{mm})$$

When the aperture of the radiating element is 56 × 8 (mm), the theoretical gain coefficient is 8.34 (cosine distribution). Considering that the aperture utilization coefficient of the ridged horn is lower than that of the optical waveguide horn and the reflection loss due to standing waves. The gain coefficient g of the unit is:

$$g = 2.5$$

The equivalent radiation power of the transmitting array can be calculated as follows:

$$p = \eta p_0 g N^2 \tag{8.67}$$

In the formula, $p_0$ is the transmit power of a single traveling wave tube amplifier, g is the gain of the radiating element, N is the number of radiating elements, and $\eta$ is the synthesis efficiency.

If the equivalent radiated power is 100Kw, when $p_0 = 50$ w, g = 2.5, $\eta = 0.7$, N = 34 can be obtained, and N = 35 is taken in the project. The principle of selecting the number of beams is that the level of the intersection point of the high-end beam of the frequency band should not be too low. In the 8–18 GHz frequency band, the intersection point level should not be lower than about 4 dB because if the intersection point level is too low, it will reduce the jamming power at the intersection level. Otherwise, the number of beams must be increased. According to the above principle, the number of beams is 32, that is: $M_O = 32$. When the scan angle is zero, the 3 dB width angle of the pattern can be calculated as follows:

$$2\theta_{0.5} = 51^{\circ} \frac{\lambda}{Nd} \tag{8.68}$$

When f = 8 GHz, $2\theta_{0.5} = 5.46°$; when f = 13 GHz, $2\theta_{0.5} = 3.46°$; when f = 18 GHz, $2\theta_{0.5} = 2.43°$. When the number of beams is 32, and the coverage angle is ± 45°, the nominal beamwidth is 2.8125°. Obviously, when N = 32 can meet the requirements, when the beam is scanned, the beamwidth this calculated by the formula (8.68) and then divided by the scan factor $\cos\theta_k$, where $\theta_k$ is the beam scan angle. When $\theta_k = 45°$, the corresponding beam widths of 8 GHz, 13 GHz, and 18 GHz are 7.72°, 4.89°, and 3.44°, respectively. Assuming that 32 beams are evenly distributed within ± 450, each beam occupies an angular width of 90°/32 = 2.8125°. This beam is located at a scan angle of 0°, and the intersection with the adjacent beam is 45°/32 = 1.406°. Considering such angle value, the intersection point level e can be obtained as follows:

$$f(\theta) = \frac{1}{N} \frac{\sin\left(\frac{N\pi d}{\lambda} \sin\theta\right)}{\sin\left(\frac{\pi d}{\lambda} \sin\theta\right)} \tag{8.69}$$

When f = 8 GHz, e = – 0.76 dB; When f = 13 GHz, e = – 2.08 dB; When f = 18 GHz, e = – 8.72 dB. One characteristic of the linear array is that when the beam is scanned, the radiated power will decrease for two reasons as the scanning angle increases. One is that the beam gain decreases with the increase of the scanning angle $\theta_k$ according to cos $\theta_k$, and the second is the value of the unit pattern at the $\theta_k$ position lower than the 0° direction, so the power drop level when the beam scans to $\theta_k$ is:

$$e_b = \left[\frac{1+\cos\theta_k}{2}\frac{\sin\left(\frac{D_2\pi}{\lambda}\sin\theta_k\right)}{\frac{\pi D_2}{\lambda}\sin\theta_k}\right]\cdot\cos\theta_k \tag{8.70}$$

In fact, as the beam scanning angle increases, the irradiation of the array aperture becomes more and more uneven, which causes a decrease in gain. And as the frequency increases, the feeder's exposure level to the edge of the array aperture will decrease, which will undoubtedly reduce the aperture utilization coefficient, which will also cause the edge beam gain to overlap. Therefore, the actual overlap of the edge beam's gain is larger than the result calculated above. The lens is the core component of the multi-beam transmit array, and the rational design of the lens has a decisive influence on the performance of the transmit array.

Introducing a new parameter m into the bootlace plane lens can make the original lens more flexible. The lens obtained in this way is called a variant bootlace lens. When m = 1, the mutated bootlace lens becomes the original bootlace lens. The mutated bootlace lens is shown in Fig. 8.18. Σ1 is the feed arc, Σ2 is the focal arc, and Σ1 and Σ2 are the same curves.

Point Fo (0, 2A–C) is the focus of the positive axis. F1 (m, A), F2 (− m, A) is the off-axis's focus. A and C are undetermined constants; m and $\theta_m$ are parameters that determine the lens structure. xoy is the rectangular scale of the lens, z is the coordinate axis of the array aperture, and the contour equation of the lens is:

$$ax^2 + bx + c = 0$$

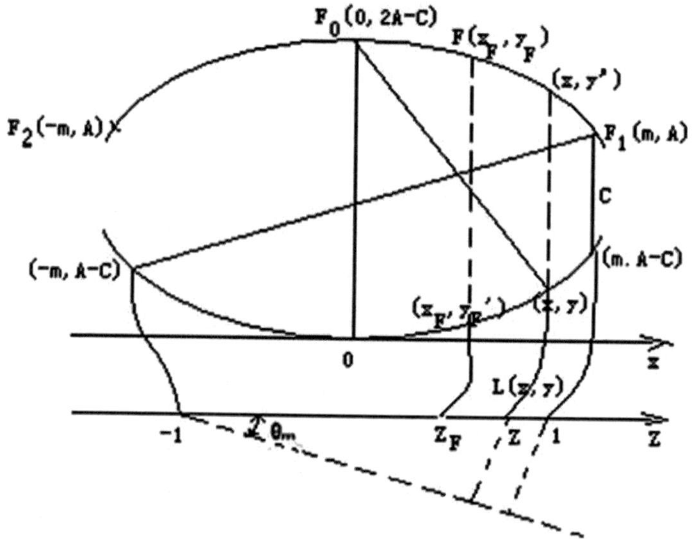


**Fig. 8.18** Variation bootlace lens

$$a = \frac{8m^2}{9p^2z^2} - 1, \quad b = \frac{4m}{9p}\left(\frac{1}{2} + z^2 - \frac{m^2}{p^2}\right) \tag{8.71}$$

$$c = \left[z^2 - \frac{4}{9}\left(\frac{1}{2} + z^2 - \frac{m^2}{p^2}\right)^2\right]p^2 - m^2 \quad L = -\frac{mx}{pz} + \frac{m^2}{p} + \frac{p}{4} \tag{8.72}$$

$$y = -\frac{mx}{3pz} + \frac{m^2}{3p} + \frac{p}{12} + \frac{2}{3}pz^2 \tag{8.73}$$

$$y' = \frac{mx}{3pz} + \frac{2m^2}{3p} + \frac{5p}{12} - \frac{2}{3}pz^2 \tag{8.74}$$

$$p = \sin\theta_m \tag{8.75}$$

where $\theta_m$ is the beam scanning angle when the feed is located at the ideal focus off-axis. The values of the constants A, C and 2A–C are:

$$A = \frac{m^2}{p} - \frac{p}{4} \qquad C = \frac{m^2}{p} - p \qquad 2A - C = \frac{m^2}{p} + \frac{p}{2} = T$$

T is called the thickness of the lens. It can be seen from Eqs. (8.71), (8.72), (8.73), (8.74) that the lens profile only depends on the parameters m and $\theta_m$. Assuming that a feed is located at point F($x_F$, $y_F$), the array aperture coordinate $Z_F$ corresponding to this point has the following relationship with the beam scanning angle $\theta$ of this point and the beam scanning angle $\theta_m$ of the off-axis ideal focus:

$$\sin\theta = z_F \sin\theta_m \tag{8.76}$$

When the beam scanning angle $\theta$ is given, $z_F$ can be obtained from the above formula, then $x_F$ is obtained by formula (8.71), and $y_F$ is obtained according to the formula (8.73), then the point F is determined. For the calibers of feeding triangle and the receiving triangle, assuming that $M_0$ feeds are evenly distributed on the arc between $F_1$ and $F_2$, then the caliber of each feed can be calculated as follows:

$$D_k = 2rtg(\psi_0/M_0)$$

$$r = \frac{3p}{8} + \frac{2m^2}{3p} \tag{8.77}$$

$$\psi = tg^{-1}\left[\frac{24mp}{16m^2 - 9p^2}\right]$$

The aperture D of the receiving element is also calculated according to formula (8.77), but the number $M_0$ of beams is changed to N. The lens described above is normalized, that is, the lens size obtained by dividing the array aperture by all the dimensions of the lens. To obtain the actual lens, the value obtained by the above formulas should be multiplied by the normalization coefficient to be $k^g$, then:

$$k_g = Nd/\sqrt{\varepsilon_r} \tag{8.78}$$

where $\varepsilon_r$ is the relative dielectric constant of the material used to make the lens, N is the number of elements, and d is the element spacing. When N = 35, d = 10; when $\varepsilon_r$ = 2.2, $k^g$ = 235.97. Whether it is a feeding delta element or a receiving delta element, the 50 Ω microstrip line from the aperture of the lens outline to the beam or array element can be regarded as a resistance transition section. In addition to considering the reflection coefficient, this transition section's length should also consider the width of the two adjacent beam openings or the minimum distance between the array element openings. This minimum distance must ensure that the cable head can be connected to the SMA of the lens. The coefficient S can be calculated as follows:

$$S = \left| \sum_{i=0}^{N-1} \frac{z_{i+1} - z_i}{z_{i+1} + z_i} e^{-j\left(2i\alpha \frac{H}{N}\right)} \right| \tag{8.79}$$

The formula $z_i$ is the characteristic impedance of the i-segment, which is equal to the N-segment of the transition section of the microstrip. H is the gradient line's length, and N is the number of segments in the transition section. The larger the N, the higher the calculation accuracy. $\alpha = 2\pi/\lambda$ is the wavenumber, λ is the operating wavelength. When the feed's triangle diameter is 10.11 mm, the width of the 50 Ω microstrip line is 2.8 mm, the length of the transform section is H = 1.6, = 60 mm, and εr = 2.2, the standing wave coefficient S = 1.13 is calculated according to the above formula. Take the length of the feed triangle of the actual lens H = 40.5 mm. When illuminating the array aperture during off-axis feeding, an asymmetry will occur. The larger the scanning angle is, the more severe this asymmetry is. Since the formula for calculating the amplitude distribution of the lens's aperture field during off-axis feeding is complicated, only the calculation results are given here. When the scanning angle $\theta_{m=}$ 450, the amplitude ratio of the left and right edges of the aperture is about 0.9–0.2. This calculation is obtained, assuming that the height of the triangle coincides with the normal of the lens outline in the triangle aperture. Obviously, changing the direction of the triangle (the center of the aperture is unchanged) can change the triangle's radiation pattern so that the exposure to the aperture can be adjusted. The results show that this method can change the ratio of uneven irradiation to about 0.6:0.4. In the lens, its aperture phase error is divided into inherent error and manufacturing error. The inherent phase error is inherent to the lens. When the lens parameters $\theta_m$ and m and the array aperture Nd are determined, the inherent phase error is a constant. When $\theta_m$ = 450, m = 1.25, Nd = 350 mm, the inherent phase error is about 40 at f = 18 GHz. The effect of such a small phase error on the pattern can be completely ignored.

For the convenience of processing, set the lens output and input port socket outline to a circle. Otherwise, the processing will be greatly complicated. According to the focusing equation of the lens, the path length of the wave from $F_0$ to $A'$ on $\Sigma_2$ and then to A is equal to that of the wave from $F_0$ to $B'$ on $\Sigma_2$ and then to B on $\Sigma_3$. When calculating the path length, the geometric lengths $A'A$ and $B'B$ are usually regarded as electrical length. In fact, due to the transition section of the triangular microstrip, their electrical length and geometric length are not equal. This is because the microstrip line has an equivalent dielectric constant, and it is a function of impedance, so it is difficult to strictly calculate the dielectric constant of the triangle. However, the numerical method can be used for approximate calculation, that is, the triangle is divided into many small sections, and the electrical length of each small section is obtained, and then the electrical length of the triangle can be obtained by adding them together. Thus, the phase error can be obtained.

The above phase error due to simplified design can be corrected. As can be seen from Fig. 8.19, without changing the position of B′, moving the center of the $\Sigma_3$ circle on the FoA axis can change the length of the triangle B′B, thereby changing the phase error. Calculations show that moving the center of the circle to a certain position can basically eliminate the above manufacturing error, that is, the maximum phase error exceeds 10. For the transmit antenna array, the power it carries depends on each component of the transmit antenna's breakdown power. For the lens-fed multi-beam antenna, since the traveling wave tube is added between the lens and the radiating element, only the coaxial line's breakdown power and the radiating element needs to be connected. The breakdown power of the same axis is:

$$p_{\max} = \frac{1}{S}\frac{V_m^2}{2Z_0} \tag{8.80}$$

where S is the standing wave coefficient, $V_m$ is the breakdown impedance, and $Z_0$ is the characteristic impedance. When SFT-50-3 semi-hard coaxial cable is used, the characteristic impedance is 50 Ω and $V_m$ is 5 kw. Assuming that the standing wave coefficient is 4, then $P_{\max} = 62.5$ kw which is higher than the traveling wave tube's transmission power. Since the radiating element adopts a ridge horn and is a

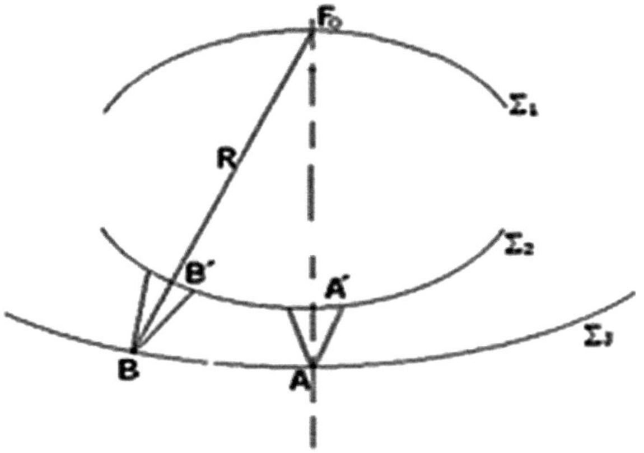


**Fig. 8.19** Lens contour line

**Table 8.4** Maximum withstand power

| F(GHz) | 110 | 11 | 12 | 13 | 14 | 15 |
|---|---|---|---|---|---|---|
| $P_{max}$(w) | 399 | 377 | 358 | 341 | 327 | 313 |

microstrip line at the feeding point, the breakdown power of the radiating element depends on the breakdown power of the feeding microstrip line. When h = 0.5 mm (the distance between the microstrip and the floor), t = 0.01 mm (t is the thickness of the strip), $\varepsilon_r = 2.6$, then the breakdown power at a temperature of 20 °C is as follows (Table 8.4):

According to the above trend of $P_{max}$ decreasing with frequency, it is estimated that when f = 18 GHz, if the standing wave coefficient is 4, then $P_{max} = 67.5$ W, which is still greater than the transmission power of the traveling wave tube 50 W. When $\varepsilon_r$ or h increases, $P_{max}$ will also increase. But as the temperature increases and the pressure decreases, the breakdown voltage will decrease. The relationship is as follows:

$$P'_{max} = P_{max}\left[2.9 \times \frac{p}{10^3 T}\right]^{3/2} \tag{8.81}$$

where p is the atmospheric pressure (in Pascals), T is the absolute temperature, and $P_{max}$ is the calculated value at a temperature of 20 °C under standard atmospheric pressure. When the environment temperature is 60 °C (p is the standard atmospheric pressure), $P_{max}$ calculated according to the above formula is 258 w (h = 0.5 mm, = 2.6, f = 15 GHz), and it can still withstand 50 w of transmitting power. Since the antenna's performance and the various parameters that affect the antenna's performance are mostly complex implicit function relationships, it is impossible to find a simple optimization design method. Therefore, the improved bootlace lens is applied to the multi-element array, which breaks through the mutual influence between the feeding isolation and solves the problem of matching the stray waves on the left and right edges and successfully developed a larger broadband beam antenna array. The typical pattern is shown in Fig. 8.20. The system with a multi-beam transmitting antenna array has the characteristics of coping with multiple targets and capable of synthesizing high power in space, operating frequency bandwidth, large instantaneous suppression range, high reliability, and fast response speed.

### 8.3.2 *Beamforming of Luneburg Lens*

The array of the beamformer by the bootlace lens can also be a curved array, which helps broaden its beam scanning range, but there is a natural level difference between the center beam and the edge beam. In order to further equalize the beam level in a wide scanning range, a double bootlace lens combination can be used, but the additional loss increases. Another method is to use an R–KR lens, a circular array

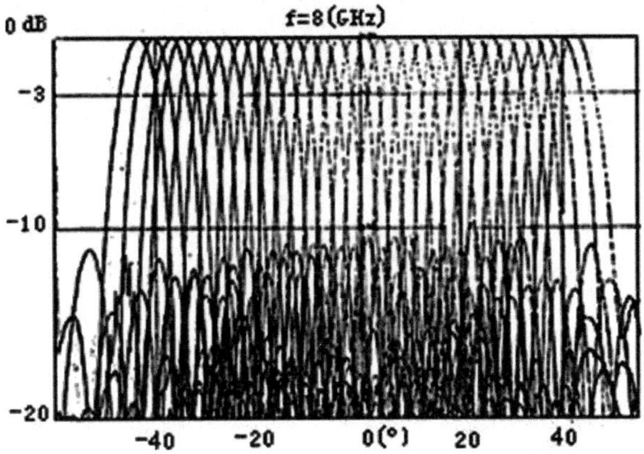


**Fig. 8.20** Typical multi-beam radiation pattern

with radius R connected to a circular lens with radius KR by a series of lines with equal length, where K is a constant. Usually, this lens does not have an ideal focus. But due to the symmetry of the circle, the same phase error at all feed points can be minimized by choosing K appropriately. Generally, the connection angle between the array and the lens port at the center of the lens cannot be the same. However, for K $\approx$ 1.9, when the circular symmetry is achieved, we have a lens with diameter approximately twice to that of the array. In this case, the array and the lens cannot be coplanar. However, if the lens is made of a high dielectric constant material, its size can be reduced, and it can be installed in the middle of the array to form a coplanar device. The combination of the traditional manufacturing process and new ideas developed the Luneburg lens to achieve the purpose of beam balancing.

As we all know, the Luneburg lens is made of non-metal dielectric material as the substrate, and the dielectric constant of the dielectric material is required to be gradual. The dielectric constant at the edge of the lens is 1, and the dielectric constant at the center is a constant, which satisfies the following formula (derived from the ray equation). The Luneburg lens is a fully-focused dielectric lens antenna. The lens is generally cylindrical. The lens uses the principle of reflection and refraction of the beam rays. Through the continuous refraction of rays in a medium with a smoothly changing dielectric constant, the lens focuses the rays in an appropriate spatial orientation so that when a feed is placed on the circumference, the spherical wave radiated from the horn at the beam port becomes a plane wave after passing through the lens. Thus, a high-gain sharp directional beam required for direction finding is obtained, as shown in Fig. 8.21.

$$n^2(r) = \varepsilon(r) = 1 + k^2 - k^2(r/R)^2 \tag{8.82}$$

In the formula, k is the ratio coefficient of the lens radius R and the lens's focus radius. The value of k can be greater than or less than 1. When k is less than 1, the focal circle radius is greater than the lens radius; k equal to 1 is a special case. At this time, the electric constant is smoothly gradual, the dielectric constant of the edge is

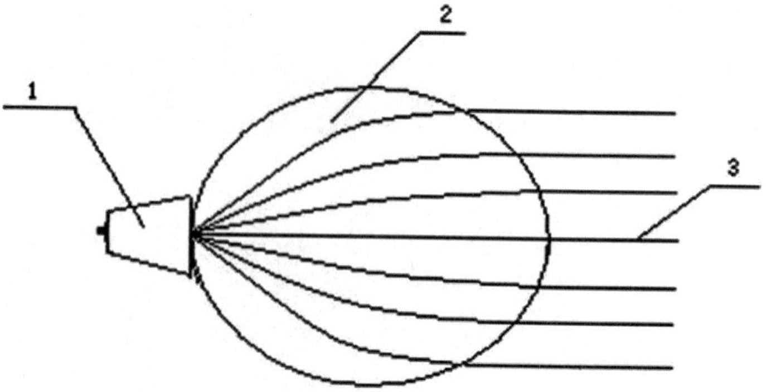


**Fig. 8.21** Schematic diagram of Longbo lens. 1—Beam mouth horn 2—Lumber lens 3—Flat wave after lens focusing

1, and the center position is 2. To achieve the gradual change of dielectric constant, there are several methods in engineering:

One is to use a continuously variable medium that satisfies the change rule of formula (8.82). From the material's manufacturing process, there is corresponding equipment to realize the gradual change of the dielectric constant through the foaming method. However, this method requires expensive special equipment, and the processing cost is very high. It is very difficult to control the processing's uniformity, so its engineering application's value is not great.

The second method is to use a combination of dielectric material layers with different dielectric constants. Each layer of dielectric has a certain thickness. The dielectric constant in one layer is constant, which is equivalent to approaching continuity with small intervals. However, this method has two major disadvantages. The dielectric constant of the glue between each layer is too large, which will have a serious impact on the lens's electrical performance. For the lens formed by the stacking method, the dielectric constant of the lens changes stepwise rather than smoothly, causing the deterioration of the lens's electrical performance.

Thirdly, the method of drilling holes in the lens dielectric body is used to realize the dielectric change, but this method also has obvious disadvantages. First, the dielectric constant of the medium is changed stepwise. The second is that the punching method is troublesome and difficult to control the dielectric value strictly. Also, for the beamforming in the azimuth plane, the lens is deformed into a curved panel to meet the equal phase, but it is cumbersome in design and requires high positioning processing. On this basis, air and medium are used, and the equivalent dielectric constant of the electric field at different heights when vertically polarized is different to continuously approach the dielectric change at different positions in the Luneburg lens to realize the Luneburg lens. This method's Luneburg lens has a smooth gradation of dielectric constant, simple design, excellent electrical performance, and convenient processing.

The working principle of the Luneburg lens multi-beam antenna developed here is exactly the same as that of the traditional Luneburg lens antenna. The only difference

**Fig. 8.22** Schematic diagram of Longbo lens. 1—feed antenna 2—metal disc 3—air 4—homogeneous medium 5—open horn

is the way of forming a gradual equivalent dielectric constant. As shown in Fig. 8.22, the lens's medium is between two metal plates and consists of two media. One is the dielectric material; the other is the air. Their interface is a specific curved surface, which can be designed by the relevant formula. The equivalent dielectric constant of the lens medium can be calculated by the formula (8.83), and its value continuously changes between 1 and 2 or a certain dielectric constant value.

$$\varepsilon_{eq} = \frac{D\varepsilon_r}{D\varepsilon_r + (1-\varepsilon_r)\cdot t} = \frac{\varepsilon_r}{\varepsilon_r + (1-\varepsilon_r)\cdot (t/D)} \tag{8.83}$$

In the formula, D is the spacing between parallel plate, t is the medium's height, and $\varepsilon_r$ is the relative dielectric constant of the medium. Available from formula (8.82) and formula (8.83)

$$(r/R)^2 = \frac{1 + k^2 - [\varepsilon_r/(\varepsilon_r + (1-\varepsilon_r)(t/D))]}{k^2} \tag{8.84}$$

In the formula, r is the distance from the center of the lens, which can be concluded that under a certain situation of k, the relationship between (r/R) and (t/D) relative to the fixed dielectric material. R is determined by the number of antenna beams and beamwidth, D is determined by the highest frequency, calculated by the following formula:

$$2R = 68\frac{\lambda}{2\theta_{0.5}} \tag{8.85}$$

$$D \le \lambda_{\min}\big/\sqrt{\varepsilon_{eq}} \tag{8.86}$$

The formula $2\theta_{0.5}$ is the beamwidth, and $\lambda_{\min}$ is the wavelength corresponding to the highest frequency in the operating band. Thus, the medium's height curve with the radius is obtained between the flat plates, and the curve is rotated 360° around the central axis to obtain the planar Luneburg lens with an equivalent dielectric constant gradient. Vivaldi feeds the lens antenna printed antennas. Each antenna element corresponds to a beam. The antennas are distributed at even intervals, and their phase centers are in the same circle. Under normal circumstances, the lens's center is on the lens's circumference, but it can also be adjusted according to the electrical performance. The polarization of the antenna element is perpendicular to the metal plate. Since the lens is an axisymmetric all-focus structure, the feed direction is the lens antenna's beam direction. Evenly spaced feeds obtain uniformly spaced beam directions, and the frequency dispersion of the beam direction is small. According to the requirements of electrical indicators, the lobes of the required beams are determined by the azimuth coverage angle and the number of beams. Then determine the size of the lens aperture, that is, the size of the metal disc and the longitudinal dimension of the lens (that is, the spacing between the two metal discs) depends on the size of the beam aperture feed aperture and formula (8.85). Simultaneously, the lens's volume and weight are also considered, and the longitudinal dimension is generally small. The problem of spatial matching of the wide beam of the pitch plane can be solved by designing a small arc on the radiation port surface to transition the opening section. The 16 beams cover 90° azimuth, and the beams are separated by 6° to form multiple beams. The Vivaldi antenna is selected because it is a microwave-printed structure with the advantages of small lateral size, good processing consistency, wide working frequency band, and relatively stable phase center. In the engineering design, the antenna impedance matching circuit, the slot gradient curve, and the radiation port surface are carefully designed to make it have good impedance matching characteristics and radiation characteristics. Due to the small distance between the feed antennas, the mutual coupling problem should be highly valued. In order to reduce the impact of coupling as much as possible, in addition to taking appropriate measures in antenna design and feed layout, it is also essential for the antenna to use a vertically polarized form. The phase center of the Vivaldi antenna is in the same circle. Generally, the circumference is selected to be near the outer edge of the lens, but appropriate adjustments can be made. There is space for properly adjust the design, so that in the wideband range, the phase center of the feed element is kept as far as possible on the focal arc of the lens. It is also quite difficult to realize the structure of the lens antenna, mainly reflected in the support structure of the two metal discs. Because there is a part of the air between the metal discs. By properly adding circular dielectric rod support, the position of the lens area between the two can be processed to form a certain foam entity for structural reinforcement.

The dielectric material used in actual development is polystyrene. The lens radius is 150 mm. The longitudinal length of the antenna aperture should be as short as possible to avoid the additional loss introduced by adding the array port and the radiating element array. The airspace covers 90° (azimuth plane) × 50° (pitch plane), and the polarization form is vertical linear polarization. The typical pattern is shown in Fig. 8.23. This Luneburg lens's multi-beam antenna has a relatively simple structure, good electrical performance, all-round focusing, good beam uniformity, no array element port and antenna array, small size, and lightweight. Such obvious advantages occupy an important position in the lens antenna. In addition, by connecting each array element in the array to a broadband circulator, a 360° compact circular array can be obtained.

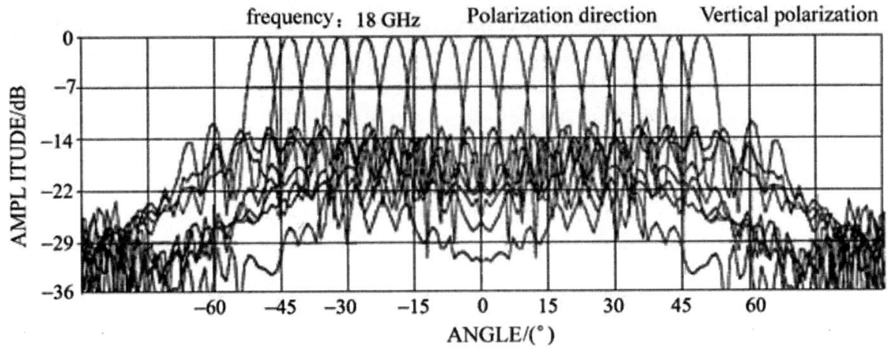


**Fig. 8.23** Typical pattern

### *8.3.3 Two-Dimensional Multi-beam Area Array Antenna*

As the electromagnetic environment becomes more complex, electromagnetic signals become denser and denser, and the types of threat targets become more and more diverse. Systems composed of conventional single-beam antennas are increasingly difficult to adapt to the needs of modern warfare. The one-dimensional multi-beam linear array antenna has multi-directional beam clusters in the azimuth plane. It can quickly correspond to various targets in different azimuths and different pitch directions, achieve dual aiming of frequency and direction, and have space synthesis. If a medium power traveling wave tube is connected after each antenna element, it can jam with multiple targets with powerful equivalent radiated power. The beam control is convenient, the beam pointing is constant, and it is not affected by the environmental humidity. The structure is compact, the insertion loss of the feed lens is small, and the reliability is high. The spacing beam is formed by the amplitude-phase transformation of the air-fed lens, which has the best cost-effective ratio, and the system formed by it effectively reduces the jamming dead zone. Simultaneously, it has the function of single-station positioning during reconnaissance, which makes it more flexible and easier to achieve the best jamming. It has great application potential in electronic warfare.

The two-dimensional multi-beam area array is composed of the beam port's curved surface, the curved surface of the array port, the planar array, and the connecting cable. The curved surface of the beam mouth and the curved surface of the array mouth form the feeding lens, which is the integration of a complex network composed of many microwave phase shifters, spatial power combiners, and power dividers. When a plane wave is an incident on the surface array from the direction of the planar array, the electromagnetic wave will be focused on the center of the beam mouth (i.e., $\theta = 0$, $\psi = 0$). When a plane electromagnetic wave is an incident on the surface array from a certain direction (the spatial angle of the direction is $\theta$, $\psi$), the electromagnetic wave will concentrate at the center point's position the curved surface of the beam mouth. The output position of the signal from the beam port depends on the incident direction of the plane wave. According to the reciprocity theorem, if a power supply is provided on the beam aperture, beams with different directions can be formed in space. It is required that the planar array system work in broadband and the planar array antenna must work in broadband. The lens maintains its focusing characteristics within the required operating frequency band. At present, there are several ways in engineering. The one-dimensional multi-beam lens introduced above has a feed source on the focal arc and constitutes an FT system. In contrast, a two-dimensional scanning lens has a focal plane and is made into a 0, 1, or 2 order FT system. It is very convenient to classify, according to FT. The zero-order Fourier transform lens system usually includes a large feed array and a very large aperture surface, such as a Sperry circular antenna, with a wide scanning angle.

Usually, a first-order FT system such as Ruze, Rotman, and R-KR lenses can be expanded to obtain a two-dimensional beam using several methods. For example, each row of a planar array is fed by a single horizontal lens, and it is, in turn, fed by a series of vertical lenses. Similarly, the stereo lens has either 4 or 5 focal points. This design phase is very similar to the use of the bootlace lens and Rotman lens in one dimension. The selection of the location of the focus can lead to a very different performance. In a stereo lens system, if both the lens's inner and outer surfaces are simplified as planes and a printed circuit unit is used, although this will increase the phase error, this method is still used. Beam scanning can also be accomplished by adding a diode switch to a constrained lens, just like a Radiant lens consisting of multiple diodes arranged in multiple layers. The three-dimensional Luneburg lens is a dielectric lens. The entire lens is a sphere. Multiple antenna elements are placed on the left sphere of the dielectric sphere. The dielectric constant of the medium in the sphere is spherically symmetric, and as the radius of the sphere becomes smaller, the dielectric constant εr becomes larger. εr = 1 on the medium's spherical surface, through the action of the non-uniform medium, the beam emitted by the beam port unit. After passing through the dielectric sphere, they are parallel to the ray passing through the center of the sphere by this beam port unit, forming a plane wave. If the Luneburg sphere is used for reception, the dielectric sphere narrows the beam to the beam port unit and improves the gain. At the same time, it is used for all beam port units to be multiplexed, so it is fully aggregated. The airspace coverage is extensive, the beam is uniform, and it is easy to achieve broadband operation. When used as a transmitting array, it will reduce the efficiency of beam distortion, and it is extremely difficult to achieve gradual $\varepsilon_r$ in practical engineering. If a layered and gradual dielectric layer is used to approximate the continuously changing dielectric layer, manufacturing and processing will produce uncontrollable errors, which will be costly, and the dielectric ball will have additional energy loss. It has great prospects of application as a corner reflector in passive jamming. Multiple two-dimensional lenses are combined to form a two-dimensional multi-beam. It is relatively easy to realize two-dimensional multi-beam by combining multiple such lenses. As shown in Fig. 8.24, on the right is a planar antenna array. This method of forming a spatial beam uses a set of azimuth beamformers (each row of the array has a two-dimensional lens), followed by a set of pitch plane wave formers, as shown in the figure. In theory, the antenna array has good focusing characteristics in azimuth

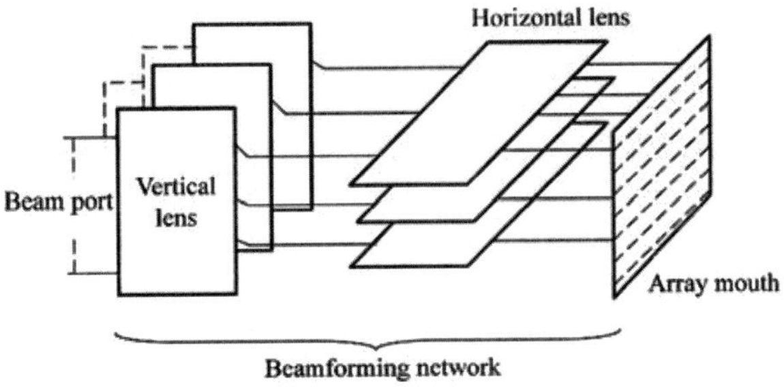


**Fig. 8.24** Combined lens two-dimensional multi-beam

and pitch. In actual engineering, the increase in the number of lenses increases the loss of the system. Simultaneously, connecting cables and connectors causes additional reflections, causing a new phase imbalance, making the accumulated phase difference larger, and aggravating the distortion of the beam. Moreover, the volume of the feeding part becomes large. On the other hand, when the planar array element antennas are arranged in a triangle or other forms, and the beam positions of the beam ports appear alternately in space, the types of two-dimensional lenses increase, and the interconnection becomes complicated.

The cylindrical lens with a phase shifter is used to form a two-dimensional multi-beam, as shown in Fig. 8.25. A phase shifter is connected to each planar array element antenna. The center of the cylindrical lens is air or medium, which is a combination of phase array and multi-beam. The azimuth plane forms a multi-beam, and the pitch plane is a phase shifter focused on forming a beam. The working mode is flexible, but the cost is high. The system has a long reaction time, and the debugging workload is large. The phase shifter is a controllable device, especially in the case of broadband. The complexity of the system increases. A compact two-dimensional multi-beamforming capability can be formed by adopting a broadband air-fed three-dimensional lens and a two-dimensional planar array antenna.

The two-dimensional multi-beam antenna formed by the lens is mainly composed of three parts, namely a three-dimensional lens, a connecting cable, and a planar

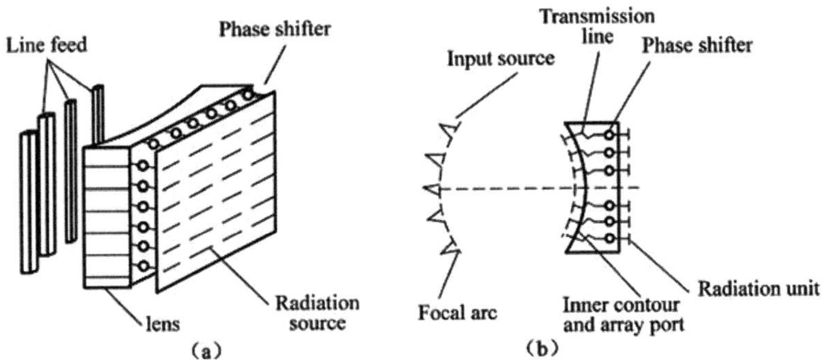


**Fig. 8.25** Cylindrical lens with phase shifter

array antenna. Through the curved surface of a special lens, the spatial amplitude and phase transformation is realized. According to the different geometric shapes, the lens is divided into three main forms: Rotman lens, Ruze lens, and Bootlace lens. The Rotman lens has a remarkable feature. The curved surface of the beam port is a spherical surface with a wide scanning angle. All the dimensions are normalized to the off-axis focal length, which is very suitable for three-dimensional processing. At the same time, the length of each connecting line between the planar array and the array port is the same, which is beneficial to the development in the project. In the broadband range, the planar array is directly fed by a three-dimensional lens of the Rotman air-feeding method with a curved array. This structure is compact, and the distribution and number of beams are predetermined. It can be used as both a transmitting array and a receiving array.

It is known from electromagnetic theory that a single focused beam needs only one focus in space. A typical example is a parabolic antenna. To focus the multi-beam linear array, the two-dimensional lens must have at least two or three focus and forming a gathering curve. Obviously, the beam quality of the three-focus curve is better. To generate a spatial beam cluster and make the beam port a curved surface, at least four focus is required. The coordinate selection of the entire lens is shown in Fig. 8.26. If you consider a conformal surface array, then the lens has five focus. In the figure, the coordinate system of the lens is analyzed in the case of a planar array. The four focus can be placed arbitrarily, and mathematical processing is more complicated. Compared with the symmetrical distribution, there will be a large phase difference. In this way, one focal point G1 is selected to be located on the lens axis, and the other three focal points F1, F2, and F3 are symmetrically located at three vertices of an equilateral triangle. Assuming that the beam passes through the spherical surfaces of the four focuses, the radius of curvature of the focal plane is controlled by the ratio of G and F (g = G/F), and the angle $F_1OG_1 = \alpha$ is variable. So, the change in the position of focus can be restricted by g and $\alpha$. It can be seen from Fig. 8.26 that the focus is defined by the following equation.

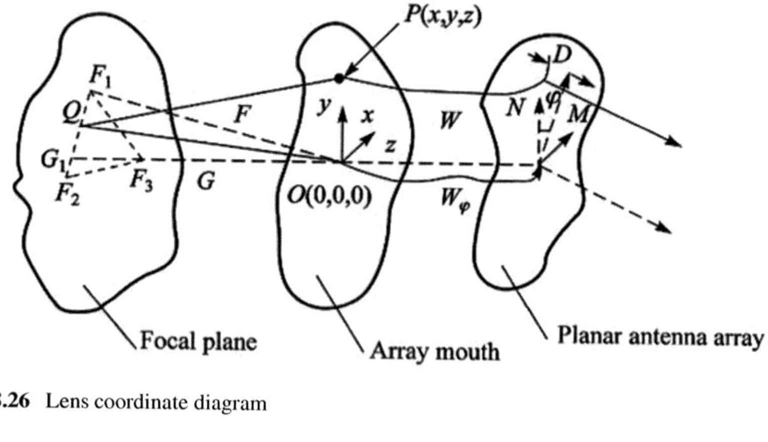


**Fig. 8.26** Lens coordinate diagram

$$\overline{F_1P} + w(M, N) + D_1 = \overline{F_1O} + w(0, 0)$$
$$\overline{F_2P} + w(M, N) + D_2 = \overline{F_2O} + w(0, 0)$$
$$\overline{F_3P} + w(M, N) + D_3 = \overline{F_3O} + w(0, 0)$$
$$\overline{G_1P} + w(M, N) = \overline{G_1O} + w(0, 0) \tag{8.87}$$

In the formula, (M, N) is the Cartesian coordinate of the planar antenna array. The length of the transmission line between the element antenna at the (M, N) point and its corresponding array curved port is w (M, N). D1 is the channel difference corresponding to the beam pointing of the focus F1. Assuming that the internal angle of the lens $\theta \neq \psi$, that is, $S2 \neq 1$, great flexibility can be obtained. Let:

$$x = X/F, \quad y = Y/F, \quad z = Z/F$$

$$w = (W - W_0)/F, \quad m = M/F, \quad n = N/F \tag{8.88}$$

Through derivation, the following formula is obtained:

$$x = A + Bw$$
$$y = C + Dw$$
$$z = E + Fw \tag{8.89}$$

In the formula:

$$A = [m + (m \cdot nS_1/2)] \cdot S_2 \quad B = -m \cdot S_2$$

$$C = \left[n + \left(m^2 - n^2\right)S_1/4\right] \quad D = -nS_2$$

$$E = m^2S_1^2/4(c_0 - g) + m^2S_1^2/4 \quad F = (g - 1)/(c_0 - g) \tag{8.90}$$

$$S_1 = \sin\beta \quad c_0 = \cos\alpha \quad S_2 = \sin\psi/\sin\theta$$

$\alpha$ is the internal flare angle of the source at the focus. β is the corresponding beam pointing angle. The length of the normalized transmission line is given by the following formula:

$$aw^2 + bw + c = 0$$

In the formula: $a = B^2 + D^2 + F^2 - 1$

$$b = 2AB + 2CD + 2EF + 2gF + 2g$$

$$c = A^2 + C^2 + E^2 + 2gE \tag{8.91}$$

If the parameters g, $\alpha$, $S_2$ are given, the above equation can be used to observe the change the error $\sigma_e$ of each normalized m, n channel of the source at any point on the focal plane with. When q = 30° and g = 0.87–1.30, the focal plane changes from an approximate plane to an approximately spherical surface, and the change of $\sigma_e$ is very small, which is less than 10%. Considering the case of g = 1, $S_2 = 1$, take $\alpha = 30°$, so that the center of the focal plane and the center of the array surface are the same, and their phase errors can be found. The formula is:

$$\sigma_e = \sqrt{(x_1 - x)^2 + (y_1 - y)^2 + (z_1 - z)^2} - \sqrt{x_1^2 + y_1^2 + z_1^2} + D(m, n) + w - w_0 \tag{8.92}$$

In the formula, x, y, z are the coordinates of the array port. $x_1$, $y_1$, $z_1$ is the coordinates of the beam port. D(m,n) is the spatial relative path difference required by the coordinate points of the antenna array on the surface of the array port corresponding to the co-location at the center of the array, and its expression is:

$$\mathrm{D(m, n)} = \mathrm{m} \cdot \sin\theta\cos\phi + \mathrm{n} \cdot \sin\theta\sin\phi \tag{8.93}$$

w is the length of the connecting cable between points m, n and x, y, z on the surface of the array. $w_0$ is the length of the cable between the center unit of the array aperture and the center of the surface of the array where x = y = 0. The calculation result of the normalized path difference is shown in Fig. 8.27.

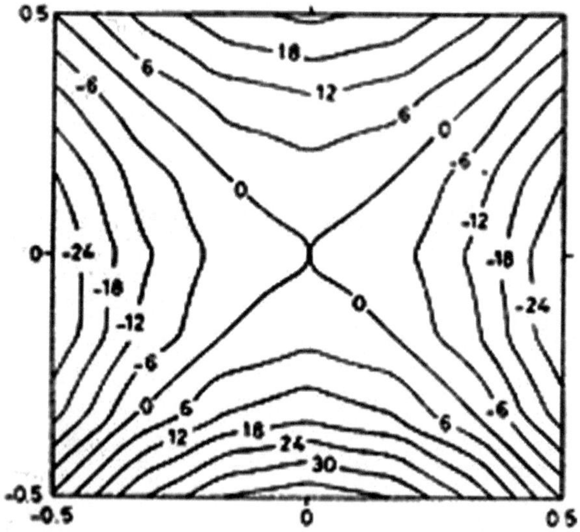


**Fig. 8.27** Normalized channel error ($\sigma e \times 10^3$)

Parameter $\alpha$ is an implicit variable, which defines $S_0 = \sin\beta_0/\sin\psi_0$ as the convergence degree of the focus. When g = 1, $S_2 = 1$, $S_0$ changes from 1 to 0 (all four focuses are closed together). In the range of m, n < 0.5, $\sigma_e$ is still a quadratic in two principal planes and plays an important role. A similar size error is introduced through F1, F2, and F3 to change the maximum channel error. However, the change of $S_2 = \sin\theta/\sin\psi$ has little effect on the normalized channel error. For the two-dimensional Rotman lens, the spacing of the array port near the center of the array port is about $S_2 \times$ array pitch. For waveguide-fed lenses, $S_2$ is required to be greater than 1 to avoid the cutoff of waveguides with lower frequency. This is also applicable to the three-dimensional Rotman lens. As $S_2$ increases, the size of the array port surface corresponding to the array also increases. If y (or z) is much greater than 0.5, distortion of the array port surface is required, which is difficult to achieve. Thus, if $S_2 = 5/3$, m (or n) < 0.3 is required. For a specific array, the minimum length F can be determined. The maximum should also be considered. We choose S for the required value of the array port.

$$F \geq S_2 \cdot P\lambda_0$$

($P\lambda_0$ is the maximum size of the array)

The maximum phase difference is set to $\lambda_0/16$, where m $\leq 0.5$, and the largest normalized channel error $\sigma_m$ is $1.8 \times 10^{-2}$, which is the main $\sigma$ for the change of the quadratic term:

$$\sigma_m \approx 7 \times 10^{-2}\,\mathrm{m}^2$$

$$\text{Then } \mathrm{F} \geq 0.3\mathrm{P}\,\lambda_0 \tag{8.94}$$

The off-axis focal length F can be determined according to the maximum scanning angle, array size, etc. As long as the focal length F is determined in the design, the antenna array's specific physical dimensions and the lens can be determined according to the given required index. For the beam distribution quality, it is also necessary to calculate the amplitude distribution of each port of array element under the above parameters. Based on the design formula given above, a planar array two-dimensional multi-beam fed by a three-dimensional Rotman lens is given. Its array is arranged in a triangle. The entire aperture is divided into eight columns with four elements in each column. It is a rectangular form. The entire antenna array must maintain a high gain of the element antenna and prevent the grating lobe from appearing in the visible area. Simultaneously, the two aperture surfaces of the lens itself are curved surfaces, especially the curved surface of the array port is a general quadric surface. In addition to the theoretical design to ensure a good amplitude distribution with small channel differences, extracting electromagnetic energy from two curved surfaces is a major engineering difficulty. Moreover, to ensure that the theoretical design can be realized in the existing conditions, within the allowable error range, the method of successive approximation of regular surfaces is adopted to ensure the project's feasibility. It carried out the research of the feed-back horn feed source of special feeding form. This research completes the transfer of radio frequency information and completes the debugging, positioning, and fixing of the unit. It solves the problem of the realization of the lens unit from both electrical and structural aspects. The distribution of beam ports is shown in Fig. 8.28.

Figure 8.29 shows the radiation pattern (orientation map and pitch plane) of the beam port. The design considers uniform illumination (theoretically, the sidelobe

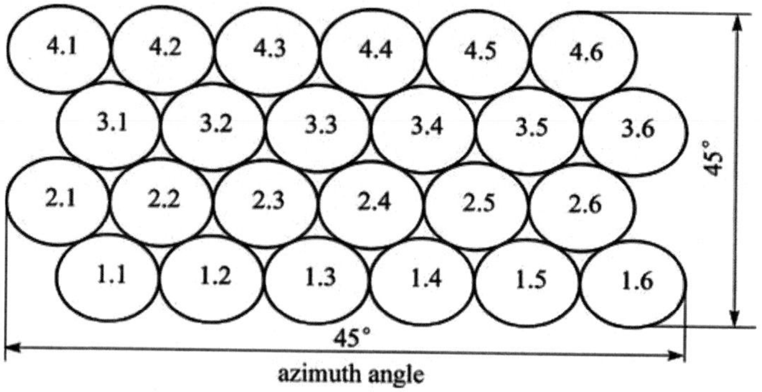


**Fig. 8.28** Beam distribution diagram

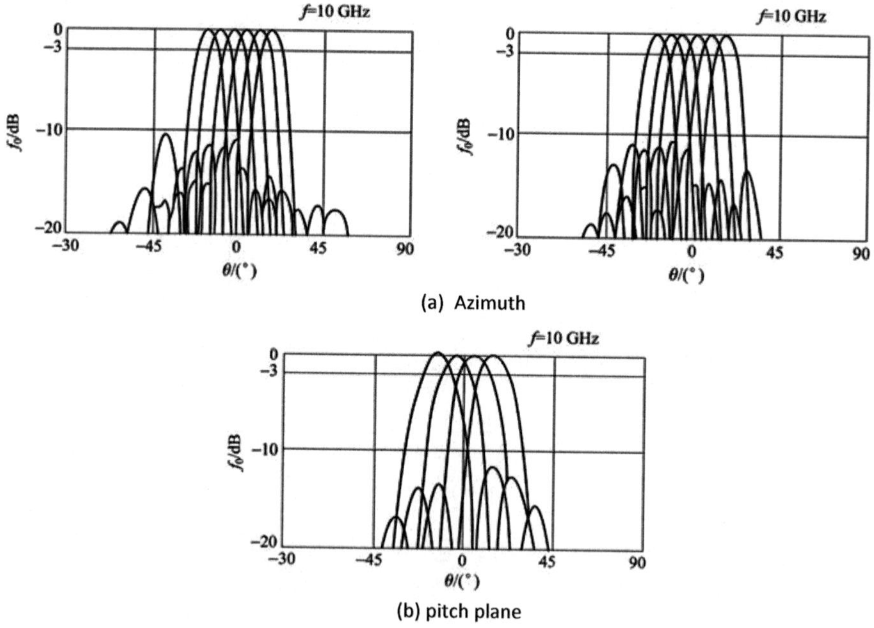


**Fig. 8.29** Two-dimensional radiation beam

level is – 13 dB), and the sidelobe level is below – 10 dB. The two-dimensional multibeam array antenna can be used alone to form a system, which can be used for both transmission and reception. It can also be multiplexed with a large-diameter parabolic antenna to produce a high gain, spatial coverage of multiple beams, large-area selection of mechanical scanning and rapid response, flexibility, and multi-function of electrical scanning. The antenna array adopts the elliptical aperture of the triangular arrangement, which will be the best selection of array arrangement.

### *8.3.4 Dielectric Lens Antenna with Matching Layer*

In the design of a specific two-dimensional antenna array, as the number of beams increases and the operating frequency becomes higher, it will be found that it is difficult to achieve in structure. This is because as the frequency becomes higher, the volume ratio of the lens becomes smaller, but the adapter of the horn of the array port cannot be smaller. However, if the number of array ports is reduced, the number of area array units will be reduced accordingly, which will cause the performance of the antenna to decrease. Especially in the application of electronic systems, the receiving array is required to have a relatively good quality of the beam in the working frequency band, and there is no blur in the direction-finding range (that is, no grating lobe). However, the higher unit gain and effect of aperture focusing pursued by the transmitting array often results in the receiving array having more antenna array units than the transmitting array. Based on these circumstances, the two-dimensional multi-beam dielectric lens is a practical selection of beamforming method. No additional port of array element and the antenna's continuous aperture surface is required in the antenna. Radio waves radiate directly from the port of the dielectric lens. Different from the Luneburg lens, it uses the dielectric lens to replace the curved surface of the array port in the air feed. There also must have a distribution relationship between the spatial feed, which is mainly used for reception. As shown in Fig. 8.30, the antenna array is mainly composed of a spherical beam array, a

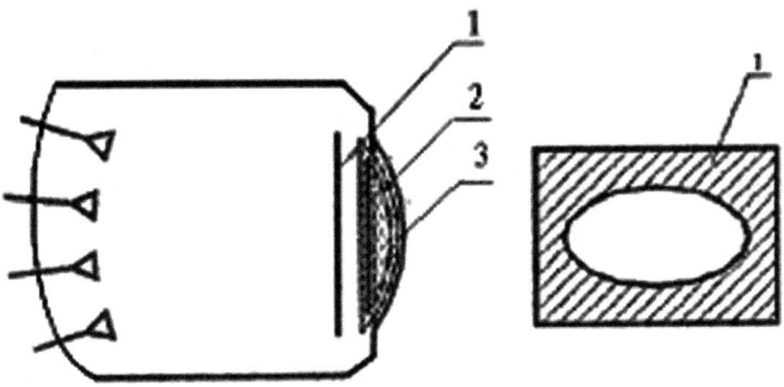


**Fig. 8.30** Schematic diagram of two-dimensional multi-beam dielectric lens structure

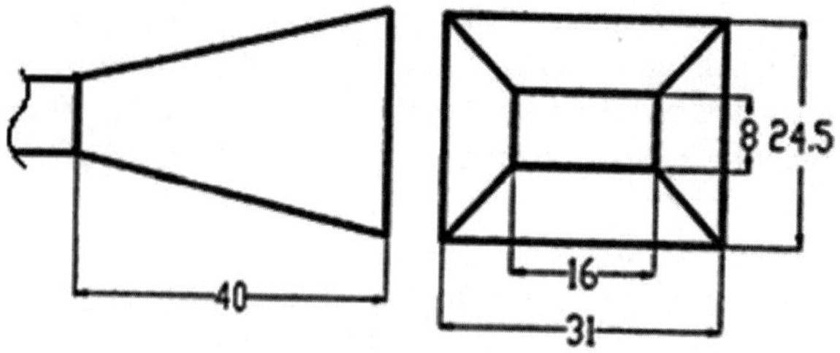


**Fig. 8.31** Dimensions of beam horn antenna

dielectric lens, and a lens cavity. The spherical beam array is composed of a horn of the beam port and a spherical crown. The horns of the beam port are arranged in a honeycomb style and are divided into four rows, each with six radiators. The phase centers of the horns of all beam ports are on a spherical surface with a certain radius, the spherical center is located in the dielectric lens, and the geometric axis of the horn points to the spherical center, as shown in Fig. 8.30. The horn of the beam port is a pyramid horn, and the size of its aperture is determined by the lobe width of the desired pattern. The external dimensions of the horn antenna are shown in Fig. 8.31.

The antenna generates multiple beams instantly in space. So, horns of multiple beam ports are needed, and each horn feed corresponds to a beam. The arrangement of horns of the beam port is shown in Fig. 8.30. The arrangement of the spatial pointing angle of the horn of each beam port is determined by the spatial distribution of the antenna beam cluster. The phase centers of the multiple horns of the beam port are located on a spherical surface with a fixed radius, and the spherical center of the spherical surface is located at a certain point in the dielectric lens. The selection of this radius is mainly subject to the insertion loss of the lens. The larger the radius is, the greater the insertion loss is, vice versa. Radius R is an important parameter in the design, which must be considered comprehensively. The output of the horn at the beam port is an SMA socket (Table 8.5).

The dielectric lens is a planoconvex lens with a transition layer on the surface of the lens body. If a convex dielectric lens is placed in the space feed cavity of the three-dimensional lens, through a certain design and selection of dielectric material and shape, the array port of the quadric surface can be turned into a plane, and the internal dielectric surface is a curved surface. In the case of broadband operation, a special matching layer is made on the surface of the dielectric lens to reduce the additional reflection caused by the medium. As shown in Fig. 8.30, in the cavity of the lens, a metal baffle is installed on the inner side of the cavity, and an elliptical window is formed on the board to form the necessary aperture surface to ensure that the sidelobe level is minimum and there is a sub-beam with a given width. The

**Table 8.5** Theoretical calculated values of antenna coverage space

| Frequency (GHz) | 12 | 13 | 14 | 15 | 16 | 17 | 18 |
|---|---|---|---|---|---|---|---|
| Azimuth plane | 47.1° | 46.4° | 45.7° | 45.2° | 44.7° | 44.2° | 43.9° |
| Pitch plane | 47.2° | 46.0° | 45.0° | 45.5° | 43.5° | 42.7° | 42.1° |

working principle of the dielectric lens can be explained by the geometric optics law. The planar electromagnetic wave of the radiation source is incident on the aperture surface and is focused on a point on the spherical surface that coincides with the phase center of the radiator of a horn through the dielectric lens. Because the axis of the radiator of the horn is consistent with the aspect of the electromagnetic wave beam. In the same way, the axis direction of the radiator that forms the horn of its main lobe is consistent.

The function of the dielectric lens is to converge the incident planar electromagnetic waves on the phase center of an array antenna at the beam port. And the incident waves in different directions. After passing through the lens, the electromagnetic waves will converge on different horns. Therefore, the key to designing a lens is to determine its focusing curve and the dielectric constant of the medium in order to ensure that it is within the operating frequency band. The planar electromagnetic waves in the required coverage airspace are better gathered, and the lens is designed as a transmitting lens according to the reciprocity principle of the transmitting and receiving antenna. According to Snell's law, its expression is:

$$\sin\theta_1 = \sin\theta_2\sqrt{\varepsilon_2/\varepsilon_1} \tag{8.95}$$

In the formula, $\theta_1$-the incident angle of the radio wave $\theta_2$-the refraction angle of the radio wave.

The spherical wave radiated by the antenna's radiating port (horn antenna) enters the inner surface (flat surface) of the lens, where it will be reflected and refracted. And the refracted electromagnetic waves will be reflected and refracted on the outer surface (convex surface) of the lens. For simplicity, the reflection of radio waves on the lens surface is not considered. After refraction by the lens, the spherical wave of the horn becomes a plane wave, which points in the axial direction of the horn. The design results show that the convex surface of the dielectric lens is a quadratic surface. Considering the difficulty in processing, when the refractive index n = 1.6, the graph with both hyperboloids on both sides becomes a plane on one side and a curved surface on the other. From the geometric relationship of the rays on the right side of Fig. 8.32, we get:

$$r_0 \sec\gamma + nS + \left(d - S\cos\beta'\right) = r_0 + nd \tag{8.96}$$

$$y = r_0 tg\gamma + S\sin\beta' \tag{8.97}$$

From the above two formulas:

$$y = \sin\gamma\left[r_0\sec\gamma + \frac{(n-1)d - r_0(\sec\gamma - 1)}{n - \cos\beta'}\right] \tag{8.98}$$

When the value of $\gamma$ is small, $\cos\beta' \approx 1$, there are:

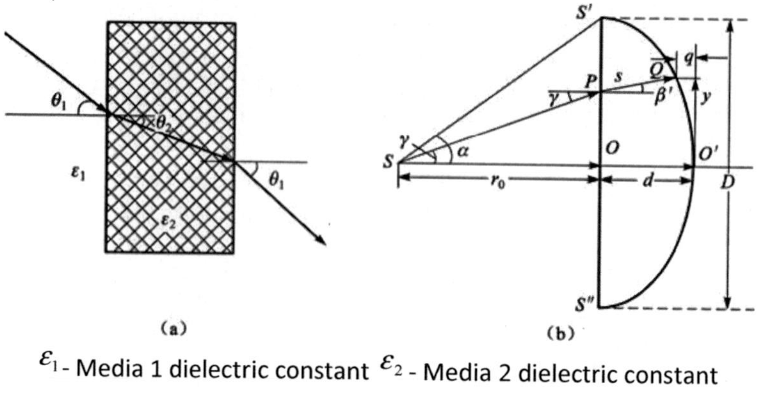


**Fig. 8.32** Schematic diagram of the refraction of radio waves in the medium

$$y \cong \sin\left(r_0 + \frac{d}{n}\right) \tag{8.99}$$

The focal length of the lens is $f = r_0 + d/n$.
For rays on edge, there is a relationship:

$$r_0 \sec\alpha + d = r_0 + nd \tag{8.100}$$

After organization,

$$y = \sin\gamma\left[f - r_0 - \frac{d}{n} + r_0\sec\gamma + \frac{r_0(\sec\alpha - \sec\gamma)}{n(n - \cos\beta')}\right] \tag{8.101}$$

From (8.100), get d = $r_0(\sec\alpha - 1)/(n - 1)$ and substitute into (8.101)

$$y = \sin\gamma\left[f + r_0\frac{n^2 - n - 1}{n - 1}(\sec\gamma - 1) - \frac{r_0(\sec\alpha - \sec\gamma)(1 - \cos\beta')}{n(n - 1)(n - \cos\beta')}\right] \tag{8.102}$$

In order to let the above formula to satisfy the sine law, its last two items must be zero

$$n^2 - n - 1 = 0 \tag{8.103}$$

Then get $n \cong 1.6$. As for the third term, when $\gamma = 0\, or\, \gamma = \alpha$, it is also 0. The value of the third term at the points between the center point O of the dielectric lens and the edges $S'$, $S''$ is not large. So,

$$y = \sin\gamma\left[f - \frac{r_0(\sec\alpha - \sec\gamma)(1 - \cos\beta')}{n(n-1)(n - \cos\beta')}\right] \tag{8.104}$$

If the sine law is accurately met, the curved surface of the lens can be represented by a circular arc. For lenses with parameter ratios D/f = 1, 3/4, 1/2, etc., the lens profile is calculated according to the above formula, and the percentage difference between the coordinate values of the two curved surfaces is not large. It is known that three points can be used as a circle. Rotate the circle to obtain a dielectric spherical crown. Obtain the width D of the aperture according to the width of the pattern, that is, determine the two edge points of the lens, and the third point must calculate the lens thickness d, which cannot be derived:

$$d = \frac{D}{n-1}\left[\frac{f}{D} - \sqrt{\left(\frac{f}{D}\right)^2 - \frac{1}{4}}\right] \tag{8.105}$$

Although this will introduce a certain phase error, as long as the spherical surface is properly selected, the phase difference can be controlled within a relatively small range so that the electrical performance of the antenna is satisfied. According to the lens formula above, the structure size of the lens is calculated, the height of the spherical crown (thickness of the medium) t = 45.6 mm, and the radius of the sphere is R = 148.5 mm. In order to improve the spatial matching of the dielectric lens, reduce the reflection and refraction of electromagnetic waves on the lens surface due to the discontinuity of the medium, thereby improving the electrical performance of the lens. The inner and outer surfaces of the lens are coated with a transition layer. The dielectric constant of the transition layer is between the dielectric constant of the medium and the air. The specific value is determined by the following formula:

$$\varepsilon_e = \sqrt{\varepsilon_1 \cdot \varepsilon_2} \tag{8.106}$$

Because it is not easy to find the medium whose dielectric constant completely meets the needs of the transition layer, it is achieved by drilling holes on the inner and outer surfaces of the lens. The depth, size, number, and distribution of the holes determine the dielectric constant of the transition layer. The greater the proportion of the sum of the volume of all holes to the volume of the entire transition layer is, the smaller the dielectric constant is, and vice versa. The variation range is 1 ~ ε (ε is the dielectric constant of the material of the medium of the transition layer before drilling). The dielectric lens is made of polystyrene, and its dielectric constant is 2.55 so that the dielectric constant of the transition layer can be determined to be 1.6. The distribution of the holes in the lens transition layer: On the side of the plane, the holes are arranged in a regular horizontal and vertical direction, the size and density of the holes are the same, the depth is one-quarter of the center wavelength, and the center axis is perpendicular to the lens plane; on the side of the lens sphere, the holes are distributed in a circle, and the radius of the circle increases gradually. The

**Table 8.6** Calculated values of dielectric lens antenna lobe width

| Frequency (GHz) | 12 | 13 | 14 | 15 | 16 | 17 | 18 |
|---|---|---|---|---|---|---|---|
| $\theta_e$ | 9.6° | 8.9° | 8.2° | 7.7° | 7.2° | 6.8° | 6.4° |
| $\theta_h$ | 15.2° | 18.7.0° | 13.0° | 12.1° | 11.4° | 10.7° | 10.1° |

number of holes increases accordingly. The holes in each circle are evenly arranged. The size and depth of all holes are the same. The center axis of the holes all points to the center of the sphere. The lens antenna does not adopt the port of array element-coaxial planar array form, but the beam horn directly radiates into space through the lens. Therefore, in order to control the beam lobe width and sidelobe level, in front of the dielectric lens, a metal partition is installed inside the metal shell of the lens, and an elliptical window is provided on the plate to form the necessary aperture surface to ensure the sidelobe level. Under the minimum condition, there is a beam of a given width, and the shape of the metal partition is an elliptical window. Its size is $D_e = 156\,\text{mm}$, $D_h = 84\,\text{mm}$:

θe-Lobe width of antenna E plane (azimuth plane) θh-Lobe width of antenna H plane (pitch plane).

$D_e$-The size of the long axis of the ellipse window, $D_h$-The size of the short axis of the ellipse window λ-Wavelength.

Using the previous aperture field formula, the 3 dB lobe width of the antenna can be calculated (Table 8.6).

In order for the lens to exert its normal characteristics, its specific structure and size are determined by engineering design. In order to reduce the reflection and scattering of the lens cavity as much as possible and cause an impact on the electrical performance of the antenna, a layer of microwave absorbing material is pasted on the surface of the cavity. The wave absorbing material uses corrugated rubber material.

The dielectric lens antenna is composed of a beam port horn, spherical metal crown, lens cavity, and dielectric lens. The installation of the metal spherical crown feeder is the same as the previous section, which is divided into four layers, and each layer has multiple holes. The beam port's horn is positioned by the normal hole and is connected to the spherical metal crown through it. The dielectric lens is installed on the lens flange, and the lens is a spherical crown. There are many small normal holes on the plane and spherical side to form a transition layer. The lens cavity is cylindrical, which provides a good electromagnetic shielding environment for the antenna, and the inner cavity wall is covered with a wave absorbing material.

The dielectric lens multi-beam antenna can simultaneously generate multiple overlapping beams in space. If divided into 4 layers, each layer has multiple beams. The entire beam can cover 450 (azimuth) × 450 (pitch) airspace. The typical radiation pattern is shown in Fig. 8.33. Due to the role of the elliptical grating hole, the width and gain of the beam can be controlled. The gain of the antenna is improved after the dielectric lens is used at the array port, which is actually equivalent to a continuous situation.

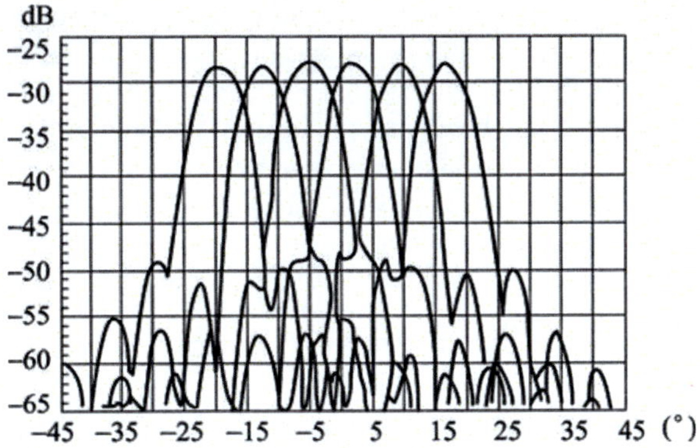


**Fig. 8.33** Typical azimuth pattern

The optical beamforming method is explained above. The use of transmission lines to link phase shifters, power splitters, and couplers to form a multi-beam network is called a circuit beamforming method. There are two main types, Blass and Butler matrix. Blass is far more flexible than the butler matrix, but because of the influence of the load of the transmission line terminal, the loss is also large. Compared with the Butler matrix, the application range of the Blass matrix is limited due to its inherent loss. The Blass matrix consists of the same series of traveling wave feeders. These feeders are connected to the line array through another set of feeders. These two sets of transmission lines are connected by directional couplers at their intersections. The signal added to the terminal port can be transmitted to the terminal load along the feeder. At each intersection, a small signal is coupled into each unit line, and the corresponding radiating unit is excited. The distance difference between the input and each unit controls the direction of the radiation beam. The coupling coefficient determines the aperture illumination. Because of the traveling wave characteristics of the network, the input is well matched and the beam position will be scanned with frequency. It is a real time-delay network with the ability to widen the frequency band. Couplers are mainly used in the network, and its directivity ensures the isolation of the beam ports. The design process of the network includes the appropriate phase and coupling degree to meet the required beam position. For a given transfer function, these can be obtained by calculation so that the terminal load consumes the least energy. In fact, the range of available coupling values is strictly limited. It corresponds to the intersection level waveform of the beam. Butler matrix is another circuit beamformer. It is composed of a series of connected fixed phase shifters and 3 dB bridge couplers. The matrix generates N orthogonally spaced beams with an intersection level of about − 3.0 dB. It has good input standing waves and beams isolation. However, the number of input ports or output ports is limited to the nth power of 2. It is difficult to optimize the beam shape. The beam will also be scanned with the frequency response. In order to maintain the orthogonality of the beam, the beam is narrowed, and the coverage is reduced. However, the use of a coupler and the terminal using asymmetric bridges and orthogonal beam synthesis can improve its flexibility. The size of the matrix is limited by manufacturing tolerances, which must be less than the size of the minimum phase shift element. Using microstrip technology on the dielectric substrate, the matrix can be composed of various transmission media, including microstrip and high-power waveguides. As the frequency increases, the shape of the beam and the level of the cross point remain unchanged, while the maximum point of the beam contracts inward, which is a specific distribution form of the general phase array antenna beam.

## 8.4 Broadband Phased Circular Array Antenna

In the wideband frequency range, the multi-beam formed by the lens system has the same beam pointing direction. As the frequency increases, the level of the intersection of adjacent beams becomes lower. In the multi-beam generated by the feeder network in the form of a Butler matrix, the beam direction changes with frequency. When the frequency increases, the same beam moves closer to the center beam, and the level between adjacent beams remains unchanged. But they all have one thing in common. Once the circuit is selected, all antenna characteristics are cured accordingly. At the same time, each beam corresponds to an output port, and how many beams correspond to how many output ports. In order to produce a more extensive beam coverage, the number of output ports is reduced, and the beam control method is flexible, which is the characteristics of a broadband phase array antenna.

### *8.4.1 Broadband Phased Circular Array Antenna*

With the development of technology, a series of new requirements are put forward for the functions of electronic warfare and the ELTNT/ECM system. It has the ability to respond quickly and accurately, the ability to simultaneously monitor and track multi-targets flying at high speed, and the ability to complete ELINT/ECM functions simultaneously and alternately. However, for a phase array antenna, this antenna changes the amplitude and phase of each element of the antenna through electronic control commands and uses the contribution of the array antenna to realize the command control of beamforming, splitting, and pointing. The main advantages are:

(1) Fast response speed: microsecond order and flexible angle coverage;
(2) Strong controllability of the beam: the beam can be divided and combined, pointing at any control;
(3) High reliability: no mechanical rotating parts. When an individual unit fails, the entire antenna can work as usual. Even if the fault element accounts for 5%, its performance is only slightly reduced;
(4) Combined work can be achieved: divided into sub-arrays to deal with the same goal or multiple goals in different forms.

Among the phase array antennas mainly used to meet the radar function, most of them use a planar array. The angular coverage area is a spherical crown, and the frequency band is generally narrow, requiring a large amount of phase change and a large and complex feed network. For engineering applications, due to the requirements of wide frequency opening and wide-angle coverage, circular phase array antennas are used. Its special advantage is all-around coverage. The array element is installed on the circumference of a cylindrical surface. It has the characteristics of full symmetry. It is excited by partitions and realizes the electric sweep in the form of stepping to cover the azimuth of 360°. The circular coplanar array only excites N/4 units in the 900 range at a time. The phase compensation of each array element and the phase shift provided by the delay line phase shifter are complementary and have nothing to do with the frequency. Therefore, the antenna is wide open in frequency. It can scan once in all directions in the order of microseconds. The diversified scanning methods can not only realize single beam scanning but also implement multi-beam scanning simultaneously. The feed network is simplified. Due to the full symmetry of the circular coplanar array, the compensation measures of partition excitation and small phase shift can be used to make the feed network more simplified than the planar array or linear array.

The N array elements of a circular coplanar phase array antenna are evenly distributed on the circumference of the same cylindrical surface. Using different feed networks, they can be divided into a single beam scanning system and a multi-beam scanning system. The array composition is shown in Fig. 8.34. The whole antenna network is composed of five parts: radiating element, switch network, phase shift network, power distribution network, and a corresponding logic control circuit. Under the control of the logic circuit, the feed network is assigned to each radiating element with a given amplitude and phase according to a certain procedure. According to the principle of array antennas, these exciting antenna elements together form a wavefront with a certain amplitude and phase relationship, thereby synthesizing the desired direction in space. The entire circular array is excited by partitions, and the excitation area can reach 180 degrees each time. Taking the unit on the quarter circle as an example, through a specially designed controllable phase shift network, each excited radiating unit is provided with the necessary phase so that the compensation of the wavefront, which is originally an arc becomes a straight wavefront. The amplitude of each radiating element is also weighted in the form of Chebyshev distribution by the controllable attenuation network designed in the phase-shift network. Therefore, the antenna element on the excited quarter circle can synthesize a pattern with strong directivity and low sidelobe level in space.

The micro-scanning of a single beam in a small range is achieved by controlling the small-scale activities of the various phasors in the phase-shifting network. The 360° scannings of the beam in a wide range is realized by controlling the switch network, and each step can move 360°/N. Because the speed of the switch is extremely fast, the entire range of 360° can be scanned in the order of microseconds. As long as the components in the array are broadband, the broadband operation of the entire antenna array can be achieved. If $\mathrm{N} = 32$ is taken in the entire beam scanning array, there are eight radiating elements on a quarter of the circle, that is, eight antenna array

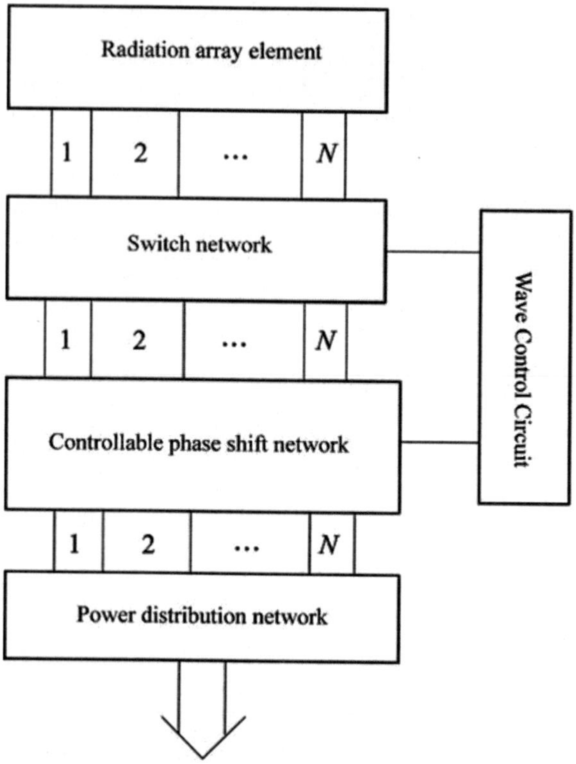


**Fig. 8.34** Block diagram of the circle array

elements are excited at a time, and 32 stepped beams can be formed in turn to cover an azimuth of 360° that each beam covers an azimuth range of 360°/32 = 11.25°. If the first beam is generated by unit 1–8, the second beam is generated by unit 2–9, the third beam is generated by unit 3–10, and the 32nd beam is generated by unit 32–7. Due to the total symmetry of the circular coplanar array, the required pattern function only needs to be derived from the N/4 elements of the circular array on the quarter circle. As shown in Fig. 8.35, suppose that the circular coplanar antenna array is composed of N elements, and the exciting elements are No. 0, 1, 2, ... (N − 1)/4 units; incoming waves enter in the θ direction; each unit is distributed on a circle with $O_1$ as the center and R as the radius, and 0, 1, 2, 3...i...(N − 1)/4 are the phase centers of the units located at direction $\alpha_0, \alpha_1, \alpha_2 \ldots \alpha_i \ldots \alpha_{(N-4)/4}$ respectively. When unit No. 0 is used as the phase reference, the lag phase of unit i in the direction θ is:

$$\Phi_i = k(Pt - Qt) = k\{[R - R\cos(\alpha_i - \theta)] - [R - R\cos\theta]\}$$

$$\Phi_i = kR[\cos\theta - \cos(\alpha_i - \theta)] \tag{8.107}$$

When N/4 units are excited, the pattern function is:

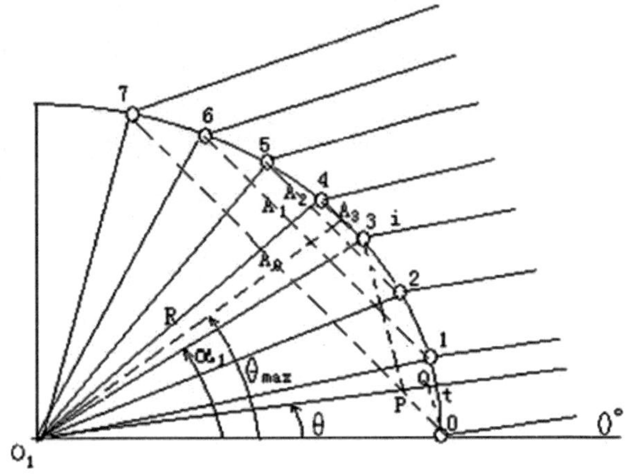


**Fig. 8.35** Coordinate relationship of direction graph function

$$F(\theta) = \sum_{i=0}^{N/4-1} f_i(\theta)e^{-jkR[\cos\theta - \cos(\alpha_i - \theta)]} \tag{8.108}$$

In the formula, $f_i(\theta)$ is the unit direction function of the i-th array element. If the E-plane horn antenna array is used:

$$f_i(\theta) = [1 + \cos(\alpha_i - \theta)]\frac{\sin\left[\frac{1}{2}ka\sin(\alpha_i - \theta)\right]}{\frac{1}{2}ka\sin(\alpha_i - \theta)} \tag{8.109}$$

In the formula, $\alpha_i = 2\pi/N \cdot i$.

a is the size of the caliber of the E-face horn, $a \leq \frac{\lambda_{\min}}{2} - 2$.

k is the number of waves of free space, $k = \frac{2\pi}{\lambda}$.

In order to make these exciting units in the direction of Q, the antenna array can obtain a larger directional gain of the antenna. It needs to use a phase shifter that is mathematical delay type to calibrate the wavefronts of these excitation elements to be flat or nearly flat. To this end, a corresponding phase shift must be added to the i-th unit:

$$\varphi_i = kR[1 - \cos(\alpha_i - \theta_{\max})] \tag{8.110}$$

So, the pattern function of the stimulated antenna array element is:

$$F(\theta) = \sum_{i=0}^{N/4-1} f_i(\theta)e^{-jkR[\cos\theta - \cos(\alpha_i - \theta)] + jkR[1 + \cos(\alpha - \theta_{\max})]} \tag{8.111}$$

The directional gain of the stimulated antenna element is:

$$\mathrm{G} = 10\,\mathrm{lgD}$$

$$D = \frac{2[F(\theta_{\max})]^2}{\int_{\theta_{\max}}^{\theta_{\max}+\pi} [F(\theta)]^2 \sin\theta d\theta} \tag{8.112}$$

The above pattern function is derived based on the ideal situation. In engineering, because each feeder channel is objectively affected and disturbed by component performance, system, or accidental factors, the required amplitude and phase cannot be obtained, which requires human intervention (weighted) and random close-up (i.e., jamming) factors. This is defined as follows: let:

AF(i) is the weighted factor for the amplitude of the i-th unit;

C(i) is the weighting factor for the phase of the i-th unit;

NF(i) is a random perturbation factor for the amplitude of the i-th unit;

NC(i) is a random perturbation factor for the phase of the i-th unit.

After considering these factors, the synthesized pattern function of the exciting unit is:

$$F(\theta) = \sum_{i=0}^{N/4-1} f_i(\theta) \cdot NF(i) \cdot AF(i) \cdot e^{j\{kR[\cos(\alpha_i-\theta)-\cos\theta]+C(i)\varphi_i+NC(i)\}} \tag{8.113}$$

Among them, fi(θ) is expressed by formula (8.109), and expressed by formula (8.110). Generally, AF(i) is weighted according to Chebyshev distribution in order to reduce the sidelobe level; C(i) is calculated according to different values of 0–1.2, and then the best value is taken; NF(i) is calculated from the values of 80–120% of different random quantities; NC(i) is calculated from the values of $-20° \sim +20°$ of different random quantities. According to the derived formulas (8.111), (8.112) and (8.113), calculations were made in different situations:

(1) The amplitude and phase are not calibrated with any weighting;
(2) The amplitude is not weighted, and the phase is weighted for calibration;
(3) The amplitude and phase are weighted and included in the random perturbation;
(4) The pattern and gain of the above three cases with different array element diameter and circular array radius.

It can be seen from the calculation results that, under a single beam scanning system, the pattern of the N/4 excited units occurs. Although the state is different, the regular main beam can still be obtained under the constraints of the array element's aperture and the array element's radius. The directional gain is above 10dBi, and the sidelobe electrical average is below $-$ 13 dB, especially if the wavefront phase is calibrated. Taking f = 2–4 GHz as an example, the simulated conditions are:

(1) Working frequency f = 2, 3, 4 GHz
(2) The caliber of the E side of the horn a = 20, 30, 35, 40 mm
(3) Weighted Ci of the calibrated phase = 1.0
(4) Weighted AF(i) of unit amplitude = 0.38, 0.58, 0.84, 1,0.84, 0.58, 0.38;
(5) Total number of units: N = 32;
(6) Root mean square value of phase perturbation, delta = 100

**Table 8.7** Calculated values for circular array antennas

| a, R | | f(GHz) | | |
|---|---|---|---|---|
| | | 2 | 3 | 4 |
| a = 20 mm<br>R = 112.23 cm | $2\theta_{0.5}$<br>SL(dB)<br>G(dBi) | 54°, 54°, 52°<br>/ / – 27<br>10.8, 10.82, 10.73 | /<br>/<br>/ | 29°, 28°, 28°<br>– 22, – 20.4, – 16<br>11.8, 11.84, 11.81 |
| a = 30 mm<br>R = 168.34 cm | $2\theta_{0.5}$<br>SL(dB)<br>G(dBi) | 40°<br>– 17.6<br>10.14 | 26°<br>– 23.0<br>12.35 | 20°<br>– 20.0<br>18.32 |
| a = 35 mm<br>R = 188.74 cm | $2\theta_{0.5}$<br>SL(dB)<br>G(dBi) | 36°<br>– 18.0<br>11.40 | 22°<br>– 22.4<br>12.7 | 18.0°<br>– 19.2<br>13.92 |
| a = 40 mm<br>R = 218.7.25 cm | $2\theta_{0.5}$<br>SL(dB)<br>G(dBi) | 32°<br>– 18.0<br>11.6 | 20°<br>– 23.0<br>18.24 | 16°<br>– 18.0<br>18.7.34 |

(7) Root mean square value of amplitude perturbation: df = 0.15.

The calculation results are shown in Table 8.7. Theoretical calculation results show that after the circular array antenna is properly weighted and calibrated for the amplitude and phase of each stimulated unit by means of partitioned excitation, even if it encounters random amplitude and phase perturbations, it can still maintain the ideal shape of the main beam, gain and sidelobe levels.

According to the above ideas, calculation methods, and results, a circular phased array working in the S-band is used. Many unit horn antennas are shared uniformly on the circumference. This kind of unit antenna is a feed-back ridge horn antenna. The horizontal polarization of the array plane is formed with its E-plane array method, as shown in Fig. 8.36. The use of multiple broadband SP4T switches to create a switch network to control the beam's switching solves the unit antenna uniform design. The phase-shifted weighted network is composed of 2 bits of the switch delay line and corresponding attenuation network. Of course, the amplitude weighting method can also be used. The power distribution network is realized by using an eight-way power splitter composed of five two-way power splitters. The antenna array is

**Fig. 8.36** Broadband circular array phase array

excited by partitioning and ingeniously combined phase compensation for delaying the phase segment to ensure the instantaneous wave array is a straight line in the azimuth plane, which provides the broadband characteristics of the array. In the network arrangement and debugging of active components, each microwave channel is optimized through engineering methods simultaneously. Its multiple channels have the same amplitude and phase characteristics under the same conditions. Under various scans, the situation is almost the same for the power distribution network in the S-band range, and their standing wave coefficients are the same. Since the beam is pointed in the radial direction, the coupling between the array elements has a relatively small effect on the radiation characteristics.

Figure 8.37 shows the reflection loss curve at the output of the antenna. From the standing wave conversion formula, the maximum standing wave is less than 2. Figure 8.38 shows the antenna beamwidth and sidelobe level values. The width of the scanning beam is 18°–36°. The pitch plane's beamwidth is 35°–70°, and the sidelobe level is greater than 13 dB. Due to its characteristics, the antenna array has all-around angle coverage, wide operating frequency, fast response speed, multiple scanning modes, and beam broadband controllable characteristics.

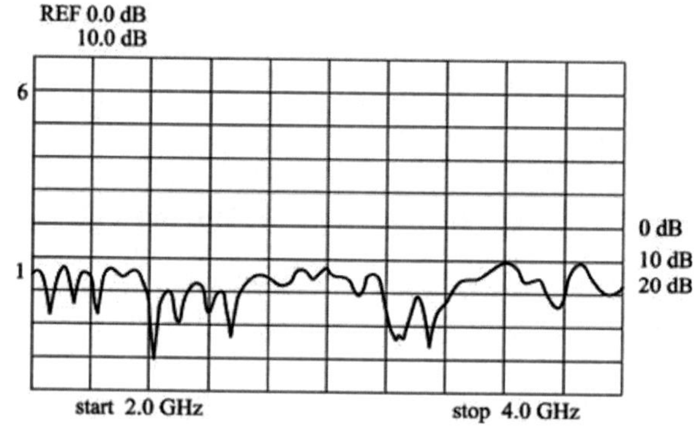


**Fig. 8.37** Reflection loss curve at the output of the antenna

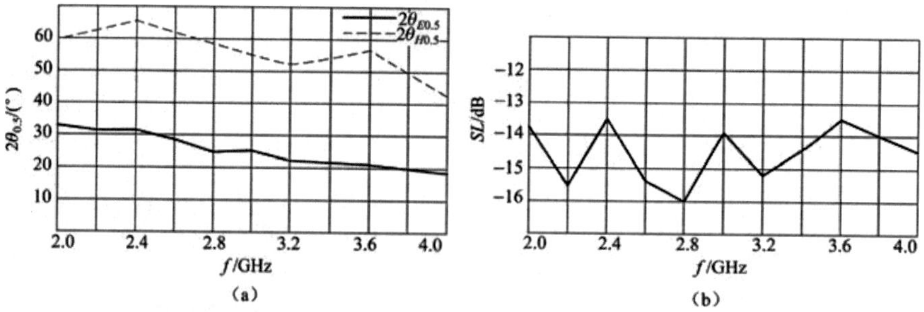


**Fig. 8.38** Radiation pattern and sidelobe level of broadband phased array antenna

### 8.4.2 Broadband Phased Transmit Array Antenna

The previous section introduced the receiving circular phase array, which is mainly determined by the device's characteristics. For the dual-ridge waveguide transmission system to transmit a wide phased array, the main transmission and conversion components must be broadband to withstand the high-power operation. It has three forms of feeding. The first type is a phased array with centralized feeding (also called a passive array). And the first type is a phased array (also called a passive array) with centralized feeding. A broadband phase shifter follows each array element antenna to form a unified input port through the power division network. Connect a high-power source or power amplifier to the input to produce a jamming beam with high equivalent radiated power. The phase shifter controls the scanning of the beam. It does not have many requirements for the consistency of the power amplifier. The second type of array is that each array element is followed by a phase shifter, which divides the entire array into several sub-arrays. Each sub-array is followed by a medium-power amplifier, followed by one or several input ports. It has certain requirements for the power amplifier's consistency, called a hybrid or semi-active phased array. The third case is that each array element is followed by a low-power amplifier and phase shifter and then forms an input terminal or several input ports. However, it has certain requirements for the consistency of the amplifier. But the power amplifier is placed directly behind the array element, and the energy loss is reduced, as shown in Fig. 8.39. This section introduces the first type, which is called a passive phased array. It can be seen from the composition of the broadband phased array system that the broadband phased transmit antenna array is the core part. By independently driving the oxygen phase shifter behind each radiating element, fast scanning of the beam in the azimuth plane is realized. The wave's speed is controlled flexibly, and a higher gain is obtained in the antenna aperture. The input port is a standard dual-ridge waveguide interface. The entire array's insertion loss is small, so it is easy to replace the simple beam scanning system of mechanical switches currently used. Simultaneously, the phased antenna array is taken as a sub-array, and multiple identical sub-arrays can make the antenna array work with several beams. It is also possible to make the array work with one beam. Combining the array is flexible, dealing with multiple targets simultaneously or against a particularly threatening

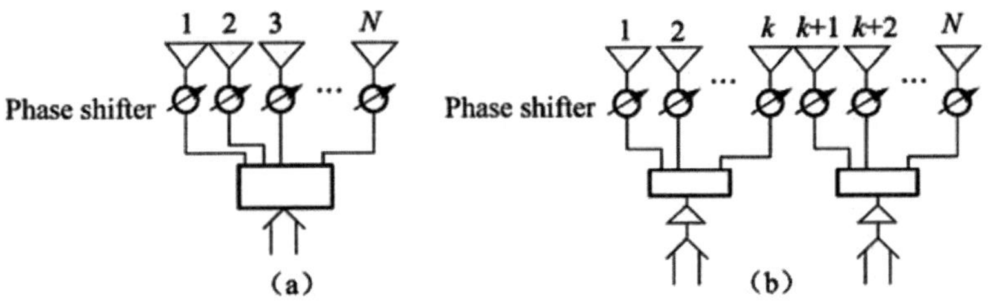


**Fig. 8.39** Several structural forms of phase array antenna

target with a very high jamming power. Multiple sub-arrays can work simultaneously on multiple different frequencies. Combining different jamming patterns can achieve effective jamming to the target. The theory of phased array antennas is well known. And the radar system of this system has been put into practical use. To achieve the array's broadband characteristics, the components of the antenna array should work with broadband. There must be broadband radiation elements, broadband connection lines, broadband phase shifters, and broadband power distribution networks.

When feeding n units in Fig. 8.39a with equal amplitude in-phase, the maximum direction of their combined field is in the direction of $\theta = 0$, that is:

$$E(\theta) = E_0 \sum_{n=0}^{N-1} e^{-jnkd\sin\theta} \tag{8.114}$$

In the formula:

| | |
|---|---|
| $k = 2\pi/\lambda$, $\lambda$ | Working wavelength. |
| d | Distance between adjacent unit antennas. |
| N | Number of elements. |

When each radiation element is given a certain amount of phase shift, the beam is scanned. The expression of this field is:

$$E(\theta) = E_0 \sum_{n=0}^{N-1} e^{-jnkd\sin\theta + j\phi_n} \tag{8.115}$$

To shift the array to the normal direction $\theta_s$, the corresponding phase shift of each array element should meet the following conditions:

$$\phi_n = \frac{2\pi}{\lambda} nd \cdot \sin\theta_s \quad (n = 0.1.2.3\ldots N-1) \tag{8.116}$$

The expression of the field is:

$$E(\theta) = E_0 \frac{\sin\frac{N}{2}(\psi - \phi)}{\sin\frac{1}{2}(\psi - \phi)} e^{-j\frac{N-1}{2}(\psi - \phi)} \tag{8.117}$$

In the formula:

$$\psi = \frac{2\pi}{\lambda}\sin\theta; \quad \phi = \frac{2\pi}{\lambda}\sin\theta_s$$

In an array antenna, as long as the phase shift amount of each phase shifter can be changed, the array antenna's beam direction can be changed. When the phase shift amount between adjacent array elements is $\phi$, the beam is directed $\theta_s$. The normalized directional function is:

$$F(\theta) = \frac{1}{N}\frac{\sin\left[\frac{N}{2}\left(\frac{2\pi d}{\lambda}\sin\theta - \phi\right)\right]}{\sin\left[\frac{1}{2}\left(\frac{2\pi d}{\lambda}\sin\theta - \phi\right)\right]} \tag{8.118}$$

Continuously changing the phase shift amount of the phase shifter behind each radiation element will form a continuous wave velocity scanning in space. When $\theta$ approaches $\theta_s$, there is the following relationship:

$$\sin\theta - \sin\theta_s \approx (\theta - \theta_s)\cos\theta_s$$

$$F(\theta) \approx \frac{1}{N}\left|\frac{\sin\left[\frac{N\pi d}{\lambda}(\theta - \theta_s)\cos\theta_s\right]}{\sin\left[\frac{N\pi d}{\lambda}(\theta - \theta_s)\cos\theta_s\right]}\right| \tag{8.119}$$

From the formula (8.119), when the beam scans to the $\theta_s$ direction, its half-power width is approximate:

$$[2\theta_{0.5}] = \frac{51\lambda}{Nd}\frac{1}{\cos\theta_s}(\circ) \tag{8.120}$$

The characteristics of the array function are discussed above. In practice, the radiating element used is directional. After considering the directional function of the element radiating antenna, the synthesized array function is:

$$F_0(\theta) = f_1 \cdot F(\theta) \tag{8.121}$$

Among them, f1 ($\theta$) represents the array element radiation's directivity function, and it is quite tedious to calculate its expression accurately. In engineering, it is approximated as a pattern function of a horn antenna with an open aperture. For wide-angle scanning, the unit antenna's radiation pattern lobe is required to be wide, and all scanning beams are within the main beam of the unit pattern. The pattern function of the phased array antenna is shown in Eq. (8.123). When the array antenna is phase-swept, the beam becomes wider after the $\theta_s$ shifts to 0°, and the gain decreases relatively. We know that the first sidelobe of an equally spaced uniformly fed array antenna is on the order of − 13 dB. To reduce the sidelobe waveguide, weighted feeding methods such as:

(1) Amplitude addend: Make the array amplitude change in the form of the slope.
(2) Density addend: the distance d between the center of the antenna array and the array element is smaller, while the distance between the outer parts gradually becomes larger.
(3) Phase addend: Change the amount of phase shift in each channel to reduce the sidelobe level.
(4) Mixed addend: weight the amplitude and phase in each channel.

$$f_1(\theta) = \frac{1+\cos\theta}{2}\frac{\sin\left(\frac{\pi a}{\lambda}\sin\theta\right)}{\frac{\pi a}{\lambda}\sin\theta} \quad (8.122)$$

In the formula:

a aperture in the direction of unit horn antenna array.

$$F(\theta) = \left|\frac{1+\cos\theta}{2}\frac{\sin\left(\frac{\pi a}{\lambda}\sin\theta\right)}{\frac{\pi a}{\lambda}\sin\theta} \cdot \frac{\sin\left(N\cdot\frac{\pi d}{\lambda}\sin\theta - N\phi\right)}{N\sin\left(\frac{\pi d}{\lambda}\sin\theta - \phi\right)}\right| \quad (8.123)$$

The scanning beam of the phased array antenna was discussed above. The continuous and constant phase shifter is very troublesome to control, and the phase shift value is easy to drift. The insertion loss is also relatively large. Digital phase shifters are generally used in engineering. Its structure is relatively simple and can carry large power. At the same time, the digital phase shifter provides great convenience for computer control. Its phase shift can only be an integer multiple of a certain minimum phase. For one n-bits phase shifters, the minimum phase shift is:

$$\Delta = \frac{360^\circ}{2^n} \quad (8.124)$$

The phase shift of basic position:

$$0, 2^0\Delta, 2^1\Delta\ldots\ldots, 2^{(n-1)}\Delta \quad (8.125)$$

It can constitute 2n phase shifts. When n is larger, $\Delta$ is smaller. That is, the difference between adjacent phase shifts becomes smaller. When $n \to \infty$, the digital phase shifter becomes a continuous analog phase shifter. Of course, the wideband phase shifter developed in the project, the more digits, the greater the insertion loss. At the same time, through the quantization of the required phase shift. In n = 4, the calculation result is compared with the calculation result in a continuous phase shift. Due to the digital phase shifter quantization, there is a minimum error in the phase shift. But it has little effect on the beam pointing. The sidelobe has a certain increase; the gain has decreased. The quantitative calculation formula is as follows:

$$E_a(\theta) = \sum_{n=0}^{N-1} \exp\left[-j\left(nkd\sin\theta + \overline{\phi}_n\right)\right]\cdot f_1(\theta) \quad (8.126)$$

Due to the use of a digital phase shifter, the phase shift of adjacent radiation elements is:

$$\varphi_n = nkd\cdot\sin\theta_0$$

$$\phi(n) = \text{mod}\left[nkd\sin\theta_0, 360^0\right] \tag{8.127}$$

$$d\phi(n) = \frac{\text{mod}\,[10\phi(n), 10\Delta]}{10}$$

In the formula:

mod(a,b) is the remainder after dividing by b.

The modified phase shift is:

$$m\phi(n) = [\phi(n) - d\phi(n)] + \Phi\left[d\phi(n) - \frac{\Delta}{2}\right] \cdot \Delta$$

$$\Phi(x) = \begin{cases} 1 & x \geq 0 \\ 0 & x < 0 \end{cases} \tag{8.128}$$

There are several basic methods for quantizing phase shifters: carry method, rounding method, rounding method, and shift weighting method. In the above calculation, the rounding method is used. Although all the array elements are different from the theoretical analog phase shift, multiple units' statistical results have the same slope of the equal phase plane. The structure of the broadband phased array antenna is shown in Fig. 8.40. Because the array antenna is used as a transmitter, the entire power transmission and radiation unit can withstand high power. The connection part uses double-ridged waveguides, and multiple units are compactly arranged in a linear subarray. Because the array antenna is used as a transmitter, the entire power transmission and radiation unit can withstand high power. The connection part uses double-ridged waveguides, and multiple units are compactly arranged in a linear subarray. To effectively solve the connection difficulties and structural realization, adjacent waveguides are alternately staggered up and down in the form of

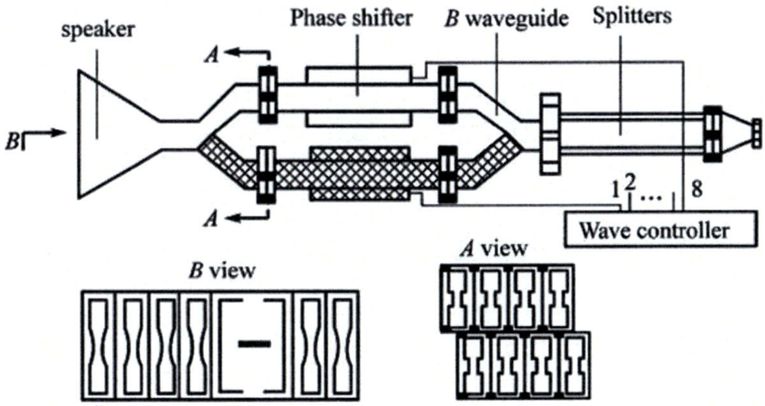


**Fig. 8.40** Structure diagram of broadband phase array sub-array antenna

elbows to make room for the connecting flange. A phase shifter is connected between each radiation unit and the power splitter port. The corresponding curved waveguide is concentrated at the output end of the power splitter. The power splitter's input end is connected with the high-frequency high-power transmitter's output waveguide port, forming a complete transmitting phased array system. The entire antenna array consists of two sub-arrays with dual-port input in the form of Fig. 8.40.

The antenna has a wide working frequency band, and each radiating unit can work under a relatively large power. The double-rib waveguide is used for the feeding part. The radiation unit is an H-plane fan-shaped horn antenna. To prevent the grating lobe from appearing in the visible region, the relationship between adjacent radiating elements must satisfy the following formula:

$$d \le \frac{\lambda_{\min}}{1 + \cos\theta_0} \tag{8.129}$$

In the formula:

$\theta_0$ Maximum scanning angle.
$\lambda_{\min}$ Minimum operating wavelength.

To make the antenna work in the frequency range of 8–16 GHz ($\lambda_{\min} = 18.75$ mm) and $\theta_0 = \pm 45°$, select: d $\le$ 10.98. The standard double-ridge waveguide 24JS7500 has an internal size of 17.55 × 8.15 mm. The E-plane size is 8.15 mm, and the wall thickness of the waveguide is 1 mm. The actual array element spacing is 10.15 mm. Considering that there is a certain pitch area coverage in the space, there is a certain requirement for the horn antenna's H-plane dimension. The E-plane of the cell is used for array formation, and the size of the H plane determines the pitch plane of the array. The in-plane dimension of the horn antenna H is a = 60 mm. Given the specific application, a ferrite phase shifter of digital waveguide type is used. The number of phase shifts is n = 4 bits. This phase shifter is composed of several sections of hollow ferrite material. They are glued into a single basic piece, with a lock wire in the center, matching transformers are added at both ends of the wire, and then it is assembled into a waveguide housing that fits closely with it. Matching sections with double waveguides are installed at both ends of the waveguide, and the peak power capacity is greater than 100 W. Provided the required information for each phase-shifting branch, the power divider used in the antenna is required to be carefully designed and manufactured. If it is implemented with a circuit in the form of a stripline, it cannot withstand large power, and the loss is large. The power divider developed in this way is space-form. Use an E-plane radial waveguide horn body to feed multiple adjacent waveguide ports assembled with the aperture to achieve power distribution. It has good electrical characteristics and a compact structure, and no other coaxial waveguide conversion is needed to reduce the power consumption of the phase shifter and power splitter. Since the waveguide space-fed power divider uses the field's principle to split, when the distance between the two sides of the apertures on the branch side is farther from the center, the difference's amplitude will change. The change of phase difference is more important in phase control. Two methods can compensate for the phase difference: using a digital phase shifter to compensate for the frequency and adding a dielectric lens between the waveguide port and the horn body so that the phase is distributed into a straight line. The TTL circuit is used for the wave controller to realize the command pulse and corresponding conduction to the phase shifter to control the scanning beam's direction, using two types of control mode, manual mode, and an automatic mode. Each phase shifter must have a control interface. The power supply voltage required by the wave controller is – 18, 24, 5, 6.5 V, and it is loaded into the EPROM through the pre-configured phase encoding. For the required frequency and azimuth, the data is automatically taken from the EPROM, and the phase shifter is driven to direct the beam to the corresponding position. One is the beam control code to control the main beam, and the other is the wave control code to control the beams of two sub-arrays. Since the array under study is a finite array element, each array element's mutual coupling environment is different, so the product relationship of the previous pattern (8.121) is approximate. As the cell spacing increases, the mutual coupling coefficient decreases. The same pitch has the same result as the frequency increases. Simultaneously, the central unit pattern is symmetrical, while the directional pattern near the edge is symmetrical, and it is more obvious at low frequency. However, the influence of mutual coupling on the array's synthetic pattern, especially the array with uniform aperture distribution, only affects the sidelobe level. For the main lobe of the array, its effect is negligible. The main influence of mutual coupling changes each unit's reflection coefficient, introduces impedance mismatch, and changes with frequency and scan angle. Therefore, it is necessary to overcome mutual coupling's influence within a wideband and a wide scanning angle. Adding a thin dielectric plate with a high dielectric constant to the antenna aperture is a practical engineering method. Considering the consistency of amplitude and phase of the channel, the factors that the performance difference of the wideband ferrite phase shifter itself and the phase difference between adjacent units during wide-angle scanning become larger affect the shape of pattern and matching and orientation. From the perspective of engineering application, using the method of correction of wideband phased array aperture, through the full-channel statistical balance of amplitude and phase, the above problems can be solved well. The correction block diagram is shown in Fig. 8.41. The figure includes a vector network analyzer, which acts as a radio frequency signal generator and amplitude-phase receiver.

Broadband waveguide phased array is shown in Fig. 8.42. While multiple units are used as correction detection elements corresponding to the array element antenna, the correction system's errors are first checked by the unit antenna and stored in the corresponding computer. The broadband phased array is placed in the correction position. According to the phase shift value required by the frequency and the scan angle, only one phase control code of a signal path can be allowed to be sent to the gated phase shifter for phase control at a time, ensuring that the phase shifter changes with the phase code issued by the controller at the same time. In practice, the phase shifter's control code is changed to obtain a phase shift value close to the theoretical requirement. And after unit 1 is debugged, connect the phase-positive path to unit 2. After debugging in the same way as the previous unit, go to the next unit, and so on, and store the actual phase shift control code of each array element at the

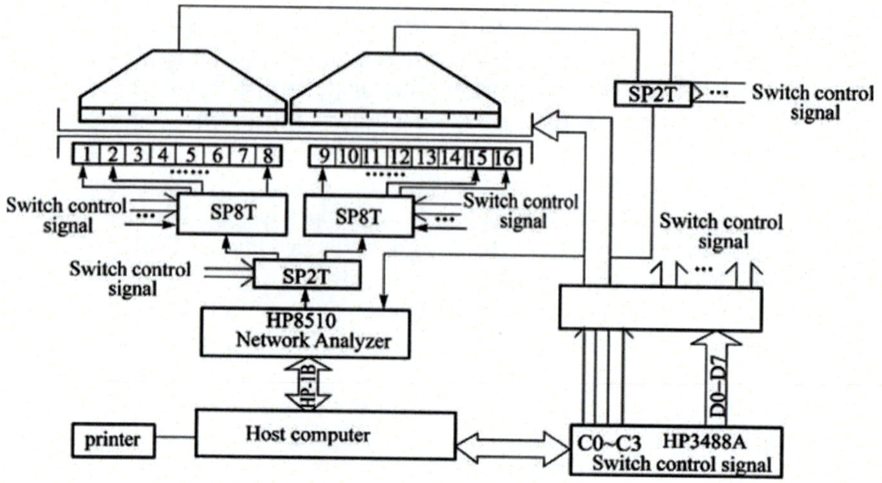


Fig. 8.41 Block diagram of wideband phased array phase correction

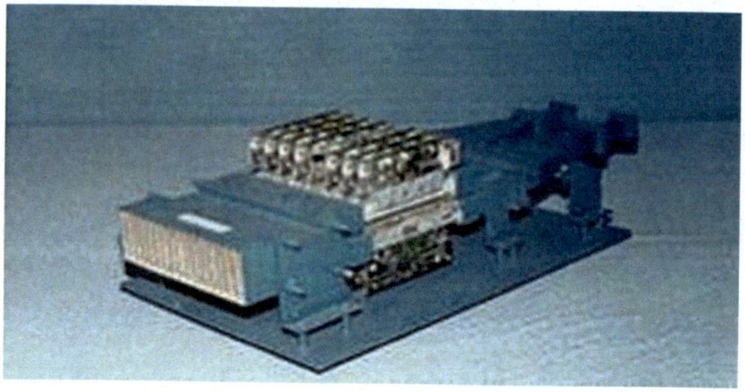

Fig. 8.42 Broadband waveguide phased array

corresponding angle and frequency into the computer. Performing the second phase correction for the same frequency, and the same angle will find that the first unit's phase shift code is adjusted. The corresponding phase shift amount is generated (at different frequencies; the phase shift code will be different). When the phase shift code of the antenna of the second unit changes, it will adversely affect the phase value of the previous unit after the fixed phase code. Since the entire calibration system is automatic, although the amount of data is large, the time taken is not long. The obtained corrected data can be stored in EPROM, and it can be used in the form of a table look-up or fitting formula. Simultaneously, a certain frequency and angle interval are allowed for correction, and a smaller frequency and angle interval can be obtained by interpolating through the theoretical formula. Dual-port input waveguide broadband phased array antenna. Its typical scanning pattern is shown in Fig. 8.43.

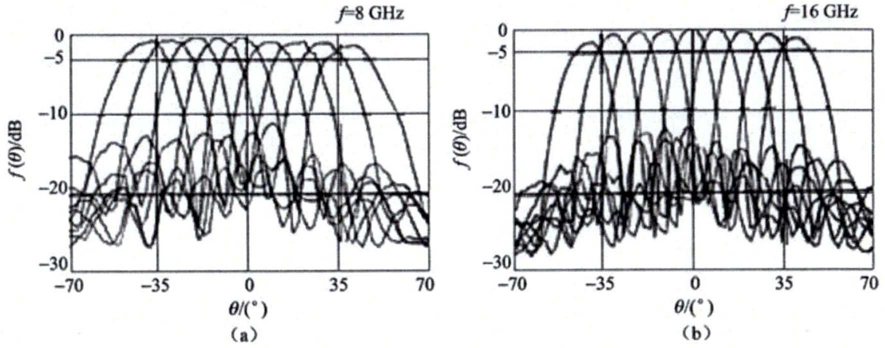


**Fig. 8.43** Sub-array scanning beam pattern

Applying the domestic digital phase shifter to the waveguide phased array, and the air feed method is used to complete the subarray feeding. The staggered arrangement realizes the ridge waveguide digital phase shifter's structural problem and the horn antenna array. The minimum angular spacing of the phase shifter scanning beam is 1°.

### *8.4.3 Broadband Active Phased Array Antenna*

Since an electric current drives the ferrite phase shifter, a single circuit can withstand larger power, but it is difficult to achieve miniaturization. Given the maturity of the low-power broadband power amplifier, PIN digital phase shifter technology, and manufacturing process technology, it is necessary to develop a solid-state broadband phase array. Figure 8.39c is a structural block diagram of phased array technology. Its characteristic is that the phase shifter's response speed becomes faster, the energy loss in the network is relatively reduced, and the difficulty in implementing the system's structure is reduced. Because the network part can be in the form of a stripline or a microstrip line, the connection line can be a general coaxial phase cable and does not require a high-voltage power supply. Simultaneously, due to small power amplifiers, the heat dissipation problem becomes easy to solve. The block diagram of a phased array antenna implemented with a solid-state microwave circuit is shown in Fig. 8.44. Each radiation element is followed by a power amplifier and a phase shifter. The power amplifier is implemented by a solid-state microwave circuit, also known as a solid-state phased array. Each solid-state power amplifier is 1 W, and its spatial equivalent radial power (ERP) is:

$$ERP = N^2 \cdot P_0 \cdot G_0 \cdot \eta \quad (8.130)$$

In the formula:

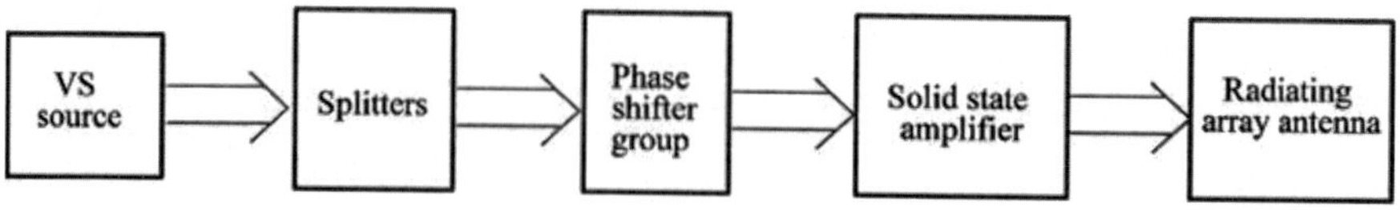


**Fig. 8.44** Block diagram of solid-state phase array

$P_0$ Amplifier output power.
$G_0$ unit gains of antenna.
$\eta$ Efficiency factor.

It can be seen that the ERP of the active phased array is proportional to the number N of the array elements. Enough units can produce very high ERP. Because the antenna array's working frequency can be extended to 18 GHz, the rear-feeding monopole horn's design is still used as the array element for the E-plane array. The narrowest diameter of the horn is 8.2 mm, and the wide-side diameter is 60 mm. The internal unit horn antenna gain is greater than 5 dB, and the array is a linear array.

The power splitter is an eight-way power splitter integrated by multiple two-way power splitters. Simultaneously, an adjustable attenuator is set in the channel to adjust the amplitude to suppress the sidelobe. The wave controller, its function is to control the scanning of the beam. Input the frequency code, azimuth code, and scanning trigger pulse to control the phase shifter through the wave control code and generate the phase scanning beam. Because it is necessary to realize the beam work without grating lobes in the wideband and wide-angle range, the distance between the array elements is measured by the minimum wavelength at the high end of the frequency. Mutual coupling is more serious at the low-frequency end. However, the array is essentially the same as the broadband phased array transmit antenna in the 8.4.2. The entire design and the compensation method and phase matching method for the array's mutual coupling are completed by a similar method in 8.4.2. As shown in Fig. 8.45, two identical element antennas are added on each side of the array, the matching load is followed by the array dummy element to improve the electromagnetic environment of the edge array element, and the number of array elements N = 8 (linear array). The phase shifter is an 8bits PIN phase shifter—beam hopping minimum distance: less than 1°. Considering the influence of the array antenna's standing wave characteristics, the power amplifier's actual output power at the low-end frequency may be less than 1 w. The final theoretical result of the equivalent radiated power is close to the measured value, and the combined efficiency is greater than 60%.

The amplitude and phase consistency requirements of the power amplifier in the phased launch array are reduced because the phase shifter can correct it. Simultaneously, because the digital phase shifter and the power amplifier are solid-state circuits, with the advancement of microelectronics technology, it is easy to achieve hybrid integration, improve reliability, and reduce size. The typical scanning emission diagram is as follows. Figure 8.46 shows the scanning pattern of f = 13 GHz. It can be seen that the jamming system with the phased array antenna is the core

**Fig. 8.45** Broadband solid-state phased array

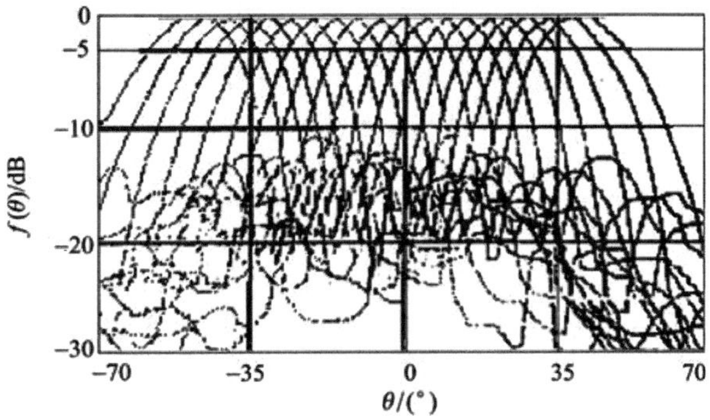


**Fig. 8.46** Scanning pattern of f = 13 GHz

because the beam spacing is very small, and almost the maximum beam is directed to the target. In the same situation, higher jamming power or jamming distance can be obtained. Each unit is preferably fed in the same amplitude to get a high array gain. For arrays with uniform amplitude distribution, the pattern's theoretical sidelobe level is – 13 dB, and uniform amplitude is also good for the power amplifier.

### *8.4.4 Broadband Planar Phased Array Antenna*

For the broadband phase array circular array and linear array, it is either used alone as a receiver or as a transmitter.

The array used as a receiver or a transmitter is mainly directly related to the broadband phase shifter, solid-state power amplifier, etc. The antenna array aperture normally works in both states. As long as the microwave switch is added, the digital phase shifter is shared. It will be transferred to the power amplifier to be radiated through the antenna unit during the transmission. During the receiving, switch from the microwave switch to the low-noise amplifier. The RF current passed from the antenna port is amplified, phase-shifted, and then interleaved to form a network to generate sum or difference or cross-direction information. To form a receipt, divergence, and azimuth, and pitch two, the broadband planar phased array with multi-dimensional function and its research can be applied to the co-curved array. In particular, the combination of optimized arrays and solid-state monolithic integrated circuit amplifiers results in efficient and reliable solid-state active arrays. The electronic warfare active array composed of solid-state power amplifier, solid-state phase shifter, low-noise amplifier, switch and excitation circuit T/R components, and high-density antenna array elements has the advantages small size, lightweight, multi-function, and high reliability. It is superior to the existing traveling wave tube/low gain antenna combined transmission array. For example, a 16-element solid-state active array with 1 W output power can obtain the same RRP as a row wave tube (25% efficiency) transmitter. The use of solid-state amplifiers can greatly improve the system's reliability and make the system have excellent "failure weakening" performance. The modular structure can be "stacked" as needed, which creates conditions for increased functionality. The active solid-state array is also suitable for installing the platform skin, especially conformal installation, which will create conditions for realizing "smart skin." The solid-state phased array used for electronic warfare should have a wideband characteristic. 2–20 GHZ is the frequency band usually required. In this wide frequency band rang, the requirements of the jammer for effective radiated power vary greatly. It depends on many parameters, such as the main battle platform's radar (RCS), the required jamming/signal ratio, the radar's minimum intercept distance, and what kind of special jamming pattern is needed.

The block diagram of the active solid-state array is shown in Fig. 8.47. It consists of the following parts: radiation area array, T/R component network, beamforming network, beam control circuit, power supply, heat dissipation, and other support

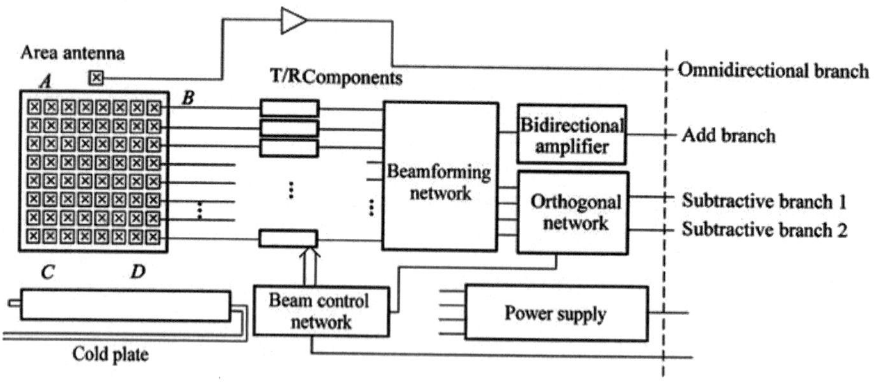


**Fig. 8.47** Broadband planar phased array composition block diagram

systems. When receiving, the planar array is divided into four identical sub-arrays, and the beamforming network generates four beams, two in the left and right, two in the upper and lower rows. The orthogonal network generates the cross beams in the upper and lower, left and right, and the sum beam in the corresponding direction. When transmitting, the four sub-arrays are combined into an array to form a higher radiated power. And the omnidirectional branch is used to cut the sidelobe during the reception. As we all know, in the design of broadband arrays, the grid lobes will appear in the scanning pattern after the larger size unit array. According to the traditional array design method, the grating-free design method is adopted, and the aperture of the unit antenna array must be relatively small, the radiation efficiency drops. Especially in the planar broadband active state phase array, the distance between the two main direction planar array elements is required, which is more prominent for the small size of the array element. At present, the better antenna elements are slot line and ridge horn antennas. Since the slot line antennas are easy to process and the consistency is easy to control, especially in the 2–6 GHz frequency band, it is very difficult to implement planar array horn antennas. So the linear antenna is used as a planar array element. The unit's radiation efficiency is at the low end of the frequency. The gain is very low, and impedance matching becomes difficult, especially the mutual coupling becomes very large, which will affect the array's beam synthesis effect. The T/R modules occupy inherent lateral dimensions. Each T/R module is directly connected to minimize the transmission path from the module to the antenna unit and effectively reduce unnecessary losses. The interval between the planar array units must be increased to meet the wideband range's low-end performance. The grid will appear in the high-end scanning pattern. How to solve the grid management problem is a key issue in design optimization.

In an array antenna, when the array pitch is large, each antenna element's far-field is in phase in multiple directions. Suppose the antenna element pattern is very wide, or the relative decrease in the antenna element's relative field strength in this direction is small. In that case, the array antenna pattern will have grating lobes. The maximum value of the grating lobe is close to the amplitude of the main lobe field strength. The magnitude difference is inversely proportional to the antenna unit beamwidth. The conditions for the presence of grating lobes are:

$$\frac{2\pi}{\lambda}[dx\sin\theta\cdot\cos\phi + dy\sin\theta\cdot\sin\phi] + \Phi_{i,k} = 2m\pi \quad m = \pm 1,\ \pm 2\cdots \tag{8.131}$$

When the beam scans to $(\theta_s, \phi_s)$:

$$\Phi_{i,k} = -\frac{2\pi}{\lambda}[dx\cdot\sin\theta_s\cdot\cos\phi_s + dy\cdot\sin\theta_s\cdot\sin\phi_s] \tag{8.132}$$

In the $\phi = 0$ main plane, the position of the grating lobe is

$$\theta x_m = \sin^{-1}\left(\frac{\lambda}{dx}\cdot m - \sin\theta_s\right) \tag{8.133}$$

In the $\phi = 90^\circ$ main plane, the position of the grating lobe is

$$\theta y_m = \sin^{-1}\left(\frac{\lambda}{dy} \cdot m - \sin \theta_s\right) \tag{8.134}$$

Therefore, the difference between the pointing angle of the main scanning beam and the grating lobe is:

$$\Delta\theta_x = \theta_s - \theta x_m$$

$$\Delta\theta_y = \theta_s - \theta y_m \tag{8.135}$$

The angle difference between the direction of the main scanning beam and the grating lobe direction can be calculated from the above formula. Among them, m = 1, f = 2–6 GHz, dx = 60 mm. It can be seen that the grating lobe is beyond 45° of the scanning main beam direction. In engineering, as long as the angle difference requirement between the main scanning beam pointing and the grating lobe pointing is given, the array spacing's maximum value range can be determined to obtain the antenna array's maximum array spacing elements. It is difficult to design without grating lobes for a wide-band phased array in a wide scanning range. Because it satisfies the aperture without a grating lobe in the high-frequency band, the aperture is too small for the low-frequency band. When scanning at a wide-angle, grating lobes are more likely to occur; otherwise, the unit antenna's aperture will be further compressed. The compression of the aperture is too small, which brings two problems: it brings difficulties to the design of the antenna unit, the array element spacing is too small, the antenna aperture is too small, the antenna voltage standing wave ratio increases, and the antenna gain is low, and the radiation efficiency is low; The array elements are too close to each other, and the mutual coupling between the antennas is enhanced. The aperture field is reflected back and forth between the antennas multiple times, making the amplitude and phase distribution errors of each channel larger, causing radiation pattern distortion, but this will cause grating lobes. The receiving array will bring multi-value ambiguity to the phase array direction finding; it will form jamming in the transmitting array's non-target direction. However, for practical applications, the above problems are not prominent. For the first problem, the relative coarse direction-finding can determine the approximate direction in the specific system. Only the accurate direction-finding of the phased array is used. For the second problem, due to the increase in the unit's aperture, the antenna gain also increases accordingly. It is possible to improve the entire phased ERP. Then the beam width, as the array element spacing becomes larger, the array beam becomes narrower, and the corresponding space becomes smaller. A part of the compressed energy is synthesized at the grating lobe. However, for the shared phase array, although the grating lobe appears, even if the direction-finding is fuzzy due to reception, the target pointing is wrong, but the transmission and reception use the same phase shift circuit, and the same phase shift value, the transmit beam just covers the receive beam, and jamming Can completely cover the target. Due to processing, material, and other errors, the inconsistency of the phase of the antenna array element is relatively large. Mutual coupling of array elements corresponds to amplitude modulation, multi-path synthesis of reflected fields between array elements, creeping waves, and the synthesis amplitude and synthesis phase change with frequency. The position of the array element and the scanning angle is different, and the errors are different.

The slot line antenna is formed by opening the metal surface of the dielectric substrate. The electric field quantity of the main wave propagating along the slit is a transverse electric wave, a traveling wave antenna whose size changes continuously and has a circular radiation aperture. The antenna pattern has a wide low frequency and a narrow high frequency, and the pattern has large ripples, which makes the scan range of the array insufficient, the sidelobe of the scan edge pattern is high, and the gain decreases. Many experimental studies and electromagnetic simulation calculations show that the circular arc opening mode and the multi-level compensation slot line antenna widen the unit beamwidth. It reduced the pattern ripple. When the slot line antenna is used as an array element, the mutual coupling between the array elements is large, and mutual coupling is a factor that seriously affects the performance of the array antenna in any form of the array antenna. It is related to the working frequency of the antenna and the scanning angle of the antenna. It is still difficult to strictly analyze and accurately calculate the mutual coupling between various units in theory. Generally, experimental methods are used to compensate for the influence of mutual coupling. Experiments and tests show that E-plane coupling is large, H-plane coupling is small, the low-frequency coupling is large, and high-frequency coupling is small. The maximum coupling of adjacent elements can reach 11 dB due to the strong E-plane surface wave coupling and the co-polarized spatial coupling between the antennas. The coupling between the antennas only limits the H-plane coupling. These coupling quantities are inversely proportional to the frequency and spacing. The special magnetic absorption materials are used between the antenna elements to reduce the mutual coupling between slot line antenna arrays. This absorption material has a large absorption volume per unit volume and a frequency bandwidth, especially a small volume; on the other hand, adding honeycomb absorbing material to the base panel between the two antennas in H-plane. Therefore, while decoupling, the radiation efficiency of the antenna array element will not be reduced. The beamwidth of the array antenna is (70°–100°) × (80°–120°), and the minimum gain is near 0 dB. In the figure, the area array antenna is a rectangular array. The array is divided into four sub-components: A, B, C, and D. The array A, B combined beam and C, D combined beam generate azimuth orthogonal difference beam through 90° bridge to achieve target pitch tracking. Similarly, A, C combined beam and B, D combined beam pass through 90° bridge. The scanning beam's sidelobe is removed through wide-beam antennas located at the four corners of the area array to ensure that the target's angle tracking is hosted. The size of the array is determined in this way to perform azimuth angle tracking. As the array element spacing becomes larger, the array element gain will increase, and the gain of the entire antenna array is:

$$G = G(\text{array}) + G(\text{unit})(\text{dB}) \tag{8.136}$$

To optimize the cell spacing, if a cosine function represents the linear pattern of 8 linear array elements: if $d/\lambda = 0.8$, the array gain will be the largest, the array element will further increase, and the array gain will gradually decrease. If $d = 72$ mm, then $d/\lambda_{min} = 1.45$ in the 2–6 GHz frequency band; If $d = 60$ mm, the grid's frequency range will be smaller in the whole frequency band, which depends on the specific engineering requirements; $dx = dy = 60$ mm can be selected in the project. This grid design method solves the expansion of the area array bandwidth, the optimized design of the entire array's effect, the mutual coupling problem, the T/R module heat dissipation problem, and the structural layout.

It is an important and indispensable component to realize the receiving and sending T/R module components of the phased array in the broadband range. It includes a digital phase shifter, microwave switch, low noise amplifier, solid-state power amplifier, equalizer, PIN attenuator, etc., as shown in Fig. 8.48. Generally speaking, the amplifier used for transmitting and receiving needs 20–30 dB of gain. As far as the power amplifier is concerned, sufficient gain is needed to compensate for the loss of any preamplifier circuit and limit the component's excitation power. The gain and noise figure should be selected for the receiving amplifier to obtain the desired receiving sensitivity. Many designs are related to the selected amplifier and other microwave components, and bandwidth, power output, and power consumption are key issues. The amplifier is required to have a uniform gain within the bandwidth. The main difficulty in obtaining broadband is to compensate for the field-effect transistor's gain decline. It is for these reasons, and the distributed amplifier method is selected. Other methods, such as the reactance equalization method, matched absorption method, and feedback method, is difficult to match the impedance. It is impossible to cascade multi-stage amplifiers to obtain the required component gain level.

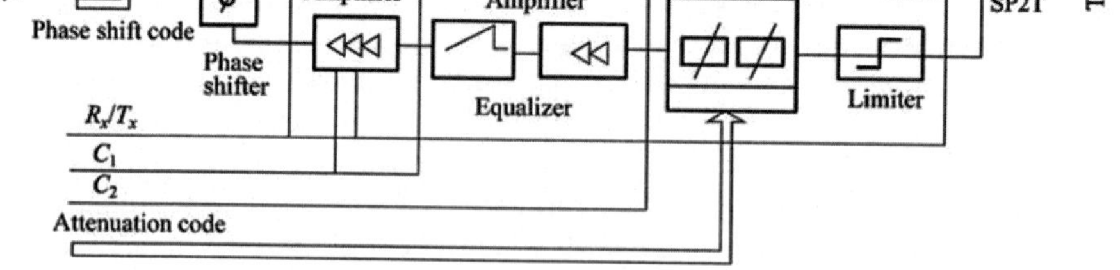


Fig. 8.48 Block diagram of broadband T/R components

The power amplifier power requirements will depend on the specific effective radiated power requirements and array size constraints. Each array element can provide the required equivalent radiated power of 1 W or higher power amplifier. Suppose the receiving channel is to have any effective gain without saturation within a wide range of input power. In that case, the receiving amplifier's output power is required to be greater than a predetermined value. If the insertion loss is within the allowable range, a broadband limiter can be added before the amplifier. Besides, power consumption is an important issue for broadband solid-state components. Broadband amplifiers have low efficiency, usually only 5–12%. Westinghouse's amplifier has an efficiency of about 10% at an output power of 0.5 watts and a relative bandwidth of 4:1. In short, each output of 1 W requires 8–20 W of input power. These amplifiers work at low drain voltages of about 7–10 V, requiring a large power supply, so the amplifier's efficiency is very important. The microwave components are packaged into small components, and the effective cross-section of each component is less than 0.16 square inches. In such a small space, the offset circuit, the control circuit, and the RF logic interface, power supply, and cooling interface must also be packaged. A non-active solid-state array uses a simple amplifier assembly. For active arrays, there are many amplifiers, and there are more complex transceiver modules. The phase shifters, attenuators, and transceiver switches must have control circuits, and the modules must be able to work in continuous wave or various types of interval pulse state. Therefore, it's internal electromagnetic jamming (especially from the high-current power bus) suppression is a problem. After taking some corresponding technical measures, the receiving and transmitting isolation is greater than 88 dB.

Multiple four-way power splitter combinations are used to obtain four basic beams. One input port and 16 output ports are connected to the T/R module behind $4 \times 4$ unit antennas, and each input port is connected to a two-way power splitter. Therefore, the array has eight input ports, four of which use the power splitter to form a beam. The other four ports use eight two-way power splitters and SP2T switches and a 90° 3 dB directional coupler to generate azimuth and the direction-finding cross beam of the pitch. In the specific design, try to use integrated technology to improve reliability. In addition to radiofrequency circuits, solid-state arrays also require a power supply, heat dissipation, and other support systems. Generally, array power supplies provide hundreds of watts, even more power for power amplifiers. Whether in pulse or continuous wave state, the power supply must provide stable power and have sufficient redundancy so that in the solid-state array, It can still work reliably under the condition of "failure weakening." A thermal management system is a big problem to be solved, especially in a solid-state array that the broadband solid-state amplifiers have lower efficiency. About 90% of power consumption is lost thermally when the power supply supplies power to each module. For example, a 32-element array of 1-W components requires a thermal management system to remove 288 W of heat from a ten cubic inch array. A heat pipe method or direct liquid cooling method can be used to dissipate the heat. The direct injection liquid cooling method has the lowest thermal resistance, which minimizes the GaAs field-effect transistor amplifier's temperature rise. Of course, with the miniaturization of T/R components and other components, the integration of broadband phase arrays has been greatly improved. It is entirely possible to use small-diameter dense broadband antenna array elements and microwave channels to form quality broadband phased radiation beam within the predetermined scanning range.

## 8.5 Broadband Array and Reflective Surface Composite Antenna

Parabolic antenna is the most commonly used high-gain reflector antenna. When the required antenna gains reach 30–50 dBi, the preferred antenna form is the reflector antenna. This antenna consists of a small illuminator (feed) and a large parabolic reflector. The small illuminator may be various other types of small antennas, such as dipoles, small horn antennas, small helical antennas, small log periodic antennas, small slot line antennas, and so on. They are mounted on the focal point of the large paraboloid. The spherical electromagnetic waves radiated by it illuminate the large parabolic reflective surface, and excite radio frequency currents on the reflective surface, which generate secondary radiation. Due to the isopath characteristic of the parabola, all rays from the focal point are reflected by the reflective surface and then reach the aperture of the reflective surface again, the path length they traverse is equal—The secondary radiation wave forms a planar equal-phase wavefront on the aperture of the reflecting surface, thereby generating a narrow beam and high gain in space. The working bandwidth of the reflecting surface mainly depends on the working bandwidth of the illuminator. For a parabolic reflector antenna with a circular aperture, when the radiating level of the illuminator on the edge of the reflector is about − 10 dB, the radiation characteristics of the reflector antenna can be estimated using the following empirical formula:

$$\mathrm{HPBW} = 70^{\circ}\lambda/D$$

$$G = 10\lg\left[\eta(\pi D/\lambda)^2\right] \tag{8.137}$$

where: λ—Working wavelength; D—Aperture size of reflecting surface; η—Radiation efficiency of antenna, including non-uniformity of illumination, shielding of illuminator and support rod, incomplete interception, spurious loss, mismatch loss, ohmic loss, etc.

According to the requirements of the system, there will be different requirements for the pattern of the reflector antenna. Therefore, there is a case of cutting a parabolic antenna, which usually cuts a part of the parabolic surface in the pitch plane to expand the coverage area of the beam in the pitch plane. In addition, a double-curved shaped reflective surface can be used to form a cosecant square beam in the pitch surface. The azimuth plane produces general focusing characteristics to cover a wide airspace with elevation angles from 0° to 40°. The lateral defocus of the parabolic reflector antenna can form sum beams and cross beams to realize the detection, tracking and positioning of long-range targets. Since the reflector antenna can generate gain to improve the sensitivity of the system, it is conducive to capturing and discovering targets. The narrow slope of the beam and the difference slope of the cross beam, the direction finding and tracking accuracy is high, which is conducive to the high-precision positioning and tracking of the target, and brings benefits to the performance improvement of other systems.

For the jamming transmitting antenna, in order to obtain a large equivalent radiated power and improve the gain of the transmitting antenna, the equivalent radiated power of the jamming system can be greatly increased. The reflector antenna can easily achieve the gain of 30–40 dBi for the microwave antenna. Assuming that the jamming transmitter can generate 1000 w of power, ERP $= 106 \sim 10 \times 106$(w) $=$ 1–10(MW). From the antenna theory, we know that the general parabolic reflector has only one focal point. Placing a broadband feed at the focal point produces an efficient beam, which has a limited lateral defocus distance and can only scan a few beam widths. To this end, a broadband electric scanning array antenna or a multi-beam antenna is used as an illuminator such as a cylindrical reflective surface or a spherical parabolic surface, to form a beam cluster of multiple spatial sub-beams. Direct each sub-beam to a different target to solve the problem of wide-angle coverage of broadband, high-gain antennas. The following discusses the case of a circular parabolic electric scanning composite array antenna.

With the development of modern satellite communication technology and monitoring technology, the antenna is required to have high gain and high-precision direction finding capability while giving other information of the target. Not only to complete the azimuth direction finding, but also to use other relevant information for spatial positioning. The planar array antenna completes a spatial coverage and has the above-mentioned functions. But its system structure is complex, there are many connecting parts, the cost is high and the passive gain is relatively low. At present, forming a multi-beam in the azimuth plane is a relatively practical method. At present, forming a multi-beam in the azimuth plane is a more practical method. The pitch plane uses other methods to perform direction finding to achieve the purpose of two-dimensional angle measurement, and there are multiple technical ways to generate the multi-beam. Here is a way to generate high-gain multi-beam—A complex multi-beam antenna system composed of a circular arc printed Vivaldi slot line antenna array and a circular parabolic reflector. It has the focusing characteristics of a parabolic antenna and the wide-angle multi-beam capability of a globe antenna. A very narrow beam is generated on the pitch plane, and a uniform beam distribution is generated on the azimuth, which improves the side lobe characteristics of the near-axis secondary pattern and the antenna has a compact structure and high reliability. A special feed network and active control circuit are used to form a fast scan, which illuminates the parabolic antenna in varying zones, so that it can work in a wide frequency range, and effectively solves the problem of the impact of the mechanical rotation scan of the high-gain antenna on the carrier.

Fig. 8.49 shows a schematic diagram of the structure of the entire antenna array. The antenna array consists of compound caliber with parabolic and spherical azimuth, circular arc book-shaped printed slot line antenna array, Connect cables and feeder

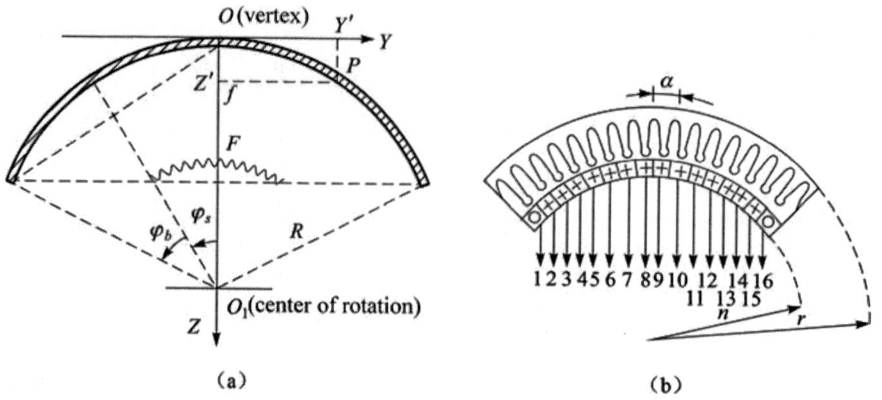


**Fig. 8.49** Schematic diagram of the structure of the antenna array

networks with good amplitude and phase consistency, and Active control circuit. At the same time, two beams are generated in space, and the rapid scanning is completed by the conversion of the control circuit. The circular arc-shaped array instantaneously in different regions, and uses a dedicated phase shifter to correct and compensate the secondary pattern. Next, it is connected to two super-heterodyne receivers with high sensitivity and good consistency. Through signal processing, we can obtain the target position and electrical index information. In the figure, the parabolic surface adopts the positive feed method. The volume of the entire antenna is smaller than that of the partial feed, and the directional pattern of the pitch plane is symmetrical, which is beneficial to the direction finding process. The occlusion of the main surface by the feed source used is a problem. Especially when the circular arc array is used as the feed array, the situation will be more prominent. In order to solve the problem of azimuth direction-finding accuracy, the distance between adjacent beams of the electric scanning multi-beam, the level of the intersection point, and the beam itself are relatively constant, thus placing stringent requirements on the size of the antenna element of the circular arc array. At the same time, it is required to work within the broadband range of the feeder array. Various factors are considered in the RF signal processing of multiple feed arrays. For a spherical reflector, when its radius is R and the radius of the illuminated area is a, the approximate expression of the optimal focal length is:

$$f_{op} = \frac{1}{4}\left[R + \sqrt{R^2 - a^2}\right] \tag{8.138}$$

For the gradual aperture field distribution, the actual focal length corresponding to the best directional pattern will be greater than the value given in Eq. (8.138). The experimental method is usually used to determine the correct position of the illuminator feed array. The radius of the irradiation area is:

$$a = 4\sqrt{14.7\frac{(\Delta/\lambda)_{total}}{(R/\lambda)}} \cdot R \tag{8.139}$$

where $\left(\frac{\Delta}{\lambda}\right)$ is Maximum allowable caliber 口径difference, for example $\Delta = \lambda/8$ or $\Delta = \lambda/16$.

The maximum path difference within the effective caliber range is:

$$\Delta\phi_{\max} = \frac{2\pi}{\lambda_{\min}} \times \frac{D^2}{2000 \cdot f^3} \tag{8.140}$$

According to the above formula, the spherical focal length on the azimuth plane can be determined. The diameter of the pitch plane is determined by the gain of the antenna beam and the broadband of the pitch plane. At the same time, taking into account the actual compromise situation:

$$\mathrm{R} = 700(\mathrm{mm}) \quad \mathrm{f} = 350(\mathrm{mm}) \quad \mathrm{D} = 870(\mathrm{mm})$$

The angular spacing between adjacent element antennas of the arc-shaped conformal array = 5°. The number of antenna elements n = 18. The outermost two element antennas are connected with matching loads to form dummy elements to ensure the electrical environment of the elements at the edges. In order to keep the radius of the circular arc changing with frequency, the width of the formed beam needs to be relatively constant. At the same time, the entire antenna array is required to work in a broadband manner to generate the required beam pattern for single pulse direction finding The feeding network of the antenna array is composed of a multi-element array, power splitter components phase shifter compensation channel phase imbalance, beam conversion SP8T switch, left and right beam synthesizer, beam controller, and etc. At least two beams should be generated simultaneously in the azimuth plane. Through the beam controller, the microwave switch is quickly switched to generate a fast beam scan. The polarization form is linear polarization and horizontal polarization along the azimuth plane. The formula of the antenna's far-field radiation field is:

$$\vec{E} = \frac{je^{-jkR}}{2f\lambda R}\iint\limits_{s}\frac{\left(J_x\vec{a}_x + J_y\vec{a}_y + J_z\vec{a}_z\right)e^{jk\Delta R}\sqrt{\left(4f^2 + y^2\right)\left(x^2 + z^2\right)}}{x}dydz \tag{8.141}$$

In the formula:

f The focal length of the compound parabolic aperture.

$$\Delta R = x\sin\theta\cos\varphi + y\sin\theta\sin\varphi + z\cos\theta$$

$$J_x = J_a R_i^x - J_b E_2^{mx}$$
$$J_y = J_a R_i^y - J_b E_2^{my} \tag{8.142}$$
$$J_z = J_a R_i^z - J_b E_2^{mz}$$

In the formula:

$$J_a = n_x E_2^{mx} + n_y E_2^{my} + n_z E_2^{mz}$$
$$J_b = n_x R_i^x + n_y E_i^y + n_z E_i^z \tag{8.143}$$

$$R_i^x = \frac{x + L\cos\alpha_m}{\sqrt{(x + L\cos\alpha_m)^2 + y^2 + (z - L\sin\alpha_m)^2}}$$
$$R_i^y = \frac{y}{\sqrt{(x + L\cos\alpha_m)^2 + y^2 + (z - L\sin\alpha_m)^2}} \tag{8.144}$$
$$R_i^z = \frac{z - L\sin\alpha_m}{\sqrt{(x + L\cos\alpha_m)^2 + y^2 + (z - L\sin\alpha_m)^2}}$$

$$n_x = \frac{-2fx}{\sqrt{(y^2 + 4f^2)(x^2 + z^2)}}$$
$$n_y = \frac{y}{\sqrt{(y^2 + 4f^2)}} \tag{8.145}$$
$$n_z = \frac{-2fz}{\sqrt{(y^2 + 4f^2)(x^2 + z^2)}}$$

In the vicinity of the mth unit antennas, the primary beam is formed by two slot line antennas at the same time as:

$$\vec{E}_2^m = \left[E_\theta(\theta'_m, \phi'_m)I_1 + E_\theta(\theta'_{m+1}, \phi'_{m+1})I_4\right]\vec{a}_x + \left[E_\theta(\theta'_m, \phi'_m)I_2 + E_\theta(\theta'_{m+1}, \phi'_{m+1})I_5\right]\vec{a}_y + \left[E_\theta(\theta'_m, \phi'_m)I_3 + E_\theta(\theta'_{m+1}, \phi'_{m+1})I_6\right]\vec{a}_z \tag{8.146}$$

$E_\theta(\theta'_m, \phi'_m)$ The radiation field of the mth unit distributed on the arc array.
$I_1 \sim I_6$ Conversion coefficient between feed array and spherical coordinate of space field

$$I_1 = \cos\alpha_m \cdot \cos\theta'_m \cos\phi'_m - \sin\theta'_m \sin\alpha_m$$
$$I_2 = \cos\theta'_m \cdot \sin\phi'_m$$

$$
\begin{aligned}
I_3 &= -\sin\alpha_m \cdot \cos\theta'_m \cos\phi'_m - \sin\theta'_m \cos\alpha_m \\
I_4 &= -\cos\alpha_{m+1} \cdot \cos\theta'_{m+1} \cos\phi'_{m+1} - \sin\theta'_{m+1} \cos\alpha_{m+1} \\
I_5 &= \cos\theta'_{m+1} \cdot \sin\phi'_{m+1} \\
I_6 &= -\sin\alpha_{m+1} \cdot \cos\theta'_{m+1} \cos\phi'_{m+1} - \sin\theta'_{m+1} \cos\alpha_{m+1}
\end{aligned} \tag{8.147}
$$

The formula for changing the rectangular coordinates of a single vivodi antenna into spherical coordinates is:

$$
\begin{aligned}
r'_m &= \sqrt{x'^2_m + y'^2_m + z'^2_m} \\
\theta'_m &= \cos^{-1}\left(\frac{z'_m}{r'_m}\right) \\
\phi'_m &= tg^{-1}\left(\frac{y'_m}{x'_m}\right)
\end{aligned} \tag{8.148}
$$

Let the coordinate of the phase reference point M of the mth vitaldi antenna in the system coordinate system be M $(x_m, y_m, z_m)$, yet:

$$
\begin{aligned}
x'_m &= r\sin\theta \cdot \cos\phi \cdot \cos\alpha_m + L - r\cos\theta \cdot \sin\alpha_m \\
y'_m &= r\sin\theta \cdot \sin\phi \\
x'_m &= r\sin\theta \cdot \cos\phi \cdot \sin\alpha_m + r\cos\theta \cdot \cos\alpha_m
\end{aligned} \tag{8.149}
$$

$\alpha_m$ is the rotation angle of the mth vitaldi antenna around the y-axis.

$$
\begin{cases}
x_m = -L\cos\alpha_m \\
y_m = 0 \\
z_m = L\sin\alpha_m
\end{cases} \tag{8.150}
$$

Vivaldi slot line unit antenna is an end-fire broadband antenna. It uses transmission line theory and dyadic Green's function to obtain the closed formula of its radiation field. Therefore, the calculated values of the radiation patterns of multiple beams in the azimuth plane can be obtained from formulas (8.141) to (8.150). Figure 8.50a shows an azimuth beam pattern with a frequency of 4 GHz. A total of 14 beams cover an angular range of 30° ~ – 30°. Figure 8.50b shows the corresponding measured beam pattern. It can be seen from the two results that they are quite consistent and have good consistency. Figure 8.51 shows the measured beam patterns of the other two frequency points (including the entire beam network). The entire array of composite multi-beam antennas works in a relatively wide frequency range and maintains good electrical characteristics. The beam pointing is almost unchanged, with a relatively

**Fig. 8.50** Azimuth beam pattern with frequency of 4 GHz

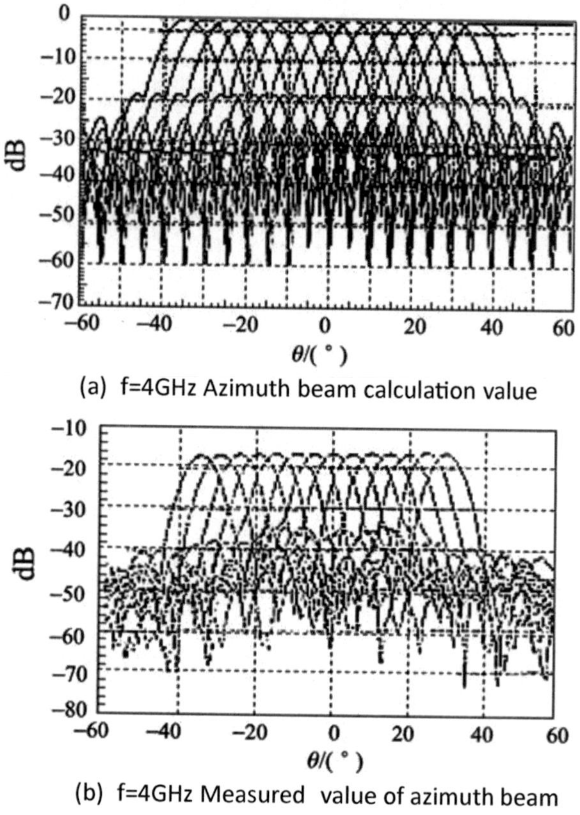


low sidelobe level, the beam is symmetrical and smooth, and the left and right beams are scanned at the same time, and there is a high gain value. One of the changes of the 3 dB width of a beam with frequency is that the pitch plane is a traditional parabolic aperture antenna. As the frequency increases, the beam becomes narrower, while the azimuth beam broadband becomes wider at the high end of the frequency.

The test results and calculation results of multiple beams in the azimuth plane are given in the figure, and they are in good agreement. At the same time, the beam quality of the entire electric scanning antenna array is relatively good. On the other hand, the angle range of the azimuth scan can be further enlarged so that the outermost beam reaches 60°. This type of antenna system that uses the array flexibility of a wide-band small array electrical scanning beam and the beam compression technology of a large-aperture antenna to form a composite function solves the technical problems of broadband, large-angle coverage, high gain and high precision. You can also use the attached surface. The intersection point of the secondary reflection deviates from the focus to form a multi-beam method. Their comprehensive consideration from the aspects of performance, reliability, cost, and usage is a trend in antenna engineering applications.

Fig. 8.51 Typical beam scanning test results of azimuth plane

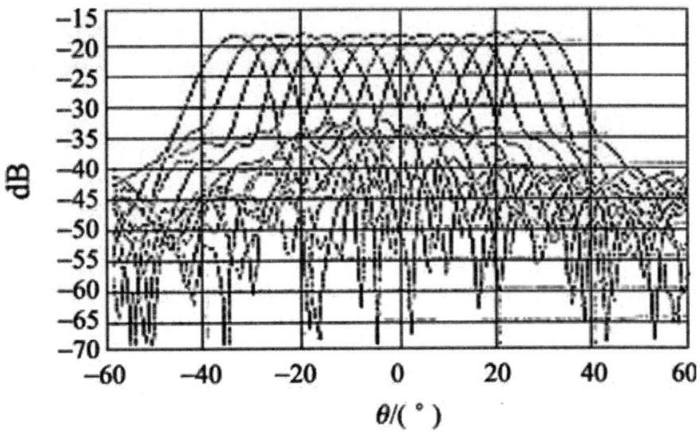


(a) f=2GHz Azimuth beam calculation value

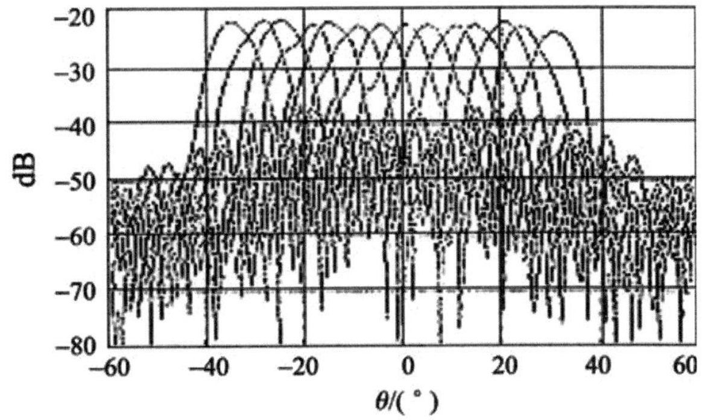


(b) f=7.5GHz Measured value of azimuth beam

## 8.6 Digital Beamforming Technology

The beam of the traditional array is formed in the microwave frequency band through the microwave network. By changing the microwave frequency band to the intermediate frequency, the amplitude and phase digital information obtained by the intermediate frequency sampling form the beam (DBF). This method is a product of the combination of digital processing technology and antenna technology in recent years. It breaks through the traditional concept of directly forming beams from RF energy on the array front. The radio frequency signals acquired by each unit on the array front are quickly transformed into digital signals through their respective receiving channels. The use of digital signal processing technology can realize the advantages of multi-beam, ultra-low sidelobes, array pattern optimization and shaping, antenna array self-correction, and super-resolution. The output signals of each element of a multi-element array are weighted and summed. The array output produces different responses to signals from different spatial directions, which is the basic concept of traditional beamforming. Mono-pulse radar and differential beamforming are one of the simplest and most typical examples. There are technical difficulties in directly converting microwaves into digital signal processing. Suppose the output signal of each array element is converted to a digital signal after appropriate frequency conversion. In that case, the processing techniques such as weighting and summation are moved to digital processing so that its output produces different responses to signals from different spatial directions. This technology converts the target information acquired by the antenna array into digitized information. The information retains the frequency, amplitude, phase, and other information of the target signal. It can construct any (shape and number) conducive to direction-finding through adaptive digital weighing technology while detecting the correlation accumulation parameters direction-finding algorithm can accurately measure the beam.

In principle, DBF technology hardware includes antenna elements, microwave front end, A/D collector, and digital beamformer (including high-speed computer). The principle block diagram of a DBF is shown in Fig. 8.52. Since the current A/D collector's working frequency cannot be too high, A/D acquisition also has requirements on analog signals. Therefore, the DBF antenna's hardware also includes radio frequency signal amplification and down-conversion, called the receiving channel or receiving front end. The converted signal is converted into a digital signal and then sent to a high-speed parallel digital signal processor for digital beam synthesis to form the required multi-beam signal. Two quadrature components must be obtained, namely the signals in-phase (I) and orthogonal (Q) components of the signal to preserve the phase and amplitude of the received signal. This method is similar to convert from polar coordinates to rectangular coordinates with the real part I and

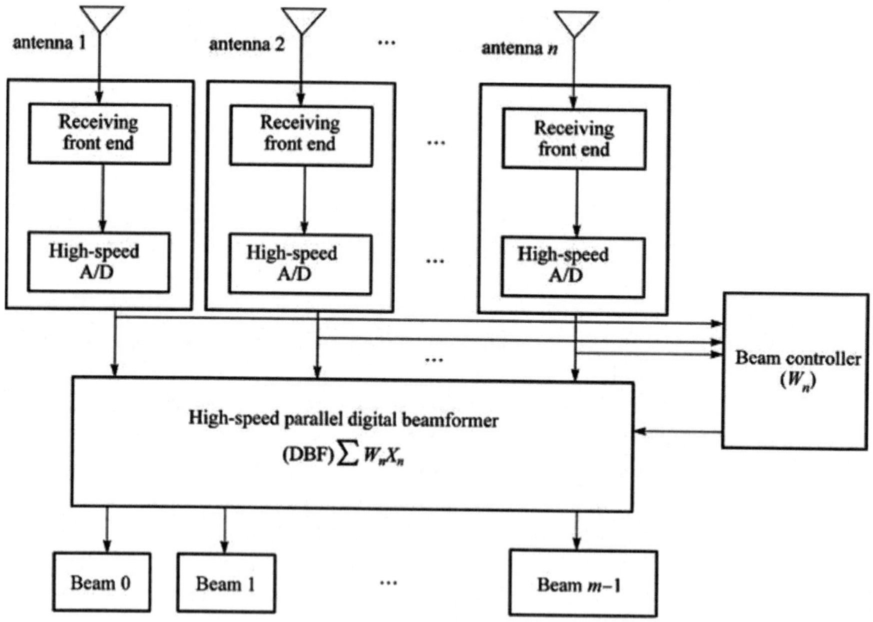


**Fig. 8.52** DBF structure diagram

imaginary part Q. After digitizing the I and Q signals, the phase and amplitude of the signal can be calculated as follows:

$$A = \sqrt{I^2 + Q^2}$$

$$\phi = \arctan(Q/I) \tag{8.151}$$

The traditional orthogonal demodulation baseband digital conversion process is as follows. First, the intermediate frequency signal extracts the I and Q baseband signals through a pair of analog mixers and low-pass filters. Next, uses two ADCs to sample the orthogonal I and Q signals, respectively. Therefore, the sampling rate of the ADC is relatively low. However, the two analog channels are required to be consistent in-phase, and amplitude and the demodulation accuracy is low. Of course, digital methods can also be used to obtain the I and Q components. When digitizing the intermediate frequency and implementing digital orthogonal demodulation using an algorithm. The intermediate frequency directly samples the digital intermediate frequency signal to obtain the I and Q baseband signals through the digital demodulation algorithm. The consistency and phase orthogonality are much better than those of the traditional demodulation method. Digital beamformers can perform fast parallel mathematical operations. Thus, the inner product between the signal vector and the weight vector received by the array element is realized to generate a beam. The beam operator is the core component of the digital beamformer. It is mainly composed of a high-speed parallel digital signal processor, which completes multiplication and addition operations. The beam controller mainly generates the optimal complex weight vector Wn according to various constraints. Using digital technology in beamforming technology can provide the following advantages:

(1) Multiple or independent controllable beams can be generated without loss of signal-to-noise ratio. In the DBF antenna array, the array element signal is digitized. The low-noise amplifier mainly determines the signal-to-noise ratio in the front-end receiving module. Digital processing of the received signal can form multiple simultaneous dense beams, providing the ability to simultaneously detect and track multiple targets. Because beam scanning control is inertial, there is minimal constraint on beam flexibility.
(2) The beam scanning control coefficient is variable, so corrections for gain and phase mismatch between channels can be combined. The receiving channel can be calibrated online, and the calibration procedure can be easily incorporated into the DBF array. Therefore, the absolute error requirement in the receiving channel can be reduced.
(3) The antenna pattern can be changed arbitrarily. Therefore, any form of direction finding beams such as differential beams and multi-beams can be formed using integrated technology during operation. Moreover, the mutual coupling calibration of the array element pattern can be conveniently performed.
(4) All signals from the array can be obtained in digital form. Therefore, array signal processing techniques can be applied, such as the adaptive algorithm of

the sampling matrix inversion (SMI) algorithm's directional pattern. Moreover, the received digital signal can also be subjected to other methods such as coherent integral.

A linear array is composed of a logarithmic periodic array element using amplitude and phase receiver experimental research. The amplitude and phase of the signals received by several units are collected, respectively. After the appropriate amplitude and phase weighting, any number of beams can be obtained simultaneously. In this way, the array's response can be formed in the digital processor, which is calculated by the weighted addition of the signals from each unit. The response of a particular beam to a set of digitized received data is given by:

$$E_r(\theta) = \sum_{k=1}^{n} w_k A_k(\theta) \exp[j P_k(\theta)] \exp[j(k-1)\varphi(r) \tag{8.152}$$

In the formula:

$w_k$ $\varphi(r)$ Amplitude and phase of the weighting function.
$A_k$, $P_k$ The amplitude and phase of the received signal from the unit.

Figure 8.53 shows the situation of beamforming. It verifies the direction-finding capability of the digital beamforming system. The incoming signal in any direction can be digitally weighted to obtain seven values using the principle of amplitude direction-finding. The seven beams formed by the digital beam and these seven values can obtain the direction of arrival of the signal. The quality of the beam obtained by the digital method is relatively good to achieve higher direction-finding accuracy.

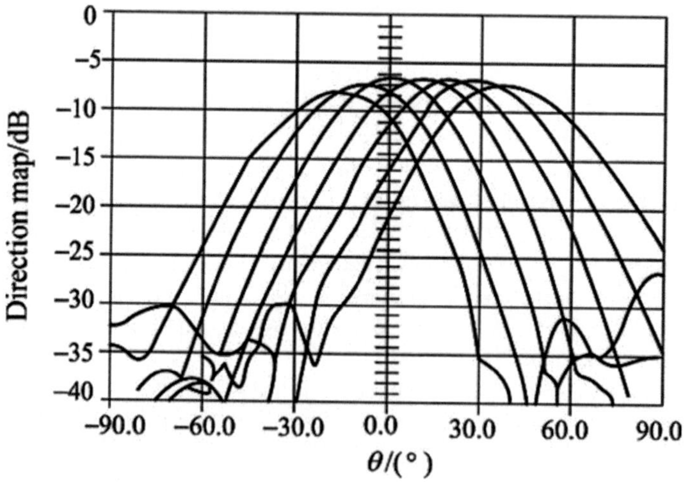


**Fig. 8.53** DBF formed beam pattern

Because each array element's output signals are properly transformed into digital signals, the frequency, amplitude, phase, arrival time, and other information it carries are all in the sampled and quantized data. Taking full advantage of these received signals and using a reasonable algorithm, you can infer the direction of arrival of the target signal in the wide-angle airspace. According to the same data set, different antenna pattern responses can be obtained after different weighting processes. It is equivalent to generating many flexible multi-beams. It performs high-precision direction finding by searching and confirming multiple target sources in the specified airspace. In terms of jamming sources, adaptive zero-adjustment or fixed band-limiting problems can be formed in a certain angular range. The real-time broadband DBF is generated using broadband RF components, high-speed sampling A/D devices, and fast beam algorithms. A power amplifier is added after each antenna array element, and the digital beamforming data is converted into an analog beam through D/A. After up-conversion to the microwave frequency band, the DBF jamming beam is generated by the power amplifier through antenna radiation. A DBF system with integrated transmission and reception is formed through the transmission and reception control switch and the common antenna array.

## 8.7 Broadband Antenna System

In actual engineering applications, to meet the requirements of the entire system's functions and indicators, many individual antennas and array antennas with certain characteristics need to be laid out in a certain way and a specific installation environment. At the same time, it is necessary to consider the mutual influence between the antennas inside the electronic warfare equipment, the mutual influence between the antennas of the electronic equipment, the mutual influence between the antenna and the installation platform, the polarization mode, the environmental adaptability, the trend of various network components inside the antenna, Installation, offset, and power supply to form the electronic information required by the system. The system involves carrier platform, radome, polarized cover, antenna arrangement, electromagnetic compatibility of the radiation field, active and passive microwave components, and beamforming methods. The system also involves radio frequency signal conversion and control methods, working environment adaptability (including temperature, vibration, shock, three-proof, etc.), heat dissipation, power supply, etc. In the following, relevant aspects will be analyzed.

### *8.7.1 Broadband Radome*

The radome is a shell that protects the antenna from the external environment. It is a special shaped electromagnetic window made of natural or artificial dielectric materials. It is used to protect the antenna or microwave system from the impact and damage of the external environment, such as wind, rain, fog, tide, sand, and high temperature. According to their different uses, different names, such as radome, feed cover, spatial filter, transparent window, microwave wall, etc. Their common feature is that they can transmit electromagnetic waves well. The radome can extend the equipment's life, reduce maintenance costs, improve the use environment, and improve its reliability. Using a radome is a good engineering measure. From the electromagnetic theory perspective, the radome shell will also affect reflection, absorption, the introduction of phase shift, aperture blocking, and parasitic scattering, which affect the antenna's directional pattern, gain, impedance, polarization, and other characteristics. The radome is often divided into aircraft radome and ground and shipborne radome according to different equipment platforms. In aircraft, missiles, and space vehicles, single-layer thick-walled covers are commonly used to withstand large aerodynamic pressure and high-temperature shocks. Aircraft radome is used in aircraft, missiles, and aerospace vehicles. A single-layer thick-walled hood is commonly used to resist large aerodynamic pressure and high-temperature shocks. Ground and shipborne radome are used in on-board equipment, fixed ground equipment, and shipborne equipment. Commonly used are A-type interlayer (three-layer), B-type interlayer (three-layer), C-type interlayer (five-layer), and multi-layer cover to improve the working bandwidth of the radome and obtain better permeability performance. There are many shapes of broadband radome used in electronic equipment, especially in bullets or airborne. The specific project is divided into an overall radome and a single radome. Generally, multiple antennas of different frequency bands are often installed on one antenna base of airborne or shipborne equipment and share a wideband radome, such as a 0.5–18 GHz wide frequency band radome.

Designing a broadband radome is not an easy task. The characteristics of the radome are not only related to all the materials used but also to the polarization, angle of incidence, operating frequency, etc. Most of the antennas used in electronic countermeasure equipment use oblique 45° linear polarization or circular polarization to adapt to various polarized waves. The radome range forms the beam pointing angle of the antenna and the angle of incidence from 0° to 90°. The operating frequency range is several octave bands. The transmission coefficient of the radome needs to be discussed first to expand the radome's bandwidth. Here, the calculation method of the transmission coefficient is discussed first. The microwave network theory transforms the rays' problem passing through the dielectric plate into a transmission line problem. Their equivalent impedances are respectively expressed as Z0 $\cos\alpha/\sqrt{\varepsilon}$ (for horizontal polarization) and Z0 $\sec\alpha/\sqrt{\varepsilon}$ (for vertical polarization). As shown in Fig. 8.54, the transmission line's characteristics are represented by a [A] matrix. If it is a multilayer dielectric radome, each layer of the radome is represented by a [A] matrix. The overall characteristics can be expressed by the product of [A] matrix, namely:

$$[A] = [A]_1 \cdot [A]_2 \cdot [A]_3 \ldots [A]_n$$

The final result of the product is A11, A12, A21, A22. Then, the voltage transmission coefficient and voltage reflection coefficient of the radome can be expressed

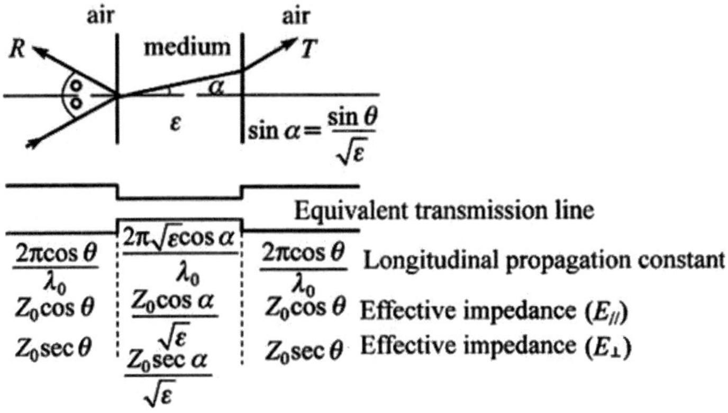


**Fig. 8.54** Electromagnetic wave transmission in a lossless medium

as:

$$T = \frac{2}{A_{11} + A_{21} + A_{12} + A_{22}}$$

$$R = \frac{A_{11} - A_{21} + A_{12} - A_{22}}{A_{11} + A_{21} + A_{12} + A_{22}} \tag{8.153}$$

According to the [A] matrix expression of the lossless transmission line, the expression of the single-layer lossless medium can be obtained:

$$A = \begin{pmatrix} \cos\phi & j\frac{Z}{Z_0}\sin\phi \\ j\frac{Z}{Z_0}\sin\phi & \cos\phi \end{pmatrix} \tag{8.154}$$

$$\phi = \frac{2\pi d\sqrt{\varepsilon - \sin^2\theta}}{\lambda} \tag{8.155}$$

In the formula:

$\varphi$ Electrical length.
$d$ Medium thickness.
$\lambda$ Working wavelength.
$\varepsilon$ Dielectric constant.
$\theta$ Angle of incidence of electromagnetic waves.

Z/Z0 is the ratio of the impedance in the medium to the impedance in free space. They are shown below:

$$\frac{Z}{Z_0} = \frac{\sqrt{\varepsilon - \sin^2\theta}}{\varepsilon\cos\theta} \quad \text{(horizontal polarization)} \tag{8.156}$$

$$\frac{Z}{Z_0} = \frac{\cos\theta}{\sqrt{\varepsilon - \sin^2\theta}} \quad \text{(vertical polarization)} \tag{8.157}$$

It can be seen from Eqs. (8.156) and (8.157) that when the incident angle $\theta = 0$, the horizontal and vertical polarization Z/Z0 are equal $1/\sqrt{\varepsilon}$. By changing (8.153) to a simple transform, the formula for the power transmission coefficient can be obtained:

$$|T_{\parallel}|^2 = \frac{4}{(2\cos\phi)^2 + \left[\sin\phi \cdot \left(\frac{\sqrt{\varepsilon - \sin^2\theta}}{\varepsilon\cos\theta} + \frac{\varepsilon\cos\theta}{\sqrt{\varepsilon - \sin^2\theta}}\right)\right]^2} \tag{8.158}$$

$$|T_{\perp}|^2 = \frac{4}{(2\cos\phi)^2 + \left[\sin\phi \cdot \left(\frac{\sqrt{\varepsilon - \sin^2\theta}}{\cos\theta} + \frac{\cos\theta}{\sqrt{\varepsilon - \sin^2\theta}}\right)\right]^2} \tag{8.159}$$

Although the above formulas are for lossless media, they are still relatively accurate for low loss and thick media. For the epoxy glass fiber reinforced plastic cover plate with a thickness of 8.79 mm, tgδ = 0.014, and ε = 8.72, the power transmission coefficients at 0° and 30° angles are shown in the table below. In Table 8.8, the frequency are7.5, 10, and 15 GHz, and the calculation formulas are (8.158) and (8.159).

According to the same calculation method, the power transmission coefficient under different parameters can be calculated. The calculation results can be seen: only when the radome's electrical thickness is k/4 wavelength (k = 0,1, ..., n), good power transmission will be obtained. N addition to the normal incidence, there is a big difference in the power transfer coefficient between horizontal polarization and vertical polarization at the same angle of incidence. Several types of broadband radomes are discussed below:

(1) The electrical thickness of the wall of the single-layer thin-walled cover is less than 0.05 wavelengths. There are generally two types of this type of radome. One is a single-layer inflatable radome; the other is a thin-skinned radome with a metal truss. These two radomes have little effect on antenna performance, high wave transmissivity, and wide operating bandwidth. It can be seen from the formulas of power transmission coefficients (8.158) and (8.159) that when

**Table 8.8** Transmission coefficients under no-loss conditions

| f(GHz) | θ | | | |
|---|---|---|---|---|
| | Vertical polarization | | Horizontal polarization | |
| | 0° | 30° | 0° | 30° |
| 7.5 | 620 | 0.536 | 0.620 | 0.708 |
| 10.0 | 0.688 | 0.593 | 0.688 | 0.753 |
| 15.0 | 1 | 0.994 | 1 | 0.997 |

the radome's electrical thickness is less than λ/20, it shows high transmission efficiency for electromagnetic waves. Table 8.10 is the minimum value of the power transmission coefficient of the thin-walled cover with $\varepsilon = 4$ when the electrical thickness is less than λ/20 (Table 8.9).

It can be seen from the calculation results that a single-layer thin-walled cover can obtain wide-band transmission performance under a small incident angle. Therefore, when developing a radome, a thin-walled radome should be used as much as possible if the structure allows. It is not only light in weight but also able to obtain broadband transmission performance. Figure 8.55 is the test result of an arc feed cover. The wall thickness of the cover is 0.5 mm.

(2) A separate foam shell is still a single-wall radome electrically, and a rigid foam radome with a very low dielectric constant. For example, its density is only 0.06–0.16 g/cm$^3$, which has been used better on shipboard equipment, but the foaming technology must be quite good to meet electrical and structural strength requirements. From the formula (8.158) or (8.159), it can be concluded that under normal incidence ($\theta = 0°$), the minimum value of the power transfer coefficient occurs periodicity when $\varphi = (n + 1/2)\pi$. So we get:

$$|T|^2_{\min} = \frac{4}{\left(\frac{\sqrt{\varepsilon}}{\varepsilon} + \sqrt{\varepsilon}\right)^2} \tag{8.160}$$

It can be seen from the above formula that for normal incidence and lossless materials, the minimum power transfer coefficient is not related to the thickness of the wall but only to the relative dielectric constant. It means that if the dielectric constant is low enough, the radome can use very thick walls. On the other hand, because the minimum value of power transmission in formula (8.160) is independent

**Table 8.9** Calculation results after considering tgδ

| f(GHz) | θ | | | |
|---|---|---|---|---|
| | Vertical polarization | | Horizontal polarization | |
| | 0° | 30° | 0° | 30° |
| 7.5 | 0.60 | 0.52 | 0.60 | 0.69 |
| 10.0 | 0.65 | 0.55 | 0.65 | 0.73 |
| 15.0 | 0.95 | 0.95 | 0.95 | 0.95 |

**Table 8.10** Minimum power transfer coefficient

| Polarization | Angle of incidence |
|---|---|
| | 0° 30° 45° 50° 60° |
| Vertical | 0.95 0.94 0.90 0.88 0.79 |
| Horizontal | 0.95 0.97 0.99 0.997 1.0 |

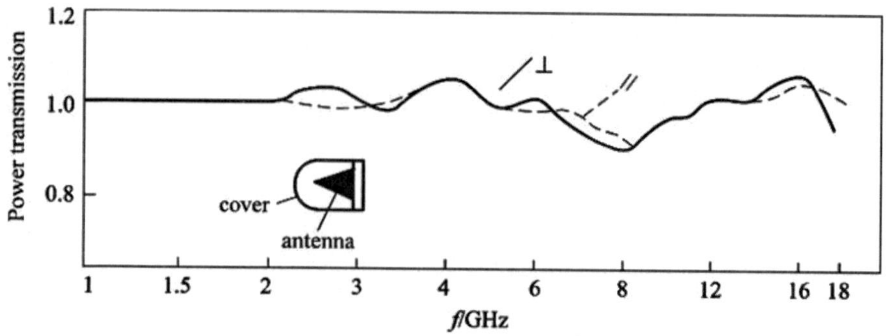


**Fig. 8.55** Test results of the arc-shaped thin-walled cover

of frequency, the foam cover has a wide-band transmission performance. The foam material is not lossless, but its tgδ is very small, so the loss of the medium itself is very small. For example, in foam with $\varepsilon = 1.2$ and $tg\delta = 0.003$, the thickness loss per wavelength is 3%. Under the condition of a large incident angle, the single layer of foam with a low dielectric constant still shows its broadband characteristics.

(3) The sandwich radome is usually composed of two high-density, high-strength skins and a low-density, low-strength core. The cascade transmission line method's calculation results show that to obtain a zero-power reflection coefficient, and it is necessary to adjust the interval between the two skins to realize the reverse phase so that the two reverse phases cancel out. Although this kind of sandwich cover only has the best characteristics in a narrow frequency band. However, for an ultra-wideband radome for electronic countermeasures, it can obtain more than 90% of the electromagnetic wave permeability within the required bandwidth range if properly designed.

Multi-sandwich radome is composed of thin skin and core with a very low dielectric constant (honeycomb core, rigid foam core). The core layer provides sufficient mechanical strength and low insertion loss. Skin layers utilize the principle of multiple wave reflection cancellations to reduce the reflection loss of the cover, thereby improving the cover's performance. As we all know, the power reflection coefficient and the radomes' power transmission coefficient have a simple relationship of energy conservation: $|T|2 + |R|2 = 1$, so when $|R|2$decreases, $|T|2$will get to promote. The most commonly used in engineering is this type A (three-layer) sandwich cover. The skin layer is epoxy glass cloth, and the core layer sandwiched between is 6–8 mm thick honeycomb material (epoxy honeycomb or Wallen honeycomb). It can generally work in the frequency range of 0.8–18 GHz, and its power transmission coefficient $|T|2$can reach 70–90%. When the radome is a flat cover, $|T|2$also has better performance. The C-type interlayer, composed of two A-type interlayers, has the advantage that it can get better transmission under a larger incident angle. There are two ways to obtain them: use two optimal A-type interlayers back-to-back together, or make the two A-type interlayers' reflections cancel each other out. The surface layer and the core are made of FRP and honeycomb-shaped

FRP. As long as it is properly designed, broadband transmission characteristics can also be obtained.

(4) For the medium with metal, adding a certain metal to the medium can increase the bandwidth. The dielectric plate's effective admittance less than 1/4 wavelength thick is the capacitance, which can be compensated by the metal grid. As can be seen from Fig. 8.54, Z0 is the impedance in free space, and Z is the impedance in the dielectric plate. Therefore, using the formula of the input impedance, we can obtain the normalized charge at the interface between air and medium:

$$B = \frac{tg\phi(k-1)}{k + tg^2\phi} \tag{8.161}$$

$$\phi = tg\left(\frac{2\pi d}{\lambda}\sqrt{\varepsilon - \sin^2\theta}\right)$$

In the formula:

$\theta$ Angle of incidence.
d Thickness of dielectric board.
λ Working wavelength.

$K = \frac{\varepsilon - \sin^2\theta}{\cos^2\theta}$ (vertical polarization); $K = \frac{\varepsilon^2\cos^2\theta}{\varepsilon - \sin^2\theta}$ (horizontal polarization).

In the case where the electric field is parallel to the metal line and perpendicular to the plane of incidence, the susceptance of the parallel metal line embedded in the medium is:

$$B = \frac{P}{\left(\frac{P}{\lambda}\right)\sqrt{\varepsilon - \sin^2\theta}\left[\ln\left(\frac{P}{\pi D}\right) + 0.6\left(\frac{P}{\lambda}\right)^2\left(\varepsilon + 2\sin^2\theta\right)\right]} \tag{8.162}$$

In the formula:

P Distance between wires.
D Wire diameter.
θ Angle of incidence.

For horizontal polarization, (ε – sin 2θ) is used instead of (ε + 2sin 2θ). In this way, the broadband characteristics of the radome can be obtained. The dielectric plate's susceptance and the susceptance of the metal grid can obtain a destructive effect at two frequency points, and a lower transmission loss is obtained within the bandwidth of 4:1.

In addition to transmitting waves, the radome also produces cross-polarization, which reduces the performance of systems that require pure polarization, such as frequency reuse satellites. It may also deteriorate the axial ratio of the circularly polarized antenna. Consider a linearly polarized wave E, which is incident on the dielectric plane at an angle I. Its polarization angle is θ, the vertical polarization component is E sin θ, and the horizontal polarization component is E cos θ. After passing through the medium, a vertical polarization component with amplitude $T_{\perp}$(E sin θ) and phase $\varphi_2$ and a horizontal polarization component with amplitude $T_{\|}$(E cos θ) and phase $\varphi_1$ is generated, where T and φ are the amplitude and phase of the transmission coefficient. These two polarization components can be decomposed into the principle plane component and the cross-plane component, respectively.

The sum of principal plane components is:

$$T_m = E\sqrt{T_{||}^2 \cos^4\theta + T_{\perp}^2 \sin^4\theta + 2T_{||}T_{\perp}\cos^2\theta\sin^2\theta \cdot \cos(\phi_1 - \phi_2)} \tag{8.163}$$

The sum of the cross-plane components is:

$$T_c = E\sqrt{T_{||}^2 \cos^2\theta\sin^2\theta + T_{\perp}^2 \sin^2\theta\cos^2\theta - 2T_{||}T_{\perp}\cos^2\theta\sin^2\theta \cdot \cos(\phi_1 - \phi_2)} \tag{8.164}$$

So we got the cross-polarization ratio:

$$\frac{T_c}{T_m} = \sqrt{\frac{1 - 2K\cos(\phi_1 - \phi_2) + K^2}{\mathrm{ctg}^2\theta + 2K\cos(\phi_1 - \phi_2) + K^2 tg^2\theta}} \tag{8.165}$$

In the formula:

K Amplitude ratio of transmission coefficient under vertical and horizontal polarization.

$\varphi_1 - \varphi_2$ For a particular angle of incidence I, the phase difference between horizontal and vertical polarization.

The circularly polarized antenna can be regarded as two kinds of linear polarization: the Ecosωt (polarization parallel to the plane of incidence) and the Esinωt (polarization perpendicular to the incidence plane). After passing through the medium, it becomes these two quantities:

$$T_{||}[E\cos(\omega \cdot t + \phi_1)] \text{ and } T_{\perp}[E\sin(\omega \cdot t + \phi_2)]$$

Then they are decomposed into left and right circular polarization components with amplitudes of A and B, respectively. Finally, the ratio of cross-polarization to primary polarization is obtained:

$$\frac{B}{A} = \sqrt{\frac{K^2 - 2K\cos(\phi_1 - \phi_2)}{K^2 + 2K\cos(\phi_1 - \phi_2)}} \tag{8.166}$$

The reduction of cross-polarization mainly depends on the geometry of the radome. The Eqs. (8.165) and (8.166) can be seen from the equations that only when

$\varphi 1 - \varphi 2 = 0$ and $K = 1$, the cross-polarization will not occur. However, except for normal incidence, $\varphi 1 - \varphi 2 = 0$ and $K = 1$ generally do not occur. Therefore, when designing the radome's geometric shape, as far as conditions permit, it should be as far as possible so that the rays can be perpendicularly incident. Generally, the flat cover has a large cross-polarization under a large incident angle. Cylindrical radomes have large cross-polarization at large incident angles. For spherical covers and conical covers, if the antenna is installed on the cover's axis, the cross-polarization is relatively small. But in the case of off-axis, its performance is worse than the flat cover. Given the specific situation in practice, especially the single antenna used in electronic warfare has a relatively small aperture, and its gain is not high. When it is installed independently, the radome's shape is not flat, and the distance between the antenna and the radome becomes very close. The radome is regarded as a part of the design's antenna, and thin and strong materials are commonly used to ensure its broadband characteristics. For the mechanical scanning system, to facilitate the manufacturing and determine the high performance of the main radiation area, the good transmission area corresponds to the radiation beam. The antenna rotates simultaneously. The idea of the local broadband radome is that one side of the radome acts as an electromagnetic window. The other side is to protect the equipment inside the hood with considerable structural and strength requirements. To complete the entire radome's structural strength and electrical performance is both laborious and laborious. Therefore, it is proposed to open a place with a radiation hole into a very good transmission window or make a radome and a large cover for compound packaging to ensure its normal performance.

There are strict restrictions on the rapid flight due to its flying speed, geometry, and installation space. Therefore, strict requirements are put forward to develop the antenna array, bringing great difficulties to the design. The antenna array is divided into an antenna array and a radome, including a circularly polarized array element antenna, feeder, a radome, an antenna array mount, and absorbing material. We know that the seeker radome first has higher flight conditions. Also, there are extremely high electrical requirements for the radome. It has wide frequency band characteristics, large angle, small phase distortion, equal amplitude, and high transmittance. The development of this type of radome is very difficult. There are several options for the materials available for the radome, such as epoxy thin-walled cover, PTFE cover, honeycomb-shaped FRP, aramid fiber, etc. Through the performance experiments of the above materials' radomes, it is shown that the wall thickness of the first radome is less than 0.5 mm, and the impact on the performance of the antenna is small. However, when the aircraft's flight speed is 500 (km/h), the huge air pressure does not allow the radome's wall thickness to be designed too small. The second type is a PTFE cover. We know that PTFE is a commonly used non-metallic material with a relatively low dielectric constant. It cannot be made into a radome that supports protection, nor can it be made into a seeker. The radome of the third material has the best structural strength and high wave transmission. It is the preferred radome for general antennas and is widely used in sea, land, and air electronic warfare equipment. However, this radome's wall thickness is generally 5–6 mm, and the size is large. And the electrical performance is worse than the PTFE cover, which cannot meet the seeker's

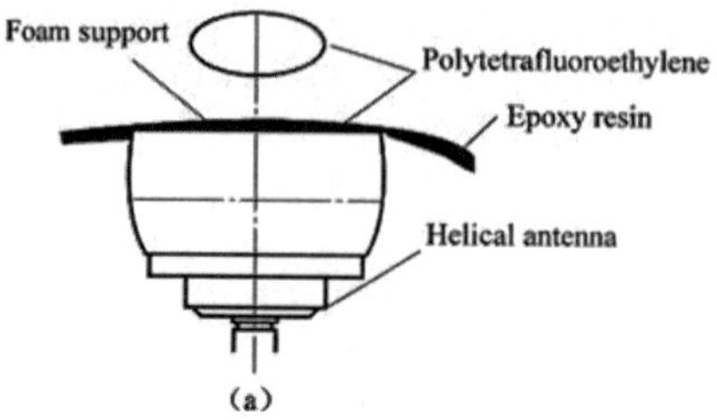


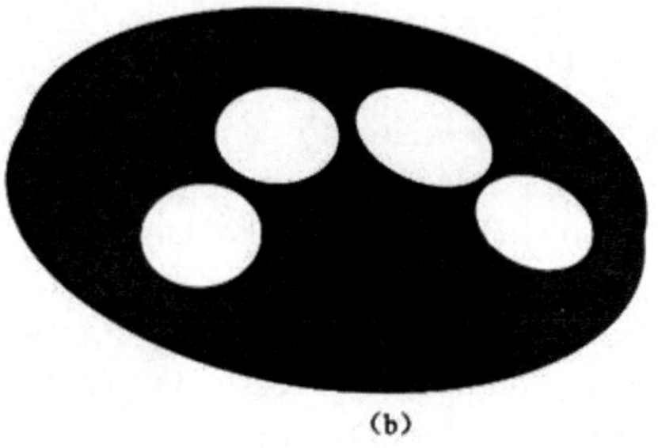


**Fig. 8.56** Composite radome

requirements for volume, quality installation, and electrical indicators. According to the above situation, design a composite radome. The cover type's design idea is based on the current processing capacity. The use of epoxy glass fiber reinforced plastics has a lightweight and good strength and considers the good electrical properties of PTFE cover. Combine the two to maximize the electrical performance of the antenna.

As shown in Fig. 8.56, it is a radome that uses two mutually orthogonal interferometer direction-finding antenna arrays. Each group of three broadband antennas with unequal spacing is distributed on a spherically curved surface. The maximum pointing of several antennas is in different directions to complete the broadband direction-finding and positioning within a limited range, resulting in a composite direction finding with specific amplitude and phase. It requires that after the antenna is installed with a radome, the wide-angle phase error is controlled at about 10 degrees, and the aperture size of the antenna is Φ40–Φ60 mm. The 2 mm thick epoxy glass fiber reinforced plastic is designed as the PTFE cover's support body so that the final cover body meets the seeker's flight conditions. Then design a small PTFE cover and buckle it on the antenna. Foam support is added between the antenna surface and the PTFE cover to enhance the Teflon cover's wind resistance. Holes and flanges position the PTFE cover and the epoxy thin-wall cover to prevent falling off, and the PTFE cover has the best electrical performance. Through the above work, the strict requirements of the antenna system on the radome are well solved. Not only is the structure strong, but also the wave transmission rate is good. It is suitable for conformal design and installation.

### 8.7.2 *Broadband Polarization Cover*

Due to electronic warfare equipment's particularity, the antenna's polarization is usually required to be oblique 45° linear polarization or circular polarization. In this way, the horizontal polarization wave of the radar can be intercepted. The vertical polarization of the radar can also be intercepted so that the system has a higher probability of responding to the electromagnetic signals in space. A linearly polarized antenna or antenna array illuminates the polarization problem of wide-bandwidth angular range and the two main planes' polarization response with different beam widths. In the vicinity of its aperture, the linear polarization is reversed to oblique 45° or circular polarization using a broadband polarization cover. The beam widths of the two polarization responses of the azimuth or elevation planes in the horizontal and vertical polarizations are close.

#### 8.7.2.1 Broadband Circular Polarization Cover

The introduction of a 90° phase shift inside the antenna can turn the linearly polarized antenna into a circularly polarized antenna. But the first condition is that the aperture must be square or round. The system's antenna includes a circularly polarized horn antenna or circularly polarized spiral antenna. The circularly polarized radome has a larger angular range. It is generally composed of five or more layers of zigzag lines, configured according to different design parameters. The orientation of the zigzag lines is at an angle of 45° to the incident wave's polarization. It is generally composed of five or more layers of zigzag lines, configured according to different design parameters. The orientation of the zigzag lines is at an angle of 45° to the incident wave's polarization. Due to the zigzag line's effect, if the incident wave is linearly polarized parallel to the X-axis direction, the zigzag line grid is equivalent to an inductor. If the incident wave is linearly polarized parallel to the Y-axis direction, the zigzag grid is equivalent to a capacitor. When the plane wave is perpendicular to the polyline grid along the Z-axis, it can be decomposed into two polarization components parallel to the X-axis and Y-axis. The two components are of equal amplitude and in the phase before entering the polarizer. After passing through the polarizer, due to the influence of the grid's equivalent inductance and equivalent capacitance, the corresponding lead and lag are generated, forming a 900 phase difference. However, the amplitude is approximately unchanged, and this conversion is achieved. When the wave leaves the circularly polarized cover, these two components are synthesized again as a circular pole wave and propagate forward. The broadband zigzag circular polarizer is composed of several layers of printed circuit boards. The distribution of each layer's zigzag lines is shown in Fig. 8.57 (only two zigzag lines are drawn in the figure), which are formed by etching by a copper-clad dielectric plate. The space between layers filled with low-loss foam is d; the dielectric constant is ε2; W1, W2, and a are the broken line dimensions. According to the boundary conditions of the broken line grid on the dielectric layer, the trigonometric function is used as the basis function to solve the integral equation of the broken line grid current distribution so as to derive a strict expression of the equivalent susceptance of the broken line grid:

$$B'' = 1/a_1 + 2\pi d(\varepsilon_2 - 1)/\lambda \tag{8.167}$$

$$B^{\perp} = 1/a_2 + 2\pi d(\varepsilon_2 - 1)/\lambda \tag{8.168}$$

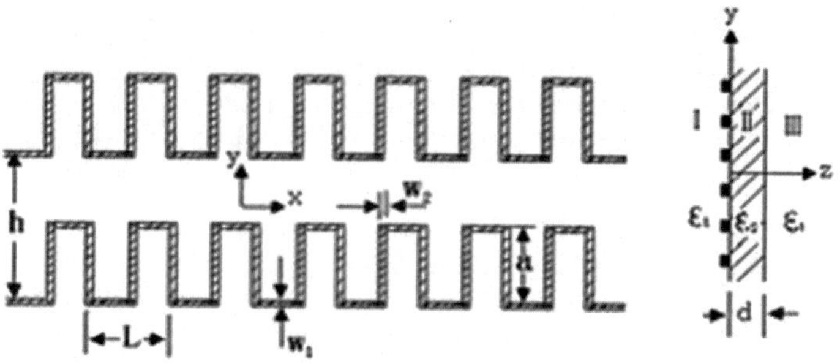


**Fig. 8.57** Schematic diagram of the meandering wire grid

In the formula:

$B''$, $B^{\perp}$ Equivalent susceptance for horizontal and vertical polarization.
$a_1, a_2$ Parameters related to the size of the zigzag wire grid.

A multilayer zigzag grid circular polarizer is used to expand the frequency band. Its equivalent network is a cascade of many basic networks. In general, the metal grid is equivalent to parallel susceptance, and the interlayer between the plates is equivalent to the transmission line. The left side of Fig. 8.58 is a side view of the symmetrical four-layer plate zigzag line circular polarizer. The structural dimensions of plates 1, 4 and plates 2, 3 are the same. The thickness of the dielectric layer between the plates is l1, l2, l3 in order. Figure 8.58 on the right is its equivalent network. According to the transmission matrix [T] of the basic network, the [T] matrix of the equivalent network can be calculated.

$$[T] = \begin{bmatrix} T_{11} & T_{12} \\ T_{21} & T_{22} \end{bmatrix} = \prod_{i=1}^{4} [T_i] \tag{8.169}$$

The scattering matrix [S] of the equivalent network is:

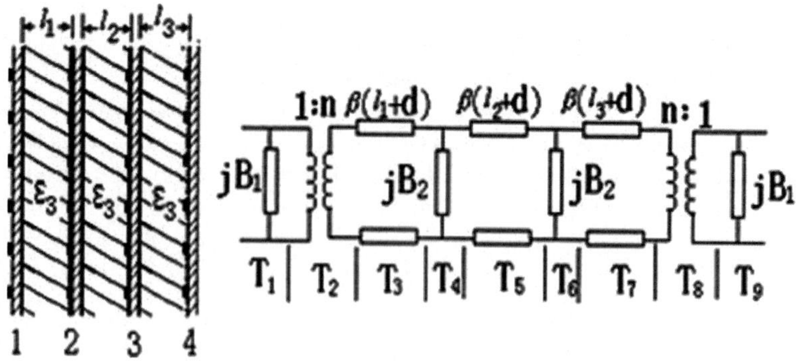


**Fig. 8.58** Four-layer curved plate broken line circular polarizer and equivalent network

$$[S] = \begin{bmatrix} S_{11} & S_{12} \\ S_{21} & S_{22} \end{bmatrix} = \frac{1}{T_{22}} \begin{bmatrix} T_{12} & T_{11}T_{22} - T_{12}T_{21} \\ 1 & -T_{21} \end{bmatrix} \quad (8.170)$$

Suppose the horizontal and vertical S parameters are $[S]^{//}$ and $[S]^{\perp}$ respectively, then the circular polarization condition is:

$$\left| \arg S_{21}^{//} - \arg S_{21}^{\perp} \right| = \frac{\pi}{2} \quad (8.171)$$

The matching conditions are:

$$\begin{cases} \left| S_{11}^{//} \right| < \delta \\ \left| S_{11}^{\perp} \right| < \delta \end{cases} \quad (8.172)$$

In the formula:

δ Allowable value of standing wave coefficient.

When the above conditions are met, the circular polarization axis ratio can be calculated using the [S] parameter at each frequency point:

$$r = \frac{\sqrt{1 + K^2 + 2K \sin \Phi} - \sqrt{1 + K^2 - 2K \sin \Phi}}{\sqrt{1 + K^2 + 2K \sin \Phi} + \sqrt{1 + K^2 - 2K \sin \Phi}} \quad (8.173)$$

In the formula:

$$\begin{cases} K = \left| S_{21}^{//} / S_{21}^{\perp} \right| \\ \Phi = \arg\left( S_{21}^{//} / S_{21}^{\perp} \right) \end{cases} \quad (8.174)$$

The decibel of the axial ratio means:

$$\mathrm{R} = 20 \log |\mathrm{r}| \quad (8.175)$$

The zigzag grid line's admittance must be obtained to calculate the axial ratio from the above formula. It is necessary to calculate the equivalent susceptance at each frequency point repeatedly. By piecewise interpolation and extrapolation. Some existing literature methods can calculate the $B\|(f)$ and $B\perp(f)$ curves. Moreover, in the ultra-wideband design, only $B\|(f)$ and $B\perp(f)$ can be processed at the center frequency, simplifying the design process. The four-layer circular polarizers were simulated and designed in the frequency range of 8–12 GHz and 2–18 GHz, respectively. The corresponding zigzag wire grid structure dimensions are listed in Tables 8.11 and 8.12.

The axial ratio calculated from the table's dimensions is $R < 0.2$ dB in the frequency range of 8–12 GHz, and $R < 1$ dB in the frequency range of 2–18 GHz. In practice, because the irradiated antenna is a non-ideal plane wave, coupled with

Table 8.11 Dimensions of 8–12 GHz four-layer circular polarizer (mm)

| Board No | L | h | $W_1$ | $W_2$ | a |
|---|---|---|---|---|---|
| 1 4 | 5 | 15 | 0.5 | 0.5 | 3.9 |
| 2 3 | 5 | 14 | 0.5 | 0.5 | 4.2 |
| d = 0.4, ε2 = 2.2, ε3 = 1.1*l*1 = 4.95*l*2 = 2.18*l*3 = 4.89 | | | | | |

Table 8.12 Dimensions of 2–18 GHz four-layer circular polarizer (mm)

| Board No | L | h | $W_1$ | $W_2$ | a |
|---|---|---|---|---|---|
| 1 4 | 5 | 16.5 | 0.5 | 0.5 | 3.6 |
| 2 3 | 5 | 12.7 | 0.5 | 0.5 | 3.6 |
| D = 0.4, $\varepsilon_2 = 2.2$, $\varepsilon_3 = 1.1$ $l1 = 9.81$ $l2 = 9.47$ $l3 = 4.68$ | | | | | |

many factors such as matching, processing, and installing the polarizing mask, the four-layer circularly polarizing mask has been studied practice has R < 3 dB in the frequency range of 8–18 GHz. Of course, it can also be deformed in the shape of the zigzag grid line, design to smooth transition at the corner of the rectangular line, and effective spacing control between the layers, and the introduction of the corresponding non-close to 1 support medium material to improve its reliability.

#### 8.7.2.2 Broadband Oblique Polarization Radome

Another widely used broadband polarizer is the broadband oblique polarizer. The multi-point Chebyshev reflection can analyze the standing wave of the center frequency oblique polarizer for its wideband characteristics. However, a complete theoretical analysis is what designers expect. The oblique polarizer comprises a multi-layer metal grid (n + 1) and N layers of air medium. As shown in Fig. 8.59, the metal grid of the 0th layer is perpendicular to the direction of the electric field

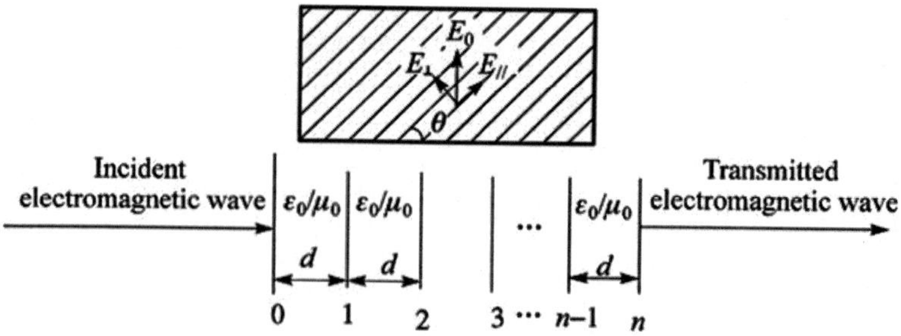


Fig. 8.59 Schematic diagram of oblique 450 polarizer

of the incident wave. The metal grid of the first layer has a deflection angle θ1 relationship with the incident electromagnetic wave. The nth layer has a 45° deflection relationship with the incident electromagnetic wave.

To analyze the polarizer, we must first analyze the propagation of electromagnetic waves on a single metal gate. When the electric field direction of the incident electromagnetic wave is perpendicular to the ideal metal gate, the electromagnetic wave will be transmitted, and the reflection will be zero. When the electric field's direction is parallel to the ideal metal grid, the electromagnetic wave will be reflected, and the transmission will be zero. For any linearly polarized electromagnetic wave, the electric field can be decomposed into vertical and parallel components. While transmission and reflection occur simultaneously, the electromagnetic waves propagating in the two media must meet their boundary conditions at the interface. That is, the tangential components of the electric field and the magnetic field must be continuous. Therefore, when any linearly polarized electromagnetic wave is perpendicularly incident on a single metal grid, its transmitted and reflected fields are:

$$
\begin{aligned}
E_{透} &= E_0 \cos\theta \\
E_{反} &= -E_0 \sin\theta
\end{aligned}
\tag{8.176}
$$

According to the definition of the [S] parameter, the [S] matrix of the single-layer metal gate is:

$$
[S]_M = \begin{bmatrix} -\sin\theta & \cos\theta \\ \cos\theta & -\sin\theta \end{bmatrix} \tag{8.177}
$$

If the oblique 45° polarizer is simply equivalent to (n + 1) [S]M matrices, it is wrong to solve it by cascading [A] or [T] matrices with n air medium [S] parameters. Because each metal gate layer and air layer participate in the reflection of electric and magnetic fields. If the 0th layer metal gate is removed (θ = 0), the standing wave of the polarization cover will become unacceptable. It is possible to find the desired [S] matrix using multiple vector superpositions through careful analysis. The process of reflecting and transmitting the electromagnetic waves simultaneously on the interface of different media is as follows: the electromagnetic wave is reflected on the interface and then transmitted. Therefore, the entire oblique polarizer can be equivalent to the structure shown in Fig. 8.60.

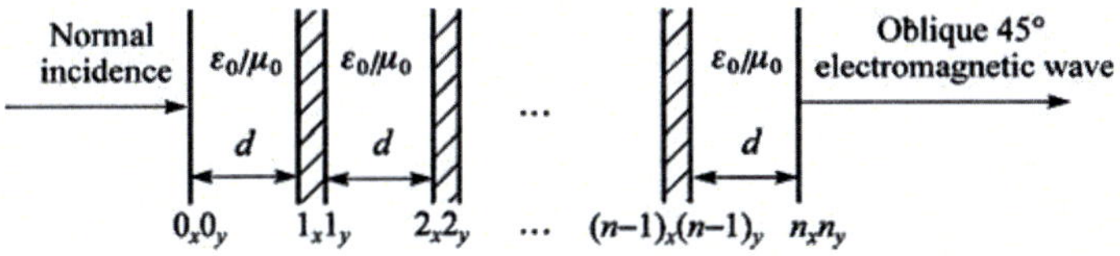


**Fig. 8.60** Equivalent structure of oblique 450 polarizer

Therefore, the entire oblique 45° polarization mask can be equivalent to n two-port networks for analysis. Figure 8.61 shows the equivalent circuit of the oblique polarizer formed by the cascade of equivalent two-port networks.

For the n-th layer polarizer, the electric field vector of the vertically incident electromagnetic wave is perpendicular to the (M − 1)y metal grid. The (M − 1)y grid has reflected the parallel component, and this incident will be reflected and transmitted multiple times in the M-layer polarizer. Table 8.13 is obtained in Fig. 8.62, according to (8.176) and Fig. 8.59 through the geometric projection relationship.

From this, the total reflected electric field ΣETR of the (M − 1)y gate layer can be obtained:

$$\begin{aligned}\sum E_{TR} &= E_{TR1} + E_{TR2} + \cdots + E_{Rn-1} + \cdots \\ &= E_{R1}\sin\theta \cdot e^{-jkd} + E_{R2}\sin\theta \cdot e^{-jkd} + \cdots + E_{Rn-1}\sin\theta \cdot e^{-jkd} + \cdots\end{aligned}$$

This is a proportional series. After taking the limit, you can get:

$$\frac{\sum E_{TR}}{E_0} = -\frac{\sin^2\theta \cdot e^{-jkd}}{1 - \cos^2\theta \cdot e^{-j2kd}} \tag{8.178}$$

According to the [S] parameter, the S11 of the polarizer in the Mth layer has been obtained:

$$S_{11} = \frac{\sum E_{TR}}{E_0} = -\frac{\sin^2\theta \cdot e^{-jkd}}{1 - \cos^2\theta \cdot e^{-j2kd}} \tag{8.179}$$

For the Mx gate layer, the total transmission field is:

$$\begin{aligned}\sum E_T &= E_{T1} + E_{T2} + \cdots + E_{Tn} + \cdots \\ &= E_1\cos\theta \cdot e^{-jkd} + \left(-E_2\sin\theta \cdot e^{-jkd}\right) + \cdots + \left(-E_n\sin\theta \cdot e^{-jkd}\right) + \cdots\end{aligned}$$

The same can be found:

$$\frac{\sum E_T}{E_0} = \cos\theta \cdot e^{-jkd} - \frac{\sin^2\theta \cdot e^{-jkd}}{1 - \cos^2\theta \cdot e^{-j2kd}} \cdot \cos\theta \cdot e^{-jkd} \tag{8.180}$$

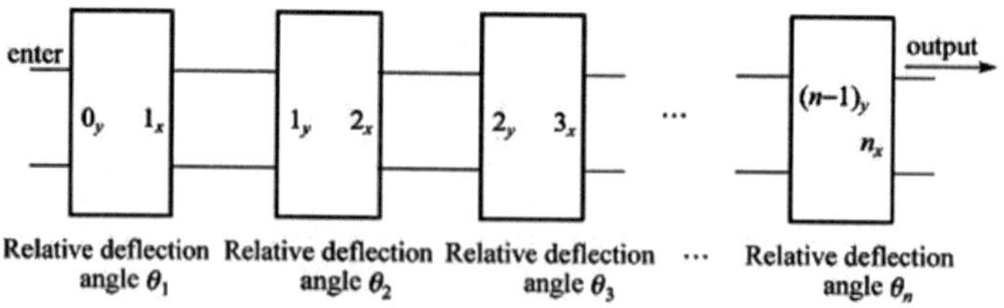


**Fig. 8.61** The equivalent circuit of the two-port network of the oblique 450 polarizer

**Table 8.13** The relationship between the incident electric field E0 and the reflected and transmitted electric fields ($k = 2\pi/\lambda$)

| Electric field $E_{TR}$ of (M − 1)y gate layer | The electric field Ei reflected by the $(M-1)_y$ grid layer is used as the incident field E1 | Electric field ER reflected by the MX gate layer | The electric field ET of the MX layer is transmitted |
|---|---|---|---|
| / | $E_1 = E_0$ | $E_{R1} = -E_1 \sin\theta e^{-jkd}$ | $E_{T1} = E_1 \mathrm{con}\,\theta e^{-jkd}$ |
| $E_{TR1} = E_{R1} \sin\theta e^{-jkd}$ | $E_2 = -E_{R1} \cos\theta e^{-jkd}$ | $E_{R2} = -E_2 \cos\theta e^{-jkd}$ | $E_{T2} = -E_2 \sin\theta e^{-jkd}$ |
| $E_{TR2} = E_{R2} \sin\theta e^{-jkd}$ | $E_3 = -E_{R2} \cos\theta e^{-jkd}$ | $E_{R3} = -E_3 \cos\theta e^{-jkd}$ | $E_{T3} = -E_3 \sin\theta e^{-jkd}$ |
| … | … | … | … |
| $E_{TR\,n-1} = E_{Rn-1} \sin\theta e^{-jkd}$ | $E_n = -E_{Rn-1} \cos\theta e^{-jkd}$ | $E_{Rn} = -E_n \cos\theta e^{-jkd}$ | $E_{Tn} = -E_n \sin\theta e^{-jkd}$ |

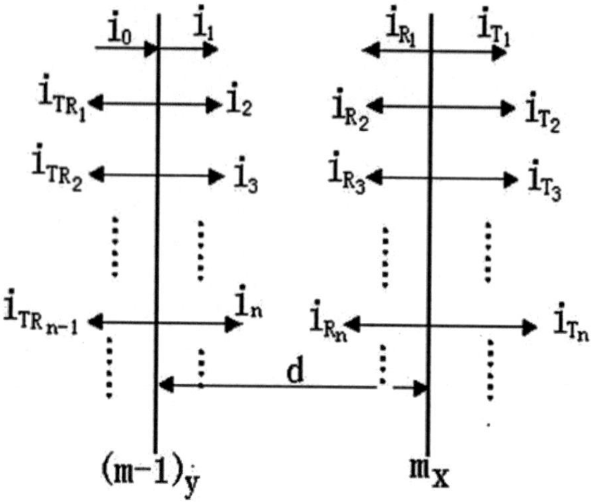


**Fig. 8.62** Schematic diagram of an electromagnetic wave i multiple reflection and transmission in the Mth layer polarizer

Then $S_{21}$ has also been found:

$$S_{21} = \frac{\sum E_T}{E_0} = \cos\theta \cdot e^{-jkd}(1 + S_{11}) \tag{8.181}$$

It is a two-port network with lossless connections: S11 = S22, S21 = S12. Therefore, the [S] matrix of the Mth polarizer has been found:

$$[S]_M = \begin{bmatrix} S_{11} & S_{12} \\ S_{21} & S_{22} \end{bmatrix} \tag{8.182}$$

Since the [S] matrix cannot be cascaded, it needs to be converted into a [T] matrix:

$$[T] = \begin{bmatrix} \frac{1}{S_{11}} & -\frac{S_{22}}{S_{21}} \\ \frac{S_{11}}{S_{21}} & S_{12} - S_{11} \cdot \frac{S_{22}}{S_{21}} \end{bmatrix} \tag{8.183}$$

Thus, the equivalent [T] m matrix of the Mth layer polarizer is obtained.

According to the [T] matrix's cascade characteristic, the equivalent [T] matrix of the entire polarizer is obtained.

$$[\mathrm{T}] = [\mathrm{T}]_1 \cdot [\mathrm{T}]_2 \cdot [\mathrm{T}]_3 \ldots [\mathrm{T}]_n \tag{8.184}$$

Then convert the equivalent [T] matrix to [S] matrix to find the equivalent [S] parameter of the oblique 45° polarized cover. According to the physical meaning of the [S] parameter, the standing wave characteristics and transmission loss characteristics of the obliquely polarized cover can be obtained:

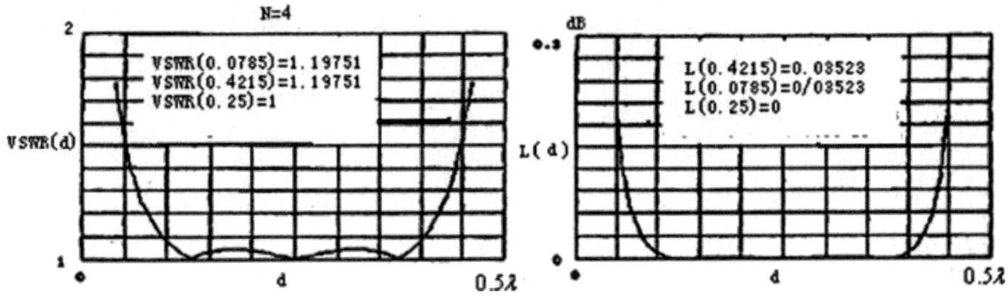


**Fig. 8.63** Standing wave and the transparent wave of the four-layer oblique polarization cover

$$VSWR = \frac{1 - |S_{11}|}{1 + |S_{11}|} \tag{8.185}$$

$$L(\mathrm{dB}) = 20 \log \frac{1}{|S_{21}|} \tag{8.186}$$

The standing wave and transmission loss of the oblique polarizer in different situations are obtained through actual calculations and compared with the actual four-layer oblique polarized cover test results (Fig. 8.63). The results show that the calculated values are in good agreement with the experimental values. The theoretical standing wave is VSWR < 1.16, and the actual result is that the worst of the standing wave of the covered antenna is 10.36 dB. The theoretical insertion loss is 0.035 dB, and the actual result is a maximum of 0.4 dB. The main reason is that the metal grid dielectric board and foam (or honeycomb) are low-loss materials. The polarizing cover's electrical properties are calculated using the above formula, for n = 9 and n = 18, which shows that the proper selection of the number of polarizer layers and grid size oblique polarizer will have a bandwidth greater than 10:1. Through theoretical analysis and calculation of the oblique polarizer and comparison with the actual results, it can be inferred that: when the dense grid condition is met, a low-loss dielectric substrate is used. The low-loss dielectric constant is close to that of air-filled materials. Proper selection of the number of layers and the deflection angle can obtain a broadband oblique polarizer with good performance. The low-loss dielectric can also make the design more flexible and reverse the polarization to any specified angle.

The supporting interlayer between the wideband oblique polarization cover described above is a foam with a low dielectric constant close to 1. It requires special installation in engineering applications, and the thickness of the polarizing cover is relatively thick, and the high-temperature resistance is poor. If the dielectric interlayer is PTFE dielectric plate and other similar materials, the ray method is used to solve this problem. If the network parameters of each unit's oblique polarization layer port can be solved, the problem will be simplified. Each unit's oblique polarization layer is composed of 0° metal gate layer (A), dielectric interlayer (B), and metal gate layer (C) with deflection angle θ. Relative to the ABC oblique polarization layer, the incident electromagnetic wave's polarization direction is always perpendicular to the A shed. After the electromagnetic wave is thrown into the ABC oblique polarization layer, a series of transmission and reflection phenomena will occur on the metal gate's surface, the metal gate dielectric interface, and the dielectric layer. It is an analysis process of multiple reflection problems, as shown in Fig. 8.64. Under the boundary conditions of energy conservation and satisfying the electromagnetic wave transmission at the interface of the same medium, through the electromagnetic wave vector projection and vector superposition, the sub-collection reflection field and the sub-collection transmission within the B medium, between the AB medium layer and between the BC medium layer are obtained field respectively. Finally, the total collective reflection field and total collective transmission field of the entire ABC oblique polarization layer are obtained. The role of electromagnetic waves in each medium layer is a typical damped oscillation process. The equivalent transmission coefficient of the electromagnetic wave in the dielectric layer B is $T_t$, and the equivalent reflection coefficient is $T_r$. Because of $I_{1BCR2}/I_{1BCR1} = I_{1BCR(n+1)}/I_{1BCRn} = -\Gamma_{\varepsilon_r}$, the sum of infinite series is:

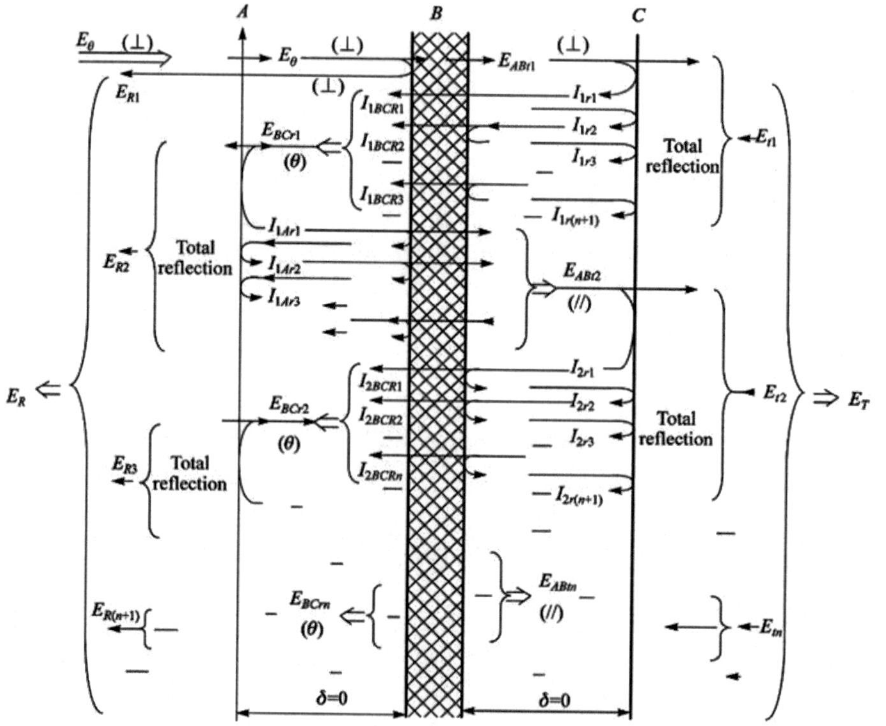


Fig. 8.64 Schematic diagram of ABC oblique polarization layer electromagnetic wave transmission

$$E_{BCR1} = \sum_{j=1}^{\infty} I_{1BCR\cdot j} = I_{1r1}T_{\varepsilon_r}/\left(1+\Gamma_{\varepsilon_r}\right) = -E_{AB\cdot t1}\sin\theta T_{\varepsilon_r}/\left(1+\Gamma_{\varepsilon_r}\right) \quad (8.187)$$

At the same time,

$$E_{t1} = I_{1BC\cdot t1} = E_{AB\cdot t1}\cos\theta \quad (8.188)$$

$$E_{AB\cdot t(i+1)} = \sum_{j=1}^{\infty} I_{iAB\cdot tj} = I_{iAr1}T_{\varepsilon_r}/\left(1+\Gamma_{\varepsilon_r}\right) = -E_{BC\cdot ri}\cos\theta\cdot T_{\varepsilon_r}/\left(1+\Gamma_{\varepsilon_r}\right) \quad (8.189)$$

$$E_{R(i+1)} = I_{iABR1} = E_{BC\cdot ri}\sin\theta \quad (8.191)$$

Therefore, the total reflected field of the obliquely polarized layer of ABC is:

$$\begin{aligned} E_R &= E_{R1} + E_{R2} + E_{R3} + \cdots + E_{Rn} + \cdots \\ &= E_0\Gamma_{\varepsilon_r} + E_{BC\cdot r1}\sin\theta + E_{BC\cdot r2}\sin\theta + \cdots + E_{BC\cdot rn}\sin\theta + \cdots \\ &= E_0\Gamma_{\varepsilon_r} + \sin\theta\left\{\sum_{j=1}^{\infty} I_{1BCRj} + \sum_{j=1}^{\infty} I_{2BCRj} + \cdots + \sum_{j=1}^{\infty} I_{nBCRj} + \cdots\right\} \end{aligned} \quad (8.192)$$

The total transmission field of the obliquely polarized layer of ABC is:

$$\begin{aligned} E_T &= E_{t1} + E_{t2} + E_{t3} + \cdots + E_{tn} + \cdots \\ &= E_{AB\cdot t1}\cos\theta + (-E_{AB\cdot t2}\sin\theta) + (-E_{AB\cdot t3}\sin\theta) + \cdots + (-E_{AB\cdot tn}\sin\theta) + \cdots \\ &= E_0\mathrm{T}_{\varepsilon_r}\cos\theta - \sin\theta\left\{\sum_{j=1}^{\infty} I_{1AB\cdot tj} + \sum_{j=1}^{\infty} I_{2AB\cdot tj} + \cdots + \sum_{j=1}^{\infty} I_{nAB\cdot tj} + \cdots\right\} \end{aligned} \quad (8.193)$$

According to the definition of network parameters, the equivalent 〖S〗 parameters of the ABC polarization layer can be obtained:

$$S_{11} = \Gamma_{\varepsilon_r} - \left(1+\Gamma_{\varepsilon_r}\right)T_{\varepsilon_r}^2\sin^2\theta/\left[\left(1+\Gamma_{\varepsilon_r}\right)^2 - T_{\varepsilon_r}^2\cos^2\theta\right] \quad (8.194)$$

$$S_{21} = \cos\theta\cdot T_{\varepsilon_r} - T_{\varepsilon_r}^2\sin^2\theta\cdot\cos\theta/\left[\left(1+\Gamma_{\varepsilon_r}\right)^2 - T_{\varepsilon_r}^2\cos^2\theta\right] \quad (8.195)$$

When the filled medium is air ($\varepsilon r = 1$), $\Gamma_{\varepsilon r =} 0$ and $T_{\varepsilon r} = e^{-j\alpha}$ are expressed as (8.181) and (8.182).

After obtaining the network parameters of the obliquely polarized layer of unit ABC, using the cascade characteristic of the matrix of 〖T〗, the standing wave and transmission loss of the broadband oblique 45° polarizer of the multilayer dielectric and polarizing grid interlayer can be obtained. The polarizer discussed here has been applied to a broadband multi-beam transmit array and is composed of two polarization grids and a polytetrafluoroethylene dielectric layer. The broadband characteristics agree with the similar oblique 45° polarizers (n = 1) in foreign countries. Figure 8.65 shows the calculated electrical performance of the oblique 45° polarizer. In the 8–8 GHz broadband, the standing wave VSWR of the oblique 45° polarizer is less than 1.3, and the insertion loss La = 0.3 dB.

Figure 8.66 is the actual test situation, which compares the broadband antenna with and without cover. The calculated value is consistent with the measured value. In order to obtain the same electrical performance, the conventional oblique 45° polarizer requires a five-layer polarization grid and four polystyrene dielectric layers.

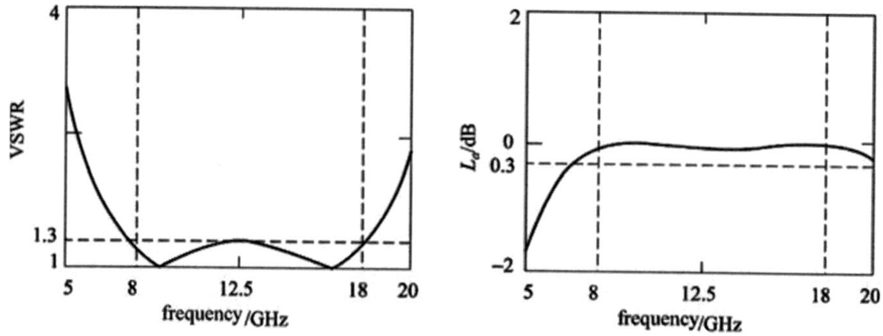


(a) Calculated value of standing wave

(b) Calculated insertion loss

**Fig. 8.65** Calculated value of single-dielectric layer polarizer

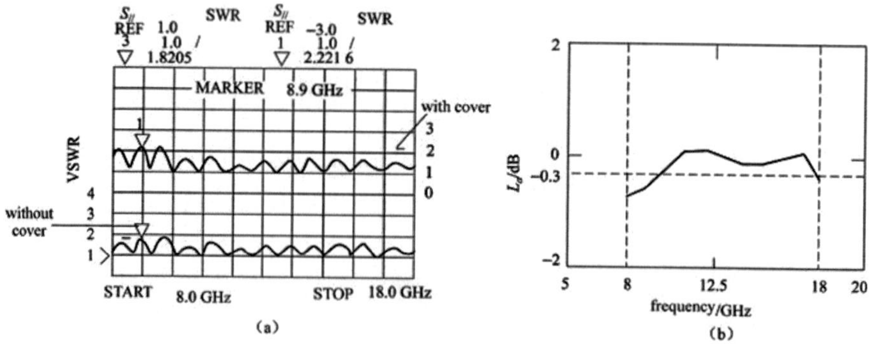


**Fig. 8.66** Measured value of single dielectric interlayer

Under the condition of meeting the same electrical index, the broadband oblique 45° polarizer designed by the above method has the advantages of the least number of layers, easy processing, low cost, high reliability, and strong environmental adaptability. Moreover, the oblique 45° polarizer's influence on the maximum beam direction and sidelobe level of the antenna pattern is much smaller than that of the ordinary oblique 45° polarizer on the antenna pattern. Cylindrical oblique polarization covers the omnidirectional oblique polarization antenna developed by special processing methods to form an integrated antenna, polarization cover, and radome.

### *8.7.3 Antenna Receiving and Transmitting Isolation Technology*

In the electronic system, due to the simultaneous operation of the receiving and transmitting parts in the wideband frequency range or the high-sensitivity receiver's time-sharing operation, the system maintains normal conditions. It uses predetermined countermeasures to deal with the target to jamming. The coupling of the radiation field caused by the transmitting antenna and the receiving antenna has become an important indicator of the broadband antenna system and an important issue to ensure electronic equipment's normal operation. The ratio of the transmitting power $P_j$ of the jamming transmitter to the sensitivity $P_{rmin}$ of the reconnaissance receiver is defined as the transceiver isolation, that is:

$$I = 10\log\left(P_j / P_{r\,\min}\right) \tag{8.196}$$

In the formula:

The $P_{rmin}$—Receiver sensitivity, which is the smallest measurable signal.

If the power radiated by the jamming transmitter reaches the receiver's input end through various channels and the power is less than the minimum measurable signal, it is said to be a good transceiver interval, which does not affect the normal reception of the signal by the receiver. When considering the system's transceiver isolation, first of all, the power that the jamming transmitter should have should be obtained according to the jamming equation (the tube produced by the manufacturer often limits the power). The reconnaissance distance equation shows the required sensitivity of the receiver to determine the required transceiver isolation requirements. Due to the difference in antenna performance, antenna installation, and the platform's surrounding environment, the jammer's actual isolation is not consistent with the required isolation. The actual isolation of the jammer can be defined as:

$$I_1 = 10\log\left(P_j / P'_{r\,\min}\right) \tag{8.197}$$

In the formula:

$P_j$ The actual transmit power of the jammer.

$P'_{r.min}$ Receiver's actual sensitivity under transmit jamming.

The actual isolation should meet to make the system reliable:

$$I_1 \geq I$$

In the broadband range, since the sensitivity of the receiver part changes in the operating frequency band, and the power level or static noise of the jammer also changes, the minimum isolation should be used as the standard. Otherwise, the system may cause self-excitation. The actual isolation must be greater than the required isolation. When the transceiver is installed in any position, the jamming signal power received by the receiver is:

$$P_r = \frac{p_j}{4\pi r^2} \cdot G_T(\theta) \frac{\lambda^2}{4\pi} \cdot G_R(\theta) \cdot \gamma A_T B_T A_r B_r S_D M_D R_D \tag{8.198}$$

In the formula:

- $\theta$ Lobe width angle.
- Pj Jammer's transmit power (w).
- r The linear distance of the wave between the phase centers of the transceiver antenna (m).
- GT($\theta$) Gain coefficient of the transmitting antenna in the coupling direction.
- GR($\theta$) Gain coefficient of the receiving antenna in the coupling direction.
- $\gamma$ Polarization coefficient between transmitting and receiving antennas.
- AT(Ar) Transmission coefficient of transmitting (receiving) antenna feeder.
- BT(Br) Reflection coefficient of the transmitting (receiving) antenna feeder.
- SD Shielding factor.
- MD Diffraction coefficient.
- $R_D$ Inductive reflection coefficient of receiving antenna.

Engineering calculations are often expressed in dB, and Eq. (8.198) becomes:

$$I_R(dB) = T_D + G_m(\theta) + \gamma_D + A_m + B_m + S_D + M_D + R_D \tag{8.199}$$

The formula (8.198) is generally called the isolation between the transmitting and receiving antennas. This formula's meaning is that the transmission power's attenuation reaches the receiver's input end in various ways. The meaning and settlement of each symbol in formula (8.199) are described as follows:

(1) Spatial isolation

$$\text{Spatial isolation is}: T_D = 20 \log \frac{4\pi r}{\lambda} \tag{8.200}$$

This formula is the relationship between the transmission loss of the free space wave and the isolation r between the transceiver antenna and the operating wavelength.

(2) Sidelobe isolation Gm($\theta$)

The subscript "m" of $G_m(\theta)$ in the symbol represents a transmitting antenna or a receiving antenna. In general, in installing the receiving and transmitting antennas (one antenna cannot be used), the two's main lobes are not expected to be opposed, so the name is "sidelobe isolation." The calculation formula is (assuming that an efficient type can approximate the antenna pattern):

$$G_m(\theta) = G(0) - 3\left(\frac{\theta_m}{\theta_{m0.5}}\right)^2 \quad (\text{dB}) \tag{8.201}$$

In the formula:

$\theta_m$ Pattern width.
$\theta_{m0.5}$ half-power beamwidth angle.

$$G(0) = 10\log\frac{20000 \sim 31800}{\theta_{E0.5}\theta_{H0.5}} \quad (\text{dB}) \tag{8.202}$$

It is the maximum gain of the main lobe of the antenna. The value of the numerator in the formula varies with the type of antenna.

(3) Polarization isolation $\gamma_D$

When both the transmitting antenna and the receiving antenna are elliptically polarized, the expression of the polarization coefficient $\gamma$ is:

$$\gamma = \frac{(e_t + e_r)^2 + \left(1 - e_t^2\right)\left(1 - e_r^2\right)\cos^2\psi}{\left(1 + e_t^2\right)\left(1 + e_r^2\right)} \tag{8.203}$$

In the formula:

et(er) Ellipticity of transmitting (receiving) antenna.
$\psi$ The angle between the long axes of the two polarized ellipses ranges from 0 to $2\pi$.

The formula (8.203) has the following three situations:

(1) Both the receiving and transmitting antennas are linearly polarized, i.e., $e_t = e^{\gamma} = 0$. Equation (8.203) becomes:

$$\gamma = \cos^2\psi$$

Thus, when $\psi = 0$ or $180°$, $\gamma = 1$, and the polarization is matched. When $\psi = 90°$ or $270°$, $\gamma = 0$, and the polarization are completely mismatched.

(2) When the transmitting antenna is linearly polarized, the polarization coefficient is:

$$\gamma = \frac{e_r^2 + \left(1 - e_r^2\right)\cos^2\psi}{1 + e_r^2}$$

If the receiving antenna is circularly polarized, that is $e\gamma = 1$ then $\gamma = 0.5$.

(3) When the receiving antenna is linearly polarized, the polarization coefficient is:

$$\gamma = \frac{e_t^2 + \left(1 - e_t^2\right)\cos^2\psi}{1 + e_t^2}$$

If the transmitting antenna is circularly polarized, $\gamma = 0.5$.

(4) Feeder transmission coefficient Am

$$A_m = -10\log(1 + 2\beta l) \tag{8.204}$$

In the formula:

$\beta$ Attenuation constant.
$l$ Feeder length.

Equation (8.204) has assumed $\beta << 1$.

(5) Feeder reflection coefficient $B_m$

$$B_m = 20\log\frac{S-1}{S+1} \tag{8.205}$$

In the formula:

S Voltage standing wave coefficient.

(6) Shielding factor $S_D$

The calculation of the shielding coefficient is relatively complicated. The following approximate calculation is used. When the transceiver antenna is placed on the cylinder, then:

$$S_D = 10\log\left[\frac{A}{\eta A + \xi}\right] \quad (8.206)$$

In the formula:

$$A = \rho\theta_s^2\sqrt{\frac{2\pi}{\lambda r_c}} \quad \eta = \begin{cases} 5.476 \times 10^{-3} & A < 26 \\ 3.340 \times 10^{-3} & A \geq 26 \end{cases} \quad \xi = \begin{cases} 0.5083 & A < 26 \\ 0.5621 & A \geq 26 \end{cases}$$

ρ Cylinder radius (m).
$\theta_S$ The angle of the transmission path around the cylinder (radian).
$r_c$ Transmission path length (m).

Within the normal use range of A < 100, the calculation error is less than 2%.

(7) Plane diffraction coefficient $M_D$

$$M_{D\cdot(TEM)} = -20\left[\log\left(\frac{a}{\lambda}\right) + \frac{1}{2}\log\left(\frac{d}{\lambda}\right) + 2.048\right]$$

$$M_{D\cdot(TE01)} = -20\left[\log\left(\frac{a}{\lambda\cos A_0}\right)\left(\frac{1}{2}\log\left(\frac{d}{\lambda}\right)\right) - 2\log\left(-\mathrm{Sec}\frac{270 - A_0}{2} + \mathrm{Sec}\frac{270 + A_0}{2}\right) + 2.348\right]$$

In the formula:

a Antenna aperture size.
d Antenna spacing.

(8) Induction reflection coefficient

When there is a symmetrical reflector in the antenna aperture, the approximate calculation of the maximum value of the degree of deterioration of the transmission and reception compartment caused by this reflector is:

$$R_D = -20\log\left|\frac{\Gamma(1 - b^2)}{1 + b^2}\right| \quad (8.208)$$

In the formula:

Γ Reflection coefficient for reflection point.
b Amplitude ratio of two orthogonal circular polarized components constituting elliptical polarization.

The value of this item is very small, usually can be ignored. In practice, the system must ensure good isolation to prevent self-excitation. The methods of increasing the antenna transceiver isolation are usually:

(1) Increase the distance between the receiving and transmitting antennas

It is obvious to increase the distance between the transmitting and receiving antennas. However, the degree of isolation is inversely proportional to the square of the distance. When r is doubled, spatial isolation only increases by 6 dB. In applications where the space size is limited, it is difficult to meet the requirements by increasing r.

(2) Reduce the relative gain of the receiving and transmitting antennas

If the transceiver antenna is installed in the same direction, the sidelobe level in the 90° direction should be reduced. Adding a cylinder and a wave absorber to the transceiver antenna's aperture surface can reduce the sidelobes by about 10 dB, thereby increasing the isolation by 20 dB.

(3) Polarization isolation

Polarization isolation is used to orthogonalize the receiving and transmitting antennas' polarization, increasing the isolation by 3–6 dB. In theory, the polarization coefficient can be zero. But in fact, it is impossible. There is always a cross-polarization component. For example, a linearly polarized receiving antenna uses a linearly polarized antenna orthogonal to it for transmission. The attenuation of its signal power is only about 10 dB.

(4) Technical measures to improve transceiver isolation

Two factors affect transceiver isolation. One is a direct coupling between antennas, and the other is the environmental coupling. The above methods (1)–(3) for increasing isolation are considered from the factor of "direct coupling between antennas." The technical measures mentioned below refer to modifying the surrounding environment that jamming with the transmitting antenna. For example, the use of metal convex plates, spikes (made of absorbers), metal grids, and metal plates to block and slow down the radiation between the transmitting and receiving antennas and destroy the energy carried by the rays. That is to diffract the rays to other directions or absorb them, as shown in Figure 8.67a–c, as follows:

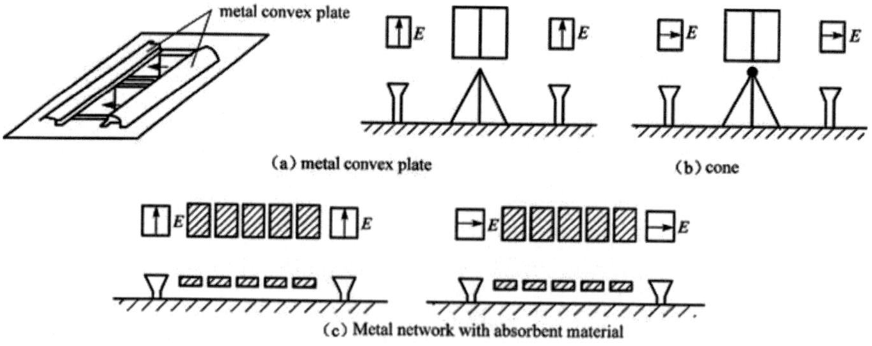


**Fig. 8.67** Various isolation measures

(a) Add a partition between the transmitting and receiving antennas. Laying microwave absorption material on the board can obtain 10–20 dB isolation.

(b) Laying microwave absorbing materials on and over the relevant shielding objects between the transmitting and receiving antennas, especially the absorbing materials in the shape of a small cone, is better. It is also beneficial to improve the receiving and to send isolation.

(c) The jamming high-frequency transmitter oscillator/amplifier shielding box. Microwave absorbing material or paint is stuck inside the shielding box. The high-frequency plug seats are also covered with microwave absorbing materials to prevent high-frequency power leakage.

(d) Transceiver antennas use impedance surface technology to reduce and eliminate creeping waves, optimized selection of antenna form, and low-side-lobe technology. If the transmitting and receiving antenna's installation distance is r = 8 m, f = 7.5 GHz (λ = 4 cm), the spatial isolation TD = 68 dB is calculated by the formula (8.199). If other technical measures are added, 20 dB can be obtained. For short-range low-power jammers, this generally meets the requirements. However, for high-power jammers and high-sensitivity receivers, the distance between the transmitting and receiving antennas cannot be too large. The above measures alone cannot meet the requirements. The following two methods can be used in the project: the time division method, also called "time isolation," to distinguish it from "energy isolation (space division)" and frequency division. This kind of method can be understood as infinite receiving and sending isolation. Therefore, they are one of the most effective measures to solve the transmission and reception isolation. This method involves the reduction of circuits and other performance of the whole machine. Besides, reducing receiver sensitivity is also away. But this method is a last resort and generally not suitable. The design example of transceiver isolation engineering is given below.

$$\text{Suppose } P_j = 1500\,\text{W}$$

$$P_{rmin} = -50\,\text{dBm} = -80\,\text{dbw}$$

The required transceiver isolation calculated by the formula (8.197) is:

$$I = 111.8\,\text{dB}$$

(1) Polarization isolation. The transmitting antenna and the frequency measuring antenna use circularly polarized antennas with opposite directions of rotation. They are taking the polarization coefficient = $\gamma 0.5$, then $\gamma_D = -\,3$ dB.

(2) Feeder loss. If the transmitting antenna uses a waveguide, the feed length is set to 6 m. In the second and third centimeters, if the waveguide loss data is 0.6 dB/m, then AT = − 3.6 dB. In the second and third centimeters, if the

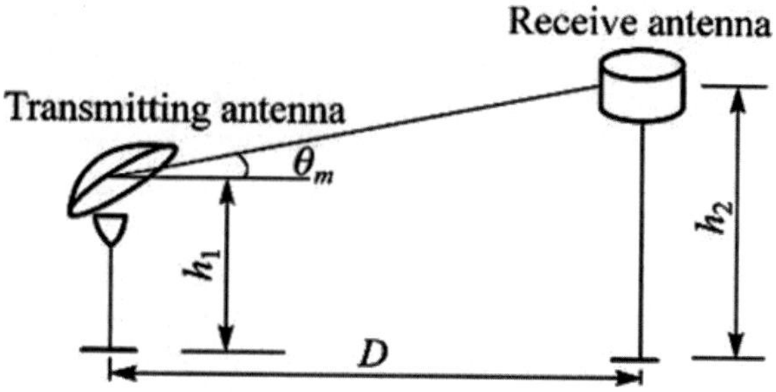


**Fig. 8.68** Installation diagram of transceiver antenna

waveguide loss data is 0.6 dB/m, then AT = – 3.6 dB. It is assumed that the reflection loss has been included. The total loss of the frequency measuring antenna feeder also takes the same value, then Am = – 7.2 dB. This value will be exceeded.

(3) Spatial isolation. The installation distance between the transmitting and receiving antenna is r = 8 m. On the 7.5 GHz frequency, TD = – 68 dB is calculated.

(4) Sidelobe isolation. The installation diagram of the transmitting antenna and frequency measuring antenna is shown in Fig. 8.68. The sidelobe level of the transmitting antenna facing the frequency measurement antenna's direction can be approximated by Eq. (8.201).

$$G_j(\theta_j) = 24 - 3\left(\frac{\theta_j}{\theta_{0.5}}\right)^2$$

In the formula:

$$\theta_j(\theta_m) = tg^{-1}\left(\frac{h_2 - h_1}{D}\right)$$

The calculation result is used as a reference, and the actual measured value shall prevail. If the sidelobe of the transmitting antenna facing the frequency measuring antenna is – 20 dB, then the gain in this direction is Gj $(\theta_j)$ = 24–20 = 4 dB.

The gain of the frequency measuring antenna is $G_r(\theta_r)$ =– 10 dB. If MD, SD, and KD are omitted, the isolation obtained from the above items (1)–(4) is I1=3+7.2+68 – 4+10 = 84.2 dB.

(5) Other technical measures adopted

It is difficult to add a partition wall between the transmitting antenna and the frequency measuring antenna. Adding a circular metal edge and absorbing material to the transmitting antenna's side can reduce the sidelobe by – 15 dB. Take the transmitter's oscillator, amplifier, connector, and other leak-proof measures to obtain – 5 dB isolation. In this way, – 15 dB transceiver isolation can be obtained, plus 84.2 dB of (1)–(4), 7.6 dB different from the required value of − 111.8 dB. By increasing the height of the frequency measuring antenna and adopting a method on the frequency measuring antenna, the above index requirements can be met under the conditions of ensuring the directivity of the antenna, reducing the side lobe of the pitch plane, and making the pattern have a certain up warping. In the broadband system, the isolation is based on the worst point in the entire frequency band between the receiving and transmitting antennas as a reference for design work. The worst isolation in practice should be greater than this value. Under normal circumstances, the isolation difficulty is most likely to occur at the low end of the broadband operating frequency range. Besides, the transmitting antenna and other receiving antennas of this system, such as multi-beam direction-finding antennas, jamming receiving antennas, etc., also have transceiver isolation. Therefore, the worst receiving and transmitting antenna isolation in the system should be taken as one of the necessary conditions for satisfying the system operation. It should also be considered from the whole machine's perspective to improve the transceiver separation, such as appropriately reducing the receiver sensitivity or reducing the power amplifier's output power, using time division, frequency division, space division, etc. It involves a wide range. The calculation method also has its points. Therefore, the problem should be considered comprehensively. Through the actual study of the isolation of the receiving and transmitting antennas, it is found that how to reduce the sidelobe of the antenna in a broadband frequency to effectively reduce the radiation coupling of the antenna and the impact of creeping current is a critical issue in engineering implementation. At the same time, when possible, comprehensive methods such as adding absorbing materials and impedance surface technology are used to meet the design requirements.

### *8.7.4 System Architecture of Broadband Antenna*

The broadband antenna system is an RF function block. It is formed by carrier platform, radome, polarizing cover, antenna arrangement, support and fixation, active and passive microwave components, beamforming, signal transformation control mode, and power supply. In this way, the larger electronic system can be divided into functional top-level designs. For example, the reconnaissance system can be divided into four major parts: the antenna, the receiver part, the information processing, and display part, the system's energy supply and the cabinet and loading platform for the structural realization, plus the transmitter part (including high-voltage power supply, heat dissipation, etc.) The generation of jamming patterns and system control requires a transmitting antenna (array) to form an electronic system. Different wideband antenna systems appear for different working platforms and tactical purposes in engineering practice.

On the other hand, the differences between antennas and different layouts must be adapted to the manufacturing technology and material conditions. The components that make up an antenna system are generally not simple combinations. It needs to

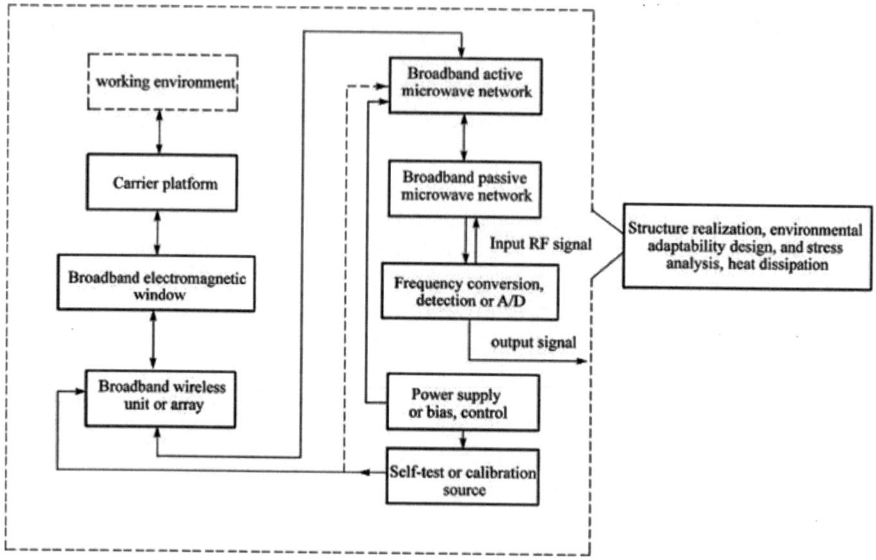


**Fig. 8.69** Block diagram of the broadband antenna system

complement each other organically. Notably, the outsourced active devices in the system may change the model and some indicators for some reason. At this time, it is necessary to adjust accordingly to achieve the required index. In the actual project, the broadband antenna's performance requirements have been transferred to the performance shown by many broadband antennas and microwave channels together with the specific loading platform.

The architecture and modules of the broadband antenna system are shown in Fig. 8.69. It mainly includes ten components. The entire broadband antenna system may only use reception, transmission, or both. It involves the radiation field design, the design of microwave components, and structural engineering design.

(1) For shipboard platforms and vehicle-mounted equipment, the radiation part should be reduced according to the specific situation where the equipment is located on the sea, ground, or mountain. The essential requirement is to use a height of at least ten wavelengths while optimizing the pitch pattern to reduce the energy radiated to the medium within a sufficient distance. Technical measures can also be adopted in the system. The media for the working environment of airborne or ballistic and satellite-based electronic equipment is relatively ideal. Still, more practical issues need to be considered in terms of environmental adaptability.

(2) The carrier platform is a big difference between engineering design and ideal design. Applicable radio equipment is always installed on a certain platform (airplanes and pods, warships, automobiles, balloons, satellites, missiles, etc.) These platforms have a relatively large impact on the performance of the antenna. Therefore, the appropriate position should be selected on the

platform, such as the aircraft's vertical tail and the head of the aircraft, the mast's height on the warship. However, some installations have a considerable distance between the antenna radiator and the microwave components below. Extended feeder losses are inevitable. In addition to the platform's preferred installation position, the resulting influence on the carrier's movement must also be considered jointly. It is best to design the carrier as part of the antenna. The layout of the antenna on the carrier is an essential technology in the design of airborne antennas. It depends on the antenna distribution, beam pointing, etc.

(3) The electromagnetic window is a radome. It has three functions: it plays the role of transmitting waves, protecting the antenna, and the role of wide-angle angular polarization twisting of the radiation port, the circular polarization transformation, the 45° oblique polarization, and the modulation of the radiation beam pattern. According to the specific engineering design, such windows have two types: centralized and distributed. Each antenna has a cover called distributed. The other is multiple antennas, an array, and a large aperture antenna sharing a cover. Sometimes the integrated design of radome and the polarized cover is used to reduce the volume. It is the junction between the antenna and the outside. Its shape, appearance, interface, environmental suitability, materials, and processing are more important.

(4) Broadband antenna unit/array section. The antennas' interface on the platform will be solidified as long as the interface is fixed on the platform. At this time, the relationship between the antenna and the platform is very close. Suppose the antenna is installed in a concentrated place on the platform. In that case, the influence between the antenna unit, the influence of the antenna and the bracket body, and the carrier platform's influence will become larger. At the same time, it also involves the polarization mode of the antenna, the pointing and beam coverage, the isolation between the transmission and reception channels and the isolation screen and surface impedance measures, and the interface relationship between the system self-test or correction components. An array antenna with a wide range of functions is used as much as possible. A certain number of virtual elements are added at the array's edges if necessary to improve the radiated field's electromagnetic compatibility. Antennas are used for frequency measurement, direction finding, warning, continuous-wave guidance, self-checking, zoned antennas, side-lobe antennas, high-power jamming antennas, and space power combined jamming. Generally, the antenna is linearly polarized, and some antennas are radiated or received by circular polarization.

(5) Broadband active network components. Including microwave switches (including mechanical ridge waveguide switches), they are limiting amplifiers, low noise amplifiers, digital phase shifters, controllable attenuators, solid-state power amplifiers, broadband T/R modules, TWT phase-consistent traveling wave tubes, related heat dissipation structure, and calibration self-test interface. These components are combined and controlled in a certain way to form a phase gradient and amplitude amplification mode that can be

scanned. The consistency debugging of these active devices is difficult due to the closed field between the channels, their electromagnetic, and easy control.

(6) Broadband passive network components. Broadband passive network components. Including microwave power divider or combiner, directional coupler, Attenuator with fixed phase shifter (including coaxial adjustable phase shifter), filter, difference device, Butler matrix feed composed of sum and difference, monopulse beam generator, Bootlace lens produced by phase detector and optical beamforming technology (the number of elements and the number of beams are the same), Rotman lens, dielectric Rotman transmission. The broadband pattern generated by the Longbo lens points to a constant beam pattern, and cables with good phase consistency are used for each connection channel. In the combination of specific networks, passive components and active components may be used together. The miniaturization and high reliability of microwave network components are required to realize multiple functions in the project.

(7) Frequency conversion/detection, and A/D converter. This part converts each antenna channel's received signals through the microwave network amplitude and phase conversion or directly output digital signals for high-precision radar parameter testing (including direction-finding accuracy). It involves system synchronization and signal pairing. The local oscillator signal required for frequency conversion is supplied by the back end of the receiver. While the video signal is obtained through the detector, the multi-channel detector system requires the detector to be a matched detector. In practice, some microwave bars have integrated the detector into a module.

(8) Power supply, offset and control part. This part provides the offset for the microwave network's active components and the primary input for the TWT traveling wave tubes' high-voltage power supply. This part provides the offset for the microwave network's active components and the primary input for the TWT traveling wave tubes' high-voltage power supply. These power supplies require stability and a low ripple. At the same time, the control level is provided by the control circuit in the module. It generates a sequence code through a certain convention, makes the switch on and off, and controls the phase shifter to generate a broadband digital phase shift value. The technical maturity of this module is relatively high.

(9) Self-test or correction source. The signal is fed to the self-test antenna through the amplifier or the self-test interface of the active microwave element through the point frequency oscillation source. The signal is output to the test instrument or receiver through each antenna and RF channel to determine whether the connection parts work normally. Also, feed the broadband signal level from the reserved self-test input port. Test its amplitude and phase from the output port, record each channel's amplitude and phase errors into the correction database through the receiver, complete the extraction of each channel's initial RF data, and perform the dynamic correction. The module is powered and controlled by the power supply, offset, and control modules.

(10) Structural design part. It includes stress analysis of the broadband antenna system, heat dissipation of active devices and structural reinforcement and guarantee of each connecting part, environmental adaptability design, consideration of three prevention measures, adjustment of the state of the antenna, cable direction, and fixing of each channel part, the realization of the antenna structure, the optimal design of the bracket, the selection of materials, the adjustment of the center of gravity of the antenna, etc. If electromagnetic compatibility characteristics are more important for large-scale broadband antenna systems, the transmitting and receiving antennas are designed in blocks.

The functions of the broadband antenna system architecture have been described above. It can be divided into two large parts, the radiation field part and the beamforming network part. In the wideband frequency range, the range of each part's mutual influence is very wide. When solving specific problems, first consider choosing a direction-finding system or jamming working method when meeting the overall equipment index requirements. It includes phase interferometer direction finding, multi-channel amplitude comparison method, maximum beam method or cross beam method, time difference direction-finding method, and hybrid method. The jamming is wide beam coverage, large area switches selection, large aperture mechanical rotation, and multi-beam or phase array power synthesis. Antenna design should also consider using the array's synthetic aperture technology as much as possible to reduce the number. Simultaneously, the problem of introducing a large radar scattering cross-section by the antenna should also be considered. The antenna can also adopt the reduction method (after certain technical measures, the antenna is restored to the original performance state) and the comprehensive method (as far as possible, in practice, the new design is adopted to obtain the performance that meets the requirements). Both methods are used in engineering. For the anti-radiation missile seeker radome, because of its limited size, it is a circular axis-symmetric structure. If the symmetrical antenna array is used, the pressure on the ultra-wideband radome will be reduced to some extent. Reducing the impact of discontinuity on the RF path and integrating the microwave network with the antenna in some ways will bring benefits in terms of improving performance.

The rapid development of microelectronic technology, digital technology, manufacturing and process, and material science has brought greater vitality to broadband beamforming technology and antenna systems. The improvement of microelectronic means has made the development of broadband microwave active antenna technology possible, resulting in a small-caliber, high-gain, multi-functional system. The emergence of new materials and new devices has promoted the exploration of plasma antennas, superconducting antennas, and nanomaterial antenna technologies. Due to the demand for military electronic equipment, the application research of time-domain technology has been accelerated. As a result, broadband and ultra-wideband antenna technologies have moved forward into new areas.

# Chapter 9
# Jamming Against Phased Array Radar

## 9.1 Overview of Phased Array Radar

The array antenna of phased array radar to switch beam quickly and accurately enables the radar to interleave in time and simultaneously perform various functions. This type of radar can also be used as a communication system. By changing the beamwidth of the antenna, certain areas can be searched more quickly with smaller gains. Random variations in the transmitting frequency between pulses or within coding in pulses can cause frequency agility. Multiple amplifiers located in the aperture can produce a large amount of power. The electronically controlled array antenna can provide the required flexibility for the radar to perform various special functions that are most suitable for specific tasks.

With the continuous development of electronic technology, the phased array antenna has been greatly improved. For example, the low side lobe can be less than −40 dB. Better components, such as radiation units, phase shifters, and power differentiators, are now available. More economical solid-state devices and memory chips have ensured precise phase control with modified frequency and temperature. Solid-state microwave devices have a broad prospect in the future system of combining solid-state modules with radiation units. A phased array can realize self-adaptive control and can be used for sidelobe cancellation.

With the development of microwave integration technology, the cost of phased array radar is decreasing. However, the lower side lobe and the better performance of the wideband will increase the cost.

A phased-antenna is an aperture composed of the same radiating elements, independently controlled in phase and amplitude. Predictable radiation pattern and beam direction can be obtained.

According to a pen-shaped beam of half-wavelength spaced, the relationship between the number of radiating elements N and beamwidth is:

$$N \approx \frac{10000}{(\theta_H)^2} \tag{9.1}$$



X. Zhang et al., *Theory to Countermeasures Against New Radars*,
https://doi.org/10.1007/978-981-16-6715-2_9

Or

$$\theta_B \approx \frac{100}{\sqrt{N}}(^\circ) \tag{9.2}$$

In the formula,
$\theta_B$ is 3 dB beamwidth.

$$G_O = \pi N n \approx \pi N n_L n_a$$

The n dropped $n_a$ due to the antenna loss $n_L$ and gain caused by uneven amplitude distribution caused by unequal weighting elements.

$$G(\theta_O) \approx \pi N n Cos\theta_O \tag{9.3}$$

### 9.1.1 Phase Scanning

The antenna beam points are in a direction perpendicular to the phase wavefront. In a phased array, this phase wavefront is adjusted by controlling each radiation element's excitation phase, thereby controlling the beam, as shown in Fig. 9.1. The phase shifter uses electrons to meet the fast scanning requirements and adjust the phase between 0 and 2π. If the distance between cells is d, when the scan is $\theta_O$,

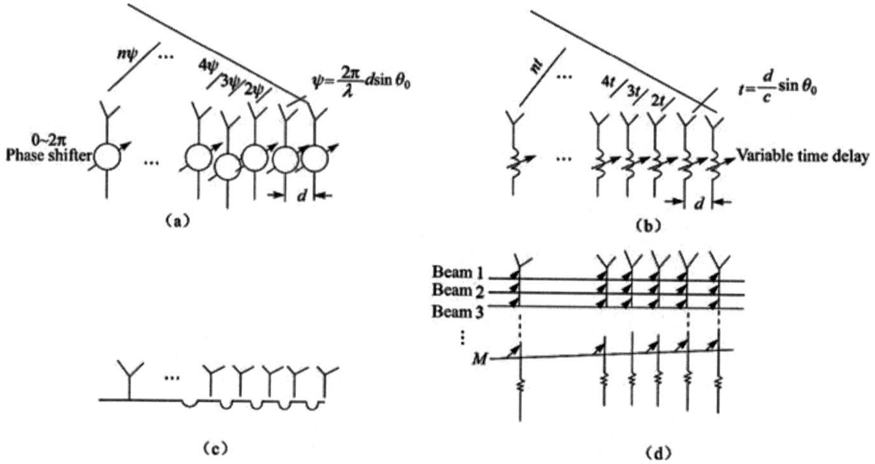


Fig. 9.1 Schematic diagram of electrical scanning beam generation. **a** phased array; **b** delay array, **c** frequency scanning array, **d** glass type array

the phase shift increment between adjacent cells is $\psi = \left(\frac{2\pi}{\lambda}\right)d\text{Sin}\theta_0$. If the phase $\psi$ does not change with frequency, the scan angle $\theta_0$ is related to frequency.

Phase scanning is very sensitive to frequency, while delay scanning is independent of frequency, as shown in Fig. 9.1b. The delay line is used to delay increment $t = \left(\frac{d}{c}\right)\text{Sin}\theta_0$ between the units instead of the phase shifter, where C is the light speed. Adding a delay network to each unit with a phase shifter can reasonably meet all aspects' requirements.

### 9.1.2 Frequency Scanning

The frequency-sensitive characteristics of phase scanning can be used to make frequency a more effective parameter. As shown in Fig. 9.1c, all radiators are in phase at a specific frequency. When the frequency changes, the beam is scanned due to the linear shift in the aperture phase.

For the reception, each radiating element's output can be amplified and digitized so that the signal can be sent to a computer for processing, including simultaneous multi-beam formation. The digital part at the subarray level is achievable. With the lens's proper design and reflecting surface, the converging feed-in plane can form many independent beams. Each beam generally has the gain and beamwidth of the entire antenna. As shown in Fig. 9.1d, the structure for providing broadband multi-beam 2 is to use parallel metal plates. This group of parallel plates contains a wide-angle microwave lens. Each port corresponds to an independent beam, and the lens provides an appropriate time delay for the aperture, giving a constant scan with frequency. The beam is selected through a switch matrix, which requires (M − 1) single-pole double-throw switches to select one of the M beams. These beams are fixed in space and overlap at 4 dB, which is very different from the scanning method.

All beams can be connected to independent receivers to replace the beam switch to provide multiple beams' simultaneous reception. The radiation pattern of the transmitter must be wide to cover all receive beams. This multi-beam system has been used in three-coordinate radar combined with mechanical rotation.

A single beamformer can generate multiple independent beams by modifying the amplitude and phase on the aperture. As shown in Fig. 9.2, two independent beams are generated. The two beams have the same amplitude (voltage) distribution F(X) but have differently inclined linear phase wavefronts. The total aperture excitation of the two beams is

$$\begin{aligned} F(X,\psi) &= F(X)e^{j2\varphi_1(x/a)} + F(X)e^{j\varphi_2(x/a)} \\ &= 2F(X)\Big[\cos(\varphi_1 - \varphi_2)\frac{x}{a}\Big]e^{j(\varphi_1+\varphi_2)(x/a)} \end{aligned} \tag{9.4}$$

The aperture amplitude distribution required to generate two independent beams varies according to cosine. The two-phase distribution is linear and has an average slope.

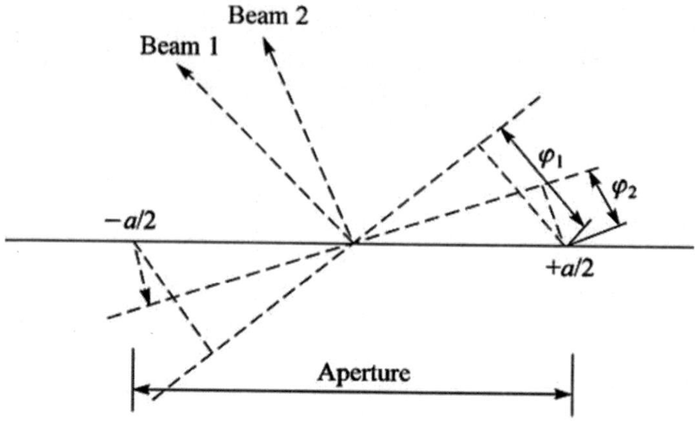


**Fig. 9.2** Schematic diagram of the aperture distribution of the dual-beam

In most phased array systems, only the phase can be controlled. Adding various required phase-shifting settings (modulo $2\pi$), regardless of the required amplitude change, can still obtain a good approximation suitable for forming multiple beams.

If multiple functions are required, the phased array system can be greatly simplified. The fire control system requires that the beam must be able to point in any specified direction at any time. Suppose the array is scanned only in the vertical plane, and the mechanical rotation provides azimuthal scanning. In that case, the number of phase control points will be reduced to horizontal lines. Scanning can be phase scanning or beam switching, or simultaneous multiple beams can be used in reception, and wide antenna patterns can be used in transmission. Many of these types of systems have been applied at sea and on land.

### *9.1.3 Typical Phased Array Radar*

The S-band phased-array radar AN/SPY-1 is part of the Navy's "Aegis" weapon system. It has four phased array apertures to provide unimpeded hemispherical coverage. The early structure used a simple feed system of 68 sub-arrays during the reception, and each sub-array included 64 waveguide-type radiators for a total of 4352 units. When transmitting, the sub-arrays are combined in pairs, and 32 sub-array pairs give an emission aperture of 4096 radiations. The phase shifter has a 5-bit, non-reversible, magnetic field line excitation, locking garnet structure, which directly feeds the waveguide radiator.

The synthesized phased array has an aperture with a forced feed structure and 4350 waveguide-type radiators. Mono-pulse and differential receive lobes and transmit lobes are optimized, respectively.

The antenna aperture includes a circularly polarized spiral radiator (used to suppress raindrops) arranged on a concentric ring around a tulip-shaped dielectric structure. The unit also contains a diode phase shifter for beam steering. The phase shift exciter is installed along the periphery of the aperture.

The feed network has a very simple structure. It consists of a flat disc-shaped medium behind the radiation aperture. The medium is powered by a single pulse feed at its center. Since the same circularly polarized antenna is used for transmission and reception, single reflection targets (including raindrops) will be suppressed.

Patriot (PATRIOT)

It is a multi-functional phased array radar system developed for the Army. This system uses an optically fed lens array, as shown in Fig. 9.3. A mono-pulse feed respectively optimizes the sum and difference lobes. The aperture is circular and contains approximately 5000 cells. It uses 4-bit magnetic field lines on both apertures to excite irreversible ferrite phase shifters and waveguide radiators. The antenna is shown in Fig. 9.3. It is installed on the vehicle and can be folded to facilitate transportation.

Electronic Agile Radar

Westinghouse Electric Company has developed an X-band phased array radar for airborne applications. Later, the system evolved into the AN/APQ-164 radar on the B1-B bomber. As shown in Fig. 9.4, it has an elliptical aperture of 1526 elements, a diameter of 39in, and is composed of 1818 radiation elements, each with its own

**Fig. 9.3** Patriot multi-function phased array radar antenna

**Fig. 9.4** AN/APQ-164 phased array radar antenna being assembled

reversible iron gas lock phase shifter. The system has many flexible features. The aperture phase can be changed to change the beam shape from a pencil shape to a cosecant square and then to a vertical fan shape. And the polarization can also be changed from vertical polarization to unidirectional circular polarization.

Airborne Warning and Control System (AWACS)

Westinghouse Electric Company developed the first combat radar antenna with an ultra-low sidelobe antenna for the AWACS system. It uses a waveguide slot and waveguide array with more than 4000 slots, which can be used for aerial surveillance, as shown in Fig. 9.5. It can be rotated together in a circular radome for mechanical rotation.

Pave Paws

As shown in Fig. 9.6, the pave paws were used to provide early warning of ballistic missiles and achieve satellite tracking. It is an ultra-high frequency (UHF) solid phased array radar. A set of systems includes radars at both ends with the apertures tilted at 120° so that it can provide a total observation range of 240°, and it can detect a $10m^2$ target at 3000 miles.

Cobra Dane

The Cobra Dane is a huge L-band phased array radar developed and deployed to collect information from foreign intercontinental missile tests, as shown in Fig. 9.7.

It is a sparse array with a diameter of 95 ft and a total of 34,768 units, of which 15,360 units are active, and the rest are passive. Active units may be used later. The active unit is divided into 96 sub-arrays, and each sub-array has 160 radiators. The traveling wave tube is used for transmission. The peak power on the antenna is 15.4 MW, the average power is 0.92 MW, the instantaneous bandwidth is 200 MHz, and the range resolution is 2.5 ft. It can detect the size and shape of the target. Since

**Fig. 9.5** AWACS radar antenna

**Fig. 9.6** Pave paws UHF solid-phase phased array radar antenna

the effective antenna aperture filling time is much longer than the pulse width, the time-delay beam's control becomes necessary even at a medium scan angle.

"Judy" Cobra

Used to collect test data of foreign ballistic missile, as shown in Fig. 9.8, it is installed on the turntable of the US ship "Lookout Island." The array has a diameter of 22.5 ft, including 12,288 units, used to transmit 16 traveling wave tube feeds.

**Fig. 9.7** Cobra Dane L-band phased array radar antenna (provided by Raytheon)

**Fig. 9.8** "Judy" Cobra large rotatable phased array radar antenna on "Lookout Island"

For an aperture with a constant illumination area of A, its normal beam gain is $G_{(0)}$

$$G_{(0)} = 4\pi \frac{A}{\lambda^2}\eta \tag{9.5}$$

Suppose the aperture is considered as a matched receiver. In that case, the sum of the energy from the direction is proportional to the aperture's projected area. And the gain during scanning is:

$$G(\theta)_0 = 4\pi \frac{A\mathrm{Cos}\theta_0}{\lambda^2}\eta \tag{9.6}$$

The actual beam width can express the gain:

$$G(\theta)_0 = \frac{32000}{\theta_B \phi_B}\eta \tag{9.7}$$

In the formula,

$\theta_B \phi_B$ is the beam width (degrees) on the two main surfaces when the beam is scanning $\theta_0$.

$\eta$ is the gain efficiency coefficient.

If the aperture is composed of N same radiation sources and the incident power is received in a matched manner, all the units contribute the same to the total gain, then

$$G(\theta) = NG_e(\theta)\eta \tag{9.8}$$

The $G_e$ is the gain of a unit, and its matched unit power lobe is

$$G_e(\theta) = 4\pi ACos\theta/N\lambda^2 \tag{9.9}$$

For a given cell spacing S, the total number $N = A/S^2$ of radiation elements in the area, then

$$G_e(\theta) = 4\pi(S/\lambda)^2 Cos\theta \tag{9.10}$$

In the scan angle direction $\theta_0$, the maximum antenna gain is:

$$G(\theta_0) = \pi N\eta Cos\theta_0 \tag{9.11}$$

For the normal beam $\theta_0 = 0$, we get

$$G_0 = \pi N\eta \tag{9.12}$$

And the unit gain $G_e = \pi$.

### 9.1.4 Planar Array Beam

The axis of the plane is the direction cosine Cosαx, Cosαy. For any direction on the hemisphere, the cosine is (Fig. 9.9)

$$\mathrm{Cos}\,\alpha_x = \mathrm{Sin}\,\theta\mathrm{Cos}\,\phi \tag{9.13}$$

$$\mathrm{Cos}\alpha_y = \mathrm{Sin}\theta\mathrm{Sin}\phi \tag{9.14}$$

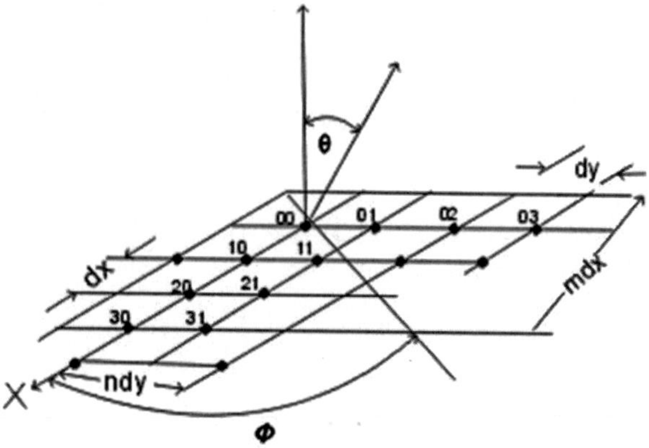


Fig. 9.9 Position and phase of planar array elements

The angle measured when the Cos $\alpha_x$-axis reflects when rotates is

$$\phi = \tan^{-1}\left(\frac{\mathrm{Cos}\,\alpha_{ys}}{\mathrm{Cos}\,\alpha_{xs}}\right) \tag{9.15}$$

Within the unit circle (Fig. 9.10).

$$\mathrm{Cos}^2\alpha_x + \mathrm{Cos}^2\alpha_y \leq 1$$

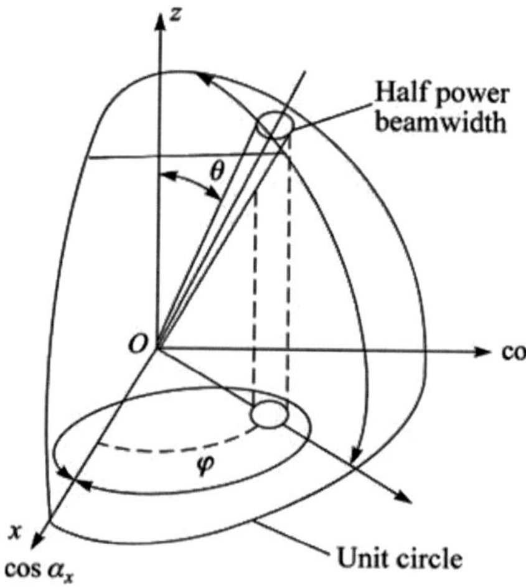


Fig. 9.10 Projection of points on the hemisphere on the array plane

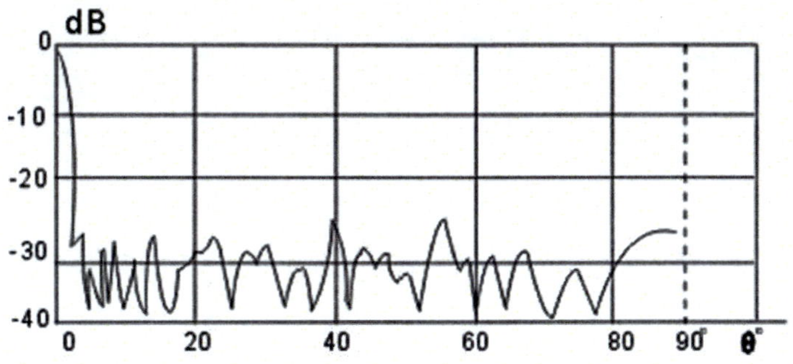


**Fig. 9.11** Typical lobe diagram of a sparse array with 4000 elements

## *9.1.5 Sparse Array and Small Array Pattern*

The side lobes in the planar array deteriorate proportionally with the number of removed units. The cell density can be thinned so that the amplitude distribution can be effectively gradual so that the thinning will not cause grating lobes due to coherent addition, as shown in Fig. 9.11.

The unit lobe has nothing to do with the array excitation. Whether it is a sparse array, a gradual array, or a uniformly illuminated array, the same part of the power is lost. The unit lobe concept is equally applicable to every unit, but it is correct when independent feeding is used, and edge effects are ignored.

Irregular cell spacing and cell gain vary from cell to cell with the surrounding environment of the cell. The array gain should be the sum of all the different unit gains $G_{en}(\theta)$.

$$G(\theta) = \sum_{n}^{N} G_{en}(\theta) \tag{9.16}$$

The pattern of a typical sparse array is shown in Fig. 9.11.

The pattern of the planar array is shown in Fig. 9.12 (small array).

## *9.1.6 Low Sidelobe Phased Array*

The low sidelobe phased array has a great influence on the jamming effect of the shield-type jammer. Antenna designers are very interested in low sidelobe. The requirement of low sidelobe for clutter suppression in AWACS radar has led to low sidelobe phased array technology production. The current technique can lower the sidelobe level by more than 50 dB from the main lobe's peak. There are many costs to obtain these low sidelobe characteristics, which include: (1) reduced gain, (2) increased beam width, (3) stricter tolerance, (4) increased cost, and (5) need to

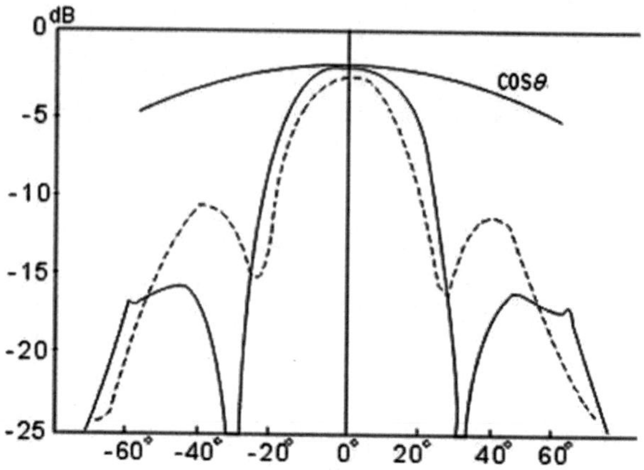


**Fig. 9.12** H-plane lobe diagram of the central unit experiment of the waveguide array

operate in a barrier-free environment. Otherwise, it is easy to increase the sidelobe. Despite these shortcomings, low sidelobe antennas have developed rapidly due to their ability to provide good resistance to electronic jamming.

Through the amplitude distribution of the aperture, the antenna sidelobe can be controlled. For a phased array, each unit's amplitude can be controlled separately so that good sidelobe characteristic can be obtained.

The steps of designing a low sidelobe antenna can be divided into two steps:

Step 1 Select the correct illumination function to obtain the desired non-difference sidelobe level;

Step 2 Control the phase and amplitude errors that affect the random sidelobes.

Only error control will fundamentally limit the effects of sidelobe performance, irradiation function, and error in these two steps.

Uniform illumination (constant amplitude) will produce the highest gain and narrowest beam width but at the cost of high sidelobes. When the amplitude decreases gradually, the gain decreases, the beam widens, and the sidelobes decrease. An

**Table 9.1** Linear illumination table

| Irradiation function | Efficiency η | Peak side lobe (dB) | Beamwidth factor (K) |
|---|---|---|---|
| Uniform | 1 | – 13.3 | 50.8 |
| Cosine | 0.81 | – 23 | 68.2 |
| Cosine squared | 0.67 | – 32 | 82.5 |
| Square of cosine on 10 dB base | 0.88 | – 26 | 62 |
| Square of cosine on 20 dB base | 0.75 | – 40 | 73.5 |
| Cosine squared | 0.73 | – 43 | 74.2 |
| D01Ph-chebyshev | 0.72 | – 50 | 76.2 |
| D01Ph-chebyshev | 0.66 | – 60 | 82.5 |
| Taylor $\overline{n} = 3$ | 0.9 | – 26 | 60.1 |
| Taylor $\overline{n} = 5$ | 0.8 | – 36 | 67.5 |
| Taylor $\overline{n} = 8$ | 0.73 | – 46 | 74.5 |

appropriate illumination function is chosen to minimize gain loss and provide low sidelobes.

For linear illumination function: beam width is $K\lambda/a$ (degree), where $a$ is the antenna length. The linear illumination function table is shown in Table 9.1

For circular illumination function: beam width is $K\lambda/D$. D is the antenna diameter. The circular illumination table is shown Table 9.2

**Table 9.2** Circular illumination table

| | | | |
|---|---|---|---|
| Uniform | 1 | – 17.6 | 58.2 |
| Taylor $\overline{n} = 3$ | 0.91 | – 26.2 | 64.2 |
| Taylor $\overline{n} = 5$ | 0.77 | – 36.6 | 70.7 |
| Taylor $\overline{n} = 8$ | 0.65 | – 45 | 76.4 |

The relationship between Taylor irradiated gain loss and sidelobe level is shown in Fig. 9.13.

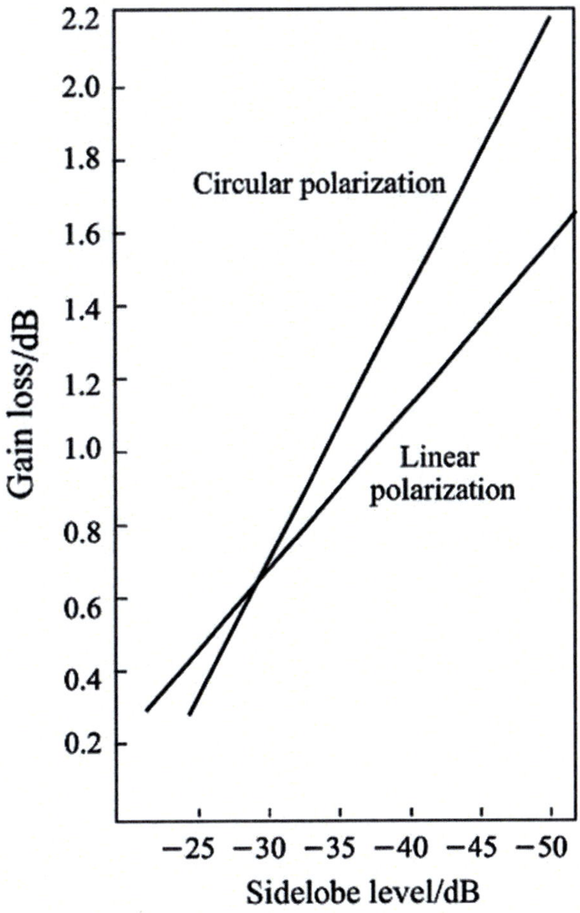


**Fig. 9.13** Taylor radiation gain loss and side lobe relationship

When there is an error in the phase and amplitude, the energy will be transferred from the main beam and distributed to the side lobes. If the error is random, the sidelobes are also random. If the errors are correlated, the sidelobe energy will be concentrated in discrete positions in the far-field, so the correlated errors can cause higher sidelobes, but only in a limited number of positions.

Independent random error is

$$\sigma_T^2 = \sigma_\phi^2 + \sigma_A^2 \tag{9.17}$$

$\sigma_\phi$ is the root mean square phase error, and $\sigma_A$ is the root mean square amplitude error.

Mean Square Sidelobe Error (MSSL) is

$$MSSL = \frac{\sigma_T^2}{\eta_a N\left(1 - \sigma_T^2\right)} \tag{9.18}$$

For example, for an array with an aperture efficiency of 0.7, 5000 elements, $\sigma_a = 0.1, \sigma_\phi = 0.1rad, \sigma_T^2 = (0.1)^2 + (0.1)^2 = 0.02$, we can get

$$MSSL = \frac{\sigma_T^2}{\eta_a N\left(1 - \sigma_T^2\right)} = \frac{0.02}{0.7 \times 5000 \times 0.98} = 5.8 \times 10^{-6} = -52\,\text{dB}$$

There are many phase errors and amplitude errors, which are caused by the mechanical structure of the phase shifter, feeder network, and radiation unit. For manufacturing a low sidelobe antenna, the amplitude error is required to be reduced to a few tenths of a decibel, and the phase error to be reduced to a few degrees. The fewer the number of units is, the stricter the error tolerance is. See random errors cause Fig. 9.14, random sidelobes.

Mean square sidelobe level (MSSL) of phased array radar

$$MSSL = \frac{(1 - P) + \sigma_A^2 + P\sigma_\phi^2}{\eta_a PN} \tag{9.19}$$

In the formula, 1 − P is the probability of a failed unit. For example, when P = 1, there is no failure, and the equation becomes

$$MSSL = \frac{\sigma_A^2 + \sigma_\phi^2}{\eta_a N} = \frac{\sigma_T^2}{\eta_a N}$$

Example: If you want to keep the side lobe 40 dB lower than the main lobe's peak at a given point in space with a probability of 0.99, you can draw a straight line from − 40 dB on the abscissa until it intersects the curve of P = 0.99. Draw a horizontal line from this intersection and read the value of the root mean square sidelobe, which is − 47 dB, namely $MSSL = -47\,\text{dB}$

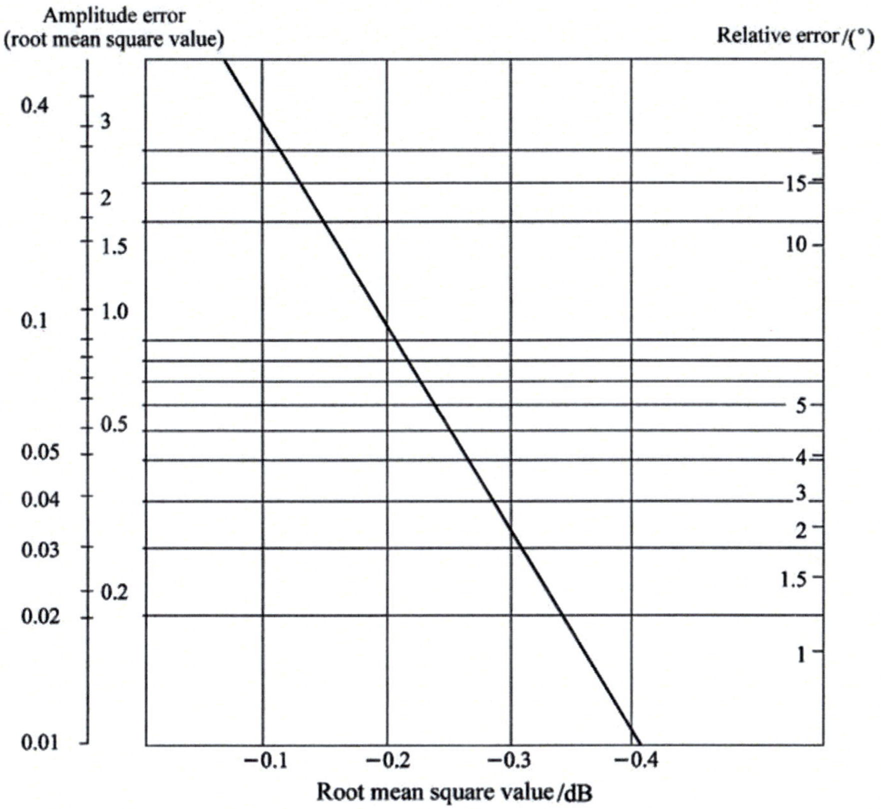


**Fig. 9.14** Random error and root mean square side lobe

$$MSSL = 2 \times 10^{-5} = \frac{(1-P) + \sigma_A^2 + P\sigma_\phi^2}{\eta_a PN}$$

As for an array with 10,000 elements, namely N = $10^4$, we can get

$$0.2 = \frac{(1-P) + \sigma_A^2 + P\sigma_\phi^2}{\eta_a P}$$

For $\eta_a = 1$, this array may allow $P = 0.83$或$\sigma_a = 3.2$或$\sigma_\phi = 25.6°$, and each type of error should be estimated. A certain distribution estimate should be given for the failed unit, amplitude error, and phase error (Fig. 9.15)

$R_T$ is the probability of a given sidelobe level. For an array of 5000 units, the probability that a single sidelobe level does not exceed $R_T$ at any single position is 0.999.

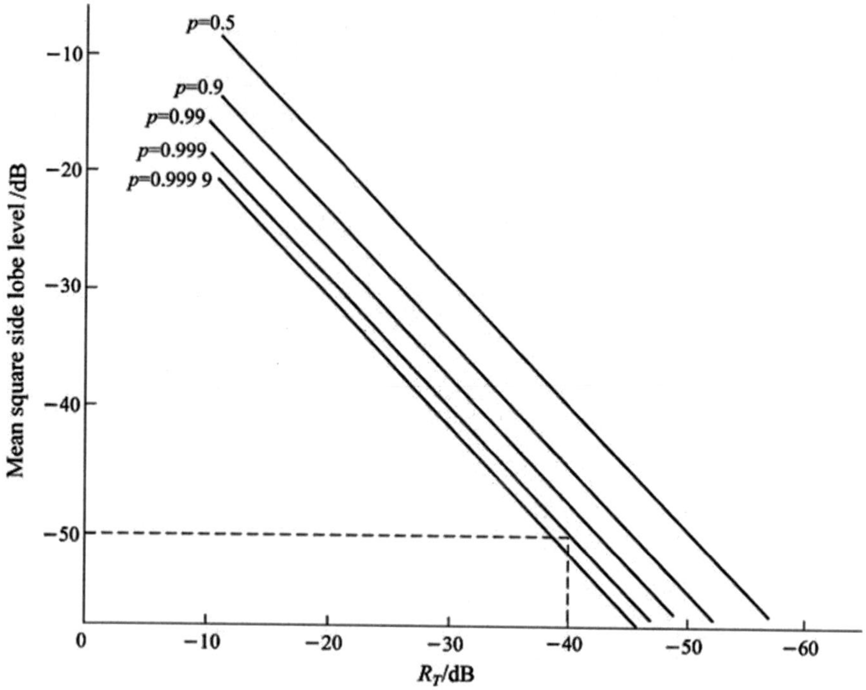


**Fig. 9.15** Relationship between sidelobe level and RT under different probability conditions

For extremely low sidelobe arrays, considering the rationality of random changes, a small number of sidelobes can exceed the mean square sidelobe value, reaching 10–12 dB, which can be regarded as the difference between the probability $P = 0.5$ and $P = 0.999$. If these tolerances are not allowed, some probability calculations must be done before specifying the sidelobe's precise requirements.

In phased array radar, many factors can cause phase error and amplitude error, and these errors will affect the sidelobe level of phased array radar. Therefore, it is very difficult to develop a phased array radar with an extremely low sidelobe level.

## 9.2 The Advantages of Phased Array Radar and Jamming Countermeasures

The phased array radar comprehensively embodies the advantages of various modern new radar systems. It has fast scanning speed, low peak power, high average power, low sidelobe antenna, and is compatible with pulse compression and frequency agility. Therefore, modern radars are developing different types of phased radar systems for different purposes on land, sea, air, and space platforms.

## 9.2.1 The Advantages of Phased Array Radar

Low peak power

Its peak power can be three to four orders of magnitude lower than conventional peak power radars. To probe long-distance targets, conventional pulse radars often use electric vacuum devices with high peak power, and the peak power is generally several megawatts. In contrast, the phased array radar can use multi-element arrays and their equivalent power ($P = N^2 P_e \eta$). Where N is the number of array elements, which can be several thousand or tens of thousands; $\eta$ is the efficiency generally less than 1, and $P_e$ is the peak power of each element. When the same equivalent peak power is reached, if the number of elements is 20000, $\eta$ is 0.5. The peak power $P_e$ can be reduced by 10,000 times so that the transmitter can avoid the difficulties of manufacturing a high-voltage pulse source and some problems of high-power heat dissipation.

High pulse energy

The maximum detection range of the radar is proportional to the pulse energy $(P\,\tau)^{1/4}$. Although the phased array radar has low peak power, its pulse width can be very wide. Generally, the pulse width of electric vacuum devices with peak power is 0.5 to several microseconds, while that of solid devices with low peak power is tens of microseconds to milliseconds. In addition to its multi-array array, it can obtain large pulse energy and provide a large pulse energy source for remote warning radar.

High average power

This is a necessary condition for remote detection radar, which can use multiple arrays to form hundreds, thousands, or tens of thousands of wide pulse sources or high heavy frequency to obtain high average power. The average power of conventional pulse radar is hundreds of watts, kilowatts, or more than ten kilowatts, while the average power of phased array radar forms the average power of tens of kilowatts or hundreds of kilowatts, which provides a powerful radiation source for radar with a detection range of thousands of kilometers.

Antenna array with high gain and low sidelobe

Phased array radar, depending on the array, can obtain antenna characteristics of high gain and low sidelobe. The side lobe of the general parabolic antenna is on the order of − 20 to − 30 dB. However, the sidelobe level of the phased array antenna ranges from − 40 to − 50 dB. Its low sidelobe level can make the radar less subject to the mutual jamming of adjacent radars in the same frequency band. It can also increase the equivalent jamming power of the human-made shield jammer, increase the difficulty of jammer manufacturing, and improve the radar's anti-jamming ability.

Low reception sensitivity and wide instantaneous bandwidth

The receiving amplifier system of phased array radar uses solid devices with low noise and high sensitivity. Meanwhile, the instantaneous bandwidth is relatively wide, which makes the receiving system of phased array radar have good reception performance.

Multi-band operation

Phased array radars are easier to achieve multi-frequency operation. Some of the phased array radars currently installed is work in two frequency bands, such as the L and X frequency bands of spaceborne radars. Some airborne radars work in three frequency bands, L, S, and X frequency bands. Multi-band operation can improve the radar's target detection performance, and at the same time, increase the radar's anti-jamming capability, which can also cause trouble for the effective jamming of the jammer.

It is easy to realize multi-band array and multiple-in multiple-out (MIMO) array radar system to enhance anti-jamming ability in a complex electromagnetic environment.

The transmitter adopts solid components with low peak power, high average power, high working frequency, small size, and lightweight, which reduces the requirements of the heat dissipation system and power supply system of high-power radars. It makes the whole machine easier, stable, and reliable.

The phased array antenna can realize fast search and multi-beam, and realize multi-direction and multi-frequency operation. The beam nulling technology (which reduces the antenna sidelobe gain very low in some directions) can easily counter the jamming of the shield jammer against the radar antenna sidelobe.

Phased array radar can integrate new technologies such as high power, low side-lobe, broadband, multiband, frequency agility, sidelobe zeroing, related reception, and imaging processing. In this case, a new type of radar system and system with multiple functions and excellent anti-jamming performance is formed, which greatly enhances the radar's performance and functions and enhances the anti-jamming capability of the radar.

### *9.2.2 Basic Jamming Countermeasures Against of Phased Array Radar*

Although phased array radar has many advantages, its radar's three inherent weaknesses have not been overcome. Therefore, it allows radar jamming to be still exploited, and effective jamming can be implemented.

Three weaknesses of phased array radar.

(1) Signals in the passband of radar or radio equipment can be received regardless of whether they are enemy or not.

(2) Regardless of the radar's signal processing method, as long as the jamming to signal ratio reaches a certain value, it cannot extract useful signals from the mixture of jamming and useful signals.
(3) Although the sidelobe of the phased array radar antenna is low, and measures such as sidelobe zeroing can be taken, it is still impossible to zero the sidelobe of its antenna, and there is still a sidelobe level. This allows the sidelobe jamming of the jammer to be used, but the required jamming power increases.

Technicians studying jammers rely on these inherent weaknesses of phased array radars to make a fuss about jamming. If we work hard on the six elements of jamming effectiveness, phased array radars can still be jammed effectively.

Six elements of effective jamming to phased array radar.

(1) The jamming frequency must be aimed at accurately. For the Pulse-Doppler system, the aiming frequency error is required to be in the order of kilohertz. For radars of other systems, the aiming frequency error is in the order of megahertz.
(2) The antenna direction of the jammer should be aimed at the radiation source, and its angular tracking error is less than $\theta_B/3$ ($\theta_B$ is the beamwidth of the point at the radar's half power).
(3) The suppression duration of noise suppressing jamming signal $\geq 90\%$.
(4) Jamming suppression coefficient $\geq$ required value of various radar systems.
(5) The jamming spectrum width is controlled within the range of $\Delta F/\Delta f = 1.2 \sim 3$. $\Delta F$ is the spectrum width of the jamming noise signal, and $\Delta f$ is the bandwidth of the radar receiver.
(6) For variable-polarization and multi-polarization radars, random polarization jamming should be used, particularly in self-defense jammers. The radar uses variable polarization to resist jamming and uses different polarizations to improve the radar's detection performance. No matter its purpose to improve polarization, for jamming, corresponding measures must be taken for radar polarization characteristics.

Several difficulties in the development of phased array radar jammer.

(1) Receiving system with high sensitivity and wide instantaneous bandwidth

The phased array radar adopts a solid transmitter. Its peak power is low, and two orders of magnitude, on average, reduce the antenna sidelobe gain. Therefore, the reconnaissance receiver's sensitivity in the jammer needs to be increased by 4–5 orders of magnitude. At the same time, the instantaneous bandwidth is wider (about tens of megahertz to gigahertz), which increased the difficulty of jamming system development.

(2) The equivalent power of the jammer needs to be larger

According to the radar jamming equation, a certain amount of decibels are reduced by the side lobe of the radar antenna to achieve the original jamming effect to make up for the low sidelobe of the phased array antenna. The same proportion of decibels should increase the equivalent power of jamming. This ratio is unchanged in principle.

(3) The isolation of the transceiver antenna of the jammer should be improved

Due to the increased sensitivity of the reconnaissance receiver and the increase in the jamming transmitter's power, the isolation between the jammer's transceiver antennas needs to be increased. If the original receiver sensitivity is increased by 40 decibels, the transmit power is increased by 10 decibels. Under the same conditions, the isolation between the transmitting and receiving antennas needs to be improved by 50 decibels, which adds new difficulties to jammers'.development.

(4) The jamming parameters can be adjusted timely.

It should be able to adaptively adjust the jammer's required jamming parameters for different jammed objects. It should detect the parameters of different jammed objects, distinguish its radar system, extract characteristic constants, and distinguish the radar system based on the three characteristic constants.

Spectrum characteristics of radar signal.

(a) Conventional pulse radar, generally 0.1–3 MHz;
(b) Pulse compression radar, generally 3–10 MHz;
(c) Synthetic aperture radar, generally 10–1000 MHz;
(d) Pulse-Doppler radar, generally 1–3 kHz.

The pulse repetition rate of the radar signal.

(a) Conventional pulse radar, generally 300–2000 MHz;
(b) Pulse compression radar, generally 200–1000 MHz;
(c) Synthetic aperture radar, generally 1000–2000 MHz;
(d) Pulse-Doppler radar, generally 5–300 kHz.

The pulse width of the radar.

(a) Conventional pulse radar, generally 0.4–4 $\mu$s;
(b) Pulse compression radar, generally 5–10000 $\mu$s;
(c) Synthetic aperture radar, generally 10–50 $\mu$s;
(d) Pulse-Doppler radar, generally 0.4–2 $\mu$s.

According to these three characteristics: pulse width, repetition frequency, and spectrum width, the system of the jammed radar is determined. And it can adaptively adjust the jamming spectrum width and aiming accuracy of the jamming transmitter.

(1) Multi-point aiming and tracking

In order to improve the anti-jamming ability, the current radar performs the work of frequency agility or multi-frequency point. In order to be able to jamming with frequency-agile radars and multi-carrier radars effectively, the jammer should be equipped with a combination of frequency-agile and multi-carrier tracking.

(2) A combination of tactics and technology
(3) Improve jammer tracking error

Improve the passive tracking error of the jammer's angle to ensure the angle tracking error $\Delta\theta \leq \theta_B/3$.

(4) Solve the high-standard antenna transceiver isolation requirements for jammers

Make sure the ratio of receiving and sending time is 1:9. The receiving time is in the order of milliseconds, and the non-transmitting and non-receiving time relationship of tens of microseconds. In addition to its transmission and reception isolation, it also needs to do electromagnetic compatibility with other electronic equipment such as radar, communication, and other electronic equipment such as platforms.

(5) Use random polarization jamming

The jammer should use random polarization jamming to deal with the multi-polarization work of radar or anti-jamming with variable polarization. Especially for self-defense jammers, the effect of variable polarization anti-jamming is remarkable, and the work of random polarization jammers should be done better.

(6) Enable jammers to achieve "one machine to multiple systems"

When designing jammers, the radar system that requires the largest jamming suppression coefficient should be selected to calculate the equivalent power of jamming required. The jammer can achieve "one machine to multiple systems" and adapt to the highly integrated phased array radar system. Only in this way can high-level radars have corresponding high-level jammers.

## *9.2.3 An Example of Jammer Designed for Intergrated Phased Array Radar*

The block diagram of the jammer of the phased array comprehensive integration radar is shown in Fig. 9.16. Comprehensive integration refers to the single or combined use of pulse compression system, Pulse-Doppler system, imaging processing synthetic aperture system, or multi-carrier frequency-agile frequency conversion system in radar. A multi-phase-array radar system is used on the antenna. This kind of radar has long detection range characteristics, many working modes, multi-function, and strong anti-jamming ability. To effectively jamming with this kind of radar, high-tech jammers are required.

### 9.2.3.1 Extension System Included in the Jammer

1. Transmitting antenna, receiving antenna, and servo tracking antenna system can adopt a phased array system or parabolic antenna.
2. High sensitivity receiver subsystem. In order to receive signals with a low probability of signal by phased array radar, the receiver should have a sensitivity of – 100 dBW and an instantaneous bandwidth of 1000 MHz.
3. Fine aiming frequency and jamming frequency guidance subsystem: the aiming frequency error should be within 1 kHz.

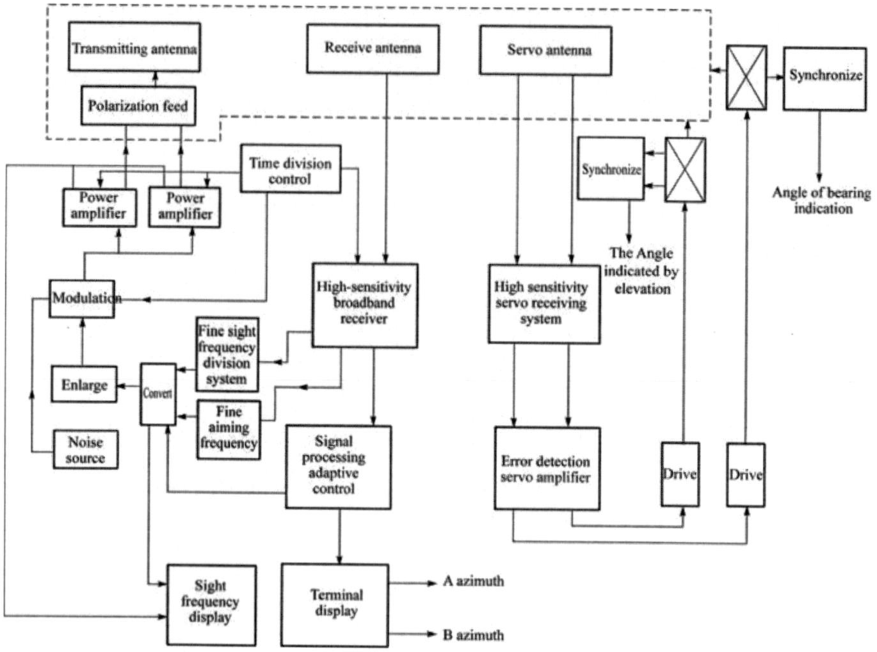


**Fig. 9.16** The principle block diagram of the jammer composition of the phased array comprehensive integration

4. Aiming frequency and jamming frequency guidance subsystem: the aiming frequency error should be in the order of megahertz; the frequency guidance time should be in the order of microseconds.
5. Self-adaptive signal detection and discrimination signal sorting processing sub-system. After confirming that the detected signal radar system is Pulse-Doppler radar, pulse compression, conventional pulsed radar, or synthetic aperture radar, the jamming is automatically guided to the corresponding jamming system.
6. The passive tracking subsystem for the radiation source can automatically make the angle tracking error less than $\theta_B/3$.
7. To make the jammer work normally, there should also be a set of adaptive time division control to ensure the transceiver antenna's isolation requirements.
8. There should be a corresponding terminal display subsystem to make the jammer easy to operate and intuitively sense.
9. For effective jamming, there should be deceptive jamming that suppresses the jamming source and auxiliary jamming.
10. There should be a supporting medium and low voltage power supply and power supply system, ventilation, and heat dissipation system for the equipment's normal operation.

These subsystems ensure that the whole machine has the six major elements of effective jamming in 9.2.2. The jamming function can effectively jam with the "comprehensive, integrated phased array radar."

#### 9.2.3.2 Quantitative Analysis of Comprehensive, Integrated Phased Array Radar Jammer

Select a typical phased array radar. Take this radar parameter as an example to quantitatively analyze the parameters of the jammer. Suppose a warning radar: the operating frequency is P-band; the peak power is 540 kW; the average power is 140 kW; the detection distance for a 10 $m^2$ target is 4800 km; the working range of elevation angle: 3°–85°; the working range of azimuth angle is 3 × 120°; Radar working system: phased array plus pulse compression; transmit pulse width is 160 μs, 1 ms, and 10 ms; pulse width after compression is 0.3–1 μs.

(1) Calculation of jamming equivalent power

The unified equation of radar jamming is:

$$P_j G_j = \frac{K_{j\Sigma} P_t G_t \sigma_t R_j^2}{4\pi R_t^4} \cdot \frac{Gt}{Gt(\theta)} \cdot \frac{L_d K_f}{L_j r_j} \tag{9.20}$$

In the formula:

| | |
|---|---|
| $P_j G_j$ | Required jamming equivalent power. |
| $P_t G_t$ | Equivalent peak power of jammed radar (96 dBW). |
| $\sigma_t$ | The equivalent reflective area of the covered target. |
| $R_j$ | Distance between jammer and radar. |
| $R_t$ | The distance between the radar and the screened target. |
| $Gt/Gt(\theta)$ | The main/sidelobe ratio of the radar antenna gain when the jammer antenna is aligned with the radar direction. |
| $L_d$ | Radar feed loss and space transmission loss. |
| $K_f$ | Jamming spectrum width and radar receiver bandwidth ratio. |
| $r_j$ | Polarization loss when jammer polarization is inconsistent with radar polarization. |
| $L_j$ | Jamming emission loss and space transmission loss. |
| $K_{j\Sigma}$ | Jamming suppression factor. |

Different radar systems require different jamming suppression coefficients to jamming effectively. The jamming suppression coefficient for pulse compression radar can be simplified to $K_{j\Sigma} = (D)^{0.8}$. D is the pulse compression coefficient of the pulse compression radar. $D = \tau/\Delta\tau$. $\tau$ is the transmitted pulse width of the pulse compression radar, and $\Delta\tau$ is the compressed pulse width (Fig. 9.17).

According to the data calculated in Table 9.3, take $P_j G_j = 55$ Dbw.

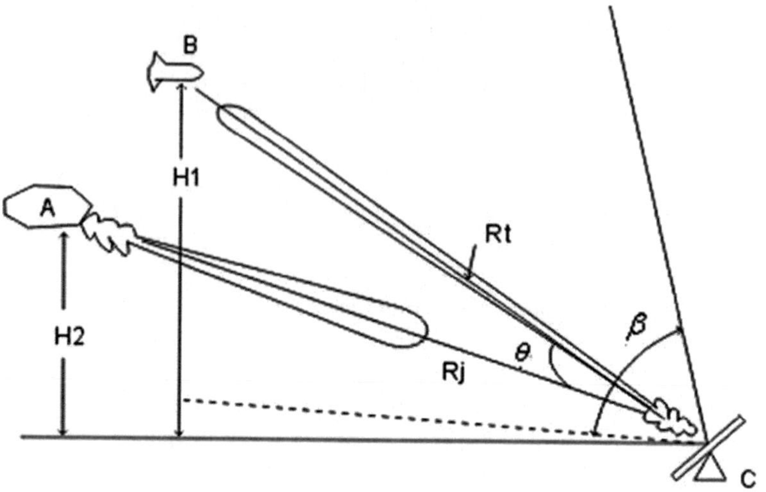


**Fig. 9.17** Schematic diagram of air-to-ground (ship) radar jamming A—Jammer carrying platform; B—Screened aircraft; C—Jamming radar position; β—Radar elevation search range

**Table 9.3** When $R_t = 200$ km and $R_j = 300$ km, the equivalent power required by the jammer

| $\sigma_t(m^2)$ | | 0.1 | 0.2 | 0.4 | 0.6 | 1 |
|---|---|---|---|---|---|---|
| $P_jG_j(dBW)$ | $R_{t\min} = 200$ km<br>$R_j = 300$ km | 39 | 42 | 45 | 47 | 49 |
| | $R_{t\min} = 200$ km<br>$R_j = 600$ km | 45 | 48 | 51 | 53 | 55 |

(2) Calculation of receiver sensitivity

The minimum signal power received by the jamming receiver from the radar is:

$$P_r = \frac{P_t G_t(\theta)}{4\pi R_j^2} \cdot \frac{\lambda^2}{4\pi} \cdot G_j \cdot L_j \tag{9.21}$$

In the formula: Radar peak power $P_t = 540$ kW; $G_t(\theta) = 0$ dB; $R_j = 600$ km; $G_j = 20$ dB; $L_j$ is the combined loss of feeder loss, space atmospheric loss, polarization loss, etc. $L_j = 0.1$.

Substituting the parameters into (9.21) formula, we get:

$$P_r = \frac{5.4 \times 10^5 \times 0.49 \times 10^2 \times 0.1}{1.56 \times 10^2 \times 36 \times 10^{10}} = -73\,\text{dBW}$$

When the actual working sensitivity of the receiver is – 85 dBW, the instantaneous bandwidth is better than 20 MHz.

(3) Isolation requirements of single-band jammer transceiver antenna

The layout of the transceiver antenna is shown in Fig. 9.18. A is the transmitting antenna, and B is the receiving antenna. When antenna A radiates a signal, the sidelobe signal of $G_A(90^\circ)$ is radiated to antenna B, and the side lobe of GB (270°) receives the signal of antenna A. The signal received by antenna B is

$$P_B = \frac{P_A G_A(90^\circ)}{4\pi d_2} \cdot \frac{\lambda^2}{4\pi} G_B(270^\circ) \tag{9.22}$$

If $P_A/P_B$ is the transmission and reception isolation coefficient of the two antennas, then:

$$D(\theta) = \left(\frac{4\pi d}{\lambda}\right)^2 \frac{1}{G_A(\theta) G_B(\varphi)} \tag{9.23}$$

The actual transmission and reception isolation required is:

$$\begin{aligned} D_N &= \frac{P_{jA} G_A(\theta)}{4\pi d^2} \cdot \frac{\lambda^2}{4\pi} \cdot G_B(\varphi) \frac{n}{P_{\min}} \\ &= \frac{P_{jA}}{P_{\min}} \cdot \left(\frac{\lambda}{4\pi}\right)^2 \cdot G_A(\theta) \cdot G_B(\varphi) \cdot n \end{aligned} \tag{9.24}$$

In the formula:

$P_{jA}$ The power radiated by the jamming transmitter.
$P_{\min}$ The minimum signal power received by the jammer receiver;
$n$ The signal/jamming ratio at which the receiver can work normally;
$\lambda$ The working wavelength;
d The distance between the transmitting and receiving antennas.

For example, the power of the transmitter

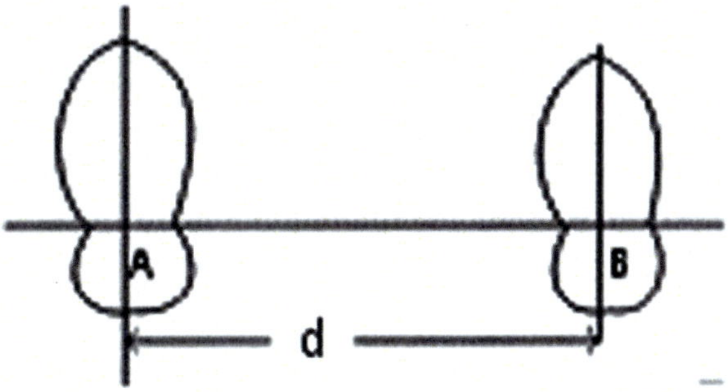


**Fig. 9.18** Schematic diagram of jammer transceiver antenna layout

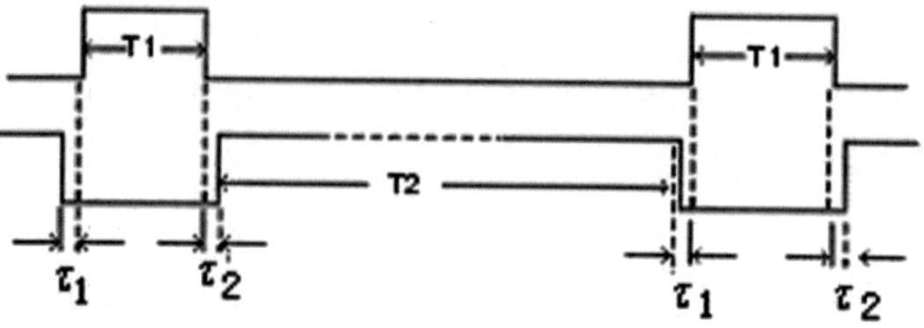


**Fig. 9.19** Time relationship of sending and receiving time separation

$$P_{jA} = 1\,\text{kW},\quad G_A(\theta) = 1,\quad G_B(\varphi) = 1,\quad P_{r\min} = -85\,\text{dBW},\quad n = 30,$$

$\lambda = 0.7\,\text{m}$, subsitute into (9.24), $d = 5\,\text{m}$.
Then, $D_N = \frac{10^3}{0.3\times10^{-8}}\left(\frac{0.7}{12.6\times5}\right)^2 \times 1 \times 1 \times 30 = 73\,\text{dB}$.
While $D(\theta) = \left(\frac{4\pi d}{\lambda}\right)^2 \frac{1}{G_A(\theta)\cdot G_B(\varphi)} = 39\,\text{dB}$.
It cannot meet the requirements of transceiver isolation. Therefore, the transceiver isolation of the jammer must be improved for the jammer to work normally. The transceiver isolation measures of jammers are generally improved in three aspects:

(a) A transceiver time separation system is adopted. When the jamming signal is transmitted, the receiver is turned off, and no signal is received. When receiving the signal, the jamming transmitter does not transmit the jamming signal. In order to avoid the reflection of ground objects, there is still a period for neither sending nor receiving. The time relationship between receiving and transmitting jamming is shown in Fig. 9.19.

$T_1$ is the receiving time, generally slightly longer than the pulse repetition period of the jammed radar. $T_2$ is the transmission time of jamming signal, generally greater than 9 times of $T_1$. $\tau_1, \tau_2$ is the time of neither sending nor receiving, generally 20–50 μs, the time separation state does not affect its jamming effect, then

$$\frac{T_2}{T_1 + T_2 + \tau_1 + \tau_2} \geq 90\%$$

The time-division control circuit shall control the control of the time-division of transmission and reception in Fig. 9.16.

(b) Measures should be taken to reduce the static noise of the transmitter. The static noise of the transmitter is produced by the terminal power amplifier of the transmitter itself. It is related to factors such as the magnification of the final stage power amplifier and the noise power at the input, such as the static noise power $P_{us}$ of the power traveling wave tube.

$$P_{us} = (K_T \Delta f N) K_r \tag{9.25}$$

In the formula, K is the Boltzmann constant, T is the absolute temperature, $\Delta f$ is the amplifier bandwidth, N is the noise figure of the input, and $K_r$ is the magnification of the power amplifier.

For example, K is $1.38 \times 10^{-23}$; $T - 300$, $\Delta f - 10\,\text{GHz}$, $K_r - 50\,\text{dB}$, $U$sually, $N = 30 - 30\,\text{dB}$ substitute into formula (9.25), $P_{us}$ is calculated to be – 14 to – 34 dBW.

If appropriate measures are taken, it can be ensured that the static noise power of the power amplifier tube is less than – 30 dBW.

(c) Adding an isolation layer between the transmitting and receiving antennas

Usually, an absorbent material is used to reduce the side lobes of $G_A(\theta)$ and $G_B(\varphi)$ under the same conditions of d.

After taking these three measures, $P_{jA}$ becomes $10^{-3}W$, $G_A(\theta)$, $G_B(\varphi)$ becomes 0.1, and other constants remain unchanged. Substitute into formula (9.24),

$$D'_N = \frac{10^{-3}}{0.3 \times 10^{-8}} \left( \frac{0.7}{12.6 \times 5} \right)^2 \times 0.1 \times 0.1 \times 30 = 13\,\text{dB}$$

Substitute the parameters into formula (9.23)

$$D'_{(\theta)} = \left( \frac{12.6 \times 5}{0.7} \right)^2 \frac{1}{0.1 \times 0.1} = 59\,\text{dB}$$

$D'_{(\theta)} > D'_N$, which meet the requirements of transceiver isolation.

(1) Transceiver isolation technology of multi-band jammer on the same platform

There are more and more demands for multi-band and multi-carrier jammers on the same platform. For example, full-time jamming aircraft can be equipped with jammers of P, L, S, C, X/Ku and other bands on this platform. There are multiple bands and corresponding multiple antennas. These antennas, in turn, are interacted with the fuselage wing, which is shielded, isolated or reflected, so the analysis is complicated. There are many antennas that need to be shielded and isolated. Some of the sidelobe characteristics of the antenna are not tested separately, but are installed at a specific location and affected by the specific surrounding environment. Therefore, when analyzing these characteristics, it needs to combine theory and actual measurement to get more realistic results. Assume that a multi-band jamming device is installed on a certain platform, as shown in Fig. 9.20.

Assuming that the optical axes of the main beams of the five antennas, $A_1$, $A_2$, $A_3$, $A_4$ and $A_i$, are parallel, antenna $A_1$ may receive radiation or coupled signal power from the other four antenna side lobes.

$$P_{\Sigma A1} = \frac{P_{A2} G_{A2}(\varphi_{A2.1})}{4\pi d_1^2} \cdot \frac{\lambda_{A2}^2}{4\pi} \cdot G_{A1}(\varphi_{A1.1}) \rho_{A1} + \frac{P_{A3} G_{A3}(\varphi_{A3.1})}{4\pi d_2^2} \cdot \frac{\lambda_{A3}^2}{4\pi} G_{A1}(\varphi_{A1.2}) \rho_{A2}$$
$$+ \frac{G_{A4} G_{A4}(\varphi_{A4.1})}{4\pi d_4^2} \cdot \frac{\lambda_{A4}^2}{4\pi} G_{A1}(\varphi_{A4.1}) \rho_{A3} + \frac{P_{Ai} G_{Ai}(\varphi_{A4.1})}{4\pi d_i^2} G_{A1}(\varphi_{A4.1}) \rho_{A4} \quad (9.26)$$

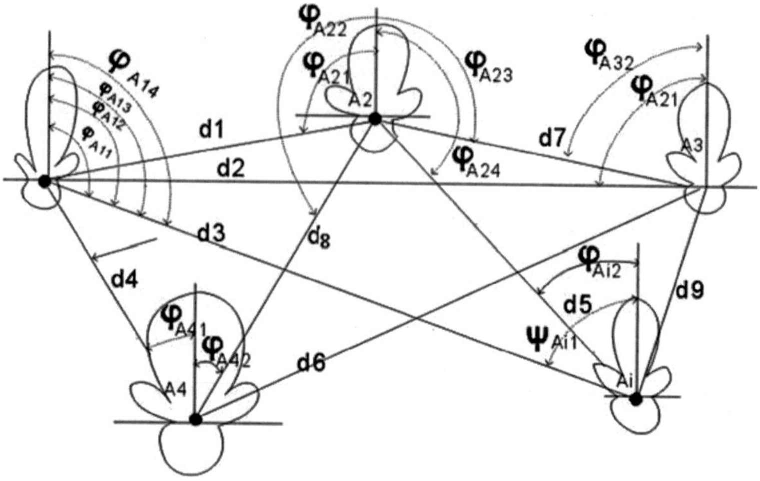


**Fig. 9.20** Schematic diagram of transceiver isolation between multi-band jamming antennas on the same platform

Antenna $A_2$ may receive radiation or coupled signal power from the other four antenna side lobes:

$$P_{\Sigma A2} = \frac{P_{A1}G_{A1}(\varphi_{A1.1})}{4\pi d_1^2} \cdot \frac{\lambda_{A1}^2}{4\pi} \cdot G_{A2}(\varphi_{A2.1})\rho_{A1} + \frac{P_{A3}G_{A3}(\varphi_{A2.3})}{4\pi d_7^2} \cdot G_{A2}(\varphi_{A2.3})\rho_{A2}$$
$$+ \frac{G_{A4}G_{A4}(\varphi_{A4.2})}{4\pi d_8^2} \cdot \frac{\lambda_{A4}^2}{4\pi} G_{A2}(\varphi_{A2.2})\rho_{A3} + \frac{P_{Ai}G_{Ai}(\varphi_{Ai.2})}{4\pi d_5^2} G_{A2}(\varphi_{A2.4})\rho_{A4} \tag{9.27}$$

By analogy, the jamming signals received by $P_{\Sigma A3}$, $P_{\Sigma A4}$和$P_{\Sigma Ai}$from the side lobe of the other antennas can be listed.

Among the mutual influences of these jammed antennas, the influence of low frequency band on high frequency band is more serious, the influence of close distance is more serious, the influence of high jamming transmission power is serious, and the influence of high antenna sidelobe level is serious.

A mutual isolation factor $\sigma_{(A1)}$, $\sigma_{(A2)}$, $\sigma_{(A3)}$, $\sigma_{(A4)}\sigma_{(Ai)}\cdots$ were added between the two antennas. This is based on an isolation factor generated after the antenna is installed in a specific location. After isolation, measures are taken between different antennas, requiring it to be as small as possible.

If there are i antennas on the same platform, it receives the jamming power $P_{\Sigma i}$ from the coupling or radiation of other antenna's side lobes.

$$P_{\Sigma i} = \sum_{i}^{i-1} P_i G_i(\varphi_i)\left[\frac{\lambda_i}{4\pi d_i}\right]^2 G_i(\theta_i)\rho_i \tag{9.28}$$

Transceiver antenna isolation of multi-band and multi-antenna array on the same platform

$$D(\theta_i, \varphi_i, \lambda_i, d_i) = \sum_i^{i-1} \left(\frac{4\pi d_i}{\lambda_i}\right)^2 \frac{1}{G_i(\theta_i) G_i(\rho_i)} \tag{9.29}$$

The isolation $D_{Ni}$ that should be met by multiple receiving antennas on the same platform

$$D_{Ni} = \frac{P_{ji}}{P_{\min i}} \left(\frac{\lambda_i}{4\pi d_i}\right) G_i(\theta_i) G_i(\varphi_i) \cdot \rho_i n_i \tag{9.30}$$

In the formula, $D_{Ni}$ is the isolation between the i-th receiving antenna and any other transmitting antenna of multiple antennas on the same platform.

$P_{ji}$ is the transmit power of multiple transmit antennas;
$\lambda_i$ is the signal wavelength radiated by the antenna;
$G_i(\theta_i)$ is the sidelobe level of the transmitting antenna aligned for the sidelobe of the receiving antenna;
$P_{\min i}$ is the sensitivity of the receiver to which the receiving antenna is transmitted;
$\rho_i$ is the isolation coefficient between the receiving antenna and the mutual coupling antenna;
$n_i$ is the signal/noise ratio of the receiver;
$D_{Ni}$ is the required value of isolation between any antenna and other multiple antennas.
$D(\theta_i, \varphi_i, \lambda_i, d_i)$ is the antenna isolation degree of any antenna in the same platform with multiple antennas. When $D(\theta_i, \varphi_i, \lambda_i, d_i) > D_{Ni}$, the isolation of multi-band and multi-antenna on the same platform can meet the requirements, that is, multi-band jammers on the same platform can work simultaneously to implement effective jamming.

To reduce the value of $D_{Ni}$, $d_i$ can be increased, $G_i(\theta_i)$ can be decreased, $G_i(\varphi_i)$ can be decreased, absorption isolation layer can be added between the two antennas, time division system can also be used to reduce $P_{ji}$, but the sensitivity of the receiver cannot be reduced, since it is the set of the requirements of the data (Fig. 9.21).

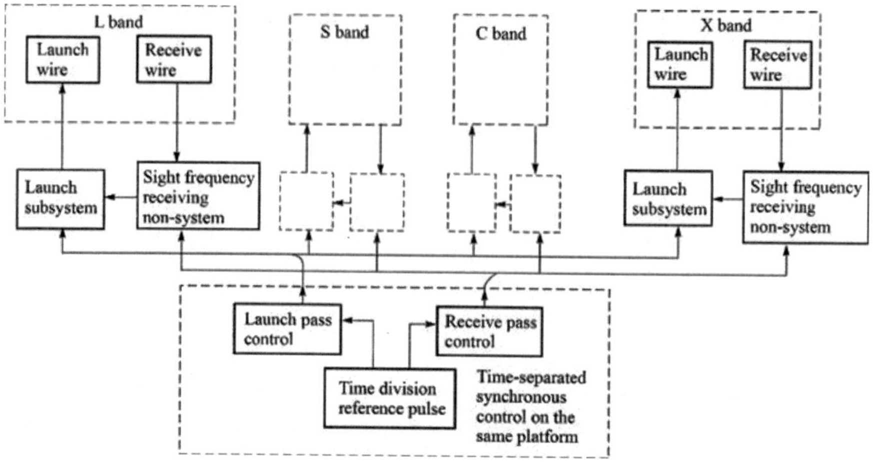


**Fig. 9.21** Block diagram of separation and synchronization control for multi-band jammers on the same platform

## 9.3 Analysis of Effective Area of Phased Array Radar Jamming

### 9.3.1 *Jamming Exposure Area of Phased Array Radar*

The unified jamming equation for radars is also suitable for jamming with phased array radars. So, Eq. (9.20) can become

$$R_{t\max} = \left[ \frac{K_{j\Sigma} P_t G_t \sigma_t R_j^2}{4\pi P_j G_j} \cdot \frac{G_t}{G_t(\theta)} \cdot \frac{L_d K_f}{L_j r_j} \right]^{1/4} \tag{9.31}$$

In the formula, $R_{t\max}$ is the jamming exposure distance, and other parameters are the same as formula (9.20). According to the typical radar parameters and jamming parameters:

Choose $P_t G_t$ to be 96 dBW, $P_j G_j$ to be 55 dBW, $K_{j\Sigma}$ to be 22 dB, $\frac{L_d K_f}{L_j r_j} = 3$, $\rho_t = 0.1$ or $1m^2$; $G_t/G_t(\theta) = 0.36(\frac{\theta_B}{2\theta})^2$; The sidelobe characteristics of the radar antenna should be calculated based on the actual measurement results, but this data is difficult to obtain and can only be calculated based on empirical formulas.

$$\frac{G_t}{G_t(\theta)} = K\left(\frac{\theta_B}{2\theta}\right)^2 \tag{9.32}$$

In the formula, K is a coefficient, which is generally 0.5–0.05;

$\theta_B$ is the beamwidth of the radar antenna at the half power point, taking 2°;

$\theta$ is the angle between the direction of the jammer and the direction of the main beam;

Taking K as 0.36, the phased array radar's sidelobe characteristics can be calculated according to Eq. (9.32).

According to Table 9.4, plot the sidelobe characteristics of the phased array radar in Fig. 9.22.

If $R_j$ is 300 km and $\sigma_t$ is $0.1\text{m}^2$ or $1\text{m}^2$, the data of $R_{t\max}^{(\theta)}$ is calculated as shown in Table 9.5. Substituting the parameters into Eq. (9.31), we get.

$R_{t\max}^{(\theta)} = \left[\frac{140\times4\times10^9\times0.1\times9\times10^{10}}{12.6\times3\times10^5}\cdot\frac{G_t}{G_t(\theta)}\times3\right]^{1/4} = 7.9\left[\frac{G_t}{G_t(\theta)}\right]^{1/4}$ km and data in Table 9.5 is calculated.

Draw the jamming effective area and jamming exposure area of a phased array radar based on the data in Table 9.5 (Fig. 9.23).

At $\sigma_t = 0.1\text{ m}^2$, the detection range of this radar is 1500 km.

Radar should be jammed to reduce its detection area, and the reduction ratio $\rho$ is

$$\rho = \frac{\pi\left[R_t^2 - \left(1-\frac{1}{6}\right)r^2\right]}{\pi R_t^2} = \frac{R_t^2 - \left(1-\frac{1}{6}\right)r^2}{R_t^2} \tag{9.33}$$

When $\sigma_t = 0.1\text{ m}^2,\ R_j = 300$ km, the radar detection area $\rho_1$ is reduced by 99%;
When $\sigma_t = 1\text{ m}^2,\ R_j = 300$ km, the radar detection area $\rho_2$ is reduced by 99%;

**Table 9.4** Necessary jamming power when Rt=200km, Rj=60km

| θ | 0° | 15° | 30° | 45° | 60° | 90° |
|---|---|---|---|---|---|---|
| $G_t/G_t(\theta)$ | 0 dB | + 26 dB | + 32 dB | + 36 dB | + 44 dB | + 45 dB |

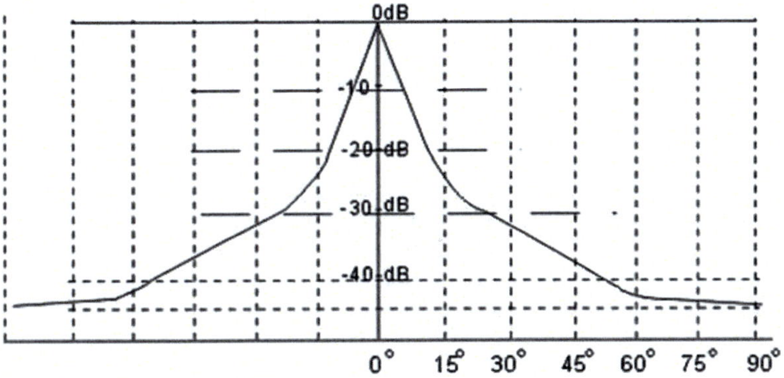


**Fig. 9.22** Schematic diagram of the side lobe characteristics of the disturbed phased array radar

**Table 9.5** Jamming Exposure areas at different angles and Rj

| θ / $R_{t\max}^{(\theta)}$ | 0° | 15° | 30° | 45° | 60° | 90° | |
|---|---|---|---|---|---|---|---|
| $R_{t\max(1)}^{(\theta)}$ | 7.9 km | 35.6 km | 48.2 km | 62.4 km | 98 km | 104 km | $R_j = 300\,\text{km}$ $\sigma_t = 0.1\,\text{m}^2$ |
| $R_{t\max(2)}^{(\theta)}$ | 14.2 km | 64.1 km | 86.8 km | 112.3 km | 176.4 km | 187.2 km | $R_j = 300\,\text{km}$ $\sigma_t = 1\,\text{m}^2$ |
| $R_{t\max(3)}^{(\theta)}$ | 11.4 km | 51.3 km | 70.0 km | 89.9 km | 141.1 km | 149.8 km | $R_j = 600\,\text{km}$ $\sigma_t = 0.1\,\text{m}^2$ |
| $R_{t\max(4)}^{(\theta)}$ | 20.5 km | 73.9 km | 100.8 km | 129.5 km | 203 km | 215.7 km | $R_j = 600\,\text{km}$ $\sigma_t = 1\,\text{m}^2$ |

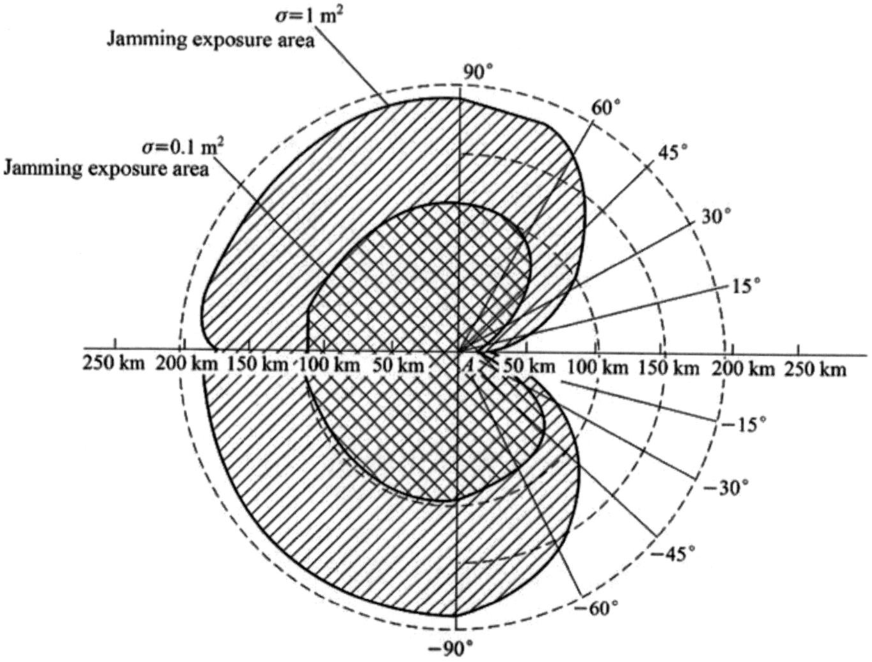


**Fig. 9.23** Schematic diagram of jamming exposure area at $R_j = 300$ km; $\sigma_t = 0.1$和$1m^2$

## 9.3.2 *The Ratio of the Radar's Detection Area Under Jamming*

Figure 9.24 is drawn according to Table 9.5.

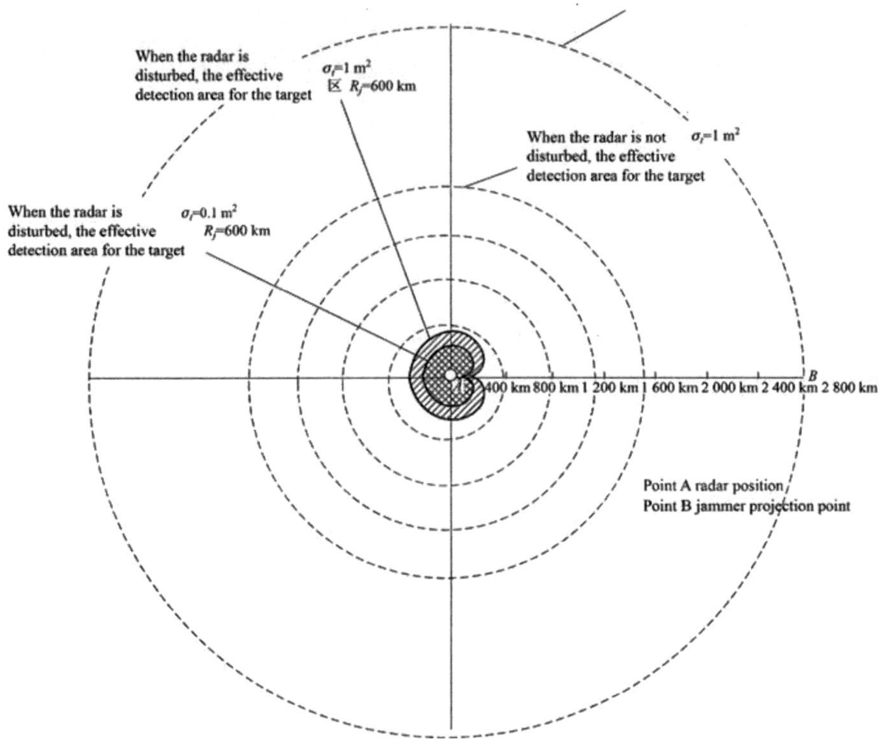


**Fig. 9.24** $R_j = 600\,\text{km}$, $\sigma_t = 0.1\,\text{m}^2, 1\,\text{m}^2$, comparison diagram of radar detection area and jamming exposure area

The horizontal line area is the jamming exposure area, the effective detection area after the radar jamming.

After the radar is jammed with, its effective detection area is reduced by 99%. It can be intuitively seen from Fig. 9.24 that when the radar is not jammed with and after jamming, its effective detection area changes significantly. It can be seen that normal suppressive jamming is a great threat to radar and almost prevents it from playing its due role.

# 9.4 The Analysis of the Jamming Effect on Phased Array SAR of Airborne (or Unmanned Aerial Vehicle)

## 9.4.1 Analysis of Jamming Effect on UAV-Borne SAR

UAV-borne SAR has broad application prospects in both military and civilian use. The military reconnaissance has a great effect, which will inevitably cause special attention to effective jamming. The jamming of UAV-borne SAR has certain difficulties. Mainly for its reconnaissance, it needs to have a higher sensitivity of the reconnaissance receiver and wider instantaneous bandwidth of the receiver. The transceiver antenna's isolation caused by the high sensitivity of the receiver is also required to be high, so it is difficult to develop the UAV-borne SAR jammer. However, according to our current technical level, it is still possible to achieve effective jamming of UAV-borne SAR.

If the UAV-borne SAR flight diagram is shown in Fig. 9.25a and the ground projection diagram is shown in Fig. 9.25b.

H is the height of the drone, $R_t$ is the distance from the drone to the imaging center, $R_j$ is the distance between the jammer and the drone, β is the elevation angle of the UAV-borne SAR antenna, d is the projection point of the drone to the imaging center, $d_i$ is the distance between the imaging centerline and the jammer, $r_i$ is the vertical distance from the jammer to the line ($d_+ d_i$), and $\theta_i$ is the sidelobe angle of the jammer to the UAV-borne SAR antenna.

According to Fig. 9.25a and b, the following equations can be listed

$$R_t^2 = h^2 + d^2; \quad d = H \tan \beta$$

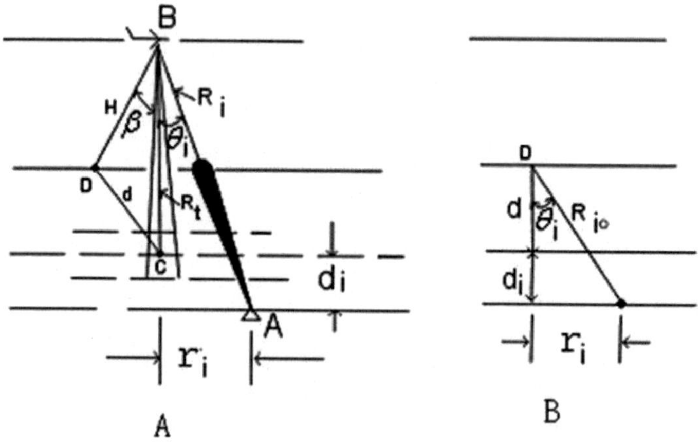


**Fig. 9.25** Schematic diagram of airborne SAR flight path and jammer configuration

$$r_i = (d + d_i)\tan\theta_i; \quad R_{j0}^2 = (d + d_i)^2 + r_i^2$$

$$R_j^2 = R_{j0}^2 + H^2 = (d + d_i)^2 + H^2 + r_i^2 = \left(d + d^i\right)^2 + H^2 + (d + d_i)^2 \tan\theta_i$$
$$= (d + d_i)^2 (1 + \tan\theta_i) + H^2$$

The main parameters of UAV-borne SAR:

The frequency band is X band, transmitter power $P_t = 3\,\text{kW}$, antenna gain $G_t = 28\,\text{dB}$.

Average sidelobe level is − 25 ~ − 30 dB, emission pulse width is 10–25 μs;

The repetition frequency is 10 kHz, the range resolution is 0.5 m, and the receiver sensitivity is − 82 dBm;

Polarized VH, the isolation of the antenna polarization is 35 dB, and the instantaneous bandwidth of the receiver is 300 MHz;

The elevation range of the antenna is 20°–70°, and the flying height is 10 km.

Required jamming equivalent power $P_j G_j$.

$P_j G_j = \frac{K_{j\Sigma} P_t G_t \sigma_t R_j^2}{4\pi R_{t\min}^4} \cdot \frac{G_t}{G_t(\theta)} \cdot \frac{L_d K_f}{r_j L_j}$, substitute the parameters into the formula and get

$$P_j G_j = \frac{140 \times 0.6 \times 10^3 \times 3 \times 10^3 \times 10^4 \times 2.2 \times 10^8 \times 0.5 \times 2 \times 10^3}{12.6 \times 1.26 \times 10^{16}} = 65\,\text{dBW}$$

According to the empirical formula of engineering calculation,

$$G_t = \frac{25,000}{\alpha_B \beta_B} = 28\,\text{dB} = 623$$

Set $\alpha_B$ to be 5°, then $\beta_B = 8^\circ$.

From $\frac{G_t}{G_t(\theta)} = -30\,\text{dB} = K\left(\frac{\alpha_B}{2\theta}\right)^2$, take K as 0.5 and $\alpha_B = 5^\circ$, it can be calculated that $\theta = \frac{0.7 \times 5 \times 31.5}{2} = 55.3^\circ$.

Example 1: $P_j G_j = 65\,\text{dBW}$, $P_t G_t = 63\,\text{dBW}$ Analysis of effective area of airborne SAR jamming.

If when $d_i = 0$, $\beta = 20^\circ$, $R_{t\min} = 10.6\,\text{km}$, $H = 10\,\text{km}$, $r_i = r_0$, now, $\theta_i = \theta_0 = 55.3^\circ$, as shown in Fig. 9.26

D is the projection point of the airborne SAR, C is the measurement center of the airborne SAR, and A is the position of the jammer.

According to Fig. 9.26, $r_0 = d\tan\theta_0 = H\tan\beta \cdot \tan\theta_0$, $\beta = 20^\circ$, $\theta_0 = 55.3^\circ$, $H = 10\,\text{km}$, $r_0 = 13\,\text{km}$.

The direction-finding distances are:

$$d_1 = H\tan\left(\beta - 4^\circ\right) = 10 \times \tan 16^\circ = 2.8\,\text{km}$$

$$d_2 = H\tan\left(\beta - 4^\circ\right) = 10 \times \tan 24^\circ = 4.4\,\text{km}$$

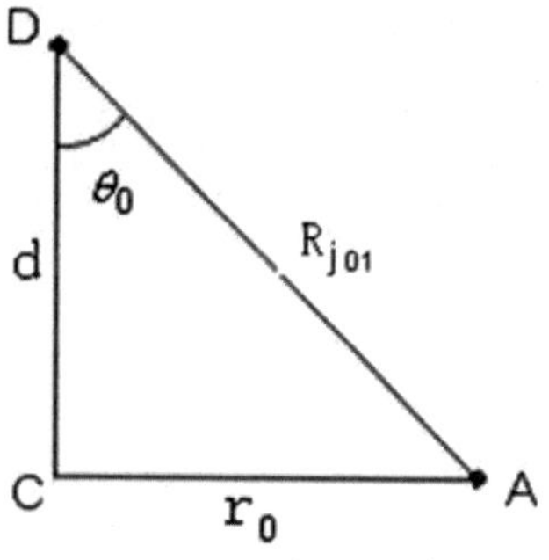


**Fig. 9.26** Airborne SAR projection

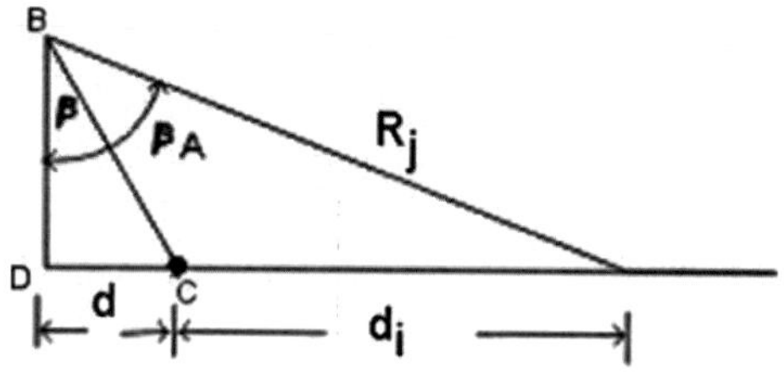


**Fig. 9.27** The side view of airborne SAR, when $r_i$ is 0

Then the direction-finding bandwidth is $\Delta R = d_2 - d_1 = 1.6\,\text{km}$.

Analysis of the longest distance of jamming: as shown in Fig. 9.27

$$r_i = 0, \quad R_{j0} = d + d_i, \quad R_j^2 = H^2 + (d + d_i)^2, \tan \beta_A = \frac{d + d_i}{H}$$

B is the airborne SAR, D is the projection point of the airborne SAR, C is the imaging center of the airborne SAR, d is the distance between the measuring center and the projection point, and $d_i$ is the distance between the jammer and the measuring center

According to the unified jamming equation: $P_j G_j = \frac{K_{j2} P_t G_t \sigma_t R_j^2}{4\pi R_{t\min}^4} \cdot \frac{G_t}{G_t(\theta)} \cdot \frac{L_d K_f}{r_j L_j}$, $R_j^2 = \frac{4\pi P_j G_j R_{t\min}^4}{K_{j\Sigma} P_t G_t \sigma_t} \cdot \frac{G_t(\theta)}{G_t} \cdot \frac{r_j L_j}{L_d K_f}$.

Solve simultaneous equations:

$$\tan \beta_A = \frac{d + d_i}{H} \tag{9.34}$$

$$R_j^2 = H^2 + (d + d_i)^2 \tag{9.35}$$

$$R_j^2 = \frac{4\pi P_j G_j R_{t\min}^4}{K_{j\Sigma} P_t G_t \sigma_t} \cdot K \left[ \frac{\beta_B}{2\beta_A} \right]^2 \cdot \frac{r_j L_j}{L_a K_f} \tag{9.36}$$

The result is:

$$\beta_A = R_{t\min}^2 \beta_B \left[ \frac{4\pi P_j G_j K}{K_{j\Sigma} P_t G_t \sigma_t \left[H^2 + (d + d_i)_2\right]} \cdot \frac{r_j L_j}{L_a K_f} \right]^{1/2} \quad (9.37)$$

In the formula,

$R_{t\min}$—the minimum detection distance of the radar, which is known to be 10.6 km;
$\beta_B$—the width of the half-power point of the beam elevation angle of the SAR antenna wave, which is 8°;
$P_jG_j$—the equivalent power of the jammer, which is 65 dBW;
$K_{j\Sigma}$—the jamming suppression coefficient, which is 140;
$P_tG_t$—the equivalent radiation power of SAR, which is 63 dBW;
H is the height of the carrier, which is 10 km;
d is the distance between the measurement zone and the projection line, which is 3.6 km;
$\sigma_t$ is the equivalent reflection area of the SAR main lobe illumination area, which is 104 $m^2$;
K is the sidelobe coefficient of the radar antenna (usually 0.5 to 0.1), but here is 0.5;

In $r_jL_j/L_aK_f$, $r_j$ is the jamming polarization loss, $r_j = 0.5$; $L_j$ is the feeder loss of the jammer, $L_j = 0.6$, $L_a$ is the radar's feeder loss, $L_a = 0.3$; $K_f$ is the ratio of the jamming spectrum width to the radar receiver bandwidth, $K_f = 2$, then $\frac{r_j L_j}{L_a K_f} = 0.5$.

Substituting formula (9.29) into formula (9.36), $\beta_A$ can be written as

$$\beta_A = R_{t\min}^2 \beta^A \left[ \frac{\pi P_j G_j K}{K_{j\Sigma} P_t G_t \sigma_t H\left(1 + \tan^2 \beta_A\right)} \cdot \frac{r_j L_j}{L_a K_f} \right]^{1/2} \quad (9.38)$$

In the formula (9.34), all except $\beta_A$ are known numbers. And $\beta_A$ is an implicit number, which is not easy to solve. Therefore, it is more convenient to use formula (9.38) and formula (9.30) to solve the curve.

Substitute the above relevant parameters and the $d_i$ value in Table 9.6 into Eqs. (9.34) and (9.42) to get the corresponding value $\beta_A$, see Table 9.6.

Draw the $\beta_A$—$d_i$ curve in Fig. 9.28 according to Table 9.6, and the intersection of the two lines is the solution of $\beta_A(d_i)$, $d_{i0} = 35$ km.

**Table 9.6** calculated βA at different ranges di (β=20°)

| $d_i$(km) | 20 | 30 | 40 | 50 | 60 |
|---|---|---|---|---|---|
| $\tan \beta_A = \frac{d+d_i}{H}$, $\beta_A$ | 67° | 73.4° | 77° | 79.5° | |
| $\beta_A = R_{t\min}^2 \beta_B \left[ \frac{\pi P_j G_j K}{K_{J\Sigma} P_t G_t \left[H^2 + (d+d_i)^2\right]} \cdot \frac{r_j L_j}{L_d K_f} \right]^{1/2}$ | 189° | 101.6° | 62.5° | 43.1° | 30.2° |

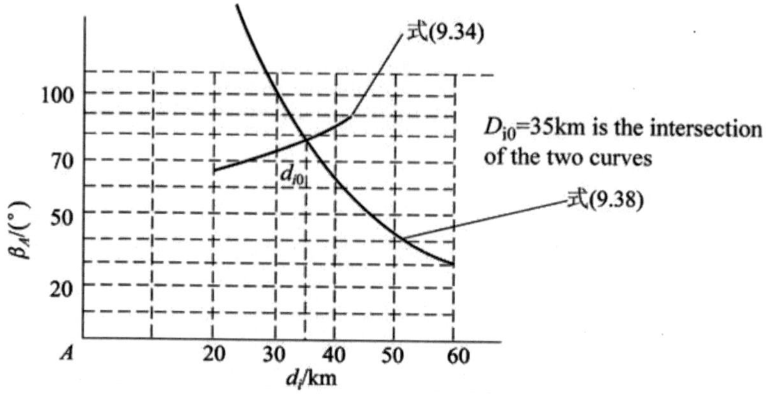


**Fig. 9.28** $\beta_A(d_i)$ inclusion curve $\left(\beta = 20^{\circ}\right)$

| | |
|---|---|
| $D_{i0} = 35$ km is the intersection point of two curves | $D_{i0} = 35$ km is the intersection of the two curves |

According to the calculated values of $r_0$, $d_{i0}$, d, the effective area of jamming to the airborne SAR at $\beta_A$ can be drawn, as shown in Fig. 9.28.

In Fig. 9.28 from A to $d_{i(1)}$, the elevation angle of the jammer is 90°. In this case, the value of $d_{i(1)}$ is taken as the value d, the value of $r_0$ is 13 km, and the value of $d_{i0}$ is 35 km. The intermediate value $r_i$ of the jamming effective area is mainly related to the sidelobe characteristics of the SAR antenna, the equivalent radiation power of SAR, the height of the machine and the equivalent power of jamming.

When $\beta$ is 60° and other parameters remain unchanged, find the effective area of the airborne SAR jamming.

When $\beta = 60^{\circ}$, $d = H\tan\beta = 17\,\text{km}$, $R_t + \left(d^2 + H^2\right)^{1/2} = 20\,\text{km}$, $k = 0.8$, $r_0 = d\tan\theta_0 = 17 \times 14 = 238^{\circ}$.

Substitute the above relevant parameters and the $d_i$ value in Table 9.6 into Eqs. (9.34) and (9.42) to get the corresponding value $\beta_A$, see Table 9.7.

**Table 9.7** Calculated βA at different ranges di (β=60°)

| $d_i$(km) | 20 | 30 | 40 | 50 | 60 |
|---|---|---|---|---|---|
| $\tan\beta_A = \frac{d+d_i}{H}$ | | 78° | 80° | 81.5° | 82.5° |
| $\beta_A = R_{t\min}^2 \beta_B \left[\frac{\pi P_j G_j K}{K_{J\Sigma} P_t G_t \left[H^2+(d+d_i)^2\right.} \cdot \frac{r_j L_j}{L_d K_f}\right]^{1/2}$ | 148° | 45° | 92° | 77° | 66° |

According to Table 9.7, the curve $\beta_A - d_i$ of Fig. 9.29 can be drawn, and the $d_{i0}$ value can be obtained from the intersection of the curves.

According to Table 9.7, the data draw a schematic diagram of the jamming effective area, as shown in Fig. 9.30.

Assuming that the sidelobe characteristics of the elevation angle of the airborne SAR are shown in Fig. 9.31

Substituting the above parameters into Eqs. (9.29), (9.30), and (9.31), the data of the jamming effective area can be calculated, and a schematic diagram of the jamming effective area can be drawn, as shown in Fig. 9.32.

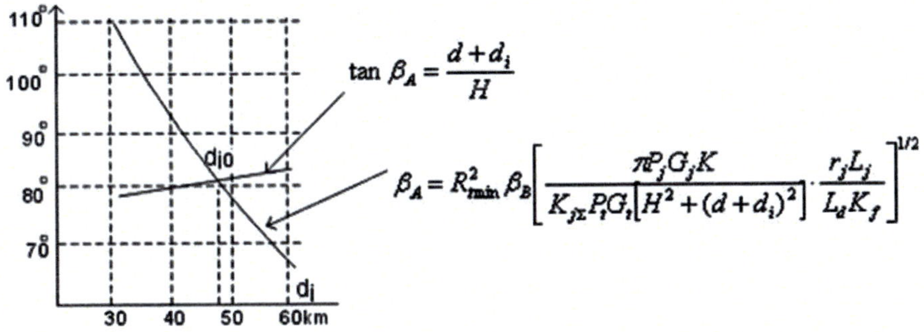


**Fig. 9.29** $\beta_A(d_i)$ inclusion curve ($\beta = 60°$)

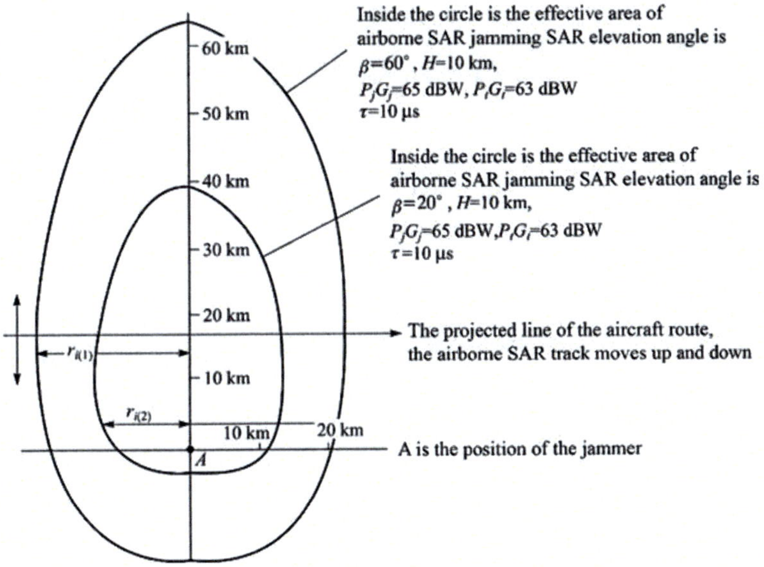


**Fig. 9.30** When the elevation angle of the airborne SAR is 20° and 60° respectively, ($P_j G_j$ = 65 dBw) Schematic diagram of the effective area of jamming ($\beta_A(d_i)$)

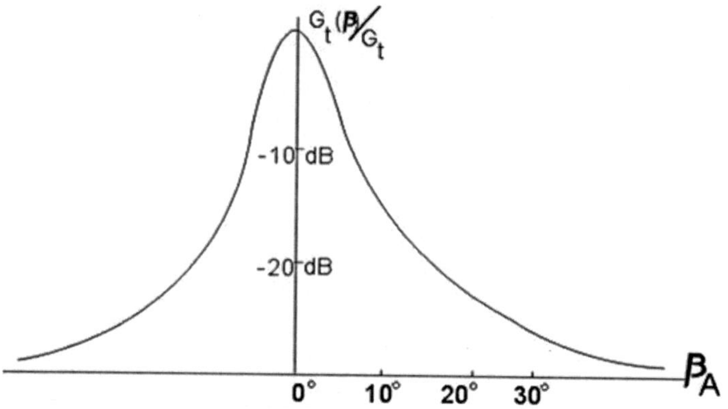


Fig. 9.31 Schematic diagram of the sidelobe characteristics of the elevation angle of the airborne SAR X-band antenna

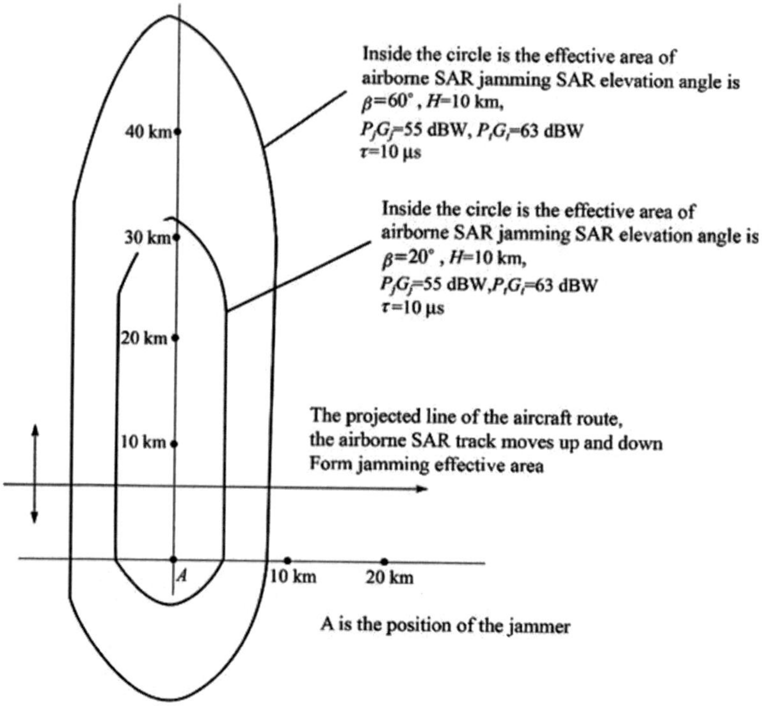


Fig. 9.32 Airborne SAR elevation angle β is 20° and 60° respectively (P $G_j$ = 55 dBW) schematic diagram of the effective area of jamming

From the perspective of anti-jamming, increasing the equivalent radiation power of the SAR, increasing the transmit pulse width, or reducing the antenna's sidelobe level will reduce the effective area of jamming. At present, radar is vigorously developing phased array radar, which is more convenient to increase the radar's equivalent radiation power, increase the transmission pulse width, and reduce the antenna sidelobe level. If the jammer wants to reach the predetermined jamming effective area, increasing the jamming equivalent power and the jamming unit array configuration are effective jamming measures.

### 9.4.2 *Analysis of the Effective Area of Point Target Jamming in the Ground Attack of Missile-Borne SAR*

A schematic diagram of a missile-borne SAR attacking a point target on the ground is shown in Fig. 9.33.

Point A—the position of the jammer.

D—the projection point of the missile on the ground.

Point B—the position of the missile, turn on the terminal guidance from point B to attack the ground target.

Point C—the attacked target.

d1—the distance between the jammer and the covered target.

H—the height at which the missile turns on the terminal guidance radar, and then the missile terminal guidance radar automatically guides the missile to the target being attacked.

d2—the horizontal distance between the missile and the target when the missile terminal guidance radar is turned on.

In Fig. 9.33b, due to the movement of the missile, which is the continuous movement of point B, $R_j$, $R_t$ (distance between the missile and the target), H, d2 are all changing. $\beta_1$ is the angle between the missile and the target line $R_t$ and H, and $\beta_2$ is the angle between $R_j$ and H. These angles also change with the movement of point B.

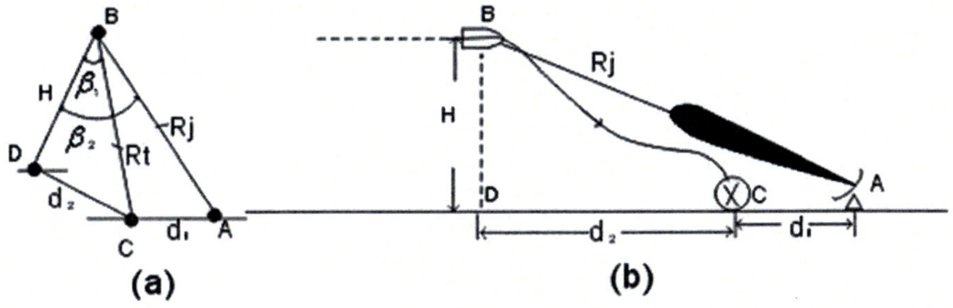


**Fig. 9.33** Schematic diagram of missile-borne SAR guided aiming to attack ground point targets

No analysis is done before the terminal guidance radar is turned on. Only analyze the situation after the missile terminal guidance radar is turned on and after being jammed with. Terminal guidance radar may be conventional pulse radar, pulse compression radar, Pulse-Doppler radar, synthetic aperture radar, etc. From the perspective of suppressing jamming, the unified equation of radar jamming can be used to describe the situation after jamming.

***Example*** Assume that the main parameters of the missile-borne radar: working in the X-band, the transmitter peak power is 200 W, the transmitted pulse width is 10 $\mu S$, the antenna gain is 28 dB; the antenna sidelobe level is – 25 dB, and it can work in pulse pressure and synthetic aperture status. When the guidance radar is turned on, H is 10 km; d2 is 10 km; the protected target is a point target (with a radius of fewer than 10 m); the distance d1 between the jammer and the protected target is 3 km, and the missile speed is 700 m/s.

When the terminal guidance radar is turned on, the distance $R_t$ is $h_2 + d_2^2)^{1/2}$, which is 14.1 km; the maximum distance $R_j$ is $\left[(d_1 + d_2)^2 + H^2\right]^{1/2}$, which is 16.4 km; $\beta_1$ is 45°; the minimum $\beta_2$ is $\tan^{-1}\left(\frac{d_2+d_1}{H}\right) = 53^\circ$.

- Calculate the equivalent power of the jammer

If the missile enters from the direction α is 60° and $R_{t\,\min}$ is 3 km, according to the jamming equation

$$P_j G_j = \frac{K_{j\Sigma} P_t G_t \sigma_t R_j^2}{4\pi R_{t\,\min}^4} \cdot \frac{G_t}{Gt(\theta)} \cdot \frac{L_d K_f}{L_j r_j}$$

In the formula, the jamming suppression coefficient is 20 dB; $P_t G_t = 6 \times 10^4$ W; The effective RCS $\sigma_t = 10^3 \text{m}^2$, $R_j \approx 3\,\text{km}$, $R_{t\,\min} = 3$ km, $G_t / G_t(\theta) = 25$ dB, $\frac{L_d K_f}{L_j r_j} = 2$.

Substitute the above parameters into the jamming equation:

$$P_j G_j = \frac{10^2 \times 1.2 \times 10^4 \times 10^4 \times 3 \times 10^2 \times 2}{12.6 \times (3 \times 103)^2} = 45\,\text{dBW}$$

- The sensitivity of the receiver

The sensitivity of the receiver is

$$P_r = \frac{P_t G_t(\theta)}{4\pi R_j^2} \cdot \frac{\lambda 2}{4\pi} \cdot G_j$$

Take $P_t = 100$ W, $G_t(\theta) = 1$, $\lambda = 3*10^{-2}$ m, $G_j = 30$, $R_j = 20$ km,
Substitute the parameters into the above formula

$$P_r = \frac{10^2 \times 1 \times 9 \times 10^{-4} \times 30}{1.5 \times 10^2 \times 4 \times 10^8} = \frac{2.7 \times 1}{6 \times 10^{10}} = -105\,\text{dBW}$$

- Angular tracking speed: 0°–45°/s, Elevation tracking range: 3°–90°;
- Calculation of jamming effective area

(1) Single jamming station suppresses jamming zone

Assuming the parameters of the missile-borne radar: working in the X-band, $P_t G_t = 0.6 \times 10^5 W$, the height is lower than 2000 m when it is 3 km close to the attacked target, the starting distance d2 is 10 km, and H is 10 km.

The jammer parameters are $P_j G_j = 45$ dBW, the configuration distance d1 = 3 km, and the working range of the antenna elevation angle is 3°–90°.

The radar's jamming equation can be written as

$$K_{j\Sigma} = \frac{4\pi P_j G_j R_{t\min}^4}{P_t G_t \sigma_t R_j^2} \cdot \frac{G_t(\theta)}{G_t} \cdot \frac{L_d K_f}{r_j L_j} \tag{9.39}$$

Substitute the parameter into the formula (9.39), when $\theta = 0°$, it is $G_t(\theta)/G_t = 10^{-3}$, $\frac{L_d K_f}{r_j L_j} = 0.5$

Then, $100 = \frac{R_{t\min}^4}{R_j^2} \cdot \frac{4\pi P_j G_j \times 0.5 \times 10^{-3}}{P_t G_t \sigma_t}$.

$\frac{R_{t\min}^4}{R_j^2} = 3 \times 10^{-8}$ Using the trial method to solve, when $R_{j(1)} = 0.8$ km and $R_{t\min} = 1.2$ km, $\frac{K_j^c}{R_{t\min}^4} \cong 3 \times 10^{-8}$.

The solution obtained that when $\theta = 0^\circ$, $R_{t\min(1)} = 1.2$ km.

When $\theta = 60°$, $G_t(\theta)/G_t = -25$ dB, $\frac{R_{j(2)}^2}{R_{t\min(2)}^4} = 1 \times 10^{-8}$, we get $R_{j(2)} = 1.7$ km, $R_{t\min(2)} \cong 1.7$ km.

(2) When the three stations are symmetrically configured, their jamming effective area and jamming exposure area are shown in Fig. 9.34.

$A_{1,2,3}$ is the position of the jammer;

C is the point target being jammed to cover;

The horizontal line area is the suppression and jamming exposure area;

Outside the horizontal line is the effective area for suppressing jamming.

The jamming suppression coefficient $K_{j\Sigma}$ outside the horizontal area is greater than 100, which is the effective prescribed value for our analysis of suppression jamming. In this case, the missile terminal guidance radar cannot observe the target and cannot keep up with the tracking target, so it is effective for the protected target. When the missile enters the horizontal zone, although $K_{j\Sigma} < 100$, the jamming comes from three different directions, and $K_{j\Sigma}$ is still greater than 1.

The angular tracking error equation caused by thermal noise is used here.

$$\Delta\theta = \frac{\theta_B}{k\sqrt{B\tau(S/N)(f_r/\beta_n)}} \tag{9.40}$$

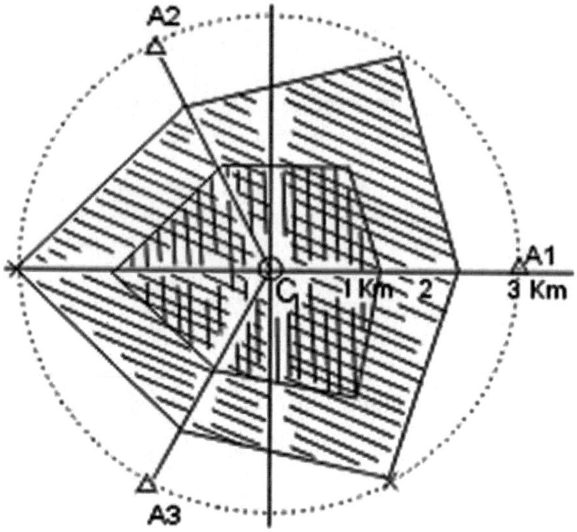


**Fig. 9.34** Schematic diagram of the jamming exposure area of the three-station jamming symmetric configuration

In the formula,

$\Delta\theta$ is the angle tracking error;
$\theta_B$ is the beam width (half power point) of the angle tracking antenna;
B is the receiver bandwidth;
$\tau$ is the radar transmit pulse width;
S/N is the signal/noise ratio and is above 15 dB during normal tracking;
$f_r$ is the repetition frequency of the radar transmit pulse;
$\beta_n$ is the loop equivalent noise bandwidth,
k is a coefficient, and the value is 1.57.

When the radar is tracking the target normally, $\Delta\theta/\theta_B \leq 0.1$. So, the error caused by thermal noise, within the allowable range of the radar's total tracking error, has a negligible impact on the missile's terminal guidance hitting the target. Under this kind of noise jamming condition, the jamming source comes from three different directions, $S/N < 1$, $B\tau$ is close to 1, $f_r/\beta_n$ is less than 1. The calculated result is $\Delta\theta/\theta_B > 1$. In the case of q, the angle tracking system will lose the target. It can be seen that the missile terminal guidance radar cannot find the target outside the horizontal line area. Although the missile terminal guidance radar may find the attacked target in the horizontal line area, it cannot track the attacked target. Therefore, under cover of a three-station symmetrical jammer, the missile terminal guidance radar cannot guide the missile to attack the target effectively. This is the result of theoretical analysis. It is best to conduct actual confrontation tests to prove it.

## 9.5 Analysis of Jamming Effect on Spaceborne Phased Array SAR

Spaceborne SAR has broad application prospects in both military and civilian applications. Because of its long detection range, the large area of action, and large equivalent radiation power, phased array radar can also be widely used in spaceborne phased array SAR. This section is devoted to analyzing the jamming principle, jamming equation, and the jamming effect of the spaceborne phased array SAR.

### *9.5.1 The Jamming Equation of Spaceborne Phased Array SAR*

The working principle of phased array synthetic aperture radar is similar to that of conventional radars, except that some measures have been taken in range resolution and azimuth resolution to improve the two-dimensional resolution to the order of meters, or less than 1 m.

In terms of distance, the principle of compression is used to transmit a wide pulse, and the echo is compressed to compress the width to $\Delta\tau$. Then the range resolution $\Delta R$ is

$$\Delta R = \frac{1}{2} C \cdot \Delta\tau \tag{9.41}$$

C is the speed of light and $\Delta\tau$ is the pulse width after compression. After range compression processing, the radar has a gain coefficient $n_r$.

$$\eta_r = \tau / \Delta\tau \tag{9.42}$$

The method to improve the azimuth resolution is to accumulate and process the signals in a beam, which is equivalent to the effect of a very narrow beam antenna, thus improving the resolution in the azimuth. As shown in Fig. 9.35,

B is the carrier of the spaceborne SAR, which moves in a straight line. The SAR is installed on the side of the carrier. The radar antenna does not rotate. The radar antenna moves when the carrier moves. If the azimuth beam of the antenna is $\theta_{0.5}$,

$$R_T \cong R \cdot \theta_{0.5} = \frac{H}{\text{Cos}\beta} \cdot \theta_{0.5} \tag{9.43}$$

The time for the carrier to pass $R_T$ is $T_r$

$$T_r = \frac{R_t}{v} = \frac{h \cdot \theta_{0.5}}{v\text{Cos}\beta} \tag{9.44}$$

Fig. 9.35 Schematic diagram of SAR operation

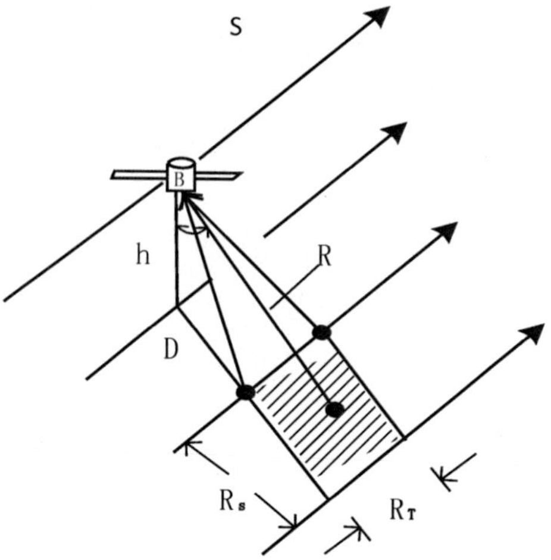


During $T_r$, the number of pulse echoes accumulated by the radar is $\eta_b$

$$\eta_b = T_r \cdot f_r = \frac{f_r \cdot h \cdot \theta_{0.5}}{v \cdot \text{Cos}\beta} \tag{9.45}$$

The radar signal echo power $P_r$ can be written as

$$P_r = \frac{P_t G_t^2 \lambda^2 \sigma L_d}{(4\pi)^3 R^4} \tag{9.46}$$

The gain of SAR is obtained through correlation receiving and two-dimensional compression processing, and the signal power converted to the input of the receiver is $S_r$

$$S_r = P_r \eta_r \eta_b = \frac{P_t G_t^2 \lambda^2 \sigma_i L_d}{(4\pi)^3 R^4} \cdot \frac{\tau}{\Delta\tau} \cdot \frac{f_r \cdot \theta_{0.5} \cdot h}{v\text{Cos}\beta} \tag{9.47}$$

In the formula:

$P_t$ is the output power of the radar transmitter;
$G_t$ is the antenna gain of the radar;
$\lambda$ is the working wavelength of the radar;
$L_d$ is feeder loss of the radar transmission and reception;
$\tau$ is the pulse width of the radar emission;
$\Delta\tau$ is the radar pulse width after receiving compression;
$f_r$ is the repetition frequency of radar pulses;
$\theta_{0.5}$ is the azimuth beamwidth of the radar antenna;
$\beta$ is the elevation angle of the radar beam;
R is the distance between the radar and the center of the detection target;

$h$ is the height of the radar carrier platform.
v is the speed of the radar carrier;
$\sigma_i$ is the effective reflection area of the target in the resolution area of the radar.

The jamming suppression coefficient of SAR is analyzed as follows.
Assume that the power of the jamming signal received by the SAR antenna is $N_j$.

$$N_j = \frac{P_j G_j}{R\pi R_j^2} \cdot \frac{\lambda^2}{4\pi} G_t(\theta) \cdot L_j r_j K_f \tag{9.48}$$

In the formula,

$P_j G_j$ is the jamming equivalent jamming power;
$\lambda$ is the working wavelength of the jammer;
$R_j$ is the distance between the jammer and the radar;
$G_t(\theta)$ is the sidelobe gain of the radar antenna aligned with the jamming antenna;
$L_j$ is the feeder loss and atmospheric loss of the jammer;
$r_j$ is the polarization loss of the jammer.

According to the definition of jamming suppression coefficient, at the minimum jamming exposure distance, the ratio of the power of the jamming signal to the power of the radar echo signal at the output end of the radar's receiving antenna is defined as the equivalent jamming suppression coefficient of the jammer, denoted by $K_{j\Sigma}$.

The passband of the radar receiver also limits the jamming signal entering the radar receiver. No matter how wide the jamming signal spectrum is, it can only allow the spectrum signal $\Delta f_j$ with the same bandwidth as the receiver to enter, and $K_f = \frac{\Delta f_r}{\Delta f_j}$ is obtained.

Define:

$$K_{j\Sigma} = \frac{N_j}{S_r} \tag{9.49}$$

Substitute Eq. (9.47) and Eq. (9.48) into Eq. (9.49).

$$K_{J\Sigma} = \frac{\frac{P_j G_j}{4\pi R_j^2} \cdot \frac{\lambda^2}{4\pi} \cdot G_t(\theta) \cdot L_j \cdot \gamma_j \cdot \Delta f_j}{\frac{P_T G_t^2 \lambda^2 \sigma_i L_d}{(4\pi) R^4} \cdot \frac{\tau}{\Delta\tau} \cdot \frac{f_\gamma \cdot \theta_{0.5} h}{v \mathrm{Cos}\beta} \Delta f} = \frac{4\pi P_j G_j R^4}{P_t G_t \sigma_i R_j^2} \cdot \frac{G_t(\theta)}{G_t} \cdot \frac{v \mathrm{Cos}\beta}{B_j \cdot \theta_{0.5} h} \cdot \frac{L_j \gamma_j K_f}{L_d} \tag{9.50}$$

In the formula,

$\overline{P}_t = \rho_t \cdot \tau \cdot f_\gamma$ is the average power of the radar;
$B_\gamma = \frac{1}{\Delta\tau}$ is the receiver's bandwidth of the radar;
$K_f$ is the ratio of the receiver's bandwidth to the width of the jamming spectrum [The rest of the parameters are the same as (9.47)].

If the conventional data is substituted into the jamming suppression coefficient,

$$K_{jn}=\frac{N_j}{S_\gamma}=\frac{\frac{P_jG_j}{4\pi R_j^2}\cdot\frac{\lambda^2}{4\pi}G_t(\theta)\cdot L_j\cdot\gamma_j\cdot K_f}{\frac{P_tG_t^2\lambda^2\sigma_iL_d}{(4\pi)R^4}\cdot\eta_r\eta_b}$$

After simplified,

$$K_{jn}=\frac{4\pi P_jG_jR^4}{P_tG_t\sigma_iR_j^2}\cdot\frac{G_t(\theta)}{G_t}\cdot\frac{L_j\cdot\gamma_jK_f}{L_d}\cdot\frac{1}{\eta_\gamma\eta_b}\tag{9.51}$$

The maximum jamming suppression coefficient $K_{j\sum\max}$ of SAR is its maximum processing gain coefficient, therefore:

$$K_{j\sum\max}=K_{jn}\eta_\gamma\cdot\eta_b\tag{9.52}$$

Substituting Eqs. (9.51) into (9.52) and after simplified, we get

$$K_{j\Sigma\max}=\frac{4\pi P_jG_jR_{\min}^4}{P_tG_t\sigma_iR_j^2}\cdot\frac{G_t(\theta)}{G_t}\cdot\frac{L_j\gamma_jK_f}{L_d}\tag{9.53}$$

In terms of form, (9.53) is similar to conventional pulse radar, but the content of some parameters is quite different from conventional radar.

$$P_jG_j=\frac{K_{j\Sigma\max}P_tG_t\sigma_iR_j^2}{4\pi R_{\min}^4}\cdot\frac{G_t}{G_t(\theta)}\cdot\frac{L_d}{L_j\cdot r_jK_f}\tag{9.54}$$

For example, the effective RCS ($\sigma_i$) of the target refers to the equivalent RCS of the target in the radar antenna beam, and the $\sigma_i$ of SAR refers to the effective RCS of the target in the resolution area of the SAR.

For another example, the peak power $P_t$ of the radar is relatively large in conventional radars. In SAR, due to the coherent receiving technology it uses, the radar receiver's signal/noise ratio is improved. In the same operating distance, the peak power of the radar transmission is greatly reduced.

In conventional pulse radar, $K_f = 0.3$; because SAR adopts pulse compression technology and the requirements of range resolution, the bandwidth of the radar receiver is relatively wide. So, $K_f \approx 0.8$ is generally adopted.

### 9.5.2 Derivation of "Unified Jamming Equation" of Phased Array Synthetic Aperture Radar

Conventional pulse radar:

$$P_r = \frac{G_a G_a^2 \lambda^2 \sigma_a k_1}{(4\pi)^3 R_a^4} \tag{9.55}$$

$$R_a^4 = \frac{G_a G_a^2 \lambda^2 \sigma_a k_1}{(4\pi)^3 P_r} \tag{9.56}$$

SAR:

$$P_s = \frac{P_t G_t^2 \lambda^2 \sigma_i k_1}{(4\pi)^3 R_t^4} \tag{9.57}$$

$$R_t^4 = \frac{P_t G_t^2 \lambda^2 \sigma_i k_1}{(4\pi)^3 P_s} \tag{9.58}$$

Assuming that the maximum detection distances of conventional pulse radar and synthetic aperture radar are equal, that is, Eqs. (9.56) and (9.58) are equal.

$$\frac{P_a G_a^2 \lambda^2 \sigma_a k_1}{(4\pi)^3 P_{r\,\min}} = \frac{P_t G_t^2 \lambda^2 \sigma_i k_1}{(4\pi)^3 P_{s\,\min}} \tag{9.59}$$

If $G_a = G_t$; $P_{r\,\min} = KT\Delta F_a N$; $P_{s\,\min} = \frac{KT\Delta F_t N}{n_d \cdot n_b}$.
In the formula,

K is the Boltzmann constant;
T is the absolute temperature;
$\Delta F_a$ is the bandwidth of the conventional pulse radar receiver;
N is the noise figure;
$\Delta F_t$ is the bandwidth of the synthetic aperture receiver.
$G_a$, $G_t$ is the radar antenna gain;
$\sigma_a$ is the equivalent reflection area of the target in the main beam of the radar;
$\sigma_i$ is the equivalent reflection area of the target in the resolution area of the synthetic aperture radar;
$K_1$ is the feeder loss coefficient.

If $G_a = G_t$, substituting the parameters into Eq. (9.59), after simplified

$$\frac{P_a G_a}{\Delta F_a} = \frac{P_t \sigma_i n_d \cdot n_b}{\Delta F_t} \tag{9.60}$$

$$P_t = \frac{P_a \sigma_a \Delta F_t}{\Delta F_a \sigma_i n_d \cdot n_b} \tag{9.61}$$

$$\Delta F_a = \frac{1}{\tau};\ \Delta F_t = \frac{1}{\Delta\tau}$$

Then

$$\frac{\Delta F_t}{\Delta F_a} = \frac{\tau}{\Delta\tau} = n_d,$$

We get

$$P_t = \frac{P_a \sigma_a}{\sigma_i n_b} \tag{9.62}$$

Substitute formula (9.62) into formula (9.54) to get

$$P_j G_j = \frac{n_d P_a G_a \sigma_a R_j^2}{4\pi R^4} \cdot \frac{G_t}{G_t(\theta)} \cdot \frac{L_a}{L_j \cdot r_j K_f} \tag{9.63}$$

It can be seen from the formula (9.63) that the calculation of the jamming equation for SAR has been calculated and simplified into a formula similar to the calculation of the pulse compression radar jamming equation. Therefore, the jamming equations for synthetic aperture radars can be unified into the jamming equations for conventional pulse radars. As long as different radars' systems, different jamming suppression coefficients can be used, which can greatly simplify the calculation of the jamming equivalent power required for the jamming of synthetic aperture radar.

For example, the jamming equivalent power required for a certain spaceborne SAR:

If $G_t = 42$ dB; $P_t = 37$ dBW; $\sigma_a = 47$ dBm$^2$, $G/G_t(\theta) = 40$ dB.

$H = 700$ km; $R_j = 1000$ km; $n_d = (\tau/\Delta\tau)^{0.8} = 26$ dB.

$\frac{L_a}{L_j r_j K_f} = 2$; $R_{\min} = H$. The protected area is 100,000 km$^2$.

Substitute the parameters into the formula (9.63)

$$\begin{aligned} P_j G_j &= \frac{n_d P_t G_t \sigma_a R_j^2}{4\pi R_{\min^4}} \cdot \frac{G_t}{G_t(\theta)} \cdot \frac{L_a}{L_j r_j K_f} \\ &= (26 + 37 + 42 + 47 + 40 + 3)\,\text{dB} \times \frac{10^{12}}{12.6 \times \left(0.7 \times 10^6\right)^4} \\ &= 70.3\,\text{dBW} \end{aligned}$$

If $G_j$ is 35 dB, then $P_j$ can be 3.3 kW.

### 9.5.3 Calculation of Equivalent Power of Jamming to Spaceborne Phased Array Synthetic Aperture Radar

Calculate with the formula (9.54):

$$P_j G_j = \frac{\left(K_{j\sum\max}\right)^{0.8} P_t G_t \sigma_i R_j^2}{(4\pi) R_{\min}^4} \cdot \frac{G_t}{G_t(\theta)} \cdot \frac{L_j r_j K_f}{L_d}$$

Set $M = \frac{P_t G_t R_j^2}{(4\pi) R_{\min}^4} \cdot \frac{G_t}{G_t(\theta)} \cdot \frac{L_j r_j K_f}{L_d}$, then

$$P_j G_j = \left(K_{j\Sigma\max}\right)^{0.8} \sigma_i M$$

Known $\left(K_{j\Sigma\max}\right)^{0.8} = 59.4dB$; $\sigma_i = 13dB$.
Calculate with formula (9.54), $P_j G_j = (594 + 13)dB \times M = 72.4dB \times M$.
Calculate with formula (9.59)

$$P_j G_j = \frac{n_d \cdot \sigma_a P_t G_t R_j^2}{4\pi R_{\min}^4} \cdot \frac{G_t}{G_t(\theta)} \cdot \frac{L_j r_j K_f}{L_d}$$
$$= n_d \cdot \sigma_a \times M$$

Known $n_d = (n_r)^{0.8} = 26.4$ dB.
Equivalent reflection area of main lobe $\sigma_a = 50{,}000$ m$^2 = 47$ dB.
So, $P_j G_j = (47 + 264)$ dB $\times M = 73.4$ dB $\times M$.

The equivalent jamming power calculated by (9.54) and (9.63) is basically the same. Therefore, the "jamming unified equation" that jamming with the new system radar is established.

### 9.5.4 Analysis of Jamming Effective Area Against Spaceborne Phased Array SAR

Set the configuration of spaceborne SAR and jammer, as shown in Fig. 9.36.

The known satellite height H is 700 km; the elevation angle β is 25°, $r_0$ is 240 km; $P_t G_t = 80$ dBW.

$$P_j G_j = 69\,\text{dBW};\ K_{j\Sigma} = 24\,\text{dB};\ G_t/G_t(\theta) = 43\,\text{dB};\ \frac{L_d}{r_j L_j K_f} = 2$$

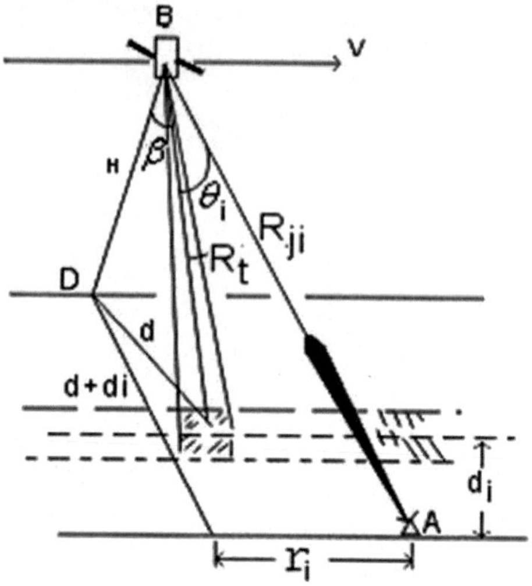


**Fig. 9.36** Schematic diagram of jamming to spaceborne phased array synthetic aperture radar

#### 9.5.4.1 Analysis of Effective Jamming Area When the Elevation Angle of Spaceborne Phased Array SAR Remains Unchanged, and the Track Is Shifted

It can be seen from Fig. 9.36 that when $d_i = 0$,

$$d = H\tan\beta = \tan 25^{\circ} \times 700 = 326\,\text{km}$$
$$R_t = \left(H^2 + d^2\right)^{1/2} = 772\,\text{km}$$
$$R_j = \left(R_t^2 + r_0^2\right)^{1/2} = 810\,\text{km}$$
$$\theta_i = \tan^{-1}\frac{r_0}{R_t} = 17^{\circ}$$

Point A is the position of the jammer; Point B is the spaceborne SAR; Point D is the projection point of B; $R_{ti}^2 = H^2 + (d + d_i)^2$.

According to formula (9.57) and Fig. 9.36 The simultaneous equations can be written as

$$\begin{cases} R_{ji}^2 = \frac{4\pi P_j G_j R_t^4}{K_{j\Sigma} P_t G_t \sigma_t} \cdot \frac{G_t(\theta_i)}{G_t} \cdot \frac{L_j r_j K_f}{L_d} \\ R_{ji}^2 = H^2 + (d + d_i)^2 + r_i^2 \\ r_i = \left[H^2 + (d + d_i)^2\right]^{12} \tan\theta_i \end{cases}$$

Solving the simultaneous equations of (9.63), (9.64), (9.65), we get:

$$R_{ti}^2\left(1 + \tan^2\theta_i\right) = \frac{4\pi R_t^4 P_j G_j}{K_{j2} P_t G_t \sigma_t} \cdot \frac{G_t(\theta_i)}{G_t} \cdot \frac{L_j r_j K_f}{L_d} \tag{9.64}$$

In the formula (9.64), only $\theta_i$ is an unknown number, and the rest are known numbers.

$$\frac{G_t(\theta_i)}{G_t} \cong K\left(\frac{\theta_0}{2\theta_i}\right)^2 \tag{9.65}$$

At a small angle of $\theta_{\mathrm{i}}$, $\tan^2\theta_i \ll 1$.

Given $K\left[\frac{\theta_0}{2\theta}\right]^2 = 0.5 \times 10^{-4}$, $K \cong 0.09$ can be obtained.

Substitute Eq. (9.65) into Eq. (9.64).

$$R_{ji}^2 = \left[\frac{4\pi R_t^4 P_j G_j}{K_{j\Sigma} P_t G_t \sigma_t} \cdot \frac{K\theta_0^2}{4\theta_i^2} \cdot \frac{K_2 K_h}{K_1 K_f}\right] \tag{9.66}$$

Simplify formula (9.66) to

$$\theta_i = \left[\frac{\pi R_t^4 P_j G_j}{K_{JL} P_t G_t \sigma_t R_{ti}^2} \cdot K\theta_0^2 \cdot \frac{L_j r_j K_f}{L_d}\right]^{1/2}$$

Which is

$$\theta_{\mathrm{i}} = \left(H^2 + d^2\right)\theta_0 \left[\left[\frac{K\pi P_j G_j}{K_{j\Sigma} P_t G_t \sigma_t \left[H^2 + (d + d_i)^2\right]}\right] \cdot \frac{L_j r_j K_f}{L_d}\right]^{1/2} \tag{9.67}$$

In the formula (9.67): $\theta_0$ is the beamwidth of the radar antenna at the half power point, $K = 0.09$; **(K È¡ 01)**

$$R_{\mathrm{t}} = \left(H^2 + d^2\right)^{1/2};\ R_{ti} = \left[H^2 + (d + d_i)^2\right]^{1/2} \tag{9.68}$$

$$r_{\mathrm{i}} = R_{ti} \tan\theta_i \tag{9.69}$$

Substituting the above parameters and the $d_{\mathrm{i}}$ value in Table 9.7 into Eq. (9.70), the result is listed in Table 9.8.

**Table 9.8** Calculated βA at different ranges di (β=60°)

| $d_{\mathrm{i}}$(km) | 0 | 163 | 326 | 514 | – 163 |
|---|---|---|---|---|---|
| $\theta_{\mathrm{i}}$(km) | 17° | 15° | 13° | 0° | 10° |
| $R_{\mathrm{ji}}(km)$ | 810 | 980 | 1005 | 1093 | 720 |
| $R_{\mathrm{toi}}(km)$ | 326 | 489 | 652 | 840 | 163 |
| $R_{\mathrm{ti}}$(km) | 770 | 850 | 956 | 1093 | 720 |
| $r_i(km)$ | 234 | 221 | 219 | 0 | 130 |

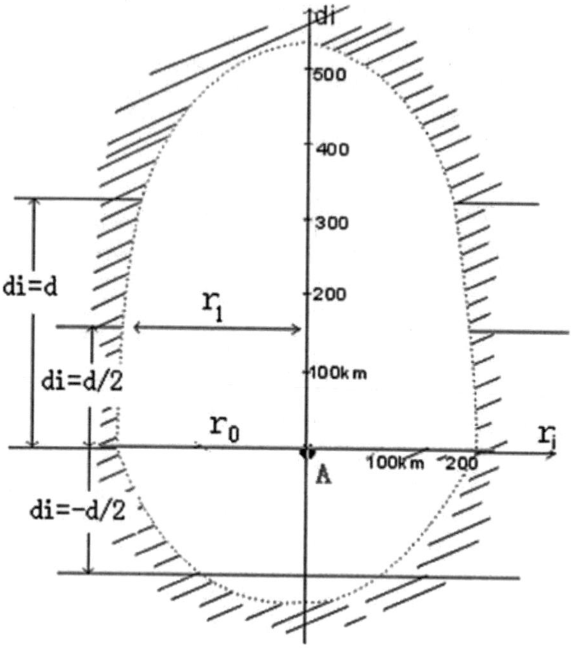


**Fig. 9.37** Schematic diagram of the effective jamming area and jamming exposure area for spaceborne SAR

According to Table 9.8, plot the effective jamming area and jamming exposure area for spaceborne SAR, and the elevation angle β of the spaceborne phased array SAR is 25. When the track moves along the jammer's vertical direction, the effective jamming area is formed, as shown in Fig. 9.37.

Inside the circle is the effective jamming zone; outside the circle is the jamming exposure zone; $P_tG_t = 80\,\text{dBw}$, $P_jG_j = 69\,\text{dBw}$. For space-borne phased array SAR: H = 700 km; β = 25°; the antenna's sidelobe level at 17° is − 43 dB; the space-borne SAR working in the L-band forms an effective jamming area (Fig. 9.38). The effective area is about 200,000 km$^2$.

Figure 9.38 is just the result of analysis and calculation. If you are interested, you can do a formal jamming test to verify the theoretical and actual compliance and provide a basis for subsequent actual combat.

Inside the circle is the effective jamming zone; outside the circle is the jamming exposure zone; point A is the jamming station, which can automatically and passively track the spaceborne SAR; H is the SAR height that remains unchanged, $D_{1,2,3,4,5}$ are the SAR projection points, β is 25° and maintains unchanged; ν is the speed of spaceborne SAR and maintains unchanged. Only the SAR's trajectory moves along the vertical direction to the jammer, forming the effective jamming area and the jamming exposure area. The jamming-effective area in Figs. 9.37 and 9.38 is the same, but the drawing method is somewhat different.

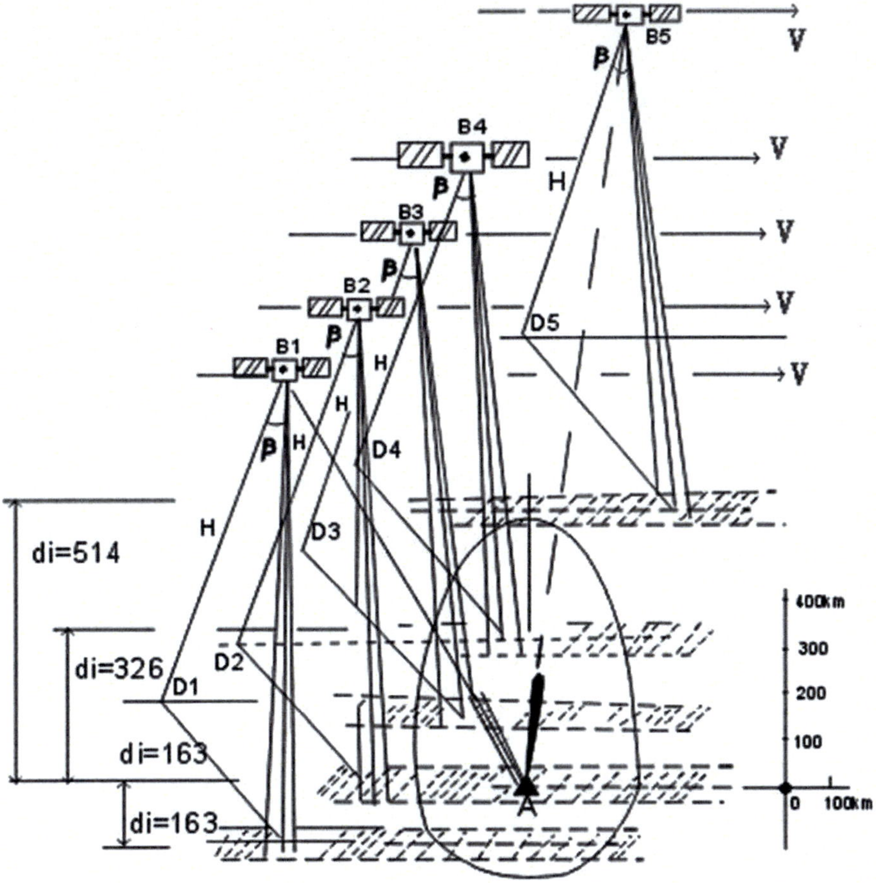


**Fig. 9.38** Schematic diagram of the jamming effective area formed by the track shift of the spaceborne phased array SAR

Substituting formula (9.68) and formula (9.69) into formula (9.67), the unified expression of $\mathrm{r_i}$ can be written:

$$r_{\mathrm{i}} = R_{ti} \tan \theta_i$$

$$r_{\mathrm{i}} = \left[H^2 + (d + d_i)^2\right]^2 \cdot \tan\left[H^2\left(1 + \tan^2 \beta\right) \cdot \theta_0 \cdot \frac{k\pi P_j G_j}{K_{J\Sigma} P_t G_t \sigma_t \left[H^2 + (d + d_i)^2\right]} \cdot \frac{1}{K_a}\right] \quad (9.70)$$

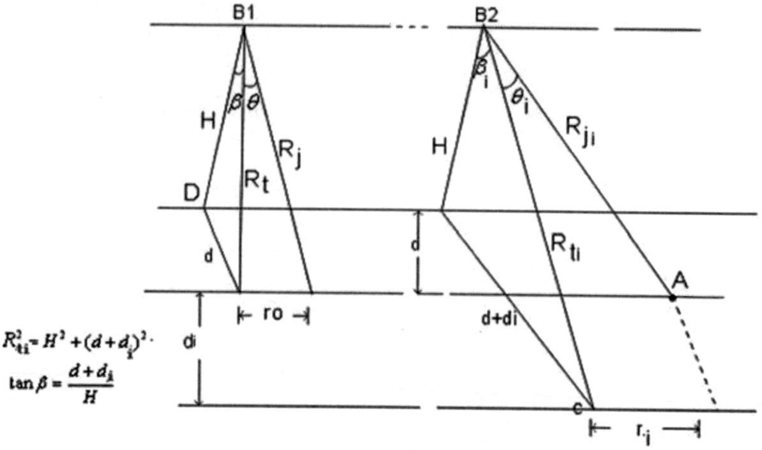


**Fig. 9.39** The trajectory of the spaceborne phased array SAR remains unchanged, and the effective area of jamming formed by β increases

#### 9.5.4.2 Analysis of Effective Jamming Area When the Track of Spaceborne Phased Array SAR Changes with Constant Elevation Angle

According to Fig. 9.39, the equation can be listed as

$$\begin{cases} R_{ji}^2 = \frac{4\pi P_j G_j R_{ti}^4}{K_{j\Sigma} P_t G_t \sigma_{ti}} \cdot \frac{G_t(\theta_i)}{G_t} \cdot K_a \\ R_{ji}^2 = H^2 + d^2 + r_i^2 \\ r_i = \left(H^2 + d^2\right)^{1/2} \tan\theta_i \end{cases}$$

Solve simultaneous equations for Eqs. (9.69), (9.70), and (9.71)

$$H^2 + d^2 + r_i^2 = \frac{4\pi P_j G_j \left[H^2 + (d+d_i)^2\right]^2}{K_{j\Sigma} P_t G_t \sigma_{ti}} \cdot \frac{G_t(\theta_i)}{G_t} \cdot K_a$$

$$H^2 + d^2 + \left(H^2 + d^2\right)\tan^2\theta_i = \frac{4\pi P_j G_j \left[H^2 + (d+d_i)^2\right]^2}{K_{j\Sigma} P_t G_t \sigma_{ti}} \cdot \frac{G_t(\theta_i)}{G_t} \cdot K_a$$

$$R_t^2\left(1 + \tan^2\theta_i\right) = \frac{4\pi P_j G_j \cdot R_{ti}^4}{K_{j\Sigma} P_t G_t \sigma_{ti}} \cdot \frac{G_t(\theta_i)}{G_t} \cdot K_a \qquad (9.71)$$

In the formula (9.71), only $\theta_i$ is an unknown number, and the others are all known numbers.

According to an empirical formula based on engineering calculation,

$$\frac{G_t(\theta_i)}{G_t} \cong K\left(\frac{\theta_0}{2\theta_i}\right)^2 \tag{9.72}$$

Substituting Eq. (9.72) into Eq. (9.71), the Eq. (9.73) can be obtained

$$R_t^2(1+\tan^2\theta_i) \cong \frac{\pi P_j G_j R_{ti}^4 K\theta_0^2 \cdot K_a}{K_{j\Sigma} P_t G_t \sigma_{ti}} \tag{9.73}$$

$$\sigma_{ti} = \sigma_t \frac{R_{ti}^2}{R_t^2}$$

When the elevation angle of SAR increases, the area of reflection area of the target increases, so

$$(1+\tan^2\theta_i)\theta_i^2 \cong \frac{\pi P_j G_j R_{ii}^4 K\theta_0^2 \cdot K_a}{K_{j\Sigma} P_t G_t \sigma_t} \tag{9.74}$$

To find the implicit number of (9.74), set a $d_i$, then a $\theta_i$ can be obtained, shown in Table 9.9. Parameters of radar and jammer: d = 326 km; Rt = 772 km; $r_0$ = 240 km; H = 700 km; $P_tG_t$ = 69 dBw; $\sigma_t = 5 \times 10^4$米$^2$; $K_a = 0.5$; K = 0.1; $\theta_0 = 0.8°$; $K_{j\Sigma}$ = 24 dB; $R_{ti}^2 = H^2 + (d+d_i)^2$; $R_t^2 = H^2 + d^2$;

Substituting the parameters into Eq. (9.74), $\theta_i$ is obtained:

When $d_i = 0$, $\theta_i = 17°$, $\beta_0 = 25°$; when $d_i = d/2$, $\beta_1 = 35°$; when $d_i = d$, $\beta = 42°$; when $d_i = 3d/2$, $\beta = 50°$; when $d_i = 2d$, $\beta = 55°$

According to Table 9.9, Fig. 9.40 can be plotted.

Comparing Figs. 9.38 with 9.40, the effective jamming area of Fig. 9.40 is larger than that of Fig. 9.38. This is because, with the increase of elevation angle, the echo signal weakens, but the jamming signal does not change much, so the effective zone of jamming increases; as the translational jamming signal of the SAR weakens, and the echo signal remains unchanged, the effective jamming area is smaller (under the condition that the SAR and jammer parameters remain unchanged).

It can be seen from Eqs. (9.70) and (9.74) that the effective area of jamming to spaceborne SAR is related to the height H of the spaceborne SAR, antenna elevation angle β, antenna beamwidth $\theta_0$ and antenna sidelobe characteristics, transmitter equivalent radiation power $P_tG_t$, pulse compression factor $\tau/\Delta\tau$; equivalent radiation power $P_jG_j$ of jammer, jamming suppression coefficient $K_{j\Sigma}$, translation distance

**Table 9.9** calculateted ri at different ranges di

| $d_i$ (km) | 0 | d/2 | 3d/2 | 2d | 5d/2 |
|---|---|---|---|---|---|
| $\beta_i$ (degree) | 25° | 35° | 42° | 50° | 55° |
| $R_{ti}$ (km) | 772 | 789 | 838 | 1010 | 1119 |
| $\theta_i$ (degree) | 17° | 17.9° | 19° | 22.6° | 24° |
| $r_i$ (km) | 240° | 247 | 262 | 316 | 339 |

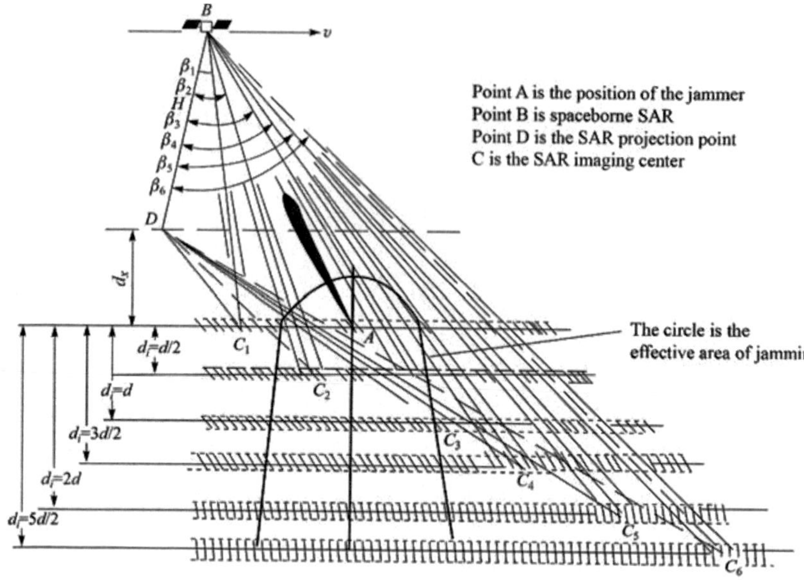


**Fig. 9.40** Diagram of the effective area of jamming when the trajectory of the spaceborne SAR is unchanged and the elevation angle changes

of the carrier's flight path $d_i$, feeder loss of SAR transmitter and atmospheric transmission loss, feeder loss of jamming signal, atmospheric transmission loss, jamming polarization loss, spectrum width of jamming, bandwidth ratio of SAR signal and many other factors such as the.

Figure 9.41 is a schematic diagram of imaging spaceborne SAR jamming in various working modes.

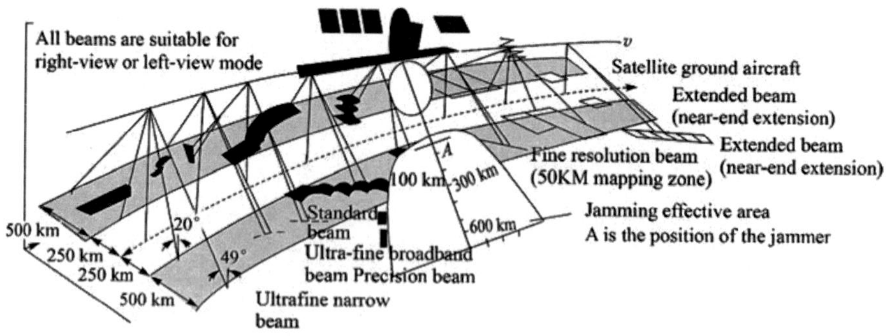


**Fig. 9.41** Schematic diagram of the imaging area in various working modes

# Chapter 10
# New Concepts and Technologies of Electronic Warfare

As one of the main operational forms of information warfare, electronic warfare will become more and more prominent in future wars. The research and development of electronic warfare technology and equipment will also receive more and more attention. Its level will continue to develop with the rapid development of electronic technology. Electronic warfare is moving from land, sea, and air to space. Electronic warfare technology is advancing at an unprecedented speed. Multi-function, full airspace, full-spectrum, and integration have become the development direction. The original batch of electronic warfare equipment will be continuously upgraded and improved, and its performance will be further improved. The development of T/R module technology and active phased array technology will realize the comprehensive integration of radar and electronic warfare functions. Distributed jamming and networking technologies for air defense radar networks are under development. As electronic warfare is changing from traditional "soft" killing to "hard" killing, new electronic warfare technology concepts have emerged. The following mainly analyzes high-power microwave and electromagnetic pulse technology and "reconnaissance, jamming, positioning" integrated technology.

## 10.1 Combat Effect and Protection Technology of Electromagnetic Pulse

In nuclear radiation weapons, people find that there are various types of radiation and shock waves that can damage various buildings and personnel. It was also found that an electromagnetic pulse wave does not damage people and buildings, but it can cause damage to some electronic devices. In the range of several kilometers to tens of kilometers, electromagnetic pulse waves can make some radio equipment not work properly or be damaged.

Nuclear radiation weapons are too destructive to humans. Therefore, countries worldwide have established treaties banning nuclear weapons and are not allowed to



X. Zhang et al., *Theory to Countermeasures Against New Radars*,
https://doi.org/10.1007/978-981-16-6715-2_10

Classification of electromagnetic pulses

- Carrier-free electromagnetic pulse
  - High repetition frequency low peak power electromagnetic pulse
  - Low repetition frequency and low peak power electromagnetic pulse
- with carrier electromagnetic pulse
  - Single high peak microwave pulse
  - Low repetition rate high peak microwave pulse

**Fig. 10.1** Classification of electromagnetic pulse

manufacture or use nuclear weapons. Although nuclear weapons are prohibited, the use of nuclear electromagnetic pulse waves is not forbidden. Because it generally does not pose a threat to personnel. Some countries have managed to develop a strong electromagnetic pulse wave to destroy the other party's radio equipment. This is the origin of the electromagnetic pulse bomb. The effect and classification of electromagnetic pulse is shown in Fig. 10.1.

### *10.1.1 High Repetition Frequency Electromagnetic Pulse with No Carrier and Low Peak Power*

This electromagnetic pulse is generally only used to jamming with the P and L band radars, and it does not damage the electronic equipment. The ultra-wideband high-repetition frequency pulse source is used as a jammer under certain conditions. The pulse waveform is shown in Fig. 10.2.

The form of the bipolar pulse:

$$f(t) = \begin{cases} \sin(\omega_0 t) & (0 \le t \le \tau) \\ 0 & (\tau \le t \le T) \end{cases} \tag{10.1}$$

$$\omega_0 = 2\pi/\tau;\ \omega = 2\pi/T$$

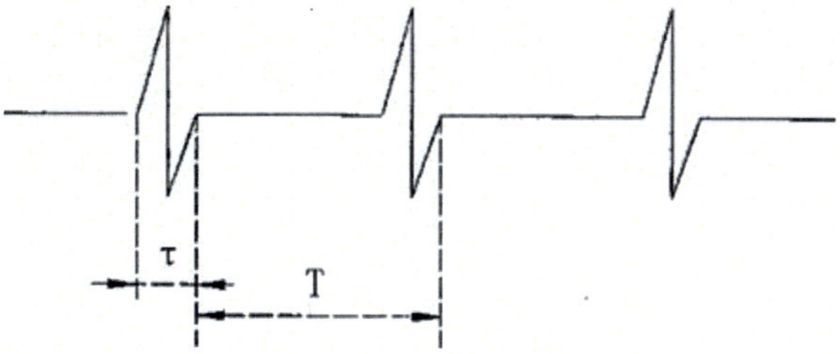


**Fig. 10.2** Bipolar no-carrier ultra-wideband pulse

In the formula:

$\tau$ Width of bipolar pulse.
T Pulse repetition period.

After Fourier transforms, we get:

$$f(t) = \sum_{K=1}^{\infty} (a_k \cdot \cos(K\omega t) + b_k \cdot \sin(K\omega t)) \tag{10.2}$$

$$a_k = \frac{\omega}{2\pi}\left\{\frac{1}{\omega_0 + k\cdot\omega}\left[1 - \cos\frac{2\pi(\omega_0 + k\cdot\omega)}{\omega_0}\right] + \frac{1}{\omega_0 - k\cdot\omega}\left[1 - \cos\frac{2\pi(\omega_0 - k\cdot\omega)}{\omega_0}\right]\right\}$$

$$b_k = \frac{\omega}{2\pi}\left\{\frac{1}{\omega_0 + k\cdot\omega}\left[\sin\frac{2\pi(\omega_0 + k\cdot\omega)}{\omega_0}\right] + \frac{1}{\omega_0 - k\cdot\omega}\left[1 - \sin\frac{2\pi(\omega_0 - k\cdot\omega)}{\omega_0}\right]\right\}$$

The spectrum of an ultra-wideband pulse is shown in Fig. 10.3. Repetition frequency $f_r = 1/T$, and the width of the spectrum is $1/\tau$.

The ultra-wideband burst spectrum is shown in Fig. 10.4.

1. The composition of high-repetition frequency ultra-wideband jammer is shown in Fig. 10.5.

   UWB jammers can be made very small. Its source is simple. After charging with an avalanche tube plus power supply, the avalanche tube automatically discharges to generate a narrow pulse when the voltage reaches a certain height.

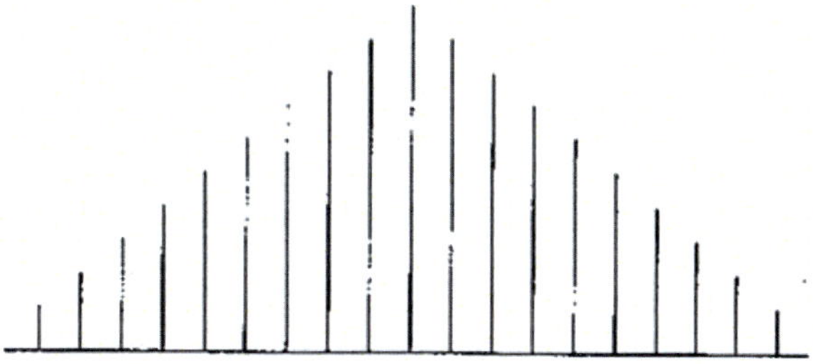

**Fig. 10.3** Narrow pulse spectrum

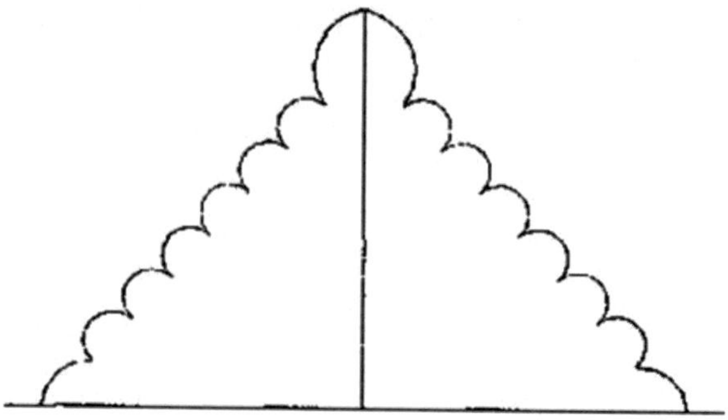

Fig. 10.4 Broadband burst spectrum

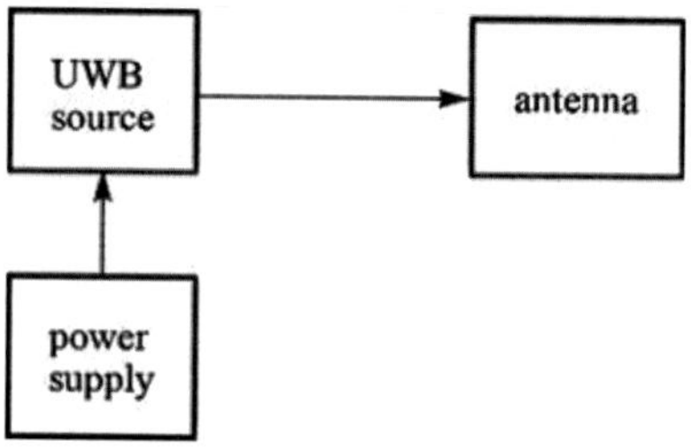


Fig. 10.5 Composition of UWB jammer

After discharging to a certain voltage, the avalanche tube ends, and the charging circuit recharges to a certain voltage and then discharges. Repeating this will produce a series of narrow pulses.

It produces a pulse width from a few nanoseconds to 3, 4 ns. The reciprocal of the pulse width is its spectral width. Therefore, its spectrum width is several hundred MHz to more than 1 GHz, and it can cover most of the radar bands of P-band and L-band.

2. The jamming effect of high repetition ultra-wideband jammer

   This jammer can be made very small, generally within one or two kilograms, and it can be placed near the jammed radar.

   For example: 1 UWB jammer pulse width 2 ns; repetition frequency 400 kHz; average power less than 0.5 W; antenna gain 3 dB. Place it at a distance of 500 m from the jammed radar to jamming with the P-band radar. The radar detection range in the 30° sector can be reduced from more than 400 km to less than 50 km.

### *10.1.2 Effect of Electromagnetic Pulse on Electronic Devices*

The effects of microwave weapons and microwave jamming are different. The difference between them can be roughly divided into: when the jamming disappears, the device can automatically resume normal work. This phenomenon is the jamming effect. If, after the jamming disappears, the device still cannot work normally, it needs to be repaired before it can work normally. This phenomenon is called the microwave weapon effect.

There is a lot of information on microwave effects.

1. The effect on the detector is shown in Table 10.1.
2. The damage effect on the mixer is shown in Table 10.2.
3. The power levels for computer and personnel damage are shown in Table 10.3.

According to theoretical analysis and experimental results, electronic devices' damage has certain rules with pulse width and pulse power, as shown in Fig. 10.6.

**Table 10.1** The effect on the detector

| Pulse width | Number of pulses | Peak power (dBW) | Damage situation | Performance changes | Remarks |
|---|---|---|---|---|---|
| 24 ns | Single | >40 | Breakdown short circuit | Cannot detect normally | Burn |
| | | 36 | Severe breakdown | Detector output drops by 50% | Burn |
| | | 25.6 | Breakdown | Small signal performance remains unchanged, large-signal backpressure <0.1 V | |
| | | 6 | No damage | Signal output drops | |
| 670 ns | Single | 18.6 | Severe breakdown | The detection output amplitude drops by 40–50% | Backpressure < 0.1 V |
| 100 ns | 100 Hz | 17.6 | Breakdown | The detection amplitude drops by 50% | |
| 1 μs | 10 Hz | 17 | Breakdown short circuit | Loss of detection ability | |

**Table 10.2** The effect on the detector

| Pulse width (ns) | Pulse repetition frequency | Peak power (dBW) | Suppress signal time | Waveform influence time | Remarks |
|---|---|---|---|---|---|
| 30 | Single 2CV3A | 16 | | | Damage breakdown |
| 370 | Single 2CV9A | 11.4 | Short circuit 500 μs | 15,000 μs | |
| | | 0.4 | | | |
| 20 | 100 Hz | 17.1 | Signal amplitude drops by 50% | | |

**Table 10.3** The effect on the detector

| Serial number | Jamming characteristics | Damaged piece name | Degree of damage | Power density | Remarks |
|---|---|---|---|---|---|
| 1 | Continuous-wave | Mixing diode | Burn out | $10^{-7}$ W/cm | |
| 2 | Continuous-wave | Linear integrated circuit | Burn out | $10^{-6}$ W/cm | |
| 3 | Continuous-wave | Bipolar integrated circuit | Burn out | $10^{-5}$ W/cm | |
| 4 | | CMOS integrated circuit | Burn out | $10^{-4}$ W/cm | |
| 5 | | Zener diode, power transistor | Burn out | $10^{-3}$ W/cm | |
| 6 | | Computer | Node is disturbed | 0.01–1 μW/cm$^2$ | |
| | | | The chip is damaged | 0.01–1 W/cm$^2$ | |
| | | | Component burnt out | 10–100 W/cm$^2$ | |
| | | | The node is burnt out | 100–1000 W/cm$^2$ | |
| 7 | | Personnel | Spasms | 10–60 mW/cm$^2$ | |
| | | | Cardiopulmonary exhaustion | 100 mW/cm$^2$ | |
| | | | Burn skin | 560 mW/cm$^2$ | |
| | | | Third-degree burn (2 s) | 20 W/cm$^2$ | |
| | | | Death (1 s) | 80 W/cm$^2$ | |

This law has great implications for those who develop electronic pulse weapons. As long as the pulse amplitude, pulse width, and pulse repetition frequency can be properly adjusted, the best electromagnetic pulse effect may be obtained.

### *10.1.3 Analysis of Combat Effect of Electromagnetic Pulse*

There are two main forms of electromagnetic pulse combat: one is the form of microwave pulse bombs; the other is the form of a microwave pulse gun.

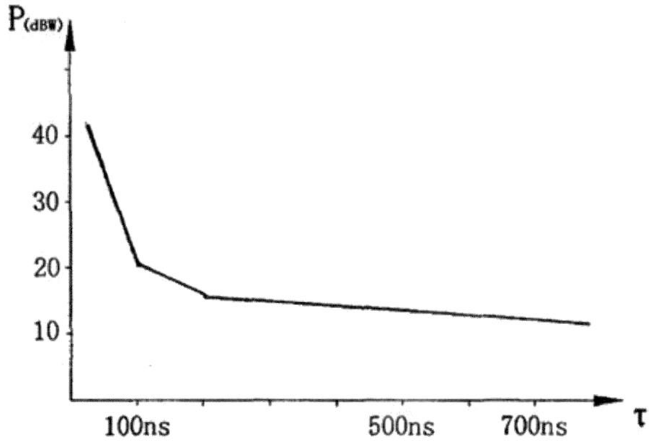


**Fig. 10.6** Schematic diagram of the relationship between the electromagnetic pulse effect $\tau$ and P

Microwave pulse bombs convert chemical energy into electrical energy to produce high-power electromagnetic pulses. Use this pulse to control a radio frequency oscillating tube to generate high-power microwave electromagnetic pulses. The pulse enters the inside of the electronic device, which damages the electronic device and destroys the electronic device's normal operation.

There are two ways for electromagnetic pulses to enter the inside of electronic equipment: one is to enter from the front door, that is, enter from the passband of the antenna receiver, and then act on the electronic device; the other is to enter from the back door, that is, from some coupling holes, slit, or enter outside the receiver, and then acts on the electronic device.

1. Enter from the front door

   The pulse power required to cause damage to the electronic device is entered from the passband of the receiver and can be expressed by (10.3):

$$P_a = \frac{P_j \cdot G_j(\phi)}{4\pi \gamma^2} \cdot \frac{\lambda^2}{4\pi} \cdot G_t(\theta) \tag{10.3}$$

   In the formula:

| | |
|---|---|
| $P_a$ | The power required for the burning of electronic devices. |
| $P_j G_j$ | Equivalent radiated power of high power pulse source. |
| $\gamma$ | Distance between high power and electronic device under attack. |
| $\lambda$ | Carrier frequency wavelength of pulse source. |
| $G_t(\theta)$ | Gain of the receiving antenna of the attacked electronic device in the direction of the electromagnetic pulse source. |

   (a) Electromagnetic pulse power required by the main antenna lobe to the main lobe

      Suppose the width of the electromagnetic pulse source is 30 ns, 100 ns, and 500 ns, respectively, and the required pulse burning power is:

      When the pulse width is 30 ns, the required power $P_{a1}$ is 40 dBW.

When the pulse width is 100 ns, the required power $P_{a2}$ is 23 dBW.
When the pulse width is 500 ns, the required power $P_{a3}$ is 20 dBW.
Suppose R = 0.5 km, $G_t = 20$ dB, $G_j = 15$ dB, the calculated data is as follows:
When the pulse width $\tau_1 = 30$ ns, the required pulse power is 140 GW.
When the pulse width $\tau_2 = 100$ ns, the required pulse power is 2.8 GW.
When the pulse width $\tau_3 = 500$ ns, the required pulse power is 1.4 GW.

(b) The electromagnetic pulse power required to align the main lobe of the jamming antenna with the side lobe of the radar
Suppose $\gamma = 100$ m, $G_t(\theta) = 4$, and other parameters remain unchanged:
When the pulse width $\tau_1 = 30$ ns, the required pulse power is still 140 GW.
When the pulse width $\tau_2 = 100$ ns, the required pulse power is still 2.8 GW.
When the pulse width $\tau_3 = 500$ ns, the required pulse power is still 1.4 GW.

(c) Electromagnetic pulse power required for attacking anti-radiation missiles
Assuming that the attack pulse repetition frequency is 100 Hz, $G_j = 40$ dB, the anti-radiation missile antenna gain is 10 dB, and $\gamma = 3$ km.
When the pulse width is 30 ns, the required power $P_{a1}$ is 35 dBW.
When the pulse width is 100 ns, the required power $P_{a2}$ is 20 dBW.
When the pulse width is 500 ns, the required power $P_{a3}$ is 18 dBW.
Using Eq. (10.3), the calculation data is as follows:
When the pulse width $\tau_1 = 30$ ns, the required pulse power is 52 GW.
When the pulse width $\tau_2 = 100$ ns, the required pulse power is 1.6 GW.
When the pulse width $\tau_3 = 500$ ns, the required pulse power is 1.06 GW.

2. Coupling entry from the back door
Coupled into electronic equipment, the damage distance to the electronic device is:

$$\gamma \cong \left[\frac{P_j \cdot G_j \cdot \rho \cdot \lambda^2}{(4\pi)^2 P_a}\right]^{\frac{1}{2}} \tag{10.4}$$

In the formula:

$P_j G_j$ The equivalent power of the electromagnetic pulse radiation source, take $P_j G_j = 10$ GW.
$\rho$ Coupling coefficient.
$\lambda$ Radiation source wavelength, take $\lambda = 3$ cm.
$P_a$ The damaged power required by the device, take $P_a = 30$ dBW.

The coupling coefficient is related to the aperture, shape, material and other factors of the coupling hole and is generally in the range of $10^{-3}$ to $10^{-6}$ (taken $10^{-4}$). Substituting the parameters into Eq. (10.4), $\gamma$ is equal to 7.5 cm.

3. Common high-power pulse source components are shown in Table 10.4.

It can be seen from calculation and analysis that the pulse power required to damage electronic devices is in GW's order to tens of GW. It can be seen from Table 10.4 that the existing high pulse power tubes can reach the GW level to tens of GW level. Therefore, high-power microwave pulse attack equipment can be achieved. But it is still in the experimental stage. This high-power microwave equipment will be used in actual combat in the near future.

### *10.1.4 Protection Technology for High Power Electromagnetic Pulse*

1. Adding a limiter on the receiver's front end can reduce the attack damage distance of high-power microwave pulses, as shown in Fig. 10.7.

   When the limiter is not added, the high-power microwave pulse from the antenna is directly applied to the amplifier's input. If the input pulse width is 100 ns and the peak power reaches 100 Wh, the amplifier may burn out. After adding an isolator at the front end, the isolator's anti-burnout power can be two orders of magnitude higher than that of the amplifier. Therefore, the high-power pulse output from the antenna is limited to the order of tens of mill watts by the isolator. The isolator will not be damaged, and at the same time, protect the subsequent amplifier from damage.

   When a wide pulse signal passes through the limiter, it needs a delay of several nanoseconds to limit the amplitude, as shown in Fig. 10.8.

   The peak power of the pulse does not decrease in $\Delta\tau$ time, but its pulse width is relatively narrow (about 5–10 ns). The amplifier's anti-burnout ability for narrow pulses is 40 dBW. Therefore, the peak power of 20 dBW of $\Delta\tau$ cannot burn out the amplifier. The following wide pulse is limited by the isolator, which reduces its amplitude and cannot burn the amplifier. Therefore, after adding an isolator, the subsequent electronic devices cannot be burned by high-power microwave pulses. In other words, after adding an isolator, the burnout distance of high-power microwave pulses is shortened by ten times.

2. Add bandpass filter or bandstop filter at the front end of the receiver

   The high-power electromagnetic pulse generator's frequency band is very narrow, which is close to the point frequency plus some frequency drift. Therefore, band-stop filtering can weaken this frequency signal at the front end of the receiver. Adjusting the receiver's operating frequency away from this band-stop frequency range or adding band-pass filtering can make the passband frequency deviate from the high-power pulsed microwave frequency.

**Table 10.4** High-power pulse source components

| Serial number | Name | Operating voltage | Working current | Pulse width | Frequency | Output power | Effectiveness (%) | Up to peak power |
|---|---|---|---|---|---|---|---|---|
| 1 | Relativistic klystron | 1 mV | 30 kA | 160 ns | 1.3 GHz | 15 GW | 50 | 100 GW |
| 2 | Virtual cathode device | 0.2–0.65 mV | 10–100 kA | | 0.5–17 GHz | 20 GW | 15 | |
| 3 | Multiwave Direnkov oscillator | 1.8–2 mV | 13–15 kA | 60–70 ns | X band | 15 GW | | 100 GW |
| 4 | Relativistic magnetron | | | | 46 GHz | 3.5 GW | | 10–20 GW |
| | | | | | S band | 7 GW | | 100 GW |
| | | | | | L band | | | |
| 5 | Relativistic echo tube | 0.7 mV | 4.5 kA | 10 ms | | 1 GW<br>100 GW | 33 | 200 GW |
| 6 | Gyrotron | | | | 140 GHz | 1 MW (Peak) | | Average power 412 kW |
| | | | | | 10 GHz | 2 MW (Peak) | | |
| | | | | | Millimeter wave | 7 GW (Peak) | | |

(continued)

**Table 10.4** (continued)

| Serial number | Name | Operating voltage | Working current | Pulse width | Frequency | Output power | Effectiveness (%) | Up to peak power |
|---|---|---|---|---|---|---|---|---|
| | | | | | 64 GHz | 3.5 GW (Peak) | | |
| 7 | Typical values of modern electronic warfare traveling wave tubes | | | Continuous wave | 2–4 GHz | 1 kW | 15–20 | |
| | | | | Continuous wave | 2.5–3.5 GHz | 5 kW | 10–15 | |
| | | | | Continuous wave | 2.5–8 GHz | 0.2 kW | 15–30 | |
| | | | | Continuous wave | 6–16 GHz | 0.2 kW | 15–30 | |
| | | | | Continuous wave | 8–18 GHz | 0.3 kW | 15–20 | |

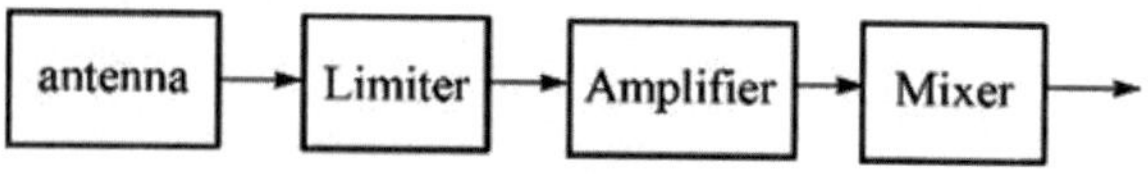


**Fig. 10.7** Schematic diagram of adding a limiter in front of the receiver

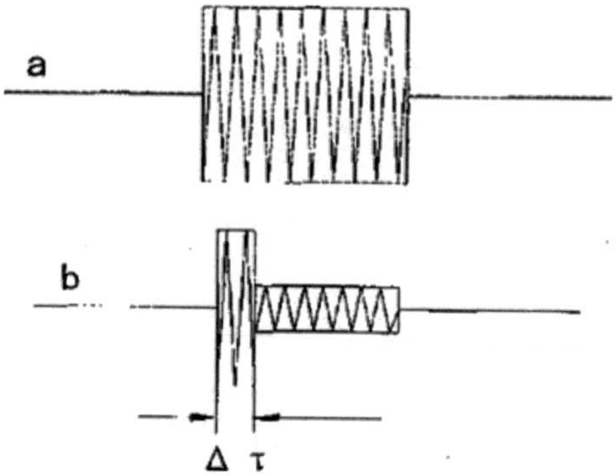


**Fig. 10.8** **a** Schematic diagram of the input pulse of the limiter, **b** schematic diagram of the output pulse of the limiter

3. Do a good job of electromagnetic shielding of electronic equipment

   For example, ground electronic equipment needs to be well shielded for carriages, cabinets, and some key components. There is a certain contradiction between shielding and ventilation. When leaving ventilation holes, choose the aperture and shape to make its leakage rate as low as possible. For components that do not allow ventilation holes, air cooling, water cooling, and other measures can be used to resolve such contradictions.
4. Improve the anti-burnout ability of electronic devices

   After many tests on the pre-stage solid-state amplifier, some inputs have not burned out with continuous-wave input of several hundred milliwatts to 1 W, and some inputs have burned out after tens of milliwatts. This shows that improving the device's anti-burnout ability has great potential to be tapped. Efforts should be made in the production process, material selection, component selection, etc. to improve the components' anti-burnout ability. This is a key link to resist the burnout of high-power electromagnetic pulses and protect electronic equipment safety.
5. The switch tube of the radar is used to protect the radar receiver from being burnt by high-power microwave pulses

   The radar switch tube function: When the radar transmits a high-power pulse, the switch tube is turned on so that the high-power pulse is radiated through the antenna. The power leaking into the radar receiver is very low and will not burn the radar receiver. Simultaneously, the small leaked signal can also be used as a radar reference pulse to measure the distance of the echo. It is precise because of the radar switch tube that the radar can transmit and receive a common antenna.
6. Adjustable conduction power level of the radar switch tube

   When the switch tube's input power reaches p (the critical power that the radar front-stage receiver burns out), the switch tube is turned on to attenuate the power. The leakage into the radar receiver is very small and cannot be burned out. In fact, it plays the role of adding a limiter to the front end of the receiver, thereby protecting the radar receiver from being burned by high-power microwave pulses.

A high-power electromagnetic pulse is a high-tech under research and development. It will play an important role in future wars. However, the emergence of new technology is followed by a strategy to deal with it. If there is a spear, there must be a shield. This is the law of historical development. With the development of high-power microwave pulse technology, the development of its protection technology will also promote.

## 10.2 Radar Countermeasures and High-Power Microwave Devices

### *10.2.1 Mechanism of High Power Microwave*

With the widespread use of electronic equipment in military equipment, electronic countermeasures have become increasingly fierce. In radar field, many anti-jamming measures have been enhanced, such as improved radar antennas and reduced radar antenna sidelobes; some carrier frequency and pulse waveform modulation can be easily carried out; in radar reception and signal processing, it has been changed to related Receiving, matched filtering, correlation accumulation, etc. improve the radar's signal-to-noise ratio performance. Coupled with other anti-jamming measures, it becomes very difficult and complicated to jamming with the radar.

Due to the enhancement of radar anti-jamming measures, radar jammers have become more complicated. For example, to deal with the frequency agility of radar, the jammer adds the "instantaneous frequency guidance sub-system"; to deal with the multi-carrier frequency agility of radar, channelized receiver and channelized aiming frequency division system are added to the jammer; to deal with For Pulse-Doppler radar related reception and matched filtering, an jamming source and fine-pointing frequency division system are added to the jammer; to cope with the radar's high transmitting power and low sidelobe antenna, a narrow beam jamming transmitting antenna is added to the jammer And precision angular tracking antenna servo subsystem; To deal with complex radar waveforms, radar networking, and multiple targets, complex signal processing, identification, and threat discrimination measures are added to the jammer. In short, with the upgrading of electronic countermeasures, radars and radar jammers have become more complex, equipment larger, and more expensive. However, in this change, the radar always dominates, and the jammer changes with the radar's change.

With the emergence and development of high-power microwave devices, radar jamming equipment can be greatly simplified. As long as there is enough microwave power, there is no need for jammers' design to require the accuracy of the aiming system, and there is no need for radar agility. Adding a complex aiming frequency system to the jammer can simplify the signal sorting and identification processing.

The way the jamming signal enters the radar equipment can enter through the front door and the back door. The "front door" mentioned here refers to the receiver's entry from the antenna of the radar, and the entry from the "rear door" refers to the coupling into the radar equipment from various gaps.

As long as there is enough power, the signal entering the radar receiving device can cause jamming or destructive effects. In short, as long as there is a large enough microwave power source, the radar jamming equipment can be greatly simplified, and the jamming effect can be greatly increased.

High-power microwave has great benefits to human society. It can be made into a "microwave oven", which can cook and grill various delicacies for people; it can be made into a microwave therapy instrument to cure people's diseases; it can be made into a high-power microwave source to extract heavy oil., Smelting oil, extracting a lot of wealth from nature for mankind; it can be made into "impact" radar to detect underground treasures; it can be made into various precision processing equipment for precision processing of workpieces. In short, microwave energy will play a huge role in human life and production activities just like "electricity" and "fire" in the future. However, if microwaves are made into weapons, they can play a hugely destructive effect. Let's make a high-power micro-source into a weapon to deal with various radars. It can burn out the microwave devices, various integrated circuits, and compute nodes in the radar equipment and even cause casualties.

The statistical data of the harm test of microwave to devices, computers, and personnel, and the relationship between energy and power required to burn various devices vary with the pulse width. The specific relationship is shown in Fig. 10.9.

Experiments have determined that when the pulse width is between 100 ns and 10, the energy required to cause the failure of the semiconductor junction is according to $t^{-\frac{1}{2}}$, and the required power is $t^{\frac{1}{2}}$. For pulse widths less than 100 ns, thermal diffusion can be ignored, and the damage of the semiconductor junction depends only on the pulse energy. The value of damage power varies with $t^{-1}$.

When the pulse width is greater than 1 ms, there is a steady-state situation where the thermal diffusivity is equal to the energy deposition rate. The temperature is proportional to power. These quantitative relationships result from the shortest pulse requiring the highest energy and the lowest power. Most HPM sources achieve between 100 ns and 10 μs. If the energy in the weapon used is reduced, a shorter pulse should be used, and if the power value is limited, a longer pulse should be used.

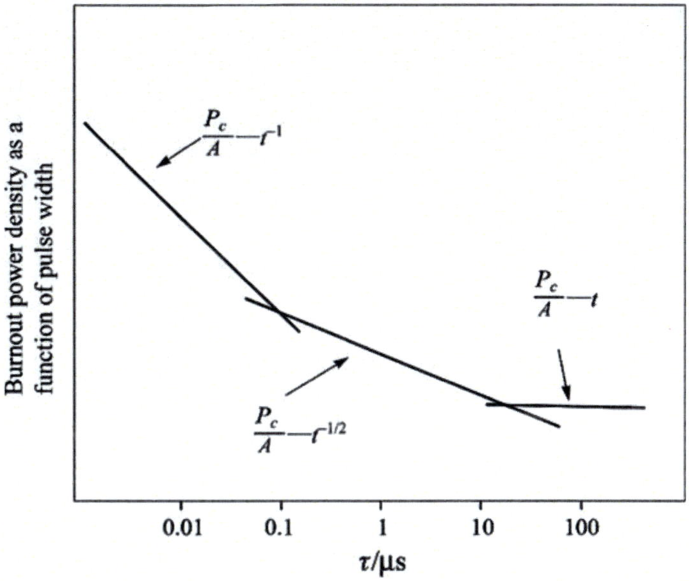


**Fig. 10.9** Burnout power density as a function of pulse width

### *10.2.2 The Power Required by Various Jammers and High-Power Microwave Weapons*

The jammer is one of the core parts of the electronic countermeasure equipment, and the jamming transmitter tube is the heart of the jammer. This is the relationship between electronic countermeasures and microwave devices. It is not excessive to evaluate the role of microwave emission sources in electronic countermeasures. People who used to work in jammer technology have a deep experience. A jamming device has not been developed for a long time. Where is the key technology stuck? Most of them are stuck in the microwave transmitter tube. When designing the jammer scheme, determine the kind of microwave transmitting tube there is, which is the key component to determine the jammer scheme.

The power levels of various types of jammers are shown in Table 10.5.

According to the use of tactics and existing microwave devices, the power levels of various existing radar jammers are listed in Table 10.5. According to the current status and development possibilities of high-power devices, the power levels of super jammers, smart high-power microwave weapons, and high-power microwave weapons are listed. We further understand that there is no essential difference between jammers and high-power microwave weapons. There is only a difference in power. The main difference between microwave jamming and microwave damage is that when the jamming signal exists, the radar loses its proper function. When the jamming signal is removed, the radar recovers its original performance without being repaired. This effect is called the jamming effect of the microwave. When there is a jamming

**Table 10.5** Power levels of various radar jammers

| Serial number | Type of jammer | Average power level | Peak power level | Equivalent radiated power |
|---|---|---|---|---|
| 1 | Throw jammer | 0.1–1 W | Several watts to one hundred watts | $10–10^2$ |
| 2 | Airborne self-defense jamming machine | Tens to hundreds of watts | Hundreds of watts to thousands of watts | $10^2–10^4$ |
| 3 | Shipborne self-defense jammer | Hundreds of watts | Kilowatt level | $10^3–10^6$ |
| 4 | Airborne support jamming aircraft | Hundreds of watts to thousands of watts | Tens of kilowatts | $10^4–10^6$ |
| 5 | Ground-to-air shielding jammer | Hundreds of watts to thousands of watts | Hundreds of kilowatts | $10^5–10^8$ |
| 6 | Super jammer | Kilowatts to tens of kilowatts | Hundreds of megawatts | $10^7–10^{12}$ |
| 7 | Smart high power microwave weapon | Kilowatt level | Thousands of megawatts | $10^{13}–10^{15}$ |
| 8 | High-power microwave weapons | Tens of kilowatts to hundreds of kilowatts | Hundreds of GW | $10^{16}–10^{18}$ |

signal, the radar loses its proper function. When the jamming signal is removed, the radar cannot restore its original performance without repair. This effect is called microwave damage. This is a process from quantitative change to qualitative change.

See Table 10.4 for the status quo and achievable output power levels of various high-power microwave devices.

### *10.2.3 Application Comparison of Solid-State Microwave Power Devices and Electric Vacuum Microwave Devices*

At present, among microwave tube power devices, "solid microwave devices" are emerging. In jammers or radars, in order to obtain the equivalent radiation power required by the whole machine, a phased array launch system is often used.

There is a theoretical analysis curve for the current solid-state devices for the achievable output power levels, as shown in Fig. 10.10a, b.

It can be seen from Figs. 10.10 and 10.11 that under conditions of tens of gigahertz, solid-state devices can only reach tens of watts, while electric vacuum devices

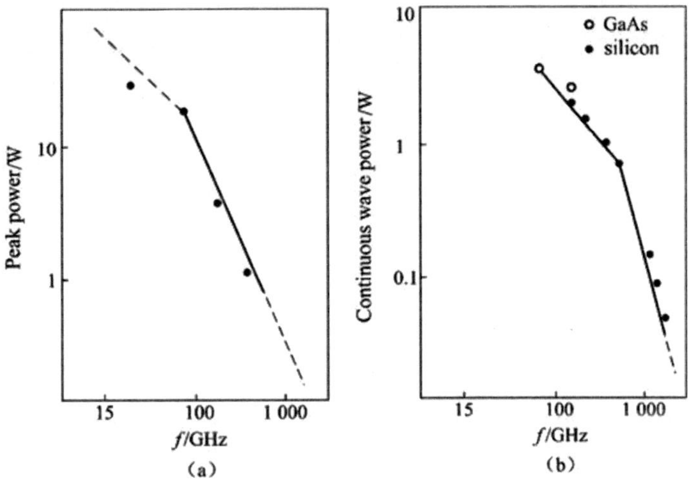


**Fig. 10.10** **a** Pulsed peak power achievable by microwave solid-state device, **b** continuous wave power achievable by microwave solid-state device

can reach hundreds of kilowatts. Some noise figure curves of several low-noise microwave amplifying devices is shown in Fig. 10.12.

The current 6–18 GHz solid-state amplifiers can actually be used on the whole machine output continuous wave power of about 0.8–1 W, with 3% efficiency.

The first type: use solid-state amplifier phased array system;

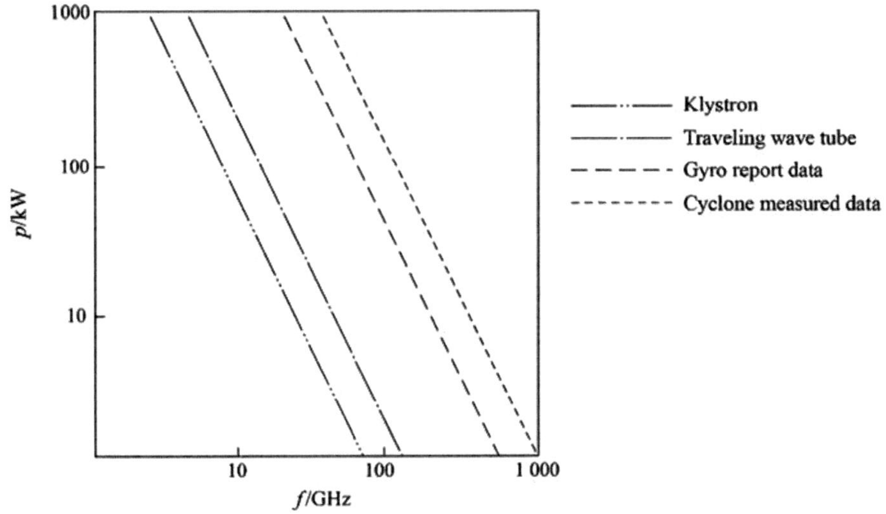


**Fig. 10.11** The achievable power value of some electric vacuum power sources

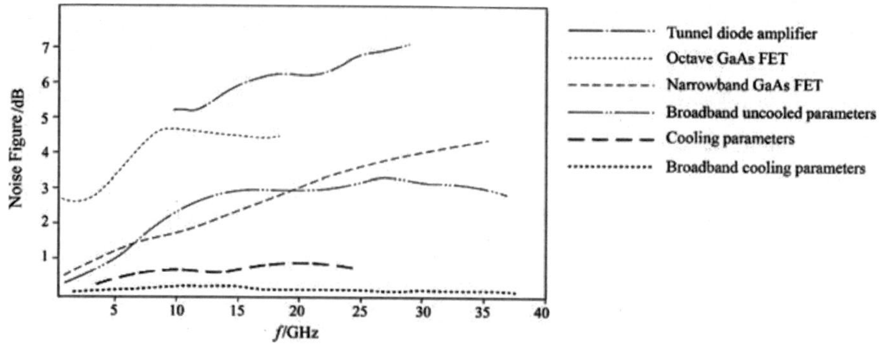


**Fig. 10.12** Noise figure curves of several low-noise microwave amplifying devices

The second type: use broadband traveling wave tube and phased array system;
The third type: use broadband traveling wave tube and single-beam system.
The equivalent radiation power of the phased array system is: $P_{\Sigma} = P_E \cdot N^2 \cdot G_E$.
The equivalent radiation power of the transistor is: $P_{\Sigma 1} = P_1 \cdot N_1^2 \cdot G_1$.
In the formula:

$P_1$ Transistor and single tube output power, $P_1 = 0.8\,\text{W}$.
$N_1$ the number of required array elements.
$G_1$ Unit antenna gain, $G_1 = 3$.

Then:

$$N_1 = \left(\frac{10^6}{0.8 \times 3}\right)^{1/2} = 640$$

The equivalent radiation power of a broadband traveling wave tube is:

$$P_{\Sigma 2} = P_2 \cdot N_2^2 \cdot G_2$$

In the formula:

$P_2$ The output power of a single traveling wave tube $P_2 = 50\,\text{W}$.
$G_2$ Unit antenna gain, $G_2 = 3$.

Then:

$$N_2 = \left(\frac{10^6}{50 \times 3}\right)^{1/2} = 81$$

The single-beam system uses a traveling wave tube output power of 200 W and an antenna gain of 37 dB.

Then, $P_j G_j = 1\,\text{MW}$.

Whether for civilian or military use, high-power microwave electric vacuum devices have broad development prospects. The comparison of the jammers of these three systems is shown in Table 10.6.

It can be seen from Table 10.6 that the phased array system for a megawatt equivalent power jammer is relatively advanced. The phased array jammer made of solid-state amplifiers has the same technical level as the same level phased array jammer made of traveling wave tube, but the cost is two or three times higher. However, a high-power jammer with a single beam has a lower cost in addition to its poor performance against multiple targets. A single beam jammer is suitable if used to jamming an early warning aircraft radar or reconnaissance satellite.

It is a more suitable solution to use transistor devices in the front stage of high-gain and high-power electric vacuum microwave devices. Low-noise microwave devices and solid-state amplifiers may gradually replace electric vacuums. However, transistor microwave power devices can't replace high-power microwave electric vacuum microwave devices from the examples listed above. It is impossible to analyze it in terms of the achieved power level, cost, efficiency, etc.

Microwave devices are widely used in civilian and military products. In people's daily life and production activities, various devices made by microwave sources are increasingly inseparable. In particular, high-power microwave sources are used to manufacture various military weapons. In future wars, it will increasingly demonstrate its power. In electronic warfare, weapons made with high-power microwave sources will greatly increase the power of electronic warfare. It can play a role in jamming and can destroy electronic equipment and have a deadly effect on operators. Solid-state microwave devices can only play a supplementary role for high-power microwave electric vacuum devices and cannot replace them. Therefore, we should see the direction and make a reasonable investment in high-power microwave devices. High-power microwave weapons are a new concept weapon. Only when we master

**Table 10.6** Comparisons between phased array launch systems and the single-beam system

| Serial number | Name | Solid phased array | A phased array of traveling wave tube | Single beam system |
|---|---|---|---|---|
| 1 | Frequency range | 8–18 GHz | 8–18 GHz | 8–18 GHz |
| 2 | Number of launch tubes | 640 | 81 | 1 |
| 3 | Control unit | Complex | More complicated | Simpler |
| 4 | Beam change time at 90° | Hundreds of milliseconds | Hundreds of milliseconds | Thousands of milliseconds |
| 5 | Number of targets that can be disturbed in the 90° range | 10–20 | 10–20 | 1 |
| 6 | Frequency lead time | Microsecond | Microsecond | Microsecond |
| 7 | Cost (estimated) | USD 10–15 million | USD 4–6 million | 1.2–1.5 million USD |

high-power microwave weapons can it be possible to ban high-power microwave weapons. In this sense, the development of high-power microwave technology and high-power microwave weapons is also very necessary.

## 10.3 Feasibility Analysis of Attacking Reconnaissance Satellites Using High-Power Microwave Weapons

The payload of reconnaissance satellites is divided into active reconnaissance and passive reconnaissance. The most widely used spaceborne active reconnaissance is the spaceborne synthetic aperture radar (SAR), which can conduct terrain mapping, military deployment detection, and find tanks and missile positions regardless of weather, day and night, and national boundaries. Spaceborne SAR can be effectively jammed in the ground to air. It has been discussed in detail in Chap. 2. This chapter only talks about how to attack high-power microwave weapons and how much equivalent radiation power is needed. Another type of satellite reconnaissance is passive reconnaissance satellites, which do not radiate signals but only receive signals radiated by various wireless devices. It can extract useful information from the reconnaissance signal and determine the radiation source's parameters and location. For example, for the reconnaissance satellite of a radar station, the carrier frequency, pulse width, pulse repetition frequency, intra-pulse characteristics, the position of the radar station, and the arrangement of various radar stations can be obtained from the reconnaissance signal to get some important military intelligence of the country.

For space-borne SAR, the ground jamming station can determine its position and aim the jamming transmitting antenna at the space-borne SAR to carry out effective jamming or high-power microwave pulse attack. For space-borne passive reconnaissance satellites, the ground jamming station itself cannot determine its direction and location. It must rely on the ground remote tracking and warning radar to guide the jamming station, and it cannot obtain useful information.

### *10.3.1 The Feasibility of Using High-Power Microwave Attacks on Passive Reconnaissance Satellites*

The composition diagram of the system is shown in Fig. 10.13.

In Fig. 10.13, A is a high-power microwave station; B is a space-borne passive reconnaissance satellite; C is a ground remote warning and tracking radar station, which guides the direction of A station; D is the ground projection point of the satellite reconnaissance station; $E_{1,2,\ldots i}$ is the target of satellite reconnaissance; and H is the height of the satellite reconnaissance station.

The satellite reconnaissance antenna continuously searches for the signal of the radar station being reconnaissance.

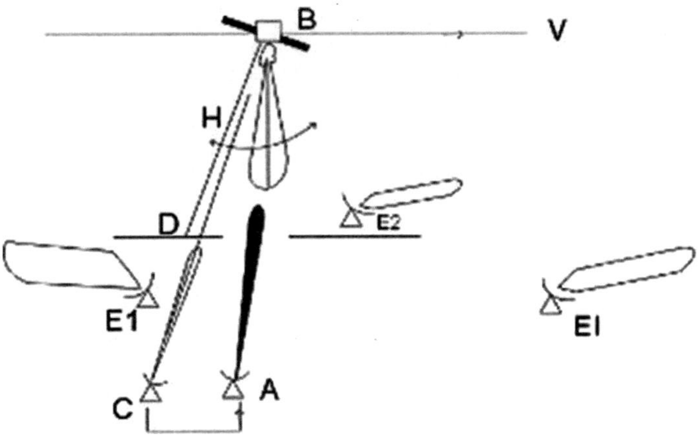


**Fig. 10.13** Schematic diagram of ground high-power microwave station attacking passive reconnaissance satellites

1. The working process of spaceborne reconnaissance station and high-power microwave attack station

   When the passive satellite reconnaissance station B reaches the space above the reconnaissance target, it turns on the radar reconnaissance antenna to search. Unknowingly, the radar station's radar parameters working on the ground are acquired by the satellite reconnaissance station and stored in their database.

   When the reconnaissance satellite station reaches the range of the long-distance guard tracking radar station C, the station C obtains the satellite B's direction and position. Transmit the coordinate information of station B to point A. After station A performs coordinate conversion according to station C's information, it automatically aligns its direction to station B. In the antenna search process of station B, at a certain moment, the main lobe of the antenna may be aligned with the main lobe of station A, so that station B receives the strongest power from station A. It is possible to reach a level sufficient to damage the B station receiver's amplifier, with reduced sensitivity and shorter detection distance. If its reconnaissance distance is reduced to less than H, the satellite has lost its reconnaissance function, which means that the A station attack is successful.
2. The equivalent power required for attacking passive reconnaissance satellite stations

   The power received by the satellite reconnaissance station from high-power microwave weapons is:

$$P_x = \frac{P_{jA} \cdot G_{jA}}{4\pi R^2} \cdot \frac{\lambda^2}{4\pi} \cdot G_B \cdot e^{-R \cdot \beta} \tag{10.5}$$

$$P_{jA}G_{jA} = \frac{\left(\frac{4\pi R}{\lambda}\right)^2 \cdot P_x}{G_B \cdot e^{-R \cdot \beta}} \tag{10.6}$$

In the formula:

$G_B$ Antenna gain of the reconnaissance receiving station.
$e^{-R \cdot \beta}$ High-power microwave station feeder loss and the loss between the microwave transmitting station and reconnaissance station.

If $P_x$ the reconnaissance receiver device's damage power, the receiving sensitivity is reduced, and the detection range is reduced to the required receiving power below H.

As shown in Table 10.1, when the pulse width is 100 ns, and the pulse power is 17.6 dBW, a single pulse can break down the detector.

As shown in Table 10.2, when the pulse width is 30 ns, and the pulse peak power is 16 dBW, a single pulse can damage the mixer and break down.

According to the above data, choose $P_x$ as 60 W; pulse width as 100 ns; single pulse $e^{-R \cdot \beta}$ as 0.03; $G_B$ as 20 dB; $\lambda$ as 3 cm. Substituting the parameters into Eq. (10.6), the relationship data between p and R is calculated, as shown in Table 10.7.

If a microwave isolator is added to the receiver of the passive reconnaissance station being attacked, the damage power may be increased by 15–20 dB. High-power microwave attacks on airborne or space-borne reconnaissance receivers have a higher probability of attack. Because the reconnaissance receiving antenna must search for various radar radiation sources. The probability that the antenna's main lobe is aligned with the high-power microwave attack station increases the probability of damage to the receiver components. However, the antenna of the spaceborne SAR generally does not search, and the probability of its main antenna lobe being aligned with the main lobe of the high-power microwave station antenna is very low. Therefore, the probability of damage to the spaceborne SAR receiver is relatively low. This is a problem in the use of some tactics. However, when we want to develop high-power microwave weapons, a series of issues from technology to tactical applications and cost must be considered in a comprehensive manner so that this equipment can play its due role. If it is technically feasible, but the cost is high, the equipment is too heavy and inconvenient. That will also limit its use-value, which must be considered comprehensively in analysis and demonstration.

The equivalent power levels required by various jammers and high-power microwave weapons are shown in Fig. 10.14.

**Table 10.7** The damage effect of microwave devices corresponds to the peak power required for different distances

| R (km) | 10 | 100 | 200 | 500 | 1000 |
|---|---|---|---|---|---|
| $P_{jA}$(dBW) | 125 | 145 | 151 | 156 | 165 |

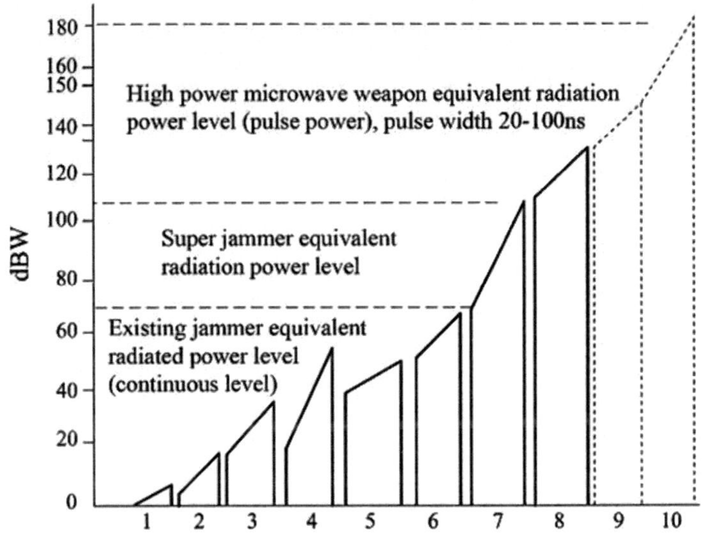


**Fig. 10.14** Statistical diagram of equivalent radiation power magnitudes of various jammers and high-power microwave weapons. (1) Deception jammers for alert guidance radars; (2) Throwing jammers; (3) Airborne self-defense jammers; (4) Shipborne self-defense jammers; (5) Airborne support jammers; (6) Ground-to-air shielding jammers; (7) Super jammers; (8) Smart high-power microwave weapons; (9) Basic microwave energy weapons; (10) Microwave energy weapons under development

### *10.3.2 The Feasibility of Strong Microwave Jamming to Reconnaissance Satellites*

Through specific reconnaissance satellite data, the importance of jamming satellites is explained. The conventional concept of jamming is used to derive using a strong microwave to jamming with reconnaissance satellites. This equation is suitable for both low-orbit reconnaissance satellites and jamming synchronous reconnaissance satellites. The use of multi-station and multi-pulse jamming saturates the reconnaissance satellite processing, making it unable to find the radiation source's location and parameters correctly, making the reconnaissance satellite lose its due role. Besides, the jamming equation for jamming the spaceborne synthetic aperture radar is derived. Calculate the strong microwave radiation power required to jamming with the reconnaissance satellite through these jamming equations.

Electronic reconnaissance has many methods, including ground reconnaissance, shipborne reconnaissance, airborne reconnaissance, etc. These reconnaissances have played an important role. With the rapid development of science and technology and the gradual balance of conventional technical forces between the two sides, airborne reconnaissance is subject to many restrictions such as airspace and territorial sea. Therefore, some technologically advanced countries have gradually turned their reconnaissance platforms into space. For example, a total of 302 military electronic reconnaissance satellites and 1090 imaging satellites have been launched worldwide since the launch of the first satellite in 1957. These reconnaissance satellites are not restricted by airspace and territorial seas, and they are always monitoring other countries' electronic radiation. Electronic radiation and the research of new technologies are a new threat from advanced countries to developing countries. In peacetime, it can be used to conduct multi-party reconnaissance and surveillance of interested countries and regions. During the war, they had a clear understanding of the combat layout in these areas, and they were in the active position of combat. On the contrary, if there are no precautions against these reconnaissance satellites and let other countries' reconnaissance satellites carry out reconnaissance and surveillance arbitrarily, they can only be passively beaten at critical moments. The Gulf War is an example of this result.

A schematic diagram of a strong microwave jamming military reconnaissance satellite is shown in Fig. 10.15.

The power received by the reconnaissance satellite from an important radiation source is:

$$P_1 = \frac{P_a \cdot G_a(\theta)}{4\pi R^2} \cdot \frac{\lambda^2}{4\pi} \cdot G_B(\beta_1) \tag{10.7}$$

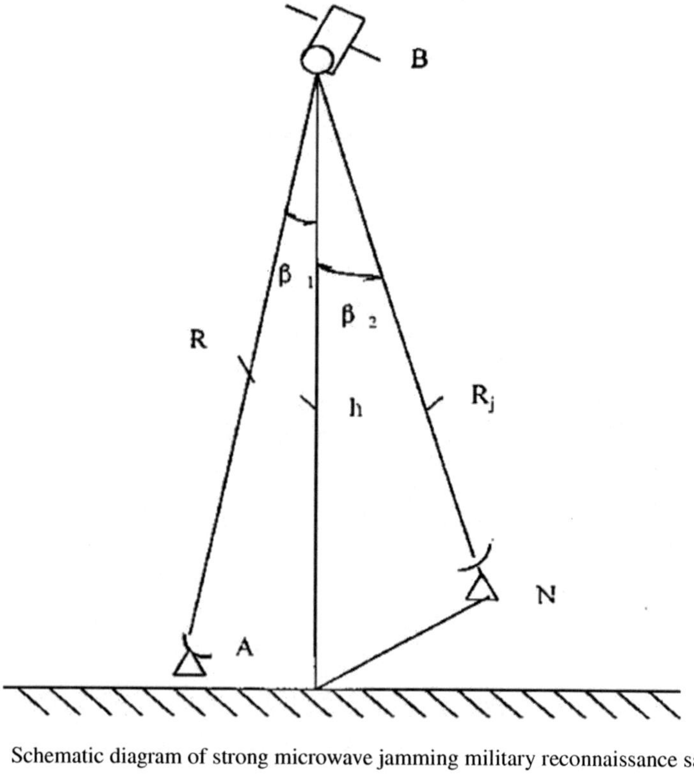


**Fig. 10.15** Schematic diagram of strong microwave jamming military reconnaissance satellite

The power received by the reconnaissance satellite from the strong microwave jammer is:

$$P_2 = \frac{P_j \cdot G_j(\phi)}{4\pi R_j^2} \cdot \frac{\lambda^2}{4\pi} \cdot G_B(\beta_2) \tag{10.8}$$

Select the jamming frequency of the jammer to be near the frequency of the shielded radiation source:

$$P_2 = \eta \cdot P_1 \tag{10.9}$$

Substituting Formulas (10.7) and (10.8) into Formula (10.9), we get:

$$\frac{P_j \cdot G_j(\phi)}{4\pi R_j^2} \cdot \frac{\lambda^2}{4\pi} \cdot G_B(\beta_2) = \eta \cdot \frac{P_a \cdot G_a(\theta)}{4\pi R^2} \cdot \frac{\lambda^2}{4\pi} \cdot G_B(\beta_1)$$

After simplification:

$$P_j \cdot G_j(\phi) = \frac{\eta \cdot P_a \cdot G_a(\theta) \cdot R_j^2 \cdot G_B(\beta_1)}{R^2 \cdot G_B(\beta_2)} \tag{10.10}$$

$$R \approx \frac{h}{\cos \beta_1}; R_j \approx \frac{h}{\cos \beta_2}$$

Substitute $R$ and $R_j$ into the Formula (10.10):

$$P_j \cdot G_j(\phi) = \frac{\eta \cdot P_a \cdot G_a(\theta) \cdot G_B(\beta_1) \cdot \cos^2 \beta_1}{G_B(\beta_2) \cdot \cos^2 \beta_2} \tag{10.11}$$

| | |
|---|---|
| $f_a$ | frequency of the radiation source. |
| P | the radiated power. |
| $G_a$ | Antenna gain. |
| $G_a(\theta)$ | Antenna sidelobe gain. |
| h | The height of the reconnaissance satellite. |
| N | The location of the strong microwave jammer. |
| B | The point where the satellite is located. |
| $P_j$ | The transmit power of the jammer. |
| $G_j$ | Jamming antenna gain (Note: To simplify the calculation, ignore the influence of the earth's curvature). |

When jamming the reconnaissance satellite, the reconnaissance satellite generally reconnoiters the blind area on the top of the ground radar emitter's radar antenna. Set: $P_a = 1$ MW; $G_a(\theta) = 1$. Under adverse jamming conditions, set: $G_B(\beta_1) \approx G_B = 10$; $G_B(\beta_2) = 0.1 G_B$; $\cos \beta_1 = 1$; $\cos \beta_2 = 0.707$; $G_j(\phi) = 10^3$. When using

**Table 10.8** The relationship between different values of n and the power of the transmitting tube

| $n$ | 1 | 5 | 10 | 15 | 20 | 30 | 50 |
|---|---|---|---|---|---|---|---|
| $P_j$ (W) | $1.4 \times 10^4$ | $7 \times 10^4$ | $1.4 \times 10^5$ | $2.1 \times 10^5$ | $2.8 \times 10^5$ | $4.2 \times 10^5$ | $7 \times 10^5$ |

continuous noise jamming, choose different values for n, and the required power of the microwave tube is shown in Table 10.8.

If double-film jamming and continuous jamming are used, n = 1, the continuous wave launch tube is 15 kW. If the pulse jamming is n = 50, the duty cycle is 10, the peak power of the pulse tube is 700 kW, and the average power is 70 kW. The area covered by four jamming stations is about 300,000 km$^2$. The relationship between signal flow density, processing time plus pulse width, and detection probability is:

$$\eta = \frac{1}{1 + t_p \cdot \lambda} \tag{10.12}$$

In the formula:

$t_p$ Pulse width plus processing time.
$\lambda$ Received signal traffic density.

The calculated data is shown in Table 10.9.

According to the data in Table 10.9, The relationship between detection probability, signal traffic density, processing time, and pulse width can be drawn in Fig. 10.16.

According to the reconnaissance processing technology, when the reconnaissance probability is less than 80%, the signal sorting process will produce large errors, which will reduce the reliability of the reconnaissance data, thereby achieving the purpose of jamming.

**Table 10.9** Relationship data between different signal flow and bit error rate

| **$t_p$ (μs)** | **λ (pulses/s)** | | | | | | | |
|---|---|---|---|---|---|---|---|---|
| | $1 \times 10^4$ | $3 \times 10^4$ | $5 \times 10^4$ | $10 \times 10^4$ | $20 \times 10^4$ | $30 \times 10^4$ | $50 \times 10^4$ | $100 \times 10^4$ |
| | η (%) | | | | | | | |
| 1 | 99 | 97 | 95.2 | 90.9 | 83.3 | 76.9 | 66.6 | 50 |
| 2 | 98 | 94.3 | 90.9 | 83.3 | 71.4 | 62.5 | 50 | 33.3 |
| 3 | 97 | 91.7 | 86.9 | 76.9 | 62.5 | 52.6 | 40 | 25 |
| 5 | 95.2 | 86.9 | 80 | 66.6 | 50 | 40 | 28.57 | 16.6 |

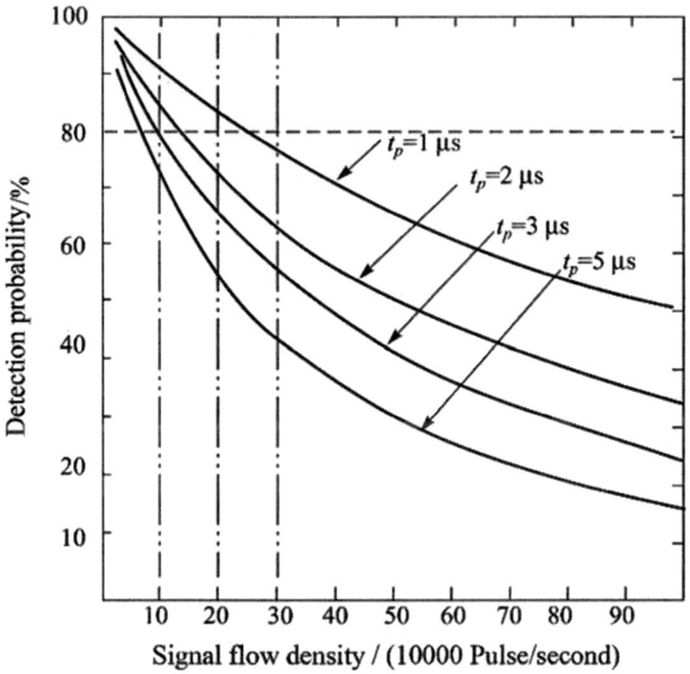


**Fig. 10.16** The relationship between detection probability, signal traffic density, processing time, and pulse width

## 10.4 Application of UWB High PRF Jammer

The meter-band radar has two special advantages: one is that it can detect stealth aircraft better and effectively; the other is that the anti-radiation missile has a low hit probability on it and has a good survival probability. Given these two advantages, the meter wave radar will continue to survive and develop. As long as the meter wave radar needs for survival and development, the jammer that jamming with the meter wave radar will be useful. Many types of jammers jamming with the meter wave radar, and the jamming effect is relatively satisfactory. However, it is necessary to find a jammer that is small in size, light in weight, better in jamming effect, very cheap, and easy to use and maintain. "Ultra-wideband high repetition frequency jammer" is one of the preferred ones.

As long as we can continuously improve the "ultra-wideband high repetition frequency jammer," make it suitable for jamming various radar systems' characteristics. Then, the "ultra-wideband and high repetition frequency jammer" will have broad development prospects and will play its role in the field of electronic warfare.

### *10.4.1 Demand Status*

Judging from the current situation, the ultra-wideband high-repetition frequency source jamming radars below 500 megabits is relatively mature. The radars that jamming with the L-band still need some work and some improvements. There are also many radars below 500 MB, currently in use (such as E-2C early warning aircraft radar AN/APY-145), and its operating frequency is 400–450 MHz. As of December 1994, Grumman had produced 169 aircraft of this early warning aircraft. Countries and regions equipped with E-2C equipment include: 139 in the United States, 13 in Japan, 4 in Israel, 5 in Egypt, 4 in Singapore, and 4 in Taiwan are renamed with E-2T. This kind of early warning aircraft radar is widely used and large in number, and the serious threat to our country is obvious. There are many ways to counter this early warning aircraft radar, and the effect is obvious. However, to find a simple and easy-to-use jammer, small in size, light in weight, low in price, and can have a certain jamming effect, the ultra-wideband high repetition frequency jammer is the most suitable.

Furthermore, there have been: the AN/APS-115, which is a submarine missile early warning radar with operating frequency between 420 and 450 MHz; the AN/the FPS-85, which is a space submarine missile target detection and distance warning radar with operating frequency 442 MHz; the AN/the SPS-29, which is a remotely Air warning radar, the frequency is 215.5–224.5 MHz; the AN/SPS-37, which is a long-range air search radar, the working frequency is 400–450 MHz, etc. In this frequency band, some of our country's meter-band radars have been equipped with many troops. Some simple, portable, and inexpensive jammers are needed for anti-jamming operation training. Therefore, as long as the "ultra-wideband high repetition frequency jammer" can be improved, it can have a better jamming effect on the radars of various systems, and its application prospects are very broad.

For the meter band and L-band radars of the United States and Taiwan province, see Table 10.10. As long as there are radars in this frequency band, Ultra-wideband jammers with high repetition frequencies will be useful. "Ultra-wideband high repetition frequency jammer" can jam some radars below the L -band. In order to narrow the range and have strong pertinence, the focus of the plan is aimed at the current E-2T early warning aircraft that has a greater impact on electronic warfare. Radar conducts analysis and demonstration. If the "ultra-wideband high-repetition jammer" is improved, it can effectively jam the E-2T early warning aircraft radar. This is an important technological breakthrough in the field of electronic warfare. This small jammer can subdue powerful radar equipment. This is the embodiment of the electronic warfare field by winning more with less and fighting the strong with the weak.

**Table 10.10** Specification of the meter band and L-band radars of the United States and Taiwan province

| Radar signal | Use | Working frequency | Range | Antenna form or gain | Lobe width | RF power | Pulse width | Pulse repetition frequency | Noise figure | Producing country |
|---|---|---|---|---|---|---|---|---|---|---|
| AN/APS-145 | E-2C(2T) Early warning aircraft radar | 400–450 | 460 | 21.5 dB | 4° | 1 MW (Peak power) 4 kW (Average power) | 13 μs compressed to 0.22 μs | 300 Hz | 3 dB | United States |
| AN/APS-115 Solid-state large-scale phased array radar | Submarine-launched ballistic missile warning | 420–450 | 4800 (Against 10 m target) | Dual array plane array, gain 38.4 dB | 2.2° | 528.4 kW (Peak power) 145 kW (Average power) | 0.25–16 μs | 25–30 Hz | 2.9 dB | United States |
| AN/FPS-85 | Space target detection and tracking and submarine-launched missile early warning | 442 | 5600–7000 | Plane array receiving, sending, and separating 29.6 × 29.6 m (transmit), 58 m (receive) | 1.4° (transmit), 0.8° (receive) | 32 MW (Peak power) 400 kW (Average power) | 1.5–250 μs | 25–30 Hz | 3 dB | United States |
| AN/FPS-118 | Strategic warning | 5–28 | 800–2880 | Dipole antenna | 2.75° | 100 kW (Peak power) 100 kW (Average power) | Continuous wave | 20–60 Hz | <3 dB | United States |
| AN/SPS-29 | Long-range air defense | 215.5–224.5 | 370 (Against 5 m target) | Rectangular grid array | 20° | 750 kW (Peak power) | 10 μs | 300 Hz | 3 dB | United States, Taiwan |

(continued)

**Table 10.10** (continued)

| Radar signal | Use | Working frequency | Range | Antenna form or gain | Lobe width | RF power | Pulse width | Pulse repetition frequency | Noise figure | Producing country |
|---|---|---|---|---|---|---|---|---|---|---|
| AN/SPS-37 | Remote air search | VHF (400–450) | | Antenna size 6 × 3.3 m$^2$ | | 180 kW (Peak power) | 200 μs, 6 μs after compression | | 3 dB | Taiwan |
| AN/SPS-40 | Mid-range air search | UHF (300–1000) | 320 (11.5 min) | | Bearing 11° Vertical 19° | 1000 kW (Peak power) 4 kW (Average power) | 0.6 μs, 60 μs before compression | 300 Hz | 3 dB | United States, Taiwan |
| AN/FPS-36 | Air search | 1220–1350 | 370 | | 1.4° | 575–615 kW (Peak power) | 2 μs | 325 Hz | Receiver sensitivity −113 dBm | Taiwan |
| AN/FPS-37 | | 1350–1450 | 450 | | 1.2° | 2.5 MW (Peak power) | 6 μs | 417–438 Hz | | Taiwan |
| AN/FPS-110B | Long-range air defense | 1250–1350 | 410 | Parabolic reflector; gain 35 dB | 1.3° | 2 MW (Peak power) 5 kW (Average power) | 3.6–6 μs | 320–330 Hz | 3 dB | Taiwan |
| AN/FPS-117 | Air vigilance | 1215–1400 | 9.27–0.370 | Flat array 7.3 × 7.3 m$^2$ | | 24.75 kW (Peak power) Coherent pulse-Doppler processing | | | | United States, Taiwan |
| AN/FPS-124 | Short-range air defense, low-altitude blinding | D band | 3.7–130 | | | | | | | United States |
| GE-592 | Air surveillance and guidance | 1215–1400 | 9.27–400 | Flat array 7.3 × 7.3 m$^2$ | 2.2° | 25 kW (Peak power) | | | 2.6 dB | Taiwan |

(continued)

**Table 10.10** (continued)

| Radar signal | Use | Working frequency | Range | Antenna form or gain | Lobe width | RF power | Pulse width | Pulse repetition frequency | Noise figure | Producing country |
|---|---|---|---|---|---|---|---|---|---|---|
| AN/SPS-6 | Remote air search | 1250–1350 | Parabolic reflector, 1.85 × 5.18 $m^2$, gain: 27.5 dB | | Bearing 3.5° Vertical 10° | 500–750 kW (Peak power) | 1, 4, 5, 20 μs | 150, 600 Hz | 8.5 dB | Taiwan |
| AN/SPS-58 | Short-range air search and target interception | 1360 | Hollow mesh reflector, 5 × 1.8 $m^2$ | | | 75 kW (Peak power) | | | | United States, Taiwan |
| DRBV-26 | Remote air search | L band | Cutting parabolic reflector, 7.5 × 3 $m^2$ | | Bearing 2.5° Vertical 50° | 2 MW (Peak power) 2 kW (Average power) | 2.5 μs | 450 Hz | | Taiwan |
| AN/TPS-63 | Balloon-mounted radar | 1250–1350 | Parabolic reflector, 4.9 × 5.5 $m^2$ | | 2.7° | 100 kW (Peak power) 300 W (Average power) | | | | United States |

### 10.4.2 Demand Analysis for High Repetition Frequency

A bright-line displayed by the radar's P-type display is formed by connecting multiple bright spots, as shown in Fig. 10.17.

If the radar detection range represented by the display radius is:

$$R = \frac{1}{2} T \cdot C \tag{10.13}$$

Then:

$$r = n \cdot \Delta r \tag{10.14}$$

In the formula:

T The minimum period of radar pulse repetition without range ambiguity.
C Speed of light.
r The distance represented by the display $r = R$.

Substitute Eq. (10.14) into Eq. (10.13):

$$\Delta r = \frac{T \cdot C}{2n} \tag{10.15}$$

$$\text{Or} \quad \Delta r = \Delta \tau_j \cdot \frac{C}{2} = \frac{1}{f_j} \cdot \frac{C}{2} \tag{10.16}$$

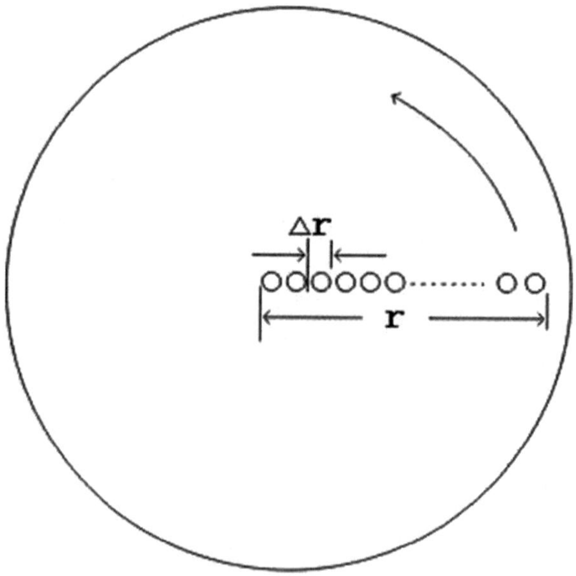


**Fig. 10.17** Schematic diagram of radar P display spot radius display. $\gamma$—Effective display radius of the display. $\Delta\gamma$—The light spot diameter of the display. n—The number of light points that can be arranged within the radius of the display

Substitute Eq. (10.16) into Eq. (10.15):

$$f_j = n \cdot f_t \tag{10.17}$$

In the formula:

$f_j$ Jamming pulse repetition frequency of high-repetition jammer.
$f_t$ Jamming radar pulse repetition.

For example:
When n = 300 and $f_t = 1.84$ kHz, $f_j = 552$ kHz.
When n = 512 and $f_t = 264$ kHz, $f_j = 135.16$ kHz.
When n = 1026 and $f_t = 2644$ kHz, $f_j = 270$ kHz.

It can be seen that the repetition frequency of the jammer with high repetition frequency, wide frequency band, and no carrier frequency is closely related to the characteristic spot radius of the radar display and the pulse repetition frequency of the radar. Only by meeting the requirements of high repetition frequency can suppress and jam effectively.

### 10.4.3 Pulse Width and Spectrum

Pulse waveform without carrier frequency, as shown in Fig. 10.18.

The video pulse spectrum width is:

$$\Delta F_j = \frac{1}{\tau_j} \tag{10.18}$$

If $\Delta\tau_j = 4$ ns,

$$\Delta F_j = 250\,\text{MHz}$$

After Fourier transforms, the mathematical expression of the bipolar narrow pulse spectrum is as follows:

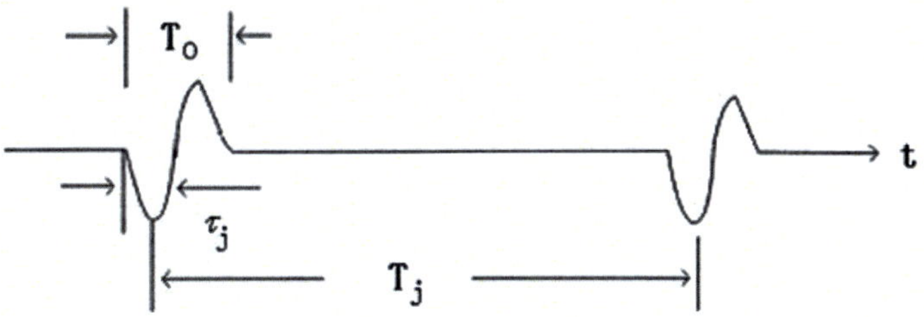


**Fig. 10.18** Pulse waveform without carrier frequency

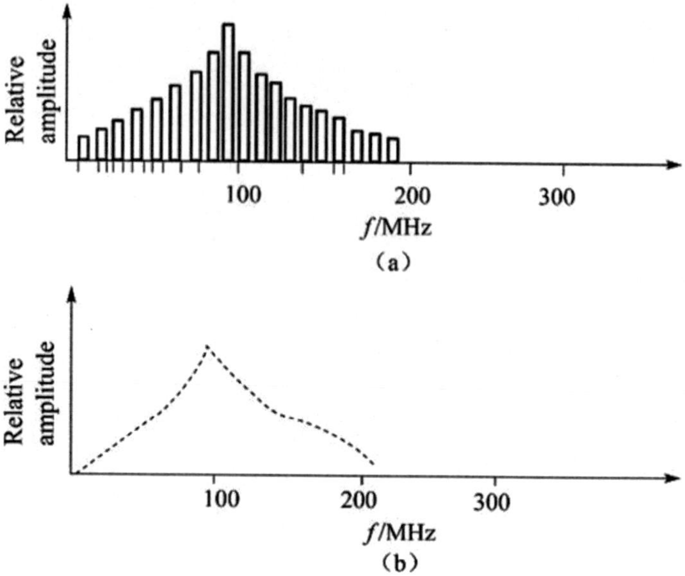


**Fig. 10.19** **a** Schematic diagram of the normalized power spectrum. **b** Schematic diagram of spectrum envelope

$$f(t) = \sum_{k=1}^{a} (a_k \cdot \cos(k\omega t) + b_k \cdot \sin(k\omega t))$$

$$a_k = \frac{\omega}{2\pi}\left\{\frac{1}{\omega_0 + k\cdot\omega}\left[1 - \cos\frac{2\pi(\omega_0 + k\cdot\omega)}{\omega_0}\right] + \frac{1}{\omega_0 - k\cdot\omega}\left[1 - \cos\frac{2\pi(\omega_0 - k\cdot\omega)}{\omega_0}\right]\right\}$$

$$b_k = \frac{\omega}{2\pi}\left\{\frac{1}{\omega_0 + k\cdot\omega}\left[\sin\frac{2\pi(\omega_0 + k\cdot\omega)}{\omega_0}\right] - \frac{1}{\omega_0 - k\cdot\omega}\left[1 - \sin\frac{2\pi(\omega_0 - k\cdot\omega)}{\omega_0}\right]\right\}$$

The corresponding normalized amplitude spectrum is shown in Fig. 10.19. The actual spectrum width is shown in Fig. 10.20.

### *10.4.4 Jamming Effect on Meter Band Radar*

It can be considered that the frequency band of the output circuit or radiating antenna is not wide enough, which attenuates the low-frequency spectrum. Therefore, the measured spectrum is quite different from the theoretical spectrum in the low-frequency band.

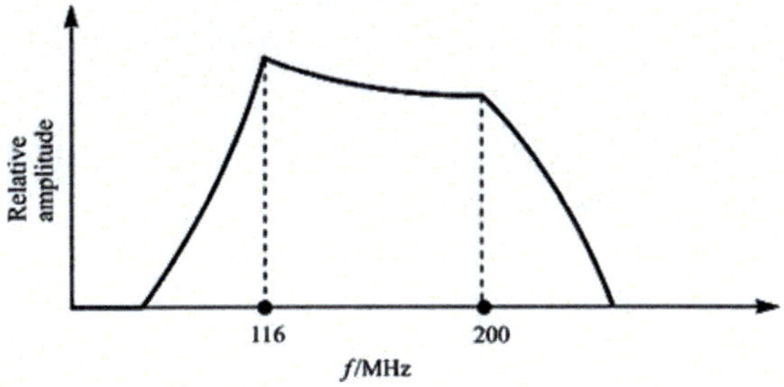


**Fig. 10.20** Measure the pulse spectrum width

The ultra-wideband high repetition frequency pulse can be equivalent to a random pointed continuous pulse. After entering the radar receiver, it is equal to increasing the noise level of the radar receiver. The radar echo is submerged in the noise, thereby reducing the detection range of the radar. This is the jamming effect of the radar without carrier frequency and high repetition frequency ultra-wideband pulse.

Suppose the power of the jamming signal received by the radar is:

$$P_1 = \frac{P_j \cdot G_j}{4\pi R_j^2} \cdot \frac{\lambda^2 \cdot K_f}{4\pi} \cdot G_a(\theta) \tag{10.19}$$

In the formula:

- $P_j$ Average power of UWB jammer.
- $G_j$ Jamming antenna gain.
- $\lambda$ Operating wavelength.

The target echo signal amplitude received by the radar is:

$$P_2 = \frac{P_a \cdot G_a^2 \cdot \sigma \cdot \lambda^2}{(4\pi)^3 R_a^4} \tag{10.20}$$

Let the jamming suppression coefficient be: $K_j = \frac{P_1}{P_2}$.

$$\text{Then: } K_j = \frac{4\pi P_j \cdot G_j \cdot R_a^4}{P_a \cdot G_a \cdot \sigma \cdot R_j^2} \cdot \frac{G_a(\theta)}{G_a} \cdot K_f \tag{10.21}$$

Substitute $P_j = \frac{\hat{P}_j \cdot \tau_j}{T}$ into (10.21), we get:

$$K_j = \frac{4\pi \hat{P}_j \cdot \tau_j \cdot G_j \cdot R_a^4}{P_a \cdot G_a \cdot \sigma \cdot R_j^2 \cdot T} \cdot \frac{G_a(\theta)}{G_a} \cdot K_f \tag{10.22}$$

In the formula:

$K_f$ The ratio of the bandwidth of the radar receiver to the width of the jamming spectrum.
$R_a$ Distance between target and radar.
$R_j$ Distance between jammer and radar.
$\sigma$ Target effective reflection area.
$P_a$ Radar peak power.
$G_a$ Radar antenna gain.
$G_a(\theta)$ Radar gain in the direction of the jammer.

Under normal circumstances, $\frac{G_a(\theta)}{G_a} = K\left(\frac{\theta_{0.5}}{2\theta}\right)^2$.

The schematic diagram of the radar antenna sidelobe characteristics is shown in Fig. 10.21.

For example:

Known parameter: $\sigma = 10\,\text{m}^2$; $P_a = 2 \times 10^5\,\text{W}$; $G_a = 100$; $G_a(\theta)$ select according to Fig. 10.20; $\tau_j = 4\,\text{ns}$; $f_j = 250\,\text{kHz}$; $\hat{P}_j = 400\,\text{W}$.

According to (10.22) formula, take different parameters to get Tables 10.10, 10.11, 10.12.

$\theta_{0.5} = 5°$ and $K_f = 0.005$.

When $R_j$ is 100 m, $K_j$ is 3, 10, 40, and the calculated data is listed in Table 10.10.

The detection range of the radar can reach 300 km when it is not disturbed. After being disturbed, its range becomes the data in Tables 10.10, 10.11, and 10.12. Throwing an ultra-wideband jammer near the radar still has a better jamming effect on the meter band radar.

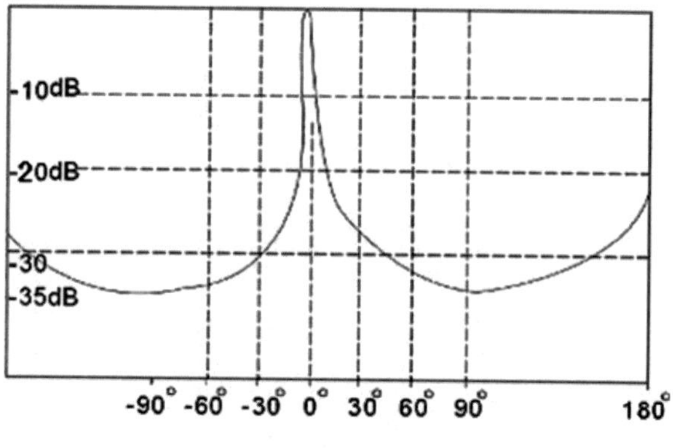


**Fig. 10.21** Schematic diagram of the sidelobes of the radar antenna

**Table 10.11** When $R_j = 100\,\text{m}$, different jamming exposure distances corresponding to different angles

| θ | 0° | 20° | 40° | 60° | 90° | 180° | $K_i$ |
|---|---|---|---|---|---|---|---|
| $G_a(\theta)/G_a$ | 1 | −28 dB | −31 dB | −33 dB | −34 dB | −25 dB | / |
| $R_{a1}$ | 4.7 km | 23.5 | 27.9 | 31.4 | 33.3 | 19.7 km | 3 |
| $R_{a2}$ | 6.3 km | 31.4 | 37.4 | 42.1 | 44.6 | 26.4 km | 10 |
| $R_{a3}$ | 8.8 km | 44.2 | 52.7 | 59.3 | 62.8 | 37.2 km | 40 |

**Table 10.12** When $R_j = 500\,\text{m}$, different jamming exposure distances corresponding to different angles

| θ | 0° | 20° | 40° | 60° | 90° | 180° | $K_i$ |
|---|---|---|---|---|---|---|---|
| $G_a(\theta)/G_a$ | 1 | −28 dB | −31 dB | −33 dB | −34 dB | −25 dB | / |
| $R_{a4}$ | 10.5 km | 52.4 | 62.2 | 70 | 74.2 | 43.9 km | 3 |
| $R_{a5}$ | 14 | 70 | 83.4 | 93.8 | 99.5 | 58.8 km | 10 |
| $R_{a6}$ | 19.6 km | 98.5 | 117.5 | 132.2 | 140 | 82.9 km | 40 |

**Table 10.13** When $R_j = 1000\,\text{m}$, different jamming exposure distances corresponding to different angles

| θ | 0° | 20° | 40° | 60° | 90° | 180° | $K_i$ |
|---|---|---|---|---|---|---|---|
| $G_a(\theta)/G_a$ | 1 | −28 dB | −31 dB | −33 dB | −34 dB | −25 dB | / |
| $R_{a7}$ | 14.8 km | 73.8 | 87.7 | 98.7 | 104.6 | 61.9 km | 3 |
| $R_{a8}$ | 19.7 | 98.7 | 117.5 | 132.3 | 140.3 | 82.9 km | 10 |
| $R_{a9}$ | 27.6 km | 138.8 | 165.7 | 186.4 | 197.4 | 116.9 km | 40 |

The jamming diagrams are drawn from Tables 10.11, 10.12, and 10.13 are shown in Figs. 10.22, 10.23, and 10.24. The dotted circle in the figure is the range that the radar should have when it is not jammed with by the UWB jammer.

When $R_j = 100\,\text{m}$, see Table 10.11 (the drawn jamming exposure area corresponds to Fig. 10.21).

When $R_j = 500\,\text{m}$, see Table 10.12 (the drawn jamming exposure area corresponds to Fig. 10.22).

When $R_j = 100\,\text{m}$, see Table 10.13 (the drawn jamming exposure area corresponds to Fig. 10.23).

### *10.4.5 Application of UWB Jammer*

The jammer is small in size and light in weight. The entire device is installed in a cylinder (including power supply but not antenna). The cylinder's diameter is 70 mm; the height is 160 mm; the total weight is less than 2 kg, and the power consumption is 3.3 W. It can work for 20–30 min when powered by a battery (it can also work with an external power supply).

The main purpose of the jammer is as follows.

It can be used as a drop-in jammer. Cannonballs, rockets, drones, etc. can be used for delivery. As shown in Fig. 10.25, the UAV is used to launch an ultra-wideband jammer. The current domestic UAV has a load of 34 kg and a range of up to 300 km. A UAV can carry up to 17 such jammers. The UAV can move around the jammed radar, and the radius around it (with the jammed radar as the center) is less than 1 km. It should be launched two minutes before our plane penetrates the defense. It

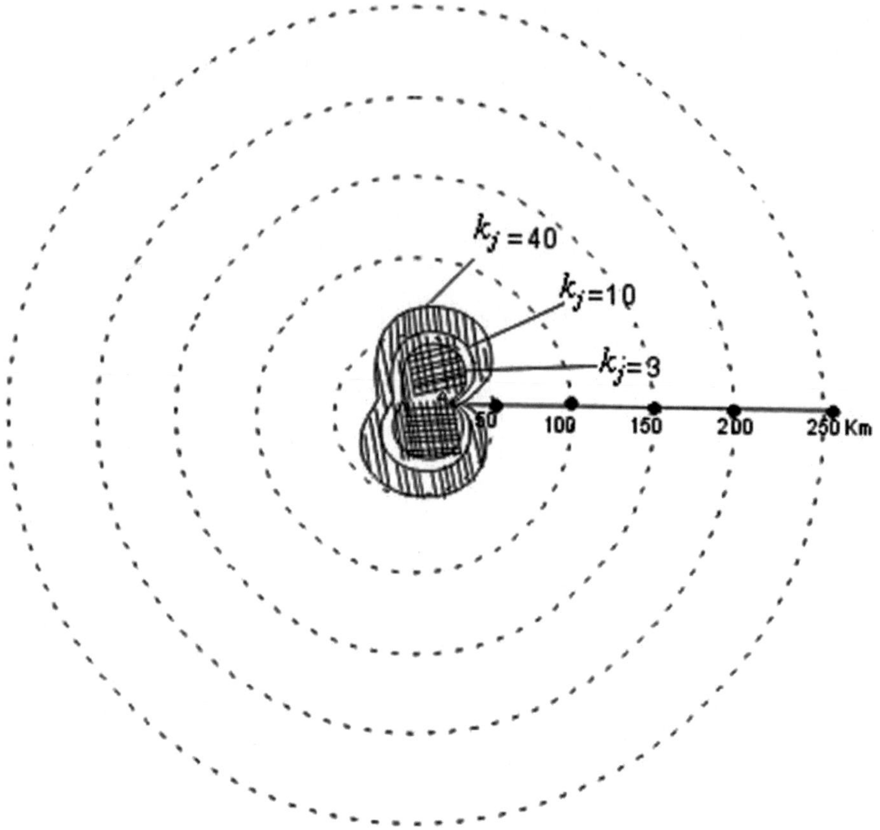


**Fig. 10.22** When $R_j = 100\,\text{m}$, the jamming exposure area and jamming effective area caused by the UWB jammer to the meter band radar. Inside the horizontal line—Jamming exposure area; outside the horizontal line—Jamming effective area

can cover the penetration of our bombers, missiles, and fighter planes to make up for some of the shortcomings of high-power support jamming aircraft.

If the deployed jammers are configured according to typical values, the distance from the radar station is 500 m on average, and the projected angle is an average of 60°. The resulting jamming exposure area is shown in Fig. 10.26.

It can be seen from Fig. 10.27 that this type of placement can make the maximum jamming exposure distance less than 50 km.

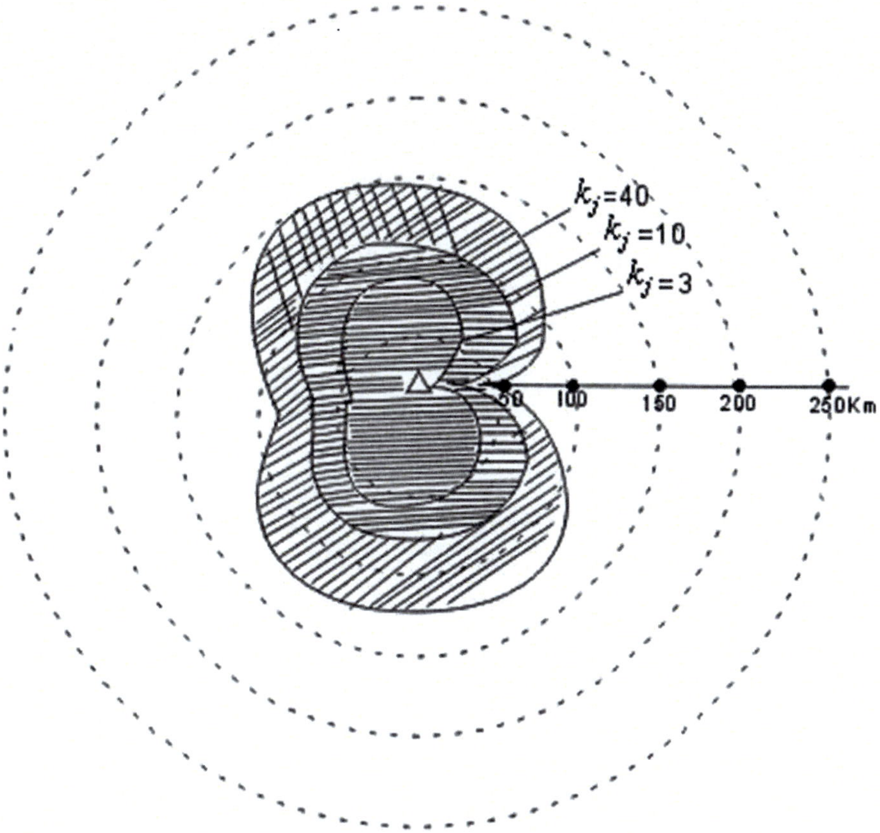


**Fig. 10.23** When $R_j = 500\,\text{m}$, the jamming exposure area and jamming effective area of the UWB jammer to the meter wave radar

### *10.4.6 Test of UWB Jammer*

1. Test layout

   The radar model of the jammed test: ××× radar.

   The distance between the UWB jammer and the radar station is 470 m.

   The UWB jamming rack is about 6 m high.

2. Basic parameters of UWB jammer

   Peak power: ** W.

   Pulse width: * ns.

   Pulse repetition frequency: 450 kHz.

   Output jamming spectrum range: 60–220 MHz.

   Power consumption: voltage 8.4 V, current 0.4 A.

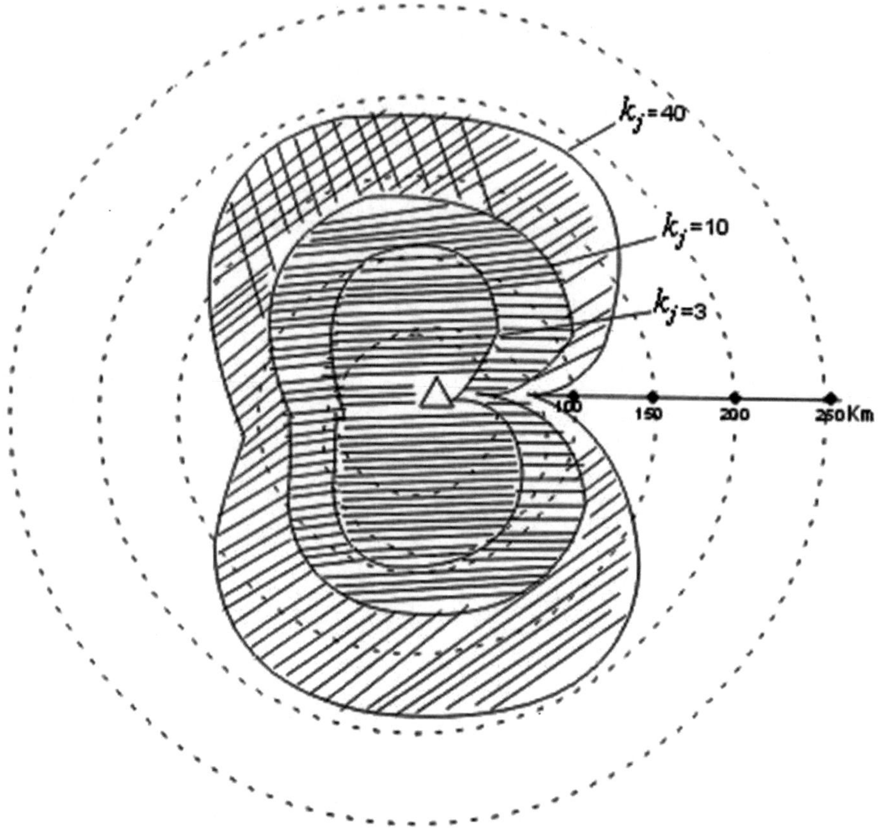


**Fig. 10.24** When $R_j = 1000$ m, the jamming exposure area and jamming effective area of the UWB jammer to the meter wave radar

3. Test method

   The radar station starts to search for targets within a range of several kilometers to tens of kilometers, and there are many ground fixed target echoes (radar 360° rotation).

   When the UWB jammer is turned on, the fixed target echoes are all suppressed on the radar's A and P displays, reaching the third level of jamming intensity.

4. Radar display image

   The image on the radar display without jamming and after being jammed is shown in Fig .10.28. On the P display, within ±30° of the direction of the jammer, the display image is dim, and the radar receiver is saturated. The frequency of the UWB jammer is not limited to the meter wave frequency band. At present, the L-band can be achieved, and the peak power can also be achieved in megawatts or more. It can be used as a drop jammer, anti-radiation missile decoy, and broadband suppressive jammer during the war. It can also be made into low

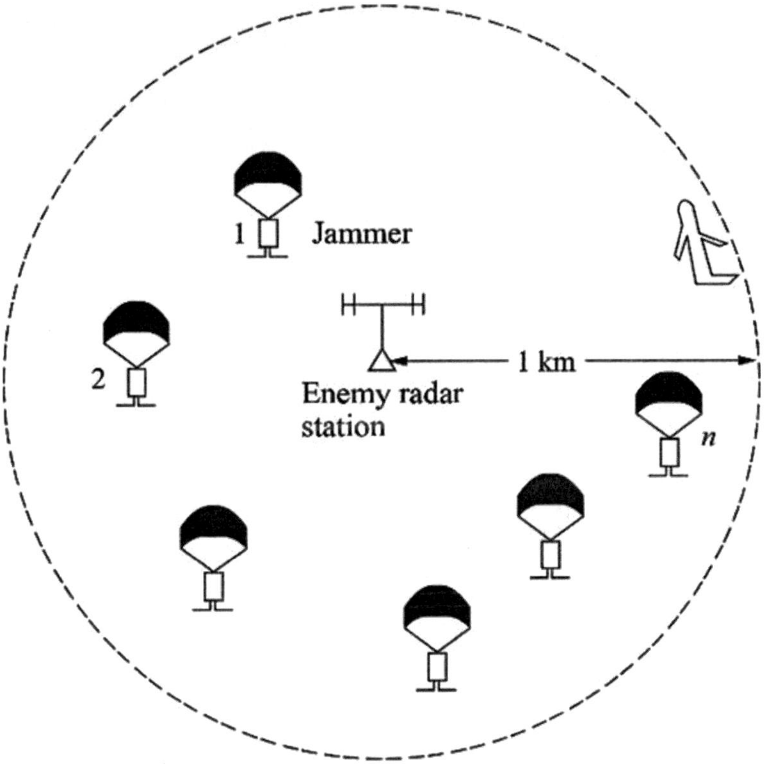


**Fig. 10.25** Schematic diagram of launching an ultra-wideband jammer

repetition frequency and super high power, damaging electronic equipment at close range and causing personal injury.

In peacetime, it can be used as a radar training simulator to train the operator's ability to find targets in the context of jamming. It is cheap and easy to use, and is a promising jammer type. Ultra-wideband jammer is a new type of jammer, which can be made of small size, lightweight, high reliability, low price, and easy to drop. It can work not only in the meter wave band but also in the L band, which can be used for simulation training.

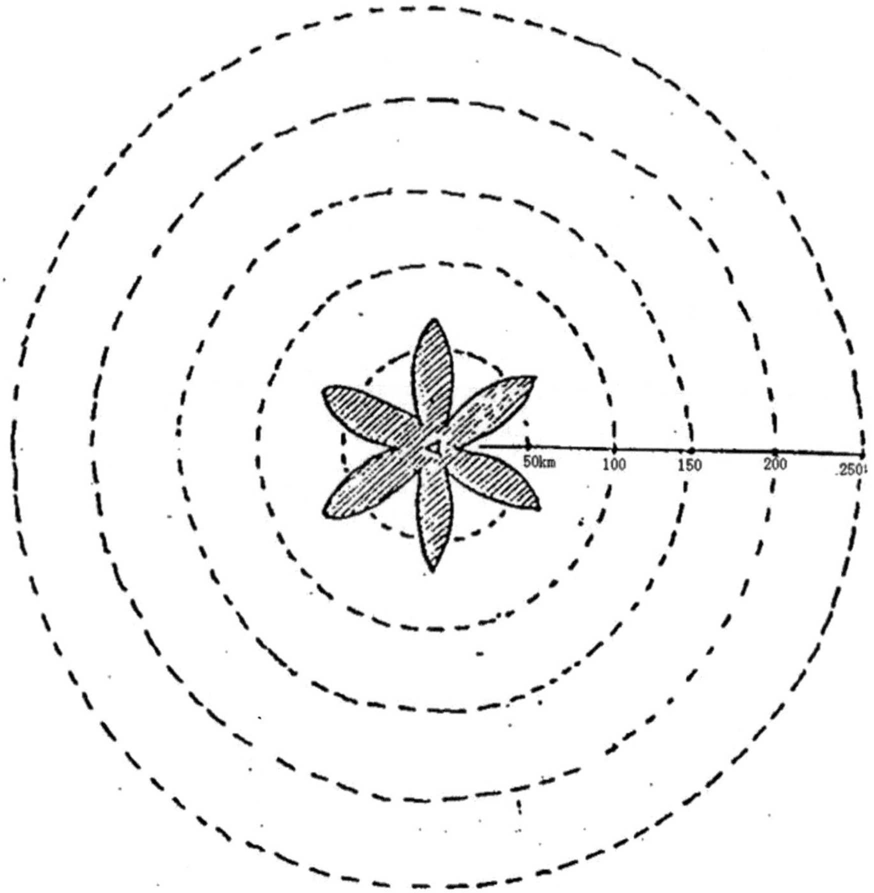


**Fig. 10.26** An jamming exposure area formed by an ultra-wideband jamming machine with an average of 60° and a distance of 500 m

## 10.5 Integrated Electronic Warfare Technology Including Reconnaissance, Jamming, and Positioning

### *10.5.1 Utilize the Integrated Technology of High-Power Continuous Wave Noise Jammer*

The basic composition of the high-power continuous-wave noise jammer is shown in Fig. 10.29.

Figure 10.29 includes receiving antennas (including reconnaissance receiving antennas and angle tracking antennas), receiving equipment (including reconnaissance receivers, aiming receivers, and angle tracking receivers), signal sorting

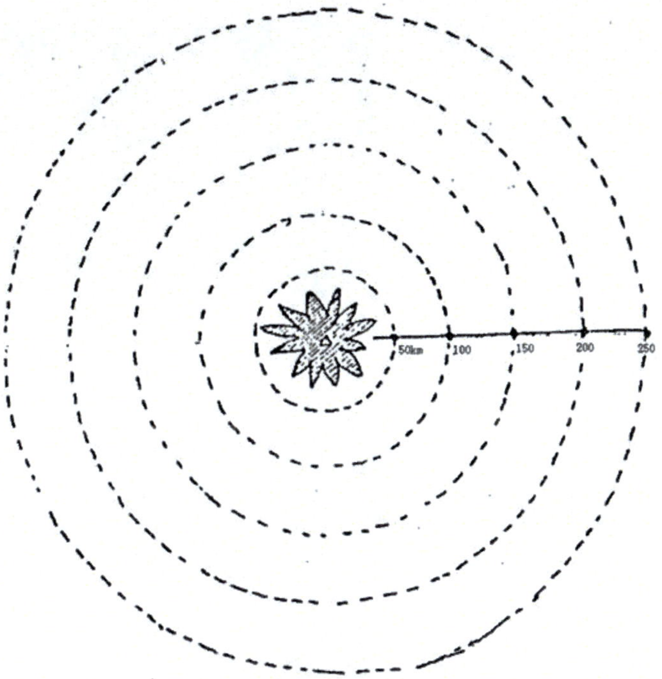


**Fig. 10.27** Disturbance exposure area formed by jammers with an average of 30° and an average distance of 500 m

(including signal sorting, identification of friend or foe, threat discrimination, signal parameter measurement display, map background display, etc.), and aiming frequency division system (including instantaneous aiming frequency, coherent jamming source generation, etc.) The jamming signal amplification link modulates the targeted continuous wave signal into a jamming signal close to white noise and then adds it to the transmitting antenna through excitation amplification and power amplification. After the radar receives this noise jamming signal, it will have a jamming-effect. The time-division control is set up to solve the spontaneous and self-receiving influence caused by installing the transceiver antenna on a platform. Its main function is to ensure that the receiver does not receive during the transmitting time, and the transmitter does not transmit during the receiving time.

The high-power noise jammer has the functions of receiving, displaying, signal processing, direction finding, servo tracking, and high-power transmission. As long as a sub-system is added to Fig. 10.29, the echo signal reflected by target can be detected, then the distance of target echo can be determined. This device can locate the target and constitute a "three-digit integrated electronic warfare device." On the basis of Fig. 10.29, the composition block diagram of the ranging function is shown in Fig. 10.30.

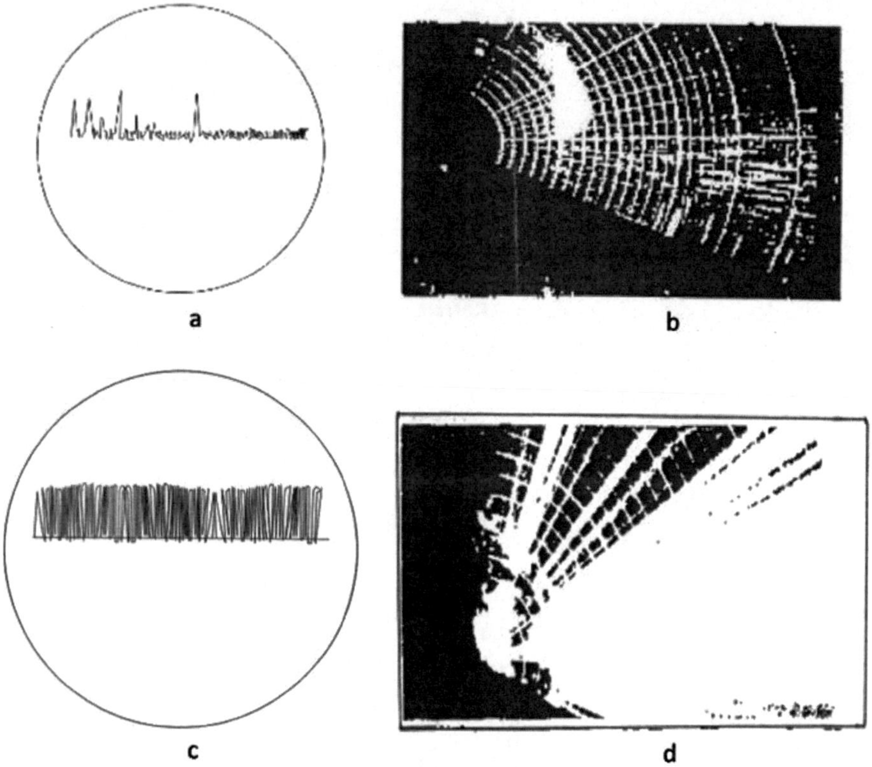

Fig. 10.28 **a**, **b** The image on the radar display without jamming. **c**, **d** The image after being jammed with by the ultra-wideband jammer

In Fig. 10.30, the original high-power noise jammer has basically maintained the receiving, display, signal sorting and processing, and transmitting and amplifying links. A "noise-related receiving system" has also been added. With this sub-system, the noise echo signal reflected by the target can be extracted, and the distance of the target can be measured from it.

The composition of the noise-related receiver and its ranging principle are discussed below.

The composition of the noise-related receiver is shown in Fig. 10.31.

The output distance signal expression after correlation integration:

$$R(\tau, \tau_0, T) = \int_0^T x(t - \tau) \cdot x \cdot (t - \tau_0)dt \tag{10.23}$$

The measurement error of distance $\Delta\tau$ depends on the spectral width of the noise signal $\Delta F$:

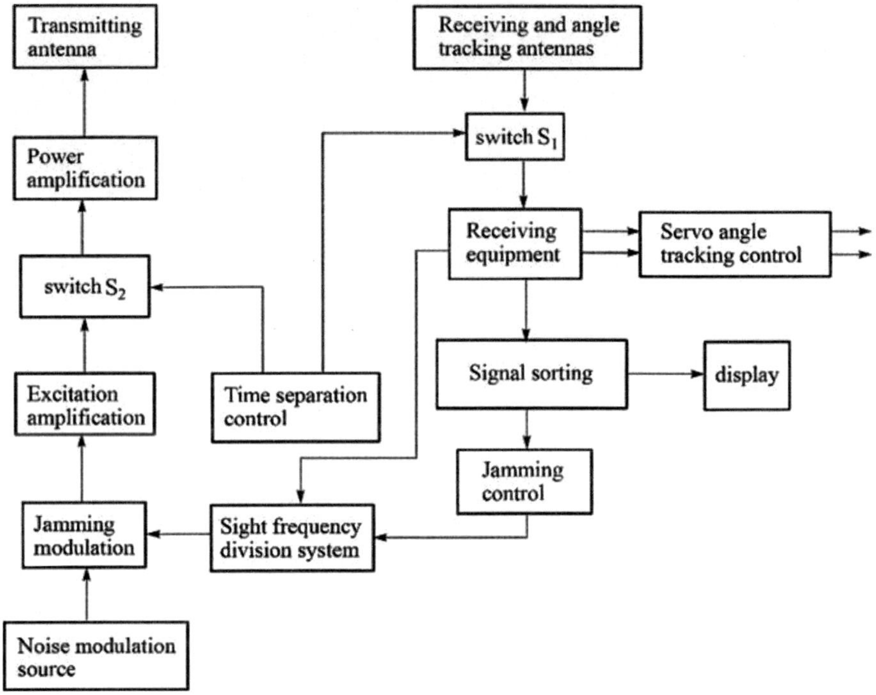


**Fig. 10.29** Basic composition block diagram of high-power noise jammer

$$\Delta\tau = \frac{1}{\Delta F} \tag{10.24}$$

The resolution of the distance is

$$\Delta L = \frac{C \cdot \Delta\tau}{2} = \frac{C}{2\Delta F} \tag{10.25}$$

If $\Delta F$ is 10 MHz, then:

$$\Delta L = \frac{3 \times 10^8}{2 \times 10 \times 10^6} = 15\,\text{m}$$

It can be seen that noise ranging has the following advantages:

1. Working with a continuous wave, the peak power is small, and the other party is not easy to detect;
2. The signal sent out is a noise spectrum, and it is difficult for the other party to measure the parameters of the signal;
3. It organically combines position measurement and distance measurement to simultaneously measure the target's distance and speed.

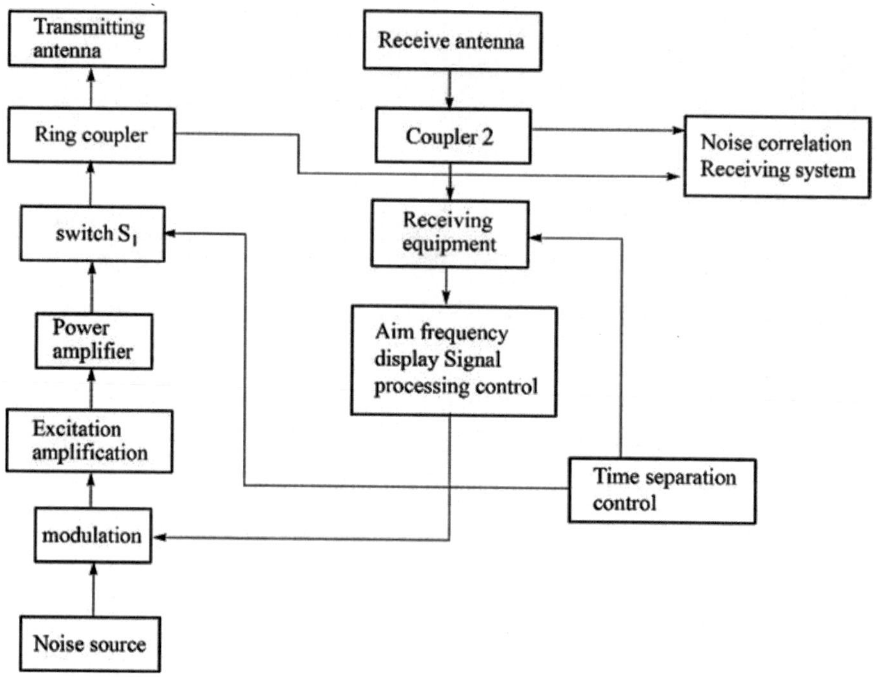


**Fig. 10.30** Schematic diagram of adding a noise-related receiving system to a high-power noise jammer to increase the ranging function

When $\tau = \tau_0$, the output $R(\tau, \tau_0, T)$ is the largest, as shown in Fig. 10.32.

The main technologies involved in the high-power noise jammer are as follows:

1. Delay technology of intermediate frequency signal.

   If the target is measured within 150 km, the delay time can be changed from 0 to 1000 μs, and the search target will be delayed automatically. When $\tau = \tau_0$, it will automatically indicate the value of $\tau$ (that is, the distance to the target).
2. High sensitivity operation of the receiver.

   Due to the low peak power of the transmission, in order to reach a certain detection distance, the receiver must be highly sensitive. Generally, the sensitivity of the relevant receiver is the ability to obtain useful information of the target echo in the case of a negative signal/noise ratio. According to some information, foreign countries can obtain useful information with a signal/noise ratio of −45 dB; domestically, they can achieve a level of −35 dB. It can be further improved in the future to increase the effective sensitivity of the receiver.
3. Isolation of the transceiver antenna.

   According to the receiver's high sensitivity requirements, there are relatively high requirements for the antenna's transceiver isolation. We can use the requirements of noise radar ranging as an example to illustrate this problem in detail.

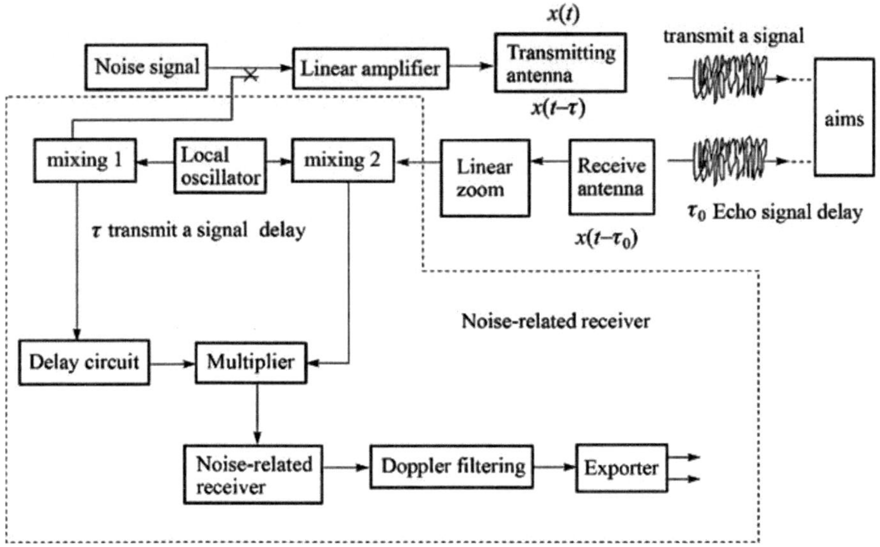


**Fig. 10.31** Principle block diagram of noise-related receivers. $x(t)$—Transmitted signal; $\tau_0$—The time it takes for the noise signal to reflect from the transmitting antenna to the target and back to the receiving antenna; $x(t-\tau_0)$—Received echo signal equation; $\tau$—Delay time of transmitting noise signal; $x(t-\tau)$—Transmit signals $x(t-\tau)$ and $x(t-\tau_0)$ are added to the multiplier at the same time

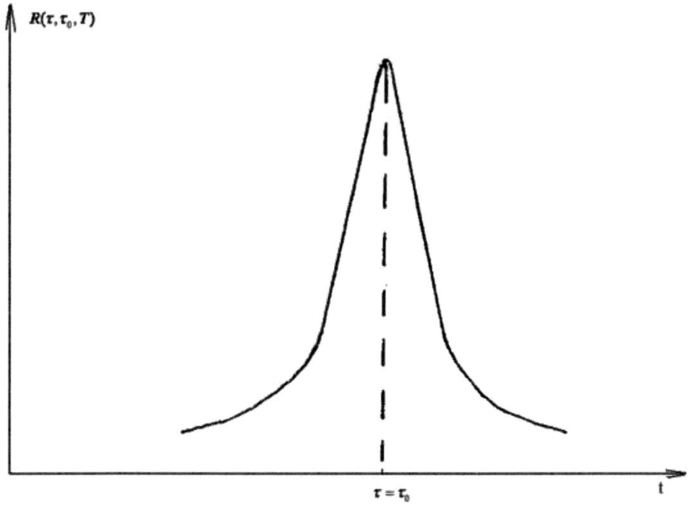


**Fig. 10.32** Schematic diagram of a maximum output of distance signal

The radar's ranging equation is:

$$R = \left[ \frac{P_t \cdot G_t \cdot G_\gamma \cdot \lambda^2 \cdot \sigma \cdot L_t \cdot L_\gamma}{(4\pi)^3 P_{\min}} \right]^{1/4} \tag{10.26}$$

In the formula:

$P_t$ Continuous wave power of radar transmitter, $P_t = 1$ kW;
$G_t$ Radar transmit antenna gain, $G_t = 40$ dB;
$G_\gamma$ Receive antenna gain, $G_\gamma = 40$ dB;
$\lambda$ Radar operating wavelength, $\lambda = 0.1$ m;
$L_t$ Transmit feed loss, $L_t = 0.3$;
$L_\gamma$ Receive feed loss, $L_\gamma = 0.3$;
$\sigma$ Target effective reflection area, $\sigma = \text{m}^2$;
$P_{\min}$ The actual working sensitivity of the receiver.

The internal noise of the radar is:

$$P_a = K \cdot T \cdot \Delta f \cdot N \tag{10.27}$$

In the formula:

K Boltzmann constant, $K = 1.38 \times 10^{-23}$;
T Absolute temperature, T = 290 K;
$\Delta f$ Receiver bandwidth, $\Delta f = 20$ MHz;
N Noise Figure, N = 3.

Substituting into the Formula (10.27), we get:

$$P_a = 1.37 \times 10^{-23} \times 290 \times 2 \times 10^7 \times 3 = 2.38 \times 10^{-13}\ \text{W}$$

If the relevant receiver can extract the useful signal under the condition of −40 dB signal/noise ratio, then:

$$P_{\min} = 2.38 \times 10^{-13} \times 10^{-4} = 2.38 \times 10^{-17}\ \text{W}$$

Substituting into the Formula (10.26), we get:

$$R = \left( \frac{10^3 \times 10^8 \times 10^{-2} \times 10 \times 0.3}{2 \times 10^3 \times 2.38 \times 10^{-17}} \right)^{1/4} = 500\ \text{km}$$

The power of the transmitter coupled to the receiver should be less than $2.38 \times 10^{-13}$ W.

That is, the isolation of the transceiver antenna $D \geq \frac{P_t}{P_a} = \frac{10^3}{2.38 \times 10^{-13}} = 4.2 \times 10^{15}$. The isolation of the transceiver antenna is required to be greater than 156.2 dB.

It is feasible to adopt the principle of noise radar to obtain three-digit integrated electronic equipment, but some key technologies need to be solved before it can be realized.

### *10.5.2 Using the Jammer's Angle Passive Tracking System to Realize the Integration of "Reconnaissance, Positioning, and Jamming"*

"Passive positioning" is a well-known term in the field of electronic countermeasures. The "passive positioning" that people know before is generally "multi-station cross positioning" or "time difference" positioning. The concept of "single-station passive positioning" sounds incredible at first, but it is not unfamiliar after careful analysis. Many of our reconnaissance and jamming devices have some of this positioning function, but the "positioning error" is too large.

"Passive positioning" has a very important position in the field of electronic countermeasures. Both the enemy and the enemy are unwilling to open the radar to expose themselves and conduct "wireless silence," They want to know whether there is an enemy in the area they protect. This is a contradiction. It is then better to use "passive positioning" without turning on the radar and determine the target position. However, in the past, "multi-station cross positioning" or "time difference positioning" was used, which required many devices to work together to determine the target position. Such positioning is relatively complicated, requires a long time for positioning, and has relatively large positioning errors, which cannot well meet tactical use requirements.

The "single station passive positioning" discussed here is to use a ground-to-air jamming station. It can not only detect the parameters of the enemy's airborne radar, but also jamming with its radar, and at the same time determine its "position." This is a kind of electronic countermeasures equipment that has three-in-one integrated functions. The so-called "three-in-one" refers to electronic countermeasures equipment capable of "reconnaissance," "jamming," and "positioning," which can simultaneously have these three functions. This kind of equipment will undoubtedly greatly improve equipment utilization and combat effectiveness.

The scope of research here is the implementation method and basic working principle of "single station passive positioning"; the mathematical expression model of "single station passive positioning"; the error of "single station passive positioning"; the possibility of achieving "single-station passive positioning" and the current equipment's realization of "single-station passive positioning."

There are three methods for single station passive positioning:

1. The first method is to obtain the position of the target by measuring the amplitude and angle of arrival of the target signal.

The second method: obtain the target position by measuring the irradiation time and the angle of arrival of the target signal.

The third type: obtain the target's position by measuring the target signal's angle and movement speed.

A reconnaissance station can usually measure the carrier frequency, azimuth angle, repetition frequency, pulse width, and pulse amplitude of the signal source.

The first method: According to the general reconnaissance equation, the signal strength received by the reconnaissance receiver can be written as:

$$P_i = \frac{P_a \cdot G_a}{4\pi R^2} \cdot \frac{\lambda^2}{4\pi} \cdot G_0 \cdot K \tag{10.28}$$

In the formula:

$P_a$ Radar transmitter peak power.
$G_a$ Radar antenna gain.
$\lambda$ Radar operating wavelength.
$R$ The distance between the reconnaissance station and the radar station under test.
$G_0$ Reconnaissance station antenna gain.
$K$ Loss factor.

Suppose the reconnaissance station is more advanced, and the measured radar parameters are relatively complete. In that case, the radar model can be obtained by comparing the measured radar parameters with the parameters stored in the database to determine the peak power of the transmitter and the antenna gain of the tested radar. After transformation, Formula (10.28) can be written as:

$$R = \frac{\lambda}{4\pi}\sqrt{\frac{P_a \cdot G_a \cdot G_0 \cdot K}{P_i}} \tag{10.29}$$

Convert the received power $P_i$ in (10.29) into voltage, and use the detection formula $V = k\sqrt{R_H \cdot P_i}$:

$$P_i = \frac{V^2}{k^2 \cdot R_H} \tag{10.30}$$

Substitute Eq. (10.30) into Eq. (10.29):

$$R = \frac{\lambda \cdot k}{4\pi V}\sqrt{P_a \cdot G_a \cdot G_0 \cdot K \cdot R_H} \tag{10.31}$$

In the formula:

k Detection coefficient.
$R_H$ Detector characteristic load.

The reconnaissance radar has its $P_a$ and $G_a$ data in the database, k and $R_H$ are the parameters of the reconnaissance station itself, and $\lambda$ are the measured parameters.

Set $A = \frac{\lambda \cdot k}{4\pi}\sqrt{P_a \cdot G_a \cdot G_0 \cdot K \cdot R_H}$.
Then A is a known number, substituting (10.31) into the formula:

$$R = \frac{A}{V_i} \tag{10.32}$$

In the formula:

$V_i$ Signal amplitude output by reconnaissance receiver.

From the Formula (10.32), we can see that the target's distance can be measured by measuring the signal amplitude. However, the amplitude of the signal received by the reconnaissance station is a function of distance, as well as a function of the radar transmitting antenna and the reconnaissance antenna. Taking aforementioned factors into account, the Formula (10.29) can be written as:

$$R = \frac{\lambda}{4\pi}\sqrt{\frac{P_a \cdot G_a(\theta) \cdot G_0(\alpha) \cdot K}{P_i}} \tag{10.33}$$

It can be seen from the Formula (10.33) that the signal amplitude varies with the irradiation point of the radar antenna and the reconnaissance antenna, and the size of the signal cannot fully indicate the distance. Therefore, the error of measuring the distance with this method is very large. In addition, using this method to measure distance requires a relatively large linear dynamic range of the receiver, which is difficult to implement on the device.

2. The second method: to achieve the purpose of distance measurement by measuring the irradiation time. According to the empirical formula of electronic warfare reconnaissance, the Formula (10.29) can be written as:

$$P_i = \frac{P_a \cdot G_a}{4\pi R^2} \cdot \frac{\lambda^2}{4\pi} \cdot K \cdot K_1 \cdot \left(\frac{\theta_{0.5}}{\theta/2}\right)^2 \cdot G_0$$

After simplification:

$$\theta = \frac{\theta_{0.5} \cdot \lambda}{2\pi R}\sqrt{\frac{P_a \cdot G_a \cdot G_0 \cdot K \cdot K_1}{P_i}} \tag{10.34}$$

In the formula:

$\theta$ Radar antenna's radiation angle range to reconnaissance antenna.
$K_1$ Sidelobe coefficient of reconnaissance antenna.

If the rotation period of the radar is T; the rotation range of the radar antenna is $\theta_F$; the irradiation time of the radar is $T_x$, according to the linear relationship:

$$T_x = T \cdot \frac{\theta}{\theta_F} \tag{10.35}$$

Substituting (10.34) into (10.35), we get:

$$T_x = \frac{T \cdot \theta_{0.5} \cdot \lambda}{2\pi\theta_F \cdot R}\sqrt{\frac{P_a \cdot G_a \cdot G_0 \cdot K \cdot K_1}{P_i}} \tag{10.36}$$

Similarly, the reconnaissance station finds the radar model in the database through the measured radar parameters, and then obtains data such as $P_a$, $G_a$,$T$, $\theta_{0.5}$, and $\theta_F$.

Set

$$B = \frac{T \cdot \theta_{0.5} \cdot \lambda}{2\pi\theta_F}\sqrt{\frac{P_a \cdot G_a \cdot G_0 \cdot K \cdot K_1}{P_i}} \tag{10.37}$$

Substituting (10.37) into (10.36), we get:

$$R = \frac{B}{T_x} \tag{10.38}$$

According to Formula (10.38), B has specific constants for specific radars. Therefore, the target distance can be obtained through the measured target irradiation time. This distance measurement method also influences the antenna sidelobe, which increases the distance measurement error. However, this method does not require a strict linear dynamic range of the reconnaissance receiver. Therefore, there are some advantages to the method of distance measurement by irradiation time.

3. The third method: Obtain the target distance by measuring the target speed and angle changes.

Suppose the target makes linear constant velocity motion within t time, as shown in Fig. 10.33.

If the target is at point B at a time $t_1$, the projection point is point A, the corresponding azimuth angle is $\alpha_1$, and the elevation angle is $\beta_1$. If the target is at point D at a time $t_2$, the projection point is point C, the corresponding azimuth angle is $\alpha_2$, and the elevation angle is $\beta_2$.

Suppose the target carrier is flying horizontally, and the speed is v. According to Fig. 10.32, we get:

$$\overline{DB} = \overline{AC} = (t_1 - t_2) \cdot V_0 \tag{10.39}$$

$$\overline{CA}^2 = \overline{OC}^2 + \overline{OA}^2 - 2 \cdot \overline{OC} \cdot \overline{OA} \cdot \cos(\alpha_2 - \alpha_1) \tag{10.40}$$

$$\tan\beta_1 = \frac{h}{\overline{OA}} \tag{10.41}$$

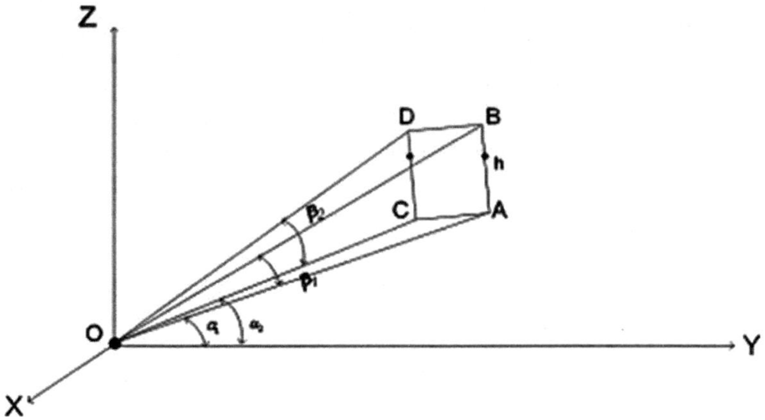


**Fig. 10.33** Reference map of target coordinates. B—Target location; A—Target projection; D—The target moves from point B to point D within $\Delta t_0$ time; C—Projection point of the target at point D; O—The location of the reconnaissance station; $\alpha_1$—The azimuth angle of the target at the projection point A; $\alpha_2$—The azimuth angle of the target at projection point C; $\beta_1$—Elevation angle of the target at point B; $\beta_2$—Elevation angle of the target at point D; h—Target height

$$\tan\beta_2 = \frac{h}{\overline{OC}} \tag{10.42}$$

After simplification:

$$\overline{OC} = \frac{\overline{CA}}{\left[1 + \left(\frac{\tan\beta_2}{\tan\beta_1}\right)^2 - 2\frac{\tan\beta_2}{\tan\beta_1}\cdot\cos(\alpha_2 - \alpha_1)\right]^{1/2}}$$

Set $\overline{OC} = R$; $\overline{CA} = V_0 \cdot (t_1 - t_2) = V_0 \cdot t_0$.
Then:

$$R = \frac{V_0 \cdot t_0}{\left[1 + \left(\frac{\tan\beta_2}{\tan\beta_1}\right)^2 - 2\frac{\tan\beta_2}{\tan\beta_1}\cdot\cos(\alpha_2 - \alpha_1)\right]^{1/2}} \tag{10.43}$$

In a reconnaissance jamming station with an automatic angle tracking system, the target's angle data, angle change data, and angle change time data can be obtained more accurately. If the target's speed is known, the target's distance can be determined so that the target's coordinate position can be determined, that is, "single-station passive positioning" can be realized.

How to obtain the angular velocity $V_0$ of the target is a key issue that requires special research. Assuming a variety of means can obtain that, according to Eq. (10.43), the distance to the target can be obtained. This positioning method is suitable for passive positioning of spaceborne radar. Since the satellite's velocity is generally known, the distance between the tracked satellite and the reconnaissance station can be calculated as long as the reconnaissance station's angle passive tracking system can obtain its elevation azimuth $\alpha_1$, $\alpha_2$, and time difference $t_0$ of the measurement angle. Therefore, the jammer of spaceborne radar can realize the integration of "reconnaissance, positioning, and jamming."

### *10.5.3 Ranging Error of Single Station Passive Positioning System Using Angle Passive Tracking System*

The range of finding formula of the "single station passive positioning" system is shown in (10.32), (10.38), and (10.43). As (10.32) and (10.38) are derived from some empirical formulas, the mathematical expressions themselves are not rigorous, and the calculation error is large, so no further analysis is done here. Here, only mathematical analysis is made for the ranging error of (10.43).

Assume that the measurement error of each variable causes the total error of the distance R, and the measurement error of each variable is independent and random. The ranging error of (10.43) can be written as:

$$\Delta R = \frac{\partial R}{\partial t_0} \cdot \Delta t_0 + \frac{\partial R}{\partial V_0} \cdot \Delta V_0 + \frac{\partial R}{\partial \beta_2} \cdot \Delta \beta_2 + \frac{\partial R}{\partial \beta_1} \cdot \Delta \beta_1 + \frac{\partial R}{\partial \alpha_1} \cdot \Delta \alpha_1 + \frac{\partial R}{\partial \alpha_2} \cdot \Delta \alpha_2 \tag{10.44}$$

According to the error of the multivariate function, the following formula is obtained.

The partial derivative of the error caused by $t_0$ is:

$$\frac{\partial R}{\partial t_0} = \frac{V_0}{\left[1 + \left(\frac{\tan \beta_2}{\tan \beta_1}\right)^2 - 2\frac{\tan \beta_2}{\tan \beta_1} \cdot \cos(\alpha_2 - \alpha_1)\right]^{1/2}} \tag{10.45}$$

The partial derivative of the error caused by $V_0$ is:

$$\frac{\partial R}{\partial V_0} = \frac{t_0}{\left[1 + \left(\frac{\tan \beta_2}{\tan \beta_1}\right)^2 - 2\frac{\tan \beta_2}{\tan \beta_1} \cdot \cos(\alpha_2 - \alpha_1)\right]^{1/2}} \tag{10.46}$$

The partial derivative of the error caused by $\beta_2$ is:

$$\frac{\partial R}{\partial \beta_2} = \frac{V_0 \cdot t_0\left(\frac{\tan\beta_2}{\tan\beta_1} \cdot \sec^2\beta_2 - \frac{\sec^2\beta_2}{\tan\beta_1} \cdot \cos(\alpha_2 - \alpha_1)\right)}{\left[1 + \left(\frac{\tan\beta_2}{\tan\beta_1}\right)^2 - 2\frac{\tan\beta_2}{\tan\beta_1} \cdot \cos(\alpha_2 - \alpha_1)\right]^{3/2}} \tag{10.47}$$

The partial derivative of the error caused by $\beta_1$ is:

$$\frac{\partial R}{\partial \beta_1} = \frac{V_0 \cdot t_0\left(\frac{\tan^2\beta_2 \cdot \sec^2\beta_1}{\tan\beta_1} - \frac{\tan\beta_2 \cdot \sec^2\beta_1}{\tan\beta_1} \cdot \cos(\alpha_2 - \alpha_1)\right)}{\left[1 + \left(\frac{\tan\beta_2}{\tan\beta_1}\right)^2 - 2\frac{\tan\beta_2}{\tan\beta_1} \cdot \cos(\alpha_2 - \alpha_1)\right]^{3/2}} \tag{10.48}$$

The partial derivative of the error caused by $\alpha_1$ is:

$$\frac{\partial R}{\partial \alpha_1} = \frac{-V_0 \cdot t_0 \cdot \sin(\alpha_2 - \alpha_1) \cdot \tan\beta_2}{\left[1 + \left(\frac{\tan\beta_2}{\tan\beta_1}\right)^2 - 2\frac{\tan\beta_2}{\tan\beta_1} \cdot \cos(\alpha_2 - \alpha_1)\right]^{3/2} \cdot \tan\beta_1} \tag{10.49}$$

The partial derivative of the error caused by $\alpha_2$ is:

$$\frac{\partial R}{\partial \alpha_2} = \frac{V_0 \cdot t_0 \sin(\alpha_2 - \alpha_1) \cdot \tan\beta_2}{\left[1 + \left(\frac{\tan\beta_2}{\tan\beta_1}\right)^2 - 2\frac{\tan\beta_2}{\tan\beta_1} \cdot \cos(\alpha_2 - \alpha_1)\right]^{3/2} \cdot \tan\beta_1} \tag{10.50}$$

If each variable has i error values, the root mean square error expression of R can be written as:

$$\begin{aligned}
\Delta R^2 = {} & \left(\frac{\partial R}{\partial t_0}\right)^2 \cdot \sum_{i=1}^{m} \Delta t_{0i}^2 + \left(\frac{\partial R}{\partial V_0}\right)^2 \cdot \sum_{i=1}^{m} \Delta V_{0i}^2 \\
& + \left(\frac{\partial R}{\partial \beta_2}\right)^2 \cdot \sum_{i=1}^{m} \Delta\beta_{2i}^2 + \left(\frac{\partial R}{\partial \beta_1}\right)^2 \cdot \sum_{i=1}^{m} \Delta\beta_{1i}^2 \\
& + \left(\frac{\partial R}{\partial \alpha_1}\right)^2 \cdot \sum_{i=1}^{m} \Delta\alpha_{1i}^2 + \left(\frac{\partial R}{\partial \alpha_2}\right)^2 \cdot \sum_{i=1}^{m} \Delta\alpha_{2i}^2 + 2\left(\frac{\partial R}{\partial t_0}\right) \cdot \left(\frac{\partial R}{\partial V_0}\right) \cdot \sum_{i=1}^{m} \Delta t_{0i} \cdot \Delta V_{0i} \\
& + 2\left(\frac{\partial R}{\partial t_0}\right) \cdot \left(\frac{\partial R}{\partial \beta_1}\right) \cdot \sum_{i=1}^{m} \Delta t_{0i} \cdot \Delta\beta_{1i} + 2\left(\frac{\partial R}{\partial t_0}\right) \cdot \left(\frac{\partial R}{\partial \beta_2}\right) \cdot \sum_{i=1}^{m} \Delta t_{0i} \cdot \Delta\beta_{2i} \\
& + 2\left(\frac{\partial R}{\partial t_0}\right) \cdot \left(\frac{\partial R}{\partial \alpha_1}\right) \cdot \sum_{i=1}^{m} \Delta t_{0i} \cdot \Delta\alpha_{1i} + 2\left(\frac{\partial R}{\partial t_0}\right) \cdot \left(\frac{\partial R}{\partial \alpha_2}\right) \cdot \sum_{i=1}^{m} \Delta t_{0i} \cdot \Delta\alpha_{2i} \\
& + 2\left(\frac{\partial R}{\partial V_0}\right) \cdot \left(\frac{\partial R}{\partial \beta_1}\right) \cdot \sum_{i=1}^{m} \Delta V_{0i} \cdot \Delta\beta_{1i} + 2\left(\frac{\partial R}{\partial V_0}\right) \cdot \left(\frac{\partial R}{\partial \beta_2}\right) \cdot \sum_{i=1}^{m} \Delta V_{0i} \cdot \Delta\beta_{2i} \\
& + 2\left(\frac{\partial R}{\partial V_0}\right) \cdot \left(\frac{\partial R}{\partial \alpha_1}\right) \cdot \sum_{i=1}^{m} \Delta V_{0i} \cdot \Delta\alpha_{1i} + 2\left(\frac{\partial R}{\partial V_0}\right) \cdot \left(\frac{\partial R}{\partial \alpha_2}\right) \cdot \sum_{i=1}^{m} \Delta V_{0i} \cdot \Delta\alpha_{2i}
\end{aligned}$$

$$+2\left(\frac{\partial R}{\partial \beta_1}\right)\cdot\left(\frac{\partial R}{\partial \beta_2}\right)\cdot\sum_{i=1}^{m}\Delta\beta_{1i}\cdot\Delta\beta_{2i}+2\left(\frac{\partial R}{\partial \beta_1}\right)\cdot\left(\frac{\partial R}{\partial \alpha_1}\right)\cdot\sum_{i=1}^{m}\Delta\beta_{1i}\cdot\Delta\alpha_{1i}$$
$$+2\left(\frac{\partial R}{\partial \beta_1}\right)\cdot\left(\frac{\partial R}{\partial \alpha_2}\right)\cdot\sum_{i=1}^{m}\Delta\beta_{1i}\cdot\Delta\alpha_{2i}+2\left(\frac{\partial R}{\partial \beta_2}\right)\cdot\left(\frac{\partial R}{\partial \alpha_1}\right)\cdot\sum_{i=1}^{m}\Delta\beta_{2i}\cdot\Delta\alpha_{1i}$$
$$+2\left(\frac{\partial R}{\partial \beta_2}\right)\cdot\left(\frac{\partial R}{\partial \alpha_2}\right)\cdot\sum_{i=1}^{m}\Delta\beta_{2i}\cdot\Delta\alpha_{2i}+2\left(\frac{\partial R}{\partial \alpha_1}\right)\cdot\left(\frac{\partial R}{\partial \alpha_2}\right)\cdot\sum_{i=1}^{m}\Delta\alpha_{1i}\cdot\Delta\alpha_{2i} \quad (10.51)$$

In order to analyze the problem further, the two sets of data are substituted into the formula to obtain the ranging error of "single-station passive positioning."

The first set of data: $V_0 = 166.6\,\text{m/s}$; $t_0 = 60\,\text{s}$; $\beta_1 = 20°$; $\beta_2 = 33°$; $\alpha_1 = 153°$; $\alpha_2 = 153°$; $\Delta V_0 = 3\,\text{m/s}$; $\Delta t_0 = 0.5\,\text{s}$; $\Delta\beta_1 = \Delta\beta_2 = \Delta\alpha_1 = \Delta\alpha_2 = 1°$, seek $\Delta R$.

Substituting the data into (10.45), (10.46), (10.47), (10.48), (10.49), and (10.50), we get: $\frac{\partial R}{\partial t_0} = 212$; $\frac{\partial R}{\partial V_0} = 76.47$; $\frac{\partial R}{\partial \beta_2} = -68{,}295$; $\frac{\partial R}{\partial \beta_1} = -44{,}455$; $\frac{\partial R}{\partial \alpha_1} = 0$; $\frac{\partial R}{\partial \alpha_2} = 0$.

Substituting these data into (10.44) equation:

$$\begin{aligned}\Delta R &= 212\times 0.5\times 76.47\times 3 - 68{,}295\times\frac{1}{57} - 44{,}455\times\frac{1}{57} + 0 + 0\\ &= 106 + 229.41 - 1198 - 779.9 = -1642.49\,(\text{m})\end{aligned}$$

The second set of data: $V_0 = 333.2\,\text{m/s}$; $t_0 = 30\,\text{s}$; $\beta_1 = 20°$; $\beta_2 = 33°$; $\alpha_1 = 153°$; $\alpha_2 = 153°$; $\Delta V_0 = 3\,\text{m/s}$; $\Delta t_0 = 0.1\,\text{s}$; $\Delta\beta_1 = \Delta\beta_2 = \Delta\alpha_1 = \Delta\alpha_2 = 0.3°$, seek $\Delta R$.

Substituting the data into (10.45), (10.46), (10.47), (10.48), (10.49), and (10.50), we get: $\frac{\partial R}{\partial t_0} = 424$; $\frac{\partial R}{\partial V_0} = 38.23$; $\frac{\partial R}{\partial \beta_2} = -68{,}295$; $\frac{\partial R}{\partial \beta_1} = -44{,}455$; $\frac{\partial R}{\partial \alpha_1} = 0$; $\frac{\partial R}{\partial \alpha_2} = 0$.

Substituting these data into (10.44) equation:

$$\begin{aligned}\Delta R &= 424\times 0.1\times 38.23\times 3 - 68{,}295\times\frac{0.3}{57} - 44{,}455\times\frac{0.3}{57} + 0 + 0\\ &= 42.4 + 114.69 - 359.4 - 233.9 = -436.2\,(\text{m})\end{aligned}$$

The above calculation shows that if the angle measurement error can be within 0.2°, the time measurement error is within 0.01 s, and the target speed error is within 3 m/s, the root mean square error of the distance measurement with "single station passive positioning" can reach several hundred meters.

### 10.5.4 Single Station Passive Positioning Based on a Passive Angle Tracking System

It seems difficult to use the angular passive tracking system for "single-station passive positioning." In fact, it has now become a reality. It's just that the direction-finding error of these current direction finding equipment is relatively large (about 1°), and the speed of the target is not measured in real-time, but obtained by other means.

At present, some equipment has initially possessed this "single-station passive positioning" function. Because it can accurately determine the azimuth and elevation

**Table 10.14** A set of data obtained by a ground-to-air jamming station during the design and final flight test

| Serial number | Name | | | | |
|---|---|---|---|---|---|
| | Time $t_0$ | Orientation $\alpha$ (°) | Elevation $\beta$ (°) | Use Formula (10.43) to calculate the distance R′ (km) | The distance R′ is provided by the target indicator radar (km) |
| 1 | 12:59 | 155 | 10 | 51 | 54 |
| 2 | 13:00 | 153 | 14 | 35.4 | 42 |
| 3 | 13:01 | 153 | 20 | 28.2 | 32 |
| 4 | 13:02 | 153 | 33 | 16.5 | 16.5 |
| 5 | 13:02:50 | 153 | 57 | 9.4 | 6 |

of the target. The position of the target can be determined when the target speed is obtained by other means. Table 10.14 is a set of data obtained by a ground-to-air jamming station during the design and final flight test.

If you make a slight improvement on the existing basis, that is, automatically input the azimuth and elevation data into the computer, and then input the time and speed data into the computer. While the jamming station tracks the target or the jamming target, the distance, azimuth, and elevation angle can be automatically displayed. That is, the jammed target position is determined.

In Table 10.14, $V_0$ is 166.6 m/s according to the speed of a certain aircraft. $R$ and $R'$ are close to each other, but there is a difference. This difference is not the ranging error of the "single station passive positioning". Because $R'$ is also inaccurate. The target indicating radar is about 5 km away from the jamming station. Without an automatic coordinate conversion device, the target indicating radar itself has a relatively large-ranging error.

It can be seen from Table 10.14 that the distance trend calculated by Eq. (10.43) is correct.

### *10.5.5 Typical Application of Single Station Passive Location Based on Angle Passive Tracking System*

The schematic diagram of the typical application scenario of this "single-station passive positioning" is shown in Fig. 10.34.

The configuration in Fig. 10.33 is a typical configuration of a ground-to-air jamming system. Point A is the center of the protected target; inside the dotted line is the jamming exposure area; outside the dotted line is the effective jamming area. The jamming exposure distance is centered at point A, and the radius is about 6–10 km. The jamming station can detect and jamming and has the function of "single-station

**Fig. 10.34** Application diagram of "single-station passive positioning" using angle passive tracking system. A—protected target; B—the target that is jammed with, tracking and positioning; $C_{1,2,3,...}$—"integrated" jamming station; D—"integrated" jamming system command station

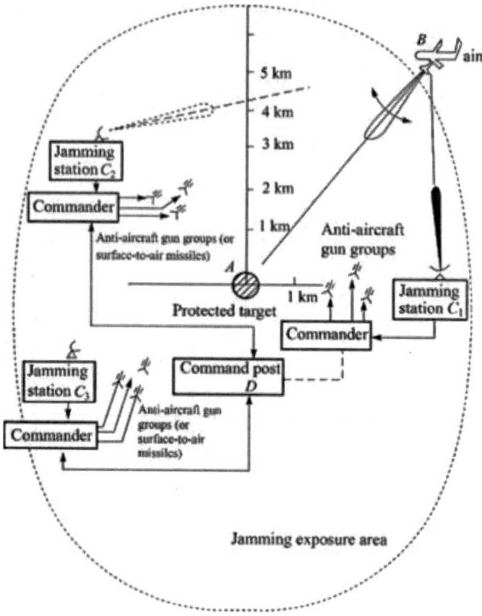


passive positioning." It is an electronic countermeasure with the "integration" of reconnaissance, jamming, and positioning.

Normally, when the enemy aircraft is outside the jamming exposure area, the jamming is the main factor, and the effective distance of the jamming is 10–200 km.

When the enemy aircraft is in the jamming exposure area, the jammer cannot effectively jamming. Fire control weapons on the ground must be used to kill enemy aircraft. Surface-to-air fire control weapons include antiaircraft artillery and surface-to-air missiles. These weapons are usually controlled by radar. Radar control has its advantages, such as high control accuracy and real-time. But there are also many shortcomings, such as being easy to expose yourself, being vulnerable to jamming, or being attacked by anti-radiation missiles. Using "single station passive positioning" can overcome these shortcomings, and the equipment can also be used for multiple purposes.

Within a short distance of tens of kilometers, it can also control surface-to-air missiles to jamming and kill enemy aircraft while launching jamming. Generally speaking, an enemy plane reaches a predetermined area after a long-distance flight. To find a specific attacker, radar is usually used to detect the target. This creates a favorable opportunity for us to target the enemy aircraft without opening any radiation source accurately. If it is true that the enemy plane cannot use the radar, from the perspective of electronic countermeasures, we have already completed the mission. This is due to the existence of our electronic countermeasure equipment, forcing the enemy's radar to dare not turn on easily and unable to complete military missions.

The ideal situation of this "single station passive positioning" equipment is its application in ground-to-air jamming stations. Because in the ground-to-air jamming stations, there are receiving systems, angle passive tracking systems, computer

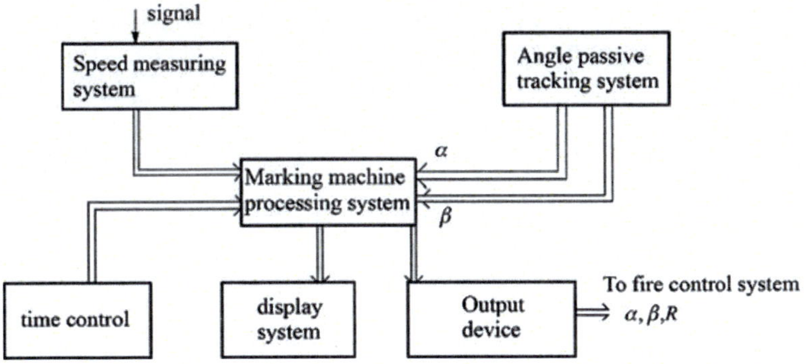


**Fig. 10.35** Block diagram of a single-station passive positioning system

processing control systems, digital display systems, etc., only a small amount of equipment can be added to take into account the "single station passive positioning" function.

Its composition block diagram is shown in Fig. 10.35.

The tracked signal passes through the speed measurement system to measure the target's moving speed (or obtain the target speed by other means) and send it to the computer storage unit. The angle passive tracking system sends the target's angle data to the computer storage unit. Take $\alpha_1$ and $\beta_1$ at $t_1$. Take $\alpha_2$ and $\beta_2$ at $t_2$. The computer uses the (10.43) mathematical model to calculate the R value, and then divides it into two outputs. One is output to the display system. The other is output to the fire control system through the output device to control antiaircraft artillery or surface-to-air missiles for accurate shooting.

According to the above discussion, single-station passive positioning using angle passive tracking system can be realized. However, in order to achieve a relatively high positioning accuracy, a lot of work needs to be done. The most important thing involves two aspects: one is to improve passive angle tracking; the other is to solve the speed measurement method and accuracy of the target. This "three-in-one" electronic countermeasure equipment of reconnaissance, jamming, and positioning can also be installed on the ship. It can enhance the comprehensive level of the ship's electronic equipment and countermeasures and bring many benefits to strengthening the fleet's air defense capabilities.

## 10.6 Analysis of Jamming Effect of "Signal Hiding" Anti-jamming Radar

The traditional jammer design first determines the maximum reconnaissance distance of the jamed radar to it, and the sensitivity of the reconnaissance receiver is higher, so that the receiver receives more signals, which increases the difficulty of sorting and recognition processing, and at the same time increases the transceiver. Therefore, the sensitivity of most jammers' reconnaissance receivers can not be too high, but also can not be too low. It is too low to receive the jammed radar signal, and also can not jam it.

In general, the sensitivity of the jammer reconnaissance receiver is 1.2 times of the maximum detection distance of the jammed radar, and this value provides an available ambush for the "signal hiding" anti-jamming of the radar.

Radar against radar:

$$R_{\max}^4 = \frac{p_t G_t^2 \lambda^2 \sigma}{(4\pi)^3 p_{to}}$$

$$P_t G_t = \frac{(4\pi)^3 R_{\max}^4 P_{to}}{G_t \lambda^2 \delta} \tag{10.52}$$

according to radar against radar:

$$R_{\max}^4 = \frac{P_t G_t^2 \lambda^2 \sigma}{(4\pi)^3 p_{to}}$$

$$P_t G_t = \frac{(4\pi)^3 R_{\max}^4 P_{to}}{G_t \lambda^2 \delta} \tag{10.52}$$

The jammer reconnaissance receiver has a sensitivity of $P_{ji}$

$$P_{ji} = \frac{P_t G_t}{4\pi R_j^2} \cdot \frac{\lambda^2}{4\pi} G_j(\theta) = \frac{P_t G_t \lambda^2 G_j(\theta)}{(4\pi)^2 R_j^2} \tag{10.53}$$

Substitute Eq. (10.52) into Eq. (10.53)

There is:

$$P_{ji} = \frac{4\pi R_{\max}^4 P_{to} G_j(\theta)}{G_t \sigma R_j^2} \tag{10.54}$$

Note: Where, $R_{\max}$ is the maximum detection range of the radar; $P_t G_t$ is equivalent power of radar radiation pulse.

$\lambda$ is the working wavelength; $R_j$ is the distance between the jammer and the radar; $P_{to}$ is the actual working sensitivity of the radar; $\sigma$ is the equivalent reflection area of the target; $G_j(\theta)$ is the gain of jammer receiving antenna.

For example: if the radar transmitter power $P_t$ is $10^5$ W; $G_t$ is $10^4$; $\lambda$ is $3 \times 10^{-2}$ m; $P_{to}$ is $10^{-13}$ W; $\sigma$ is 50 m$^2$; $R_j \approx R_{\max}$; Substituting into Eq. (10.54), $R_{\max}$ can be calculated as 220 km.

The power at which a jammer reconnaissance receiver receives a radar signal at the maximum range.

$$P_{ji} = \frac{P_t G_t}{4\pi R_j^2} \cdot \frac{\lambda^2}{4\pi} G_j(\theta) \tag{10.55}$$

Substituting the parameters into Eq. (10.55), $P_{ji}$ is calculated as $1.7 \times 10^{-6}$ W.

The sensitivity of the reconnaissance receiver for the jammer is −60 dBW (−90 dBm). According to the above equation, the radar distance $R_{ti}$, the radar echo signal power $P_{ti}$, and the actual working sensitivity of the radar jammer reconnaissance receiver $P_{ji}$ are calculated. The calculated data are shown in Table 10.15.

It can be seen from Table 10.15 that in this case, the radar does not use the "signal hiding" for anti-jamming, so the radar can be jammed from sequence 1 to sequence 5. In sequence 6, the jamming suppression coefficient to radar is the minimum jam/signal ratio required when the radar is disturbed. Different radar systems require different interference/signal ratios, which generally require $K_{jo}$ to be 3 dB for conventional pulse radar. The interference/signal ratio required by pulse compression radar is 10–15 dB. The required interference/signal ratio for SAR is between 15 and 20 dB, so whether the No. 6 in Table 10.15 could effectively interfere is determined by the radar systems.

Figure 10.36 can be drawn from Table 10.15.

In Fig. 10.36, the interference suppression coefficient and the jammer's transceiver isolation can meet the design requirements of the jammer, so the radar interference is effective from 220 to 6.8 km.

If the radar adopts "signal hiding" anti-jamming, its radar transmission power decreases as the range of action decreases, and it decreases in inverse proportion to the distance $R_{ti}^4$. The radar power received by the jammer reconnaissance receiver is inversely proportional to the square of the distance. If the sensitivity of the receiver changes to a certain distance, the reconnaissance receiver cannot receive the radar signal, and of course, the radar isn't interfered with. In this case, the radar uses the

**Table 10.15** The relationship between $R_{ti}$ and $P_{ti}$, $P_{ji}$, $P_{xi}$, $K_{ji}$

| No. | $R_{ti}$ (km) | $P_{ti}$ (dBW) | $P_{xi}$ (dBW) | $P_{ji}$ (dBm) | $K_{ji}$ (dB) | Can be jammed? |
|---|---|---|---|---|---|---|
| 1 | 220 | −130 | −88 | −88 | 42 | Yes |
| 2 | 110 | −118 | −82 | −82 | 36 | Yes |
| 3 | 55 | −106 | −76 | −76 | 30 | Yes |
| 4 | 27.5 | −94 | −70 | −70 | 24 | Yes |
| 5 | 13.7 | −82 | −64 | −64 | 18 | Yes |
| 6 | 6.8 | −70 | −58 | −58 | 12 | ? |

*Note* $R_{ti}$ is the distance between the radar and the detected target; $P_{ti}$ is the echo signal power received by radar. $P_{xi}$ is the jamming signal power received by radar. $P_{ji}$ is the power of the radar signal received by the jammer reconnaissance receiver. $K_{ji}$ is the interference/signal ratio at the input of the radar receiver

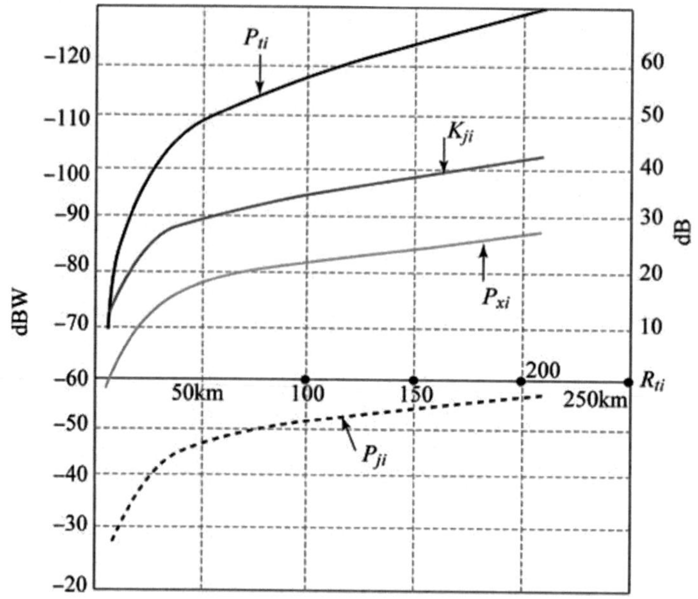


**Fig. 10.36** Relationship curve between radar detection range $R_{ti}$ and $P_{ti}$, $P_{ji}$, $P_{xi}$, $K_{ji}$

"signal hiding" method to successfully resist interference. The data are shown in Table 10.16. Fig. 10.37 is drawn according Table 10.16.

Under the condition of "signal hiding", although the interference suppression coefficient meets the requirements, the sensitivity of the jammer receiver is not enough, and the direction-finding and aiming frequency cannot be measured, so the jammer cannot play a good role. So that radar "signal hiding" anti-jamming to seize this vulnerability, then anti-jamming success.

Methods of anti-jamming against radar's "signal hiding":

The first method: The sensitivity of the interference receiver is raised to −90 dB, so that the receiver sensitivity varies with distance.

**Table 10.16** Date of $R_{ti}$ and $P_{ti}$, $P_{ji}$, ($P_{xi}$ is the radar transmit power)

| Order number | $R_{ti}$ (km) | $P_{xi}$ (dBW) | $P_{ti}$ (dBW) | $P_{ji}$ (dBW) |
|---|---|---|---|---|
| 1 | 220 | 50 | −130 | −60 |
| 2 | 110 | 38 | −130 | −66 |
| 3 | 55 | 26 | −130 | −72 |
| 4 | 27.5 | 14 | −130 | −78 |
| 5 | 13.7 | 12 | −130 | −84 |
| 6 | 6.8 | 0 | −130 | −90 |

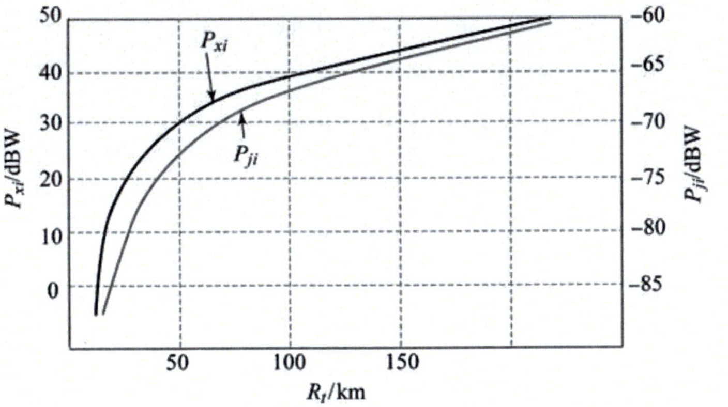


**Fig. 10.37** The relationship curve between $R_{ti}$ and $P_{xi}$, $P_{ji}$ under the condition of "signal hiding"

The interference reconnaissance receiver improves the working sensitivity, reduces the power of the interference transmitter, and maintains the "receiving, sending, and isolating" parameters of the jammer.

The power of the jammer reconnaissance receiver receiving its own jammer transmitted signal is $P_{joi}$

$$P_{joi} = \frac{P_{ji} G_j(\varphi) \cdot G_j(\varphi)\lambda^2}{(4\pi)^2\gamma^2} \tag{10.56}$$

The condition for satisfying spontaneous self-harvest is $P_{joi} < P_{jo}$.
That is:

$$1 > \frac{P_{joi}}{P_{jo}} = \frac{P_{ji} G_j(\varphi) \cdot G_j(\varphi)\lambda^2}{(4\pi)^2\gamma^2 P_{jo}} \tag{10.57}$$

The relationship between anti-jamming parameters of radar "signal hiding" and radar distance variation is shown in Table 10.17.

In Table 10.17, $P_{ti}$ is the change of radar distance and $P_{txi}$ is the change of radar transmit power; $P_{ti}$ is radar echo signal; $P_{ji}$ is radar signal power received by jammer receiver; $P_{tji}$ is interference power received by radar; $K_{ji}$ is interference/signal ratio variation.

It can be seen from Table 10.17 that the radar echo does not change, and it is still −130 dBW (the actual working sensitivity of the radar). The power of the interference signal received by the radar does not change, and the interference/signal ratio has been maintained at 72 dB, which is higher than $K_{jo}$ (interference suppression coefficient). The "transceiver isolation" coefficient of the jammer does not change. The proportion of receiver sensitivity change is proportional to the decrease of the transmitting power of the interference transmitter. Therefore, the jammer has always been able to resist the radar's "signal hiding" anti-interference. The distance has changed from

**Table 10.17** Relationship between anti-jamming parameters of radar "signal hiding" and radar distance variation

| Serial number | Radar distance $R_{ti}$ (km) | Radar's receiving echo signal $P_{ti}$ (dBW) | Radar transmitter power $P_{txi}$ (dBW) | Jamming transmitter power $P_{jxi}$ (dBW) | Received power of the jammer's receiver $P_{ji}$ (dBm) | Jamming power received by the radar $P_{tji}$ (dBW) | $K_{ji}$ (dB) |
|---|---|---|---|---|---|---|---|
| 1 | 220 | −130 | 50 | 30 | −58 | −58 | 72 |
| 2 | 110 | −130 | 38 | 24 | −64 | −58 | 72 |
| 3 | 55 | −130 | 26 | 18 | −70 | −58 | 72 |
| 4 | 27.5 | −130 | 14 | 12 | −76 | −58 | 72 |
| 5 | 13.7 | −130 | 2 | 6 | −82 | −58 | 72 |
| 6 | 6.8 | −130 | −10 | 0 | −88 | −58 | 72 |

220 to 6.8 km, and the interference has been effective in the range. Moreover, the effective sector of the interference has been expanded and the interference exposure distance has been reduced. As long as the parameters of the jammer are properly adjusted, it can effectively resist the "signal hiding" anti-interference. However, this confrontation method also increases the difficulty of the jammer adjustment. At the same time, the anti-interference of "radar signal hiding" also increases the difficulty of radar adjustment. This kind of confrontation method is feasible in theory, but it is very difficult to realize both sides in practice.

The second method:

It is to improve the "receiving, transmitting and isolating" degree of the jammer, so that the jammer has a larger margin than the "receiving, transmitting and isolating" degree of the conventional jammer, and should have a margin of 24–30 dB. The sensitivity of the general conventional jammer receiver is −60 dBW, and the jammer 'receiving, transmitting, and isolating' degree can work normally. When the receiver sensitivity is increased to −84 dBW, it can also work normally. In this case, the radar distance is reduced by 4 times, the transmission power is reduced by 24 dB, and the jamming receiver can still receive the radar signal. Effective jamming against it, the "receiving, transmitting and isolating" of the jammer still meets the requirements, and the effective jamming sector of the radar can be increased. In this case, the anti-jamming effect of radar is not increased, but reduced.

The second kind of radar 'signal hiding' is the 'hiding' of cone scanning frequency.

As early as the mid-1960s, automatic tracking of fire control radar basically used the "cone-scan" radar, one of the gun aiming radars, which is a single channel. The beam half power point of the transmitting antenna is parallel to the optical axis of the antenna and rotates around the optical axis. The rotating frequency is called cone sweep frequency, generally in 20–80 Hz. The pulse signal emitted is not equal amplitude, but modulated by the cone scanning frequency, and its shape is shown in Fig. 10.38.

**Fig. 10.38** Radar radiation pulse envelope

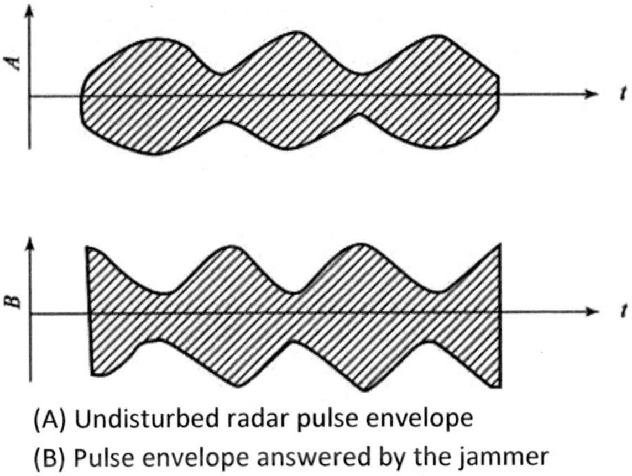


When the jammer receives the radar pulse signal, each pulse is amplified and answered. At the same time, the modulation envelope of the radar signal is amplified, and the jamming pulse is modulated by 180° of inversion, so that the jamming pulse signal envelopes the 180° of the inversion modulation waveform. The echo signal and the interference signal are received by the radar receiver, and the modulation envelope is detected. The difference between the two signals is 180°. Since the interference signal is stronger than the echo signal, the rotation of the radar servo antenna is controlled by the interference signal, so that the radar antenna turns to deviate from the target direction, thus damaging the tracking of the target by the radar.

The method of radar "hiding" cone-scan frequency is that the pulse signal transmitted is equal amplitude, so that the jammer can not receive the modulation signal, and the radar receiver is internally modulated. The radar servo receiving system can

**Fig. 10.39** Schematic diagram of "hidden" cone-scan radar signal and interference signal. **a** Transmitting equi-amplitude pulse signals for radar. **b** Modulation of echo signal for radar internal modulation. **c** It is a low-frequency sweep search for the output pulse signal of jammer in time $t_i$, and in time $t_x$, it's internal modulation frequency of radar found at low-frequency modulation time for jamming

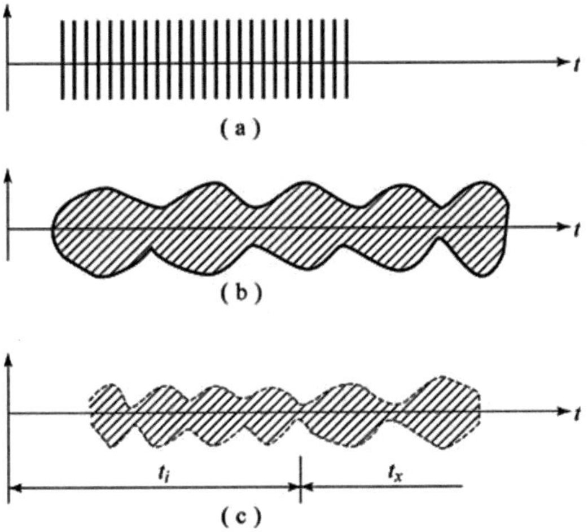

still receive the envelope of the modulation, so that the radar tracks the target, as shown in Fig. 10.39.

When the jammer finds the low-frequency modulation inside the radar, it stops at this low frequency. The radar still receives the echo signal and interference signal, and detects the echo envelope and interference envelope. Since the envelope phase and echo envelope phase of the interference is randomly changed, the radar antenna swings randomly and cannot track the target normally, the target may be lost when the swing amplitude is large. In this way, the "hidden" conical scanning radar is also disturbed. In the process of low scanning frequency, the low-frequency frequency is inconsistent with the frequency of the radar servo receiver, so the interference of low-frequency signal cannot enter the servo receiver, so the servo system still only receives the echo signal. When the low frequency is close to the radar low frequency, the servo receiver receives the low-frequency interference signal, which makes the antenna oscillate randomly.

By the late 1960s, monopulse tracking radar was widely used, and its tracking accuracy and anti-interference ability were superior to "cone-scan" radar. So the "cone-scan" radar was eliminated, jamming cone-scan radar jammers were useless, followed by the jammer that interfered with the monopulse radar.

The above several "signal hiding" radars can all be interfered with, which shows that the radar jammer is produced with the generation of radar. With the development of radars, new jammers will appear with new radars. With new anti-jamming methods, jammers will be forced to find new interference modes for these new anti-jamming radars.

With the continuous development of military radar, the jammers and jamming modes that interfere with these new radars are constantly emerging, and the jamming and anti-jamming of radars are constantly developing until human beings eliminate war.

Note: Another type of signal hiding method reduces the transmitted pulse amplitude, broadens the pulse width, and maintains the pulse energy unchanged. The pulse compression method keeps the radar's detection range unchanged, and makes the signal received by the reconnaissance receiver The amplitude is reduced.

The countermeasures taken by the jammer are to increase the sensitivity of the reconnaissance receiver, or increase the pulse receiver to improve the sensitivity of the receiver.

The final result is that the "signal hiding" anti-jamming can still be jammed, but it only increases the difficulty of the radar jammer.

# References


1. Y. Ding, *Theory and Computer Simulation of High Power Klystron* (National Defense Industry Press, Beijing, 2008)
2. Wang et al., *Precision Tracking Measurement Radar Technology* (Publishing House of Electronics Industry, Beijing, 2004)
3. Bao et al., *Radar Imaging Technology* (Publishing House of Electronics Industry, Beijing, 2004)
4. G.W. Stimson, *Introduction to Airborne Radar*. ed. by H. Wu et al., trans. (Publishing House of Electronics Industry, Beijing, 2003)
5. 38th Research Institute of China Electronic Technology Group, *Spaceborne Synthetic Aperture Radar System* (Radar Science and Technology Editorial Department, Hefei, 2002)
6. R.N. McDonough, A.D. Whalen, *Detection Signal in Noise*. ed. by Wang et al., trans. (Publishing House of Electronics Industry, Beijing, 2004)
7. Z. Zhuang, *Radar Polarization Signal Processing and its Application* (National Defense Industry Press, Beijing, 1999)
8. Z. Zhuang, N. Yuan et al., *Radar Cross Section Measurement-Compact Field Theory and Technology* (National University of Defense Technology Press, Changsha, 2000)
9. M. Skolnik, *Introduction to Radar System*, 3rd edn. ed. by Z. Qunsheng et al., trans. (Publishing House of Electronics Industry, Beijing, 2006)
10. P.E. Pace, *Detecting and Classifying Low Probability of Intercept Radar*. ed. by International Electronic Warfare, trans. (Editorial Department, 2007)
11. X. Zhang, *Modern Radar Countermeasure Technology* (National Defense Industry Press, Beijing, 1998)
12. C. Lin, Z. Nie, *Antenna Engineering Handbook* (Publishing House of Electronics Industry, Beijing, 2002)
13. M. Skolnik, *Radar Handbook*, ed. by J. Wang, Q. Lin et al., trans. (Publishing House of Electronics Industry, Beijing, 2006)




X. Zhang et al., *Theory to Countermeasures Against New Radars*,
https://doi.org/10.1007/978-981-16-6715-2